INSTRUCTOR'S SOLUTIONS MANUAL
WILLIAM B. CRAINE III

to accompany

INTRO STATS

Richard D. De Veaux

Williams College

Paul F. Velleman

Cornell University

PEARSON

Addison
Wesley

Boston   San Francisco   New York
London   Toronto   Sydney   Tokyo   Singapore   Madrid
Mexico City   Munich   Paris   Cape Town   Hong Kong   Montreal

Reproduced by Pearson Addison-Wesley from electronic files supplied by the author.

Copyright © 2004 Pearson Education, Inc.
Publishing as Pearson Addison-Wesley, 75 Arlington Street, Boston MA 02116

ISBN   0-321-16807-0

2 3 4 5 6 CRS 06 05 04 03

Contents

Chapter 2 - Data

1. **The News.** Answers will vary.

2. **Investments.** *Who* – 30 similar companies. *What* – 401(k) employee participation rates (in percent). *When* – Sometime after reading the 1992 *Fortune* article. *Where* – United States. *Why* – the company in question is concerned that its employee participation rate is lower than the rates of similar companies. *How* – Companies were "sampled", using an unspecified method. *Cases* – Each of the 30 companies is a case. *Variables* – There is one quantitative variable, 401(k) participation rate. *Concerns* – How was the sample of companies selected?

3. **Oil spills.** *Who* – 50 recent oil spills. *What* – Spillage amount (no specified unit) and cause of puncture. *When* – Sometime between 1990 (Oil Pollution Act passed) and 1995 (report made in *Marine Technology*). *Where* – United States. *Why* – To determine whether or not spillage amount per oil spill has decreased since Congress passed the 1990 Oil Pollution Act and use that information in the design of new tankers. *How* – Not specified. *Cases* – Each of the 50 recent oil spills is a case. *Variables* – There are 2 variables. Spillage amount is a quantitative variable, and cause of puncture is a categorical variable.

4. **Oscars.** *Who* – Winners of Best Actor or Best Actress Oscars. *What* – Age (in years) at the time they won Best Actor or Best Actress Oscars. *When* – From the beginning of the Academy Awards to the present. *Where* – United States. *Why* – To compare age distributions of actors and actresses. *How* – It appears that this information was collected from a census of all Best Actor and Actress Oscar winners. *Cases* – Each actor or actress is an individual. *Variables* – There is one quantitative variable, age of the actor or actress.

5. **Weighing bears.** *Who* – 54 bears. *What* – Weight, neck size, length, and sex. *When* – Not specified. *Where* – Not specified. *Why* - Since bears are difficult to weigh, the researchers hope to use the relationships between weight, neck size, length, and sex of bears to estimate the weight of bears, given the other, more observable features of the bear. *How* – Researchers collected data on 54 bears they were able to catch. *Cases* – Each bear is a case. *Variables* – There are 4 variables; weight, neck size, and length are quantitative variables, and sex is a categorical variable. No units are specified for the quantitative variables. *Concerns* – The researchers are (obviously!) only able to collect data from bears they were able to catch. This method is a good one, as long as the researchers believe the bears caught are representative of all bears, in regard to the relationships between weight, neck size, length, and sex.

6. **Molten iron.** *Who* – 10 crankshafts. *What* – The pouring temperature (in degrees Fahrenheit) of molten iron. *When* – Sometime before the *Quality Engineering* report in 1995. *Where* – Cleveland Casting Plant in Cleveland, Ohio. *Why* – Cleveland Casting is interested in quality control. They want know how close to the desired pouring temperature of 2,550 degrees Fahrenheit the machine is able to keep the molten iron. *How* – Random sampling. *Cases* – Each crankshaft is a case. *Variables* – Pouring temperature is a quantitative variable.

7. **Arby's menu.** *Who* – Arby's sandwiches. *What* – type of meat, number of calories, and serving size (in ounces). *When* – Not specified. *Where* – Arby's restaurants. *Why* – These data might be used to assess the nutritional value of the different sandwiches. *How* – Information was gathered from each of the sandwiches on the menu at Arby's, resulting in a census. *Cases* – Each sandwich is a case. *Variables* – There are three variables. Number of calories and serving size are quantitative variables, and type of meat is a categorical variable.

8. **Firefighters.** *Who* – Shipboard Firefighters. *What* – Pulling force (in Newtons), weight (probably in pounds), and gender. *When* – Not specified, but sometime prior to the 1982 report. *Where* – Not specified. *Why* – The researchers wanted to compare the abilities of men and women. *How* – Not specified. *Cases* – Each firefighter is an individual. *Variables* – There are three variables. Pulling force and weight are quantitative variables. Gender is a categorical variable.

9. **Babies.** *Who* – 882 births. *What* – Mother's age (in years), length of pregnancy (in weeks), type of birth (caesarean, induced, or natural), level of prenatal care (none, minimal, or adequate), birth weight of baby (unit of measurement not specified, but probably pounds and ounces), gender of baby (male or female), and baby's health problems (none, minor, major). *When* – 1998-2000. *Where* – Large city hospital. *Why* – Researchers were investigating the impact of prenatal care on newborn health. *How* – It appears that they kept track of all births in the form of hospital records, although it is not specifically stated. *Cases* – Each of the 882 births is a case. *Variables* – There are three quantitative variables: mother's age, length of pregnancy, and birth weight of baby. There are four categorical variables: type of birth, level of prenatal care, gender of baby, and baby's health problems.

10. **Flowers.** *Who* – 385 species of flowers. *What* – Date of first flowering (in days). *When* – Not specified. *Where* – Southern England. *Why* – The researchers believe that this indicates a warming of the overall climate. *How* – Not specified. *Cases* – Each of the 385 species at each of the 47 years is a case, for a total of 18,095 cases. *Variables* – Date of first flowering is a quantitative variable. *Concerns* - Hopefully, date of first flowering was measured in days from January 1, or some other convention, to avoid problems with leap years.

11. **Fitness.** *Who* – 25,892 men. *What* – Fitness level and cause of death. It is not clear what categories were used for fitness level. *When* - Over a 10-year period prior to the article being published in May 2002. *Where* – Not specified. *Why* – To establish an association between fitness level and death from cancer. *How* – Researchers tracked the group of men over a 10-year period. *Cases* – Each of the 25,892 men was an individual. *Variables* – There are two categorical variables: fitness level and cause of death.

12. **Schools.** *Who* – Students. *What* – Age (probably in years, though perhaps in years and months), race or ethnicity, number of absences, grade level, reading score, math score, and disabilities/special needs. *When* – This information must be kept current. *Where* – Not specified. *Why* – Keeping this information is a state requirement. *How* – The information is collected and stored as part of school records. *Cases* – Each student is an individual. *Variables* – There are seven variables. Race or ethnicity, grade level, and disabilities/special needs are categorical variables. Number of absences, age, reading test score, and math test score are quantitative variables. *Concerns* – What tests are used to measure reading and math ability, and what are the units of measure for the tests?

13. **Herbal medicine.** *Who* – experiment volunteers. *What* – herbal cold remedy or sugar solution, and cold severity. *When* – Not specified. *Where* – Major pharmaceutical firm. *Why* – Scientists were testing the efficacy of an herbal compound on the severity of the common cold. *How* – The scientists set up a controlled experiment. *Cases* – Each volunteer is an individual. *Variables* – There are two variables. Type of treatment (herbal or sugar solution) is categorical, and severity rating is quantitative. *Concerns* – The severity of a cold seems subjective and difficult to quantify. Also, the scientists may feel pressure to report negative findings about the herbal product.

14. **Tracking sales.** *Who* – Customers of a start-up company. *What* – Customer name, ID number, region of the country, date of last purchase, amount of purchase (probably in dollars), and item purchased. *When* – Present time. *Where* – United States. *Why* – The company is building a data base of sales information. *How* – Presumably, the company records the information from each new customer. *Cases* – Each customer is an individual. *Variables* – There are six variables. Name, ID number, region of the country, and item purchased are categorical variables. Date and amount of purchase are quantitative variables. *Concerns* – Region is a categorical variable, and it is potentially confusing to record it as a number.

15. **Cars.** *Who* – Automobiles. *What* – Make, country of origin, type of vehicle and age of vehicle (probably in years). *When* – Not specified. *Where* – A large university. *Why* – Not specified. *How* – A survey was taken in campus parking lots. *Cases* – Each car is a case. *Variables* – There are three categorical variables and one quantitative variable. Make, country of origin, and type of vehicle are categorical variables, and age of vehicle is a quantitative variable.

16. **Wineries.** *Who* – American vineyards. *What* – Size of vineyard (in acres), number of years in existence, state, varieties of grapes grown, average case price (in dollars), gross sales (probably in dollars), and percent profit. *When* – Not specified. *Where* – United States. *Why* – Business analysts hoped to provide information that would be helpful to producers of American wines. *How* – Not specified. *Cases* – Each vineyard is a case. *Variables* – There are five quantitative variables and two categorical variables. Size of vineyard, number of years in existence, average case price, gross sales, and percent profit are quantitative variables. State and variety of grapes grown are categorical variables.

17. **Streams.** *Who* – Streams. *What* – Name of stream, substrate of the stream (limestone, shale, or mixed), acidity of the water (measured in pH), temperature (in degrees Celsius), and BCI (unknown units). *When* – Not specified. *Where* – Upstate New York. *Why* – Research is conducted for an Ecology class. *How* – Not specified. *Cases* – Each stream is a case. *Variables* – There are five variables. Name and substrate of the stream are categorical variables, and acidity, temperature, and BCI are quantitative variables.

18. **Age and party.** *Who* – 1180 Americans. *What* – Region, age (in years), political affiliation, and whether or not the person voted in the 1998 midterm Congressional election. *When* – First quarter of 1999. *Where* – United States. *Why* – The information was gathered for presentation in a Gallup public opinion poll. *How* – Phone Survey. *Cases* – Each of the 1180 Americans surveyed is an individual in this poll. *Variables* – There are four variables. Region, political affiliation, and whether or not the person voted in 1998 are categorical variables, and age is a quantitative variable.

19. **Air travel.** *Who* – All airline flights in the United States. *What* – Type of aircraft, number of passengers, whether departures and arrivals were on schedule, and mechanical problems. *When* – This information is currently reported. *Where* – United States. *Why* – This information is required by the Federal Aviation Administration. *How* – Data is collected from airline flight information. *Cases* – Each flight is a case. *Variables* – There are four variables. Type of aircraft, departure and arrival timeliness, and mechanical problems are categorical variables, and number of passengers is a quantitative variable.

20. **Fuel economy.** *Who* – Every model of automobile in the United States. *What* – Vehicle manufacturer, vehicle type, weight (probably in pounds), horsepower (in horsepower), and gas mileage (in miles per gallon) for city and highway driving. *When* – This information is collected currently. *Where* – United States. *Why* – The Environmental Protection Agency uses the information to track fuel economy of vehicles. *How* – The data is collected from the manufacturer of each model. *Cases* – Each vehicle model is a case. *Variables* – There are six variables. City mileage, highway mileage, weight, and horsepower are quantitative variables. Manufacturer and type of car are categorical variables.

21. **Refrigerators.** *Who* – 41 refrigerators. *What* – Brand, cost (probably in dollars), size (in cu. ft.), type, estimated annual energy cost (probably in dollars), overall rating, and repair history (in percent requiring repair over the past five years). *When* – 2002. *Where* – United States. *Why* – The information was compiled to provide information to the readers of *Consumer Reports*. *How* – Not specified. *Cases* – Each of the 41 refrigerators is a case. *Variables* – There are 7 variables. Brand, type, and overall rating are categorical variables. Cost, size, estimated energy cost, and repair history are quantitative variables.

Chapter 3 – Displaying Categorical Data

1. **Graphs in the news.** Answers will vary.

2. **Graphs in the news II.** Answers will vary.

3. **Tables in the news.** Answers will vary.

4. **Tables in the news II.** Answers will vary.

5. **Causes of death.**

    a) Yes, it is reasonable to assume that heart and respiratory disease caused approximately 38% of U.S. deaths in 1999, since there is no possibility for overlap. Each person could only have one cause of death.

    b) Since the percentages listed add up to 73.7%, other causes must account for 26.3% of US deaths.

    c) A pie chart is a good choice (with the inclusion of the "Other" category), since causes of US deaths represent parts of a whole. A bar chart would also be a good display.

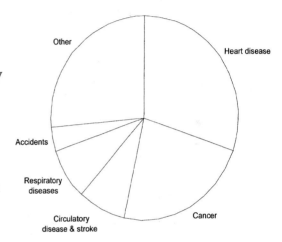

6. **Education.**

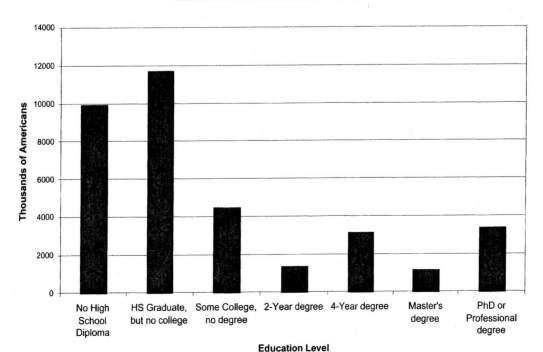

Information obtained by the Census Bureau in December 2000 reports the level of educational attainment of Americans over 65. Most older Americans have, at most, a high school diploma. Only about 15% of older Americans obtained a 4-year or higher degree.

7. **Ghosts.**

   a) It is NOT reasonable to assume the 66% of those polled expressed a belief in either ghosts or astrology. The percentages in the table add up to 185%! This tells us that we are not dealing with parts of a whole and that some respondents believe in more than one of the psychic phenomena listed. In other words, belief in ghosts and belief in astrology are not mutually exclusive. There is no way to know what percent of respondents believe in ghosts or astrology.

   b) As in 7a, since the percentages are not meant to add up to 100%, there is no way to determine what percentage of respondents did not believe in any of these phenomena. With data collected in this fashion, the only way to determine the percentage of people who did not believe in any of the psychic phenomena would be to add another category to the survey when collecting the data.

   c) Since the percentages were not intended to add up to 100%, a pie chart is not appropriate. A bar chart nicely displays the percentages as relative heights of bars. Also, we are probably interested in the most popular phenomena, it makes sense to make a Pareto Chart, with the most common belief occurring first in the chart.

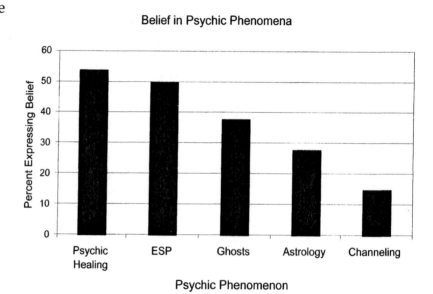

8. **Illegal guns.**

   a) *Who* – 1530 investigations into illegal gun trafficking. *What* – Percentage of cases that were the result of each of the five gun trafficking violations (straw purchase, unlicensed sellers, gun shows and flea markets, stolen from federally licensed dealers, stolen from residences). *When* – July 1996 – December 1998. *Where* – United States. *Why* – These data are being used to track the type of cases involved when gun trafficking violations occur. *How* – The BATF keeps track of this information on each case. The 1530 investigations tracked here probably represent a sample of all gun trafficking cases.

   b) The percentages listed total more than 100%. Either one of the numbers is incorrect, or some cases involved more than one type of gun trafficking violation.

c) The bar chart at the right displays the information:

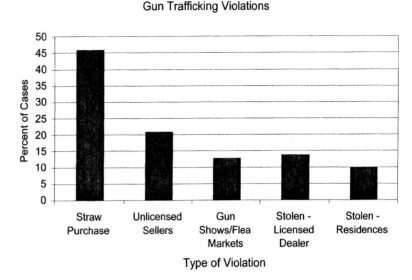

Gun Trafficking Violations

d) A survey of 1530 gun trafficking investigations conducted by the Bureau of Alcohol, Tobacco, and Firearms between July 1996 and December 1998 indicates that the greatest percentage of investigations involve straw purchases, where a legal gun buyer acts on behalf of an illegal gun buyer. Straw purchases are involved in 46% of the investigations. Unlicensed sellers are involved in 21% of investigations. Gun shows and flea markets, thefts from licensed dealers, and thefts from residences are each involved in 10-15% of investigations. It should be noted that these percentages add up to more than 100%. This indicates that either some investigations involve more than one type of violation, or that one or more of the percentages may be incorrect.

e) Corrupt licensed dealers were involved in only 9% of investigations, but were linked to almost half of the illegal firearms trafficked. One explanation for this is that each of these investigations involved large numbers of illegal firearms. Likewise, each straw purchase may have only involved a relatively small number of firearms. Analyzing these data by percentage of investigations only was obscuring crucial information. In light of this new information, corrupt licensed dealers appear to be a bigger problem.

9. Oil spills.

The bar chart shows that grounding is the most frequent cause of oil spillage for these 50 spills, and allows the reader to rank the other types as well. If being able to differentiate between these close counts is required, use the bar chart. The pie chart is also acceptable as a display, but it's difficult to tell whether, for example, there is a greater percentage of spills caused by grounding or hull failure. If you want to showcase the causes of oil spills as a fraction of all 50 spills, use the pie chart.

10. Winter Olympics 2002.

a) Here are two displays of the data:

This bar chart is confusing. There are simply too many categories!

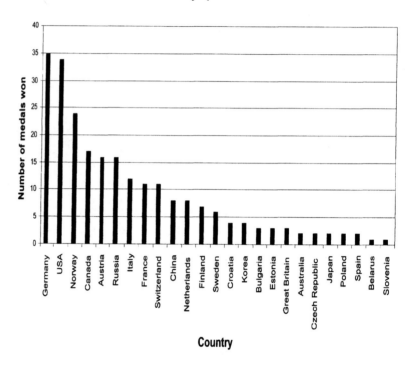

2002 Winter Olympics - Medals Won

A pie chart of the percentage of medals won by each country is even more confusing! The sections of the chart representing countries that won fewer than 5 medals are too small to even label properly.

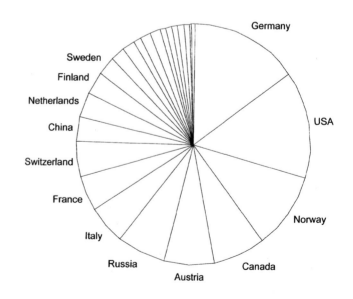

2002 Winter Olympics - Percentage of Medals Won

b) Perhaps we are primarily interested in countries that won many medals. Let's combine all countries that won fewer than 6 medals into a single category. This will make our chart easier to read. We are probably interested in number of medals won, rather than percentage of total medals won, so we'll stick with the bar chart. A bar chart is also better for comparisons.

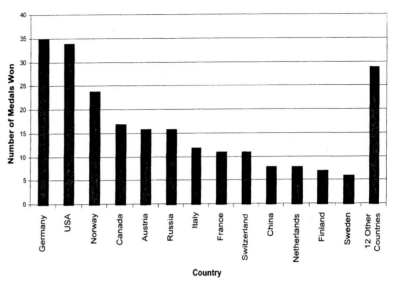

11. Teens and technology.

a) 67% of teens use a calculator and 46% of teens use an answering machine, for a difference of 21%

b) The display makes it appear that more than twice as many teens use calculators as use answering machines.

c) The vertical scale should display percentages starting at 0. This isn't simply an improvement, but rather a necessity. The display provided is misleading.

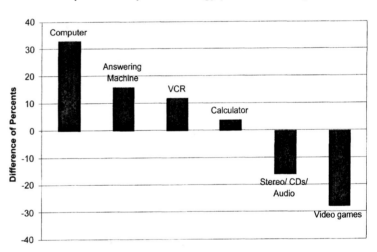

d) The display at the right is the Importance Gap in teens' opinions about technology. It displays the differences in percentages of teens who stated that the piece of technology was critical to use and the percentage of teens who actually own the piece of technology.

e) Teens tend to think that some pieces of technology are critically important to own although they don't use them daily. The Importance Gap display highlights the fact that computers and video games fall on the two extremes of this difference. 77% of teens feel computers are important to own, yet fewer than half use them daily.

12. Teens and technology II.

The main error is that this is simply the wrong type of display. The percentages of students who use each piece of technology aren't meant to represent parts of a whole, but rather overlapping categories. A bar chart would be the appropriate display. Even if a pie chart were acceptable, this one has some other problems. The three-dimensional display distorts the percentages. When using technology, always choose the two-dimensional displays. The percentages don't correspond to the angles of the pie pieces. For example,

85% looks smaller than 67% on the display. To top it all off, the display is unlabeled, giving us no context at all. Appropriate graphs are always self-explanatory.

13. Auditing reform.

a) The pie chart at the right displays the same information as the bar chart.

b) The bar chart makes it easier to compare the percentages. It also preserves the order in which the questions were originally asked.

c) Most respondents in a Gallup poll conducted February 8-10, 2002 favored some sort of reform in the way that corporations are audited. 39% of the respondents advocated major reforms, 35% said minor reforms were needed, and 17% favored a complete overhaul of the corporate auditing system. Only 4% were in favor of no reforms to the current system.

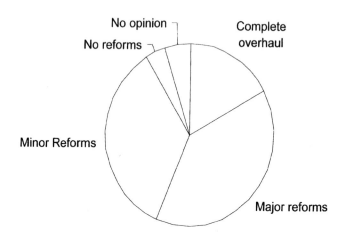

What Kind of Changes Are Needed in the Way Corporations Are Audited?

14. Cars.

a) In order to get percentages, first we need totals. Here is the same table, with row and column totals. Foreign cars are defined as non-American. There are 45+102=147 non-American cars or $147/359 \approx 40.95\%$.

Origin	Driver Student	Staff	Total
American	107	105	212
European	33	12	45
Asian	55	47	102
Total	**195**	**164**	**359**

b) There are 212 American cars of which 107 or $107/212 \approx 50.47\%$ were owned by students.

c) There are 195 students of whom 107 or $107/195 \approx 54.87\%$ owned American cars.

d) The marginal distribution of Origin is displayed in the third column of the table at the right: 59% American, 13% European, and 28% Asian.

Origin	Totals
American	212 (59%)
European	45 (13%)
Asian	102 (28%)
Total	**359**

e) The conditional distribution of Origin for Students is: 55% (107 of 195) American, 17% (33 of 195) European, and 28% (55 of 195) Asian. The conditional distribution of Origin for Staff is: 64% (105 of 164) American, 7% (12 of 164) European, and 29% (47 of 164) Asian.

f) The percentages in the conditional distributions of Origin by Driver (students and staff) seem slightly different. Let's look at a segmented bar chart of Origin by Driver, to compare the conditional distributions graphically.

The conditional distributions of Origin by Driver have similarities and differences. Although students appear to own a higher percentage of European cars and a smaller percentage of American cars than the staff, the two groups own nearly the same percentage of Asian cars. However, because of the differences, there is evidence of an association between Driver and Origin.

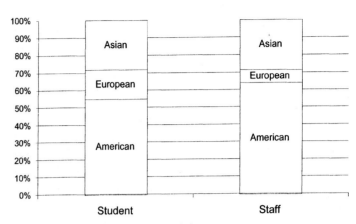

15. Class of 2000.

a) A table with marginal totals is to the right. There are 268 White graduates and 325 total graduates. $268/325 \approx 82.5\%$ of the graduates are White.

Plans	White	Minority	TOTAL
4-year college	198	44	242
2-year college	36	6	42
Military	4	1	5
Employment	14	3	17
Other	16	3	19
TOTAL	**268**	**57**	**325**

b) There are 42 graduates planning to attend 2-year colleges. $42/325 \approx 12.9\%$

c) 36 white graduates are planning to attend 2-year colleges. $36/325 \approx 11.1\%$

d) 36 white graduates are planning to attend 2-year colleges and there are 268 whites graduates. $36/268 \approx 13.4\%$

e) There are 42 graduates planning to attend 2-year colleges. $36/42 \approx 85.7\%$

f) A segmented bar chart is a good display of these data:

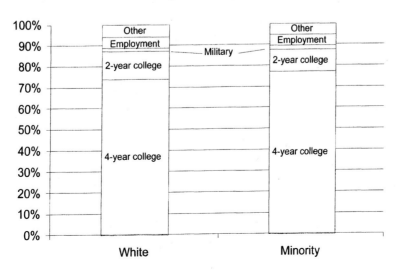

g) The conditional distributions of plans for Whites and Minorities are similar: White – 74% 4-year college, 13% 2-year college, 2% military, 5% employment, 6% other. Minority – 77% 4-year college, 11% 2-year college, 2% military, 5% employment, 5% other. Caution should be used with the percentages for Minority graduates, because the total is so small. Each graduate is almost 2%. Still, the conditional distributions of plans are essentially the same for the two groups. There is little evidence of an association between race and plans for after graduation.

16. After high school.

What graduates did	1959	1970	1980	TOTAL
Continuing education	197	388	320	905
Employed	103	137	98	338
In the military	20	18	18	56
Other	13	58	45	116
Total	333	601	481	1415

a) 56 graduates joined the military, out of 1415 total graduates. $56/1415 \approx 4.0\%$

b) In 1970, 601 students graduated. $601/1415 \approx 42.5\%$

c) In 1970, 18 graduates joined the military. $18/601 \approx 3.0\%$

d) 18 of the 56 students who joined the military were 1970 graduates. $18/56 \approx 32.1\%$

e) Convert the total column to percentages: 64.0% continued education, 23.9% employed, 4.0% in the military, and 8.2% other.

f) Convert the 1959 column to percentages: 59.2% continued education, 30.9% employed, 6.0% in the military, and 3.9% other.

g) There is evidence that the percentage of Ithaca High School graduates continuing their education has increased from 1959 to 1970 to 1980. Likewise, the percentage of students who enter the work force directly after high school appears to have decreased over these years. For example, the percentage of graduates who entered the workforce decreased from 30.9% to 22.8% to 20.4% for the years 1959, 1970, and 1980, respectively.

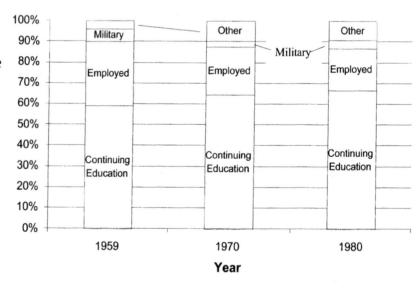

17. Canadian languages.

a) 19,134,000 Canadians speak English only. 19,134,000/28,529,000 total Canadians $\approx 67.1\%$

b) 4,078,000 Canadians speak French only and 4,843,000 speak both French and English, for a total of 8,921,000 French speakers. 8,921,000/28,529,000 total Canadians $\approx 31.3\%$

c) 3,952,000 French and 2,661,000 French and English speakers yield a total of 6,613,000 French speakers in Quebec. 6,613,000/7,045,000 Quebec residents ≈ 93.9%

d) 6,613,000 Quebec residents speak French and 8,921,000 Canadians speak French. The percentage of French-speaking Canadians who live in Quebec is 6,613,000/8,921,000 ≈ 74.1%

e) If language knowledge were independent of Province, we would expect the percentage of French-speaking residents of Quebec to be the same as the overall percentage of Canadians who speak French. Since 31.3% of all Canadians speak French while 93.9% of residents of Quebec speak French, there is evidence of an association between language knowledge and Province.

18. Tattoos.

The study by the University of Texas Southwestern Medical Center provides evidence of an association between having a tattoo and contracting hepatitis C. Around 33% of the subjects who were tattooed in a commercial parlor had hepatitis C, compared with 13% of those tattooed elsewhere, and only 3.5% of those with no tattoo. If having a tattoo and having hepatitis C were independent, we would have expected these percentages to be roughly the same.

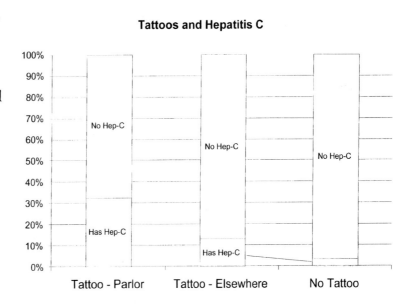

19. Weather forecasts.

a) The table shows the marginal totals. It rained on 34 of 365 days, or 9.3% of the days.

b) Rain was predicted on 90 of 365 days. 90/365 ≈ 24.7% of the days.

		Actual Weather		
		Rain	No Rain	**Total**
Forecast	Rain	27	63	90
	No Rain	7	268	275
	Total	34	331	365

c) The forecast of rain was correct on 27 of the days it actually rained and the forecast of No Rain was correct on 268 of the days it didn't rain. So, the forecast was correct a total of 295 times. 295/365 ≈ 80.8% of the days.

d) There is evidence of an association between the type of weather and the ability of the forecasters to make an accurate prediction. When Rain was forecast, it actually rained only 27 out of 90 times (30%). When the forecasters predicted No Rain, they were correct an amazing 268 out of 275 times (97.5%). The weather forecasters' predictions of Rain are much less reliable than their predictions of No Rain. However, this is hardly a ringing endorsement of the

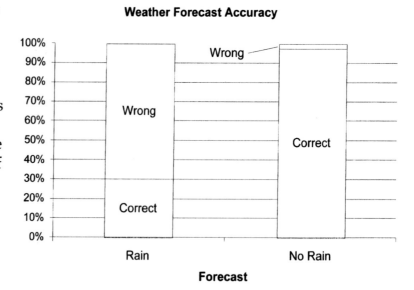

weather forecasters abilities to predict No Rain. Since the actual weather is No Rain on 90.7% of the days, these predictions are a pretty safe bet.

20. Federal prisons.

a) From 1990 to 1998, there was an increase in the percentage of the prison population that was incarcerated for drug and public order offenses. Also during this time period, there was an decrease in the percentage of the prison population incarcerated for property and violent offenses. For example, in 1990, about 52% of the prison population was incarcerated for drug offenses. By 1998, that percentage had increased to about 58%.

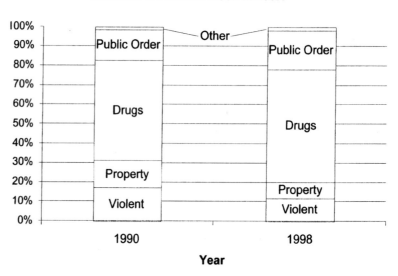

b) These data do *not* show that there has been an increase in drug *use*, merely an increase in the percentage of the prison population incarcerated for drug offenses. One more point of interest here is that the federal prison population in 1998 was almost double the 1990 population.

21. Working Parents.

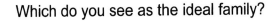

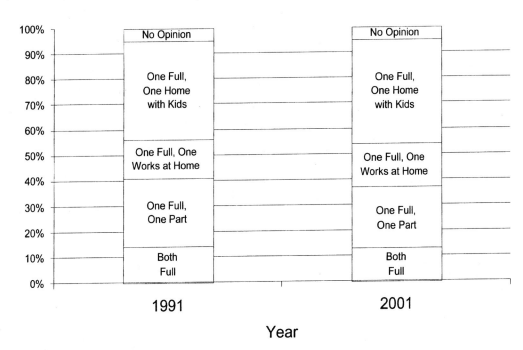

The Gallup poll doesn't provide strong evidence of a change in people's opinions regarding the ideal family in today's society between the years of 1991 and 2001. The conditional distributions of opinion by year appear roughly the same. For example, the percentage of respondents in 1991 who thought the ideal family had two parents that worked fulltime was 14%, and in 2001, the percentage was 13%.

22. Twins.

a) Of the 278,000 mothers who had twins in 1995-1997, 63,000 had inadequate health care during their pregnancies. 63,000/278,000 = 22.7%

Twin Births 1995-97 (in thousands)				
Level of Prenatal Care	Preterm (Induced or Caesarean)	Preterm (without procedures)	Term or Postterm	Total
Intensive	18	15	28	61
Adequate	46	43	65	154
Inadequate	12	13	38	63
Total	76	71	131	278

b) There were 76,000 induced or Caesarean births and 71,000 preterm births without these procedures. (76,000 + 71,000)/278,000 = 52.9%

c) Among the mothers who did not receive adequate medical care, there were 12,000 induced or Caesarean births and 13,000 preterm births without these procedures. 63,000 mothers of twins did not receive adequate medical care. (12,000 + 13,000)/63,000 = 39.7%

d)

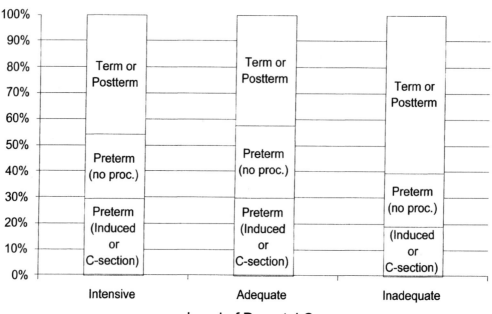

Twin Birth Outcome 1995-1997

e) 52.9% of all twin births were preterm, while only 39.7% of births in which inadequate medical care was received were preterm. This is evidence of an association between level of prenatal care and twin birth outcome. If these variables were independent, we would expect the percentages to be roughly the same. Generally, those mothers who received adequate medical care were more likely to have preterm births than mothers who received intensive medical care, who were in turn more likely to have preterm births than mothers who received inadequate health care. This does *not* imply that mothers should receive inadequate health care do decrease their chances of having a preterm birth, since it is likely that women that have some complication *during* their pregnancy (that might lead to a preterm birth), would seek intensive or adequate prenatal care.

23. Blood pressure.

a) The marginal distribution of blood pressure for the employees of the company is the total column of the table,

Blood pressure	under 30	30 - 49	over 50	Total
low	27	37	31	95
normal	48	91	93	232
high	23	51	73	147
Total	98	179	197	474

converted to percentages. 20% low, 49% normal and 31% high blood pressure.

b) The conditional distribution of blood pressure within each age category is:
Under 30 : 28% low, 49% normal, 23% high
30 – 49 : 21% low, 51% normal, 28% high
Over 50 : 16% low, 47% normal, 37% high

c) A segmented bar chart of the conditional distributions of blood pressure by age category is at the right.

d) In this company, as age increases, the percentage of employees with low blood pressure decreases, and the percentage of employees with high blood pressure increases.

e) No, this does not prove that people's blood pressure increases as they age. Generally, an association between two variables does not imply a cause-and-effect relationship. Specifically, these data come from only one company and cannot be applied to all people. Furthermore, there may be some other variable that is linked to both age and blood pressure. Only a controlled experiment can isolate the relationship between age and blood pressure.

24. Obesity and exercise.

a) Participants were categorized as Normal, Overweight or Obese, according to their Body Mass Index. Within each classification of BMI (column), participants self reported exercise levels. Therefore, these are column percentages. The percentages sum to 100% in each column, *not* across each row.

b) A segmented bar chart of the conditional distributions of level of physical activity by Body Mass Index category is at the right.

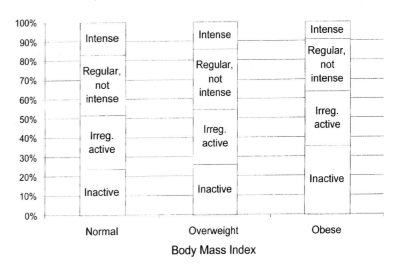

c) No, even though the graphical displays provide strong evidence that lack of exercise and BMI are not independent. All three BMI categories have nearly the same percentage of subjects who report "Regular, not intense" or "Irregularly active", but as we move from Normal to Overweight to Obese we see a decrease in the percentage of subjects who report "Regular, intense" physical activity (16.8% to 14.2% to 9.1%), while the percentage of subjects who report themselves as "Inactive" increases. While it may seem logical that lack of exercise causes obesity, association between variables does not imply a cause-and-effect relationship. A lurking variable (for example, overall health) might influence both BMI and level of physical activity, or perhaps lack of exercise is *caused by* obesity. Only a controlled experiment could isolate the relationship between BMI and level of physically activity.

25. Family planning.

The percentage of unplanned pregnancies decreased as education level increased. However, this does *not* mean that more schooling taught young women better family planning. Association between variables is not evidence of a cause-and-effect relationship. In this case, other socioeconomic variables might be related to both unplanned pregnancy and education level. Perhaps some women even had to leave school *because* of an unplanned pregnancy.

26. Pet ownership.

a) No, the income distributions of households by pet ownership wouldn't be expected to be the same. Caring for a horse is much more expensive, generally, than caring for a dog, cat, or bird. Households with horses as pets would be expected to be more common in the higher income categories.

b) These are column percentages, since each column totals 100%. Each pet was classified as belonging to a family in one of the income level categories.

c)

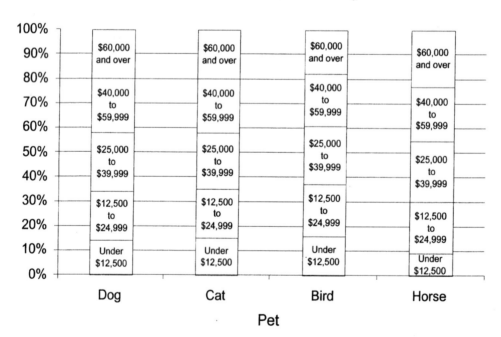

The data support the initial guess to a certain extent. The percentage of horses whose owners have income less than $12,500 is only 9%, compared to percentages in the 20s for other income levels, while the income levels of owners of other pets were distributed in roughly the same percentages. However, with the exception of those earning less than $12,500, the percentages in each income level among horse owners weren't much different.

27. Worldwide toy sales.

a) 2042.16 million dollars worth of toys are sold in European department stores, and 5987.6 million dollars worth of toys are sold in all department stores. 2042.16/5987.6 ≈ 34.1%

b) 2843.8 million dollars worth of toys are sold through catalogs. The "World" total of the table is 74395 million dollars worth of toys sold. The percentage of all toys sold through catalogs is 2843.8/74395 ≈ 3.8%.

c) Toy Chains are more popular sellers of toys in Europe than in North America, as are Toy, Hobby and Game retailers and Department Stores. In North America, General Merchandise and Food, Drug and Misc. Outlets are more popular than they are in Europe.

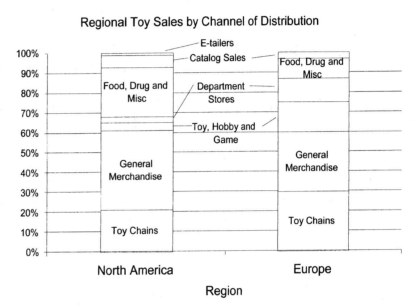

d) The distribution channel varies by region, although General Merchandise stores are either the most popular or a close second in all regions. Toy Chains are especially important in Europe and Toy, Hobby and Game Shops are important in Asia. Only North America has a significant percentage of toys sold in Food and Drug Stores. In the year 1999, E-tailing of toys was of little significance.

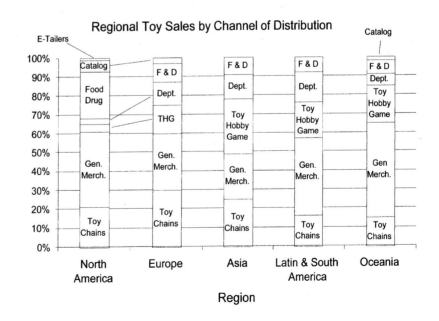

28. Driver's licenses.

a) A table with marginal totals is provided at the right. There 9,743,519 drivers under 20 and a total of 190,625,024 drivers in the U.S. That's about 5% of U.S. drivers under 20.

b) There are 95,796,069 males out of 190,625,024 total U.S. drivers, or about 50%.

Age	Male	Female	Total
19 and under	5,029,498	4,714,021	9,743,519
20-24	8,158,599	7,807,078	15,965,677
25-29	8,988,142	8,597,716	17,585,858
30-34	9,767,476	9,387,297	19,154,773
35-39	10,621,910	10,437,549	21,059,459
40-44	10,576,976	10,516,251	21,093,227
45-49	9,578,268	9,575,363	19,153,631
50-54	8,448,424	8,419,527	16,867,951
55-59	6,394,207	6,366,285	12,760,492
60-64	4,970,258	4,944,370	9,914,628
65-69	4,182,933	4,202,950	8,385,883
70-74	3,644,990	3,822,570	7,467,560
75-79	2,820,136	3,091,013	5,911,149
80-84	1,656,789	1,854,278	3,511,067
85 and over	957,463	1,092,687	2,050,150
Total	95,796,069	94,828,955	190,625,024

c) Each age category appears to have about 50% male and 50% female drivers. The segmented bar chart shows a pattern in the deviations from 50%. At younger ages, males form the slight majority of drivers. This percentage shrinks until the percentages are 50% male and 50% for middle aged drivers. The percentage of male drivers continues to shrink until, at around age 65, female drivers hold a slight majority. This continues into the 85 and over category. It should be noted that this relationship is *very* slight, and may just be a coincidence.

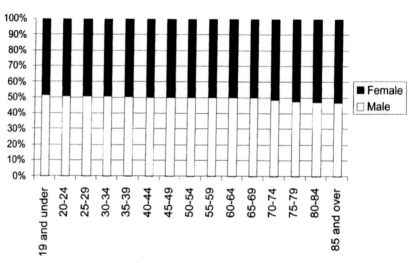

Registered U.S. Drivers by Age and Gender

d) There appears to be a slight association between age and gender of U.S. drivers. Younger drivers are slightly more likely to be male, and older drivers are slightly more likely to be female.

29. Hospitals.

a) The marginal totals have been added to the table:

		Discharge delayed		
		Large Hospital	**Small Hospital**	**Total**
Procedure	**Major surgery**	120 of 800	10 of 50	130 of 850
	Minor surgery	10 of 200	20 of 250	30 of 450
	Total	130 of 1000	30 of 300	160 of 1300

160 of 1300, or about 12.3% of the patients had a delayed discharge.

b) Major surgery patients were delayed 130 of 850 times, or about 15.3% of the time. Minor Surgery patients were delayed 30 of 450 times, or about 6.7% of the time.

c) Large Hospital had a delay rate of 130 of 1000, or 13%.
Small Hospital had a delay rate of 30 of 300, or 10%.
The small hospital has the lower overall rate of delayed discharge.

d) Large Hospital: Major Surgery 15% delayed and Minor Surgery 5% delayed.
Small Hospital: Major Surgery 20% delayed and Minor Surgery 8% delayed.
Even though small hospital had the lower overall rate of delayed discharge, the large hospital had a lower rate of delayed discharge for each type of surgery.

e) No. While the overall rate of delayed discharge is lower for the small hospital, the large hospital did better with *both* major surgery and minor surgery.

f) The small hospital performs a higher percentage of minor surgeries than major surgeries. 250 of 300 surgeries at the small hospital were minor (83%). Only 200 of the large hospital's 1000 surgeries were minor (20%). Minor surgery had a lower delay rate than major surgery (6.7% to 15.3%), so the small hospital's overall rate was artificially inflated. Simply put, it is a mistake to look at the overall percentages. The real truth is found by looking at the rates after the information is broken down by type of surgery, since the delay rates for each type of surgery are so different. The larger hospital is the better hospital when comparing discharge delay rates.

30. Delivery service.

a) Pack Rats has delivered a total of 28 late packages (12 Regular + 16 Overnight), out of a total of 500 deliveries (400 Regular + 100 Overnight). 28/500 = 5.6% of the packages are late. Boxes R Us has delivered a total of 30 late packages (2 Regular + 28 Overnight) out of a total of 500 deliveries (100 Regular + 400 Overnight). 30/500 = 6% of the packages are late.

b) The company should have hired Boxes R Us instead of Pack Rats. Boxes R Us only delivers 2% (2 out of 100) of its Regular packages late, compared to Pack Rats, who deliver 3% (12 out of 400) of its Regular packages late. Additionally, Boxes R Us only delivers 7% (28 out of 400) of its Overnight packages late, compared to Pack Rats, who delivers 16% of its Overnight packages late. Boxes R Us is better at delivering Regular and Overnight packages.

c) This is an instance of Simpson's Paradox, because the overall late delivery rates are unfair averages. Boxes R Us delivers a greater percentage of its packages Overnight, where it is comparatively harder to deliver on time. Pack Rats delivers many Regular packages, where it is easier to make an on-time delivery.

31. Graduate admissions.

a) 1284 applicants were admitted out of a total of 3014 applicants. 1284/3014 = 42.6%

Program	Males Accepted (of applicants)	Females Accepted (of applicants)	Total
1	511 of 825	89 of 108	600 of 933
2	352 of 560	17 of 25	369 of 585
3	137 of 407	132 of 375	269 of 782
4	22 of 373	24 of 341	46 of 714
Total	1022 of 2165	262 of 849	1284 of 3014

b) 1022 of 2165 (47.2%) of males were admitted. 262 of 849 (30.9%) of females were admitted.

c) Since there are four comparisons to make, the table at the right organizes the percentages of males and females accepted in each program. Females are accepted at a higher rate in every program.

Program	Males	Females
1	61.9%	82.4%
2	62.9%	68.0%
3	33.7%	35.2%
4	5.9%	7%

d) The comparison of acceptance rate within each program is most valid. The overall percentage is an unfair average. It fails to take the different numbers of applicants and different acceptance rates of each program. Women tended to apply to the programs in which gaining acceptance was difficult for everyone. This is an example of Simpson's Paradox.

Chapter 4 – Displaying Quantitative Data

1. **Statistics in print.** Answers will vary.

2. **Not a histogram.** Answers will vary.

3. **Thinking about shape.**

   a) The distribution of the number of speeding tickets each student in the senior class of a college has ever had is likely to be unimodal and skewed to the right. Most students will have very few speeding tickets (maybe 0 or 1), but a small percentage of students will likely have comparatively many (3 or more?) tickets.

   b) The distribution of player's scores at the U.S. Open Golf Tournament would most likely be unimodal and slightly skewed to the right. The best golf players in the game will likely have around the same average score, but it is easier to score 15 strokes below the mean than 15 strokes above it.

   c) The weights of female babies in a particular hospital over the course of a year will likely have a distribution that is unimodal and symmetric. Most newborns have about the same weight, with some babies weighing more and less than this average. There may be slight skew to the left, since there seems to be a greater likelihood of premature birth (and low birth weight) than post-term birth (and high birth weight).

   d) The distribution of the length of the average hair on the heads of students in a large class would likely be bimodal and skewed to the right. The average hair length of the males would be at one mode, and the average hair length of the females would be at the other mode, since women typically have longer hair than men. The distribution would be skewed to the right, since it is not possible to have hair length less than zero, but it is possible to have a variety of lengths of longer hair.

4. **More shapes.**

   a) The distribution of the ages of people at a Little League game would likely be bimodal and skewed to the right. The average age of the players would be at one mode and the average age of the spectators (probably mostly parents) would be at the other mode. The distribution would be skewed to the right, since it is possible to have a greater variety of ages among the older people, while there is a natural left endpoint to the distribution at zero years of age.

   b) The distribution of the number of siblings of people in your class is likely to be unimodal and skewed to the right. Most people would have 0, 1, or 2 siblings, with some people having more siblings.

   c) The distribution of pulse rate of college-age males would likely be unimodal and symmetric. Most males' pulse rates would be around the average pulse rate for college-age males, with some males having lower and higher pulse rates.

   d) The distribution of the number of times each face of a die shows in 100 tosses would likely be uniform, with around 16 or 17 occurrences of each face (assuming the die had six sides).

5. Heart attack stays.

The distribution of the length of hospital stays of female heart attack patients is skewed to the right, with stays ranging from 1 day to 34 days. The distribution is centered around 8 days, with the majority of the hospital stays lasting between 1 and 15 days. There are a relatively few hospital stays longer than 27 days. Many patients have a stay of only one day, possibly because the patient died.

6. Emails.

The distribution of the number of emails received from each student by a professor in a large introductory statistics class during an entire term is skewed to the right, with the number of emails ranging from 1 to 21 emails. The distribution is centered at about 2 emails, with many students only sending 1 email. There is one outlier in the distribution, a student who sent 21 emails. The next highest number of emails sent was only 8.

7. Sugar in cereals.

a) The distribution of the sugar content of breakfast cereals is bimodal, with a cluster of cereals with sugar content around 10% sugar and another cluster of cereals around 48% sugar. The lower cluster shows a bit of skew to the right. Most cereals in the lower cluster have between 0% and 10% sugar. The upper cluster is symmetric, with center around 45% sugar.

b) There are two different types of breakfast cereals, those for children and those for adults. The children's cereals are likely to have higher sugar contents, to make them taste better (to kids, anyway!). Adult cereals often advertise low sugar content.

8. Singers.

a) The distribution of the heights of singers in the chorus is bimodal, with a mode at around 65 inches and another mode around 71 inches. No chorus member has height below 60 inches or above 76 inches.

b) The two modes probably represent the mean heights of the male and female members of the chorus.

9. Wineries.

a) There is information displayed about 36 wineries and it appears that 25 of the wineries are smaller than 50 acres. That's around 69% of the wineries. (75% would be a good estimate!)

b) The distribution of the size of 36 Finger Lakes wineries is skewed to the right. Most wineries are smaller than 75 acres, with a few larger ones, from 75 to 175 acres. One winery was larger than all the rest, over 250 acres. The mode of the distribution is between 25 and 50 acres.

10. Runtimes.

The distribution of runtimes is skewed to the right. The shortest runtime was around 28.5 minutes and the longest runtime was around 35.5 minutes. A typical run time was between 30 and 31 minutes, and the majority of runtimes were between 29 and 32 minutes. It is easier to run slightly slower than usual and end up with a longer runtime than it is to run slightly faster than usual and end up with a shorter runtime. This could account for the skew to the right seen in the distribution.

11. Home runs.

The distribution of the number of homeruns hit by Mark McGwire during the 1986 – 2000 seasons is skewed to the right, with a typical number of homeruns per season in the 30s. With the exception of 3 seasons in which McGwire hit fewer than 10 homeruns, his total number of homeruns per season was between 22 and the maximum of 70.

12. Bird species.

a) The results of the 1999 Laboratory of Ornithology Christmas Bird Count are displayed in the stem and leaf display at the right. This display uses split stems, to give the display a bit more definition. The lower stem contains leaves with digits 0,1,2,3,4 and the upper stem contains leaves with digits 5,6,7,8,9.

b) The distribution of the number of birds spotted by participants in the 1999 Laboratory of Ornithology Christmas Bird Count is skewed right, with a center at around 160 birds. There are several high outliers, with two participants spotting 206 birds and another spotting 228. With the exception of these outliers, most participants saw between 152 and 186 birds.

Christmas Bird
Count Totals
1999

```
22 | 8
22 |
21 |
21 |
20 | 66
20 |
19 |
19 |
18 | 6
18 | 13
17 | 578
17 |
16 | 67        KEY:
16 | 00223     18 | 6 = 186
15 | 67        birds spotted.
15 | 233
```

13. Horsepower.

The distribution of horsepower of cars reviewed by *Consumer Reports* is nearly uniform. The lowest horsepower was 65 and the highest was 155. The center of the distribution was around 105 horsepower.

14. Population growth.

The distribution of population growth in NE/MW states is unimodal, symmetric and tightly clustered around 5% growth. The distribution of population growth in S/W states is much more spread out, with most states having population growth between 5% and 30%. A typical state had about 15% growth. There were two outliers, Arizona and Nebraska, with 40% and 66% growth, respectively. In general, the growth rates in the S/W states were higher and much more variable than the growth rates in the NE/MW states.

*Consumer
Reports*
Horsepower

```
15 | 05
14 | 2
13 | 0358
12 | 0559
11 | 00555
10 | 359
 9 | 00577      KEY:
 8 | 0058       11 | 5 =
 7 | 01158      115 horsepower
 6 | 55889
```

15. Hurricanes.

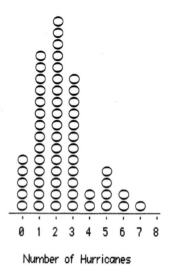

Number of Hurricanes

a) A dotplot of the number of hurricanes each year from 1944 through 2000 is displayed at the right. Each dot represents a year in which there were that many hurricanes.

b) The distribution of the number of hurricanes per year is unimodal and skewed to the right, with center around 2 hurricanes per year. The number of hurricanes per year ranges from 0 to 7. There are no outliers. There may be a second mode at 5 hurricanes per year, but since there were only 4 years in which 5 hurricanes occurred, it is unlikely that this is anything other than natural variability.

16. Hurricanes, again.

The distribution of the number of hurricanes per year before 1970 is unimodal and skewed to the right. The center of the distribution is about 2 to 3 hurricanes per year. The number of hurricanes per year ranges from 0 to 7. After 1970, the distribution of the number of hurricanes per year is also unimodal and skewed right, with center around 1 or 2 hurricanes per year. The number of hurricanes per year ranges from 0 to 6. There may be a difference in the number of hurricanes per year before and after 1970. Before 1970, there may have been a slightly greater number of hurricanes in a typical year.

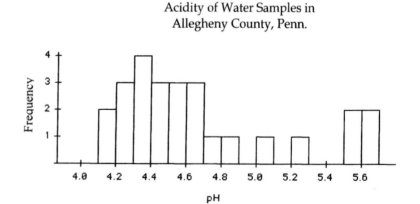

Number of Hurricanes per Year

1944 – 1969		1970 – 2000
OO	0	OOO
OOO	1	OOOOOOOOOO
OOOOOOOO	2	OOOOOOOO
OOOOOO	3	OOOOOO
OO	4	
OO	5	OO
O	6	O
O	7	

17. Acid rain.

The distribution of the pH readings of water samples in Allegheny County, Penn. is bimodal. A roughly uniform cluster is centered around a pH of 4.4. This cluster ranges from pH of 4.1 to 4.9. Another smaller, tightly packed cluster is centered around a pH of 5.6. Two readings in the middle seem to belong to neither cluster.

Acidity of Water Samples in
Allegheny County, Penn.

18. Marijuana.

The distribution of the percentage of 9th graders in 20 Western European countries who have tried marijuana is unimodal and skewed to the right. Greece, at 2%, has the lowest percentage of 9th graders who have tried marijuana. Scotland has the highest percentage, at 53%. A typical country might have a percentage of between 10% and 15%.

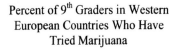
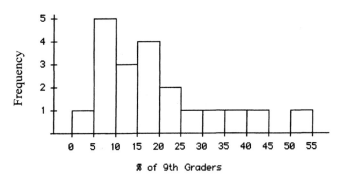

Percent of 9th Graders in Western European Countries Who Have Tried Marijuana

19. Hospital stays.

a) The histograms of male and female hospital stay durations would be easier to compare if the were constructed with the same scale, perhaps from 0 to 20 days

b) The distribution of hospital stays for men is skewed to the right, with many men having very short stays of about 1 or 2 days. The distribution tapers off to a maximum stay of approximately 25 days. The distribution of hospital stays for women is skewed to the right, with a mode at approximately 5 days, and tapering off to a maximum stay of approximately 22 days. Typically, hospital stays for women are longer than those for men.

c) The peak in the distribution of women's hospital stays can be explained by childbirth. This time in the hospital increases the length of a typical stay for women, and not for men.

20. Deaths.

According to the National Vital Statistics report in 1999, there were several key differences between the distributions of age at death for Black Americans and White Americans. The distribution of age at death for Black Americans was skewed left, with a center at approximately 65 to 75 years of age. There was a cluster of ages at death corresponding to the very young. The distribution of age at death for White Americans was also skewed to the left, although to a greater extent than the distribution of age at death for Black Americans. White Americans had a center at approximately 75 to 85 years old at death, roughly 10 years higher than Black Americans. Additionally, the cluster of ages at death corresponding to the very young was much smaller for White Americans than for Blacks, probably indicating a higher infant mortality rate for Black Americans.

21. Final grades.

The width of the bars is much too wide to be of much use. The distribution of grades is skewed to the left, but not much more information can be gathered.

22. Final grades revisited.

a) This display has a bar width that is much too narrow. As it is, the histogram is only slightly more useful than a list of scores. It does little to summarize the distribution of final exam scores.

b) The distribution of test scores is skewed to the left, with center at approximately 170 points. There are several low outliers below 100 points, but other than that, the distribution of scores is fairly tightly clustered.

23. Zipcodes.

Even though zipcodes are numbers, they are not quantitative in nature. Zipcodes are categories. A histogram is not an appropriate display for categorical data. The histogram the Holes R Us staff member displayed doesn't take into account that some 5-digit numbers do not correspond to zipcodes or that zipcodes falling into the same classes may not even represent similar cities or towns. The employee could design a better display by constructing a bar chart that groups together zipcodes representing areas with similar demographics and geographic locations.

24. CEO data revisited.

a) First of all, it must be noted that industry codes are categorical, so the use of a histogram as a display is inappropriate. Strangely enough, the investment analyst made even more mistakes! This display is really just a poorly constructed bar chart (of sorts). There are gaps in the display because all of the industry codes are integers and the widths of the bars are all less than 1. With 5 bars between 0 and 3.75, we can find the width to be 0.75. So, for example, there appears to be a gap between 2.25 and 3, simply because there are no integers between 2.25 and 3 (remember, the upper boundary is not inclusive). The gaps aren't really there, but a poor choice of scale makes them appear.

b) This question doesn't really make any sense. "Unimodal" is a vocabulary word specific to describing distributions of quantitative data. As mentioned before, the industry codes are categorical.

c) A histogram can never be used to summarize categorical data. The analyst would be better off displaying the data in a bar chart, with relative heights of the bars representing the number of CEOs involved in each industry, or a pie chart displaying the percentage of CEOs involved in each industry.

25. Productivity study.

The productivity graph is useless without a horizontal scale indicating the time period over which the productivity increased. Also, we don't know the units in which productivity is measured.

26. Productivity revisited.

The display of productivity and wages display has no scale on either axis and no units are indicated. For the vertical axis, it is unlikely that wages and productivity can be measured meaningfully in the same units, so the two are almost certainly incomparable. Also, we don't know the time scale (on the horizontal axis) over which productivity and wages were measured. In fact, given the problems on the vertical axis, it is not even apparent that the horizontal axis has comparable time periods for wages and productivity.

27. Law enforcement.

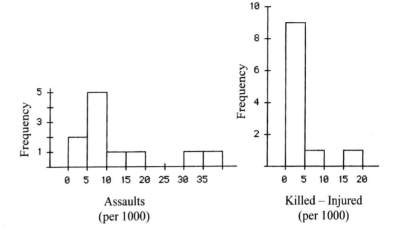

Assaults
(per 1000)

Killed – Injured
(per 1000)

a) The histograms at the right show the number of assaults per 1000 officers and number of killed or injured per 1000 officers, for eleven federal agencies with officers authorized to carry firearms and make arrests.

b) The distribution of assault rates for these federal agencies is roughly symmetric with two high outliers. The center of the distribution is between 5 and 10 assaults per 1000 officers and, with the exception of two agencies, BATF and National Park Service, with over 30 assaults per 100 officers, the distribution is tightly clustered. The distribution of killed and injured rates for these eleven law enforcement agencies is very tightly clustered with almost all of the agencies reporting rates of between 0 and 5 officers per 1000 killed and injured. Customs Service, with a rate of 5.1 officers per 1000, is essentially part of this group, as well. There is one high outlier, the National Park Service, with a rate of 15 officers per 1000 killed and injured.

c) The National Park Service is a high outlier for assaults and killed-injured, with rates of 38.7 officers per 1000 assaulted and 15 officers per 1000 killed or injured. The BATF is a high outlier for assaults, with 31.1 officers per 1000 assaulted.

28. Cholesterol.

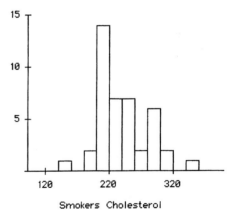

Smokers Cholesterol

The distribution of cholesterol levels for smokers is unimodal and skewed slightly to the right, with a mode around 210. Cholesterol levels vary from approximately 140 to 350, but are generally clustered between 200 and 300. There is one low cholesterol level and one high cholesterol level, but these don't seem to depart from the overall pattern.

The distribution of cholesterol levels for non-smokers is unimodal and roughly symmetric, with a center around 240. Cholesterol levels vary from approximately 120 to 340, and seem spread out. There is one low cholesterol level, but not unusually low.

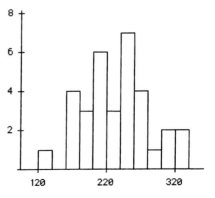

Non-Smokers Cholesterol

In general, the cholesterol levels of smokers seem to be slightly lower than the cholesterol levels of non-smokers. Additionally, the cholesterol levels of smokers appear more consistent than cholesterol levels of non-smokers.

29. MPG.

a) A back-to-back stemplot of these data is shown at the right. A plot with with tens and units digits for stems and tenths for leaves would have been quite long, but still useable. A plot with tens as stems and rounded units as leaves would have been too compact. This plot has tens as the stems, but the stems are split 5 ways. The uppermost 2 stem displays 29 and 28, the next 2 stem displays 27 and 26, and so on. The key indicates the rounding used, as well as the accuracy of the original data. In this case, the mileages were given to the nearest tenth.

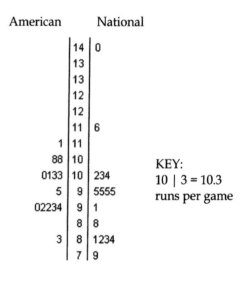

US Cars   Other Cars

```
          | 3 | 7
      44  | 3 | 45
          | 3 | 222
      01  | 3 | 01
      89  | 2 | 8
     777  | 2 | 7
          | 2 |
       2  | 2 | 222
      11  | 2 | 0
  888999  | 1 |
    6777  | 1 | 67
```

KEY:
2 | 5 = 24.5 – 25.4 mpg

b) In general, the Other cars got better gas mileage than the US Cars, although both distributions were highly variable. The distribution of US cars was bimodal and skewed to the right, with many cars getting mileages in the high teens and low twenties, and another group of cars whose mileages were in the high twenties and low thirties. Two high outliers had mileages of 34 miles per gallon. The distribution of Other cars, in contrast, was bimodal and skewed to the left. Most cars had mileages in the high twenties and thirties, with a small group of cars whose mileages were in the low twenties. Two low outliers had mileages of 16 and 17 miles per gallon.

30. Baseball.

a) The back-to-back stemplot shown at the right has split stems to show the distribution in a bit more detail than a stemplot with single stems.

American   National

```
        | 14 | 0
        | 13 |
        | 13 |
        | 12 |
        | 12 |
        | 11 | 6
     1  | 11 |
    88  | 10 |
  0133  | 10 | 234
     5  |  9 | 5555
 02234  |  9 | 1
        |  8 | 8
     3  |  8 | 1234
        |  7 | 9
```

KEY:
10 | 3 = 10.3
runs per game

b) The distribution of number of runs per game in stadiums in the American League is unimodal and slightly skewed to the right, clustered predominantly in the interval of 9 to 10 runs per game. In the National League, the number of runs scored per game is distributed symmetrically and is possibly bimodal, with clusters in the high 9s and low 10s and also in the low 8s. There are two high outliers of 11.6 and 14 runs per game. The number of runs scored per game is generally higher and more consistent in the American League.

c) The 14 runs per game scored at Coors Field is an outlier in the National League data for the first half of the 2001 season. There appear to be more runs per game scored there than in other Major League Stadiums.

31. Nuclear power.

a) The stemplot at the right shows the distribution of construction costs, measured in $1000 per mW of 12 nuclear power generators.

Construction Cost
($1000/mW)

```
8 | 0148
7 | 9
6 | 023        Key:
5 | 6            7|9 = $79000
4 |                    per mW
3 | 25
2 | 8
```

b) The distribution of nuclear power plant construction costs is skewed left, and has several modes, with gaps in between. Several plants had costs near $80,000 per mW, another few had costs near $60,000 per mW and several had costs near $30,000 per mW.

c) The timeplot at the right shows the change in nuclear plant construction costs over time.

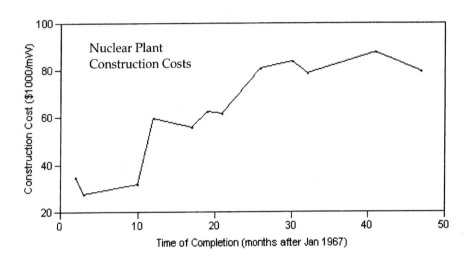

d) From the timeplot, it is apparent that the nuclear plant construction costs generally increased over time.

32. Drunk driving.

a) The stemplot (near right) shows the distribution of drunk driving deaths.

b) The timeplot (far right) shows the change in drunk driving deaths over time.

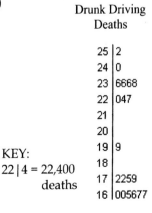

Drunk Driving
Deaths

```
25 | 2
24 | 0
23 | 6668
22 | 047
21 |
20 |
19 | 9
18 |
17 | 2259
16 | 005677
```

KEY:
22|4 = 22,400
deaths

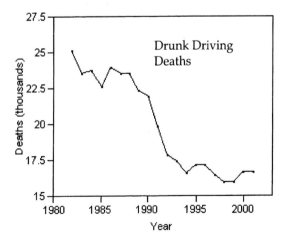

c) The distribution of the number of drunk driving deaths is bimodal, with a cluster between 22 and 25 thousand deaths and another cluster between 16 and 17 thousand deaths. The timeplot shows that this corresponds to a rapid decrease in the drunk driving deaths in the early nineties. Previously, the number of deaths was high, then decreased dramatically.

33. Assets.

a) The distribution of assets of 79 companies chosen from the *Forbes* list of the nation's top corporations is skewed so heavily to the right that the vast majority of companies have assets represented in the first bar of the histogram, 0 to 10 billion dollars. This makes meaningful discussion of center and spread impossible.

b) The distribution of logarithm of assets is preferable, because it is roughly unimodal and symmetric. The distribution of the square root of assets is still skewed right, with outliers.

c) If $\sqrt{Assets} = 50$, then the companies assets are approximately $50^2 = 2500$ million dollars.

d) If $\log(Assets) = 3$, then the companies assets are approximately $10^3 = 1000$ million dollars.

34. Rainmakers.

a) Since one acre-foot is about 320,000 gallons, these numbers are more manageable than gallons.

b) The distribution of rainfall from 26 clouds seeded with silver iodide is skewed heavily to the right, with the vast majority of clouds producing less than 500 acre-feet of rain. Several clouds produced more, with a maximum of 2745 acre-feet.

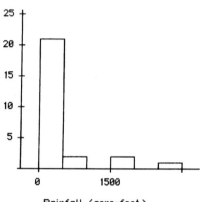

Rainfall (acre-feet)

c) The distribution of log (base 10) of rainfall is much more symmetric than the distribution of rainfall. We can see that the center of the distribution is around log 2 – log 2.5 acre-feet.

d) Since the reexpressed scale is measured in log (acre-feet), we need to raise 10 to the power of the number on our scale to convert back to acre feet. For example, if a cloud in the new scale has a log (rainfall) of 2.3, we convert back to rainfall as follows:

$$\log(rainfall) = 2.3$$
$$rainfall = 10^{2.3}$$
$$rainfall = 199.5$$

The cloud produced 199.5 acre-feet of rain.

Log(Rainfall)

Chapter 5 – Describing Distributions Numerically

1. **In the news.** Answers will vary.

2. **In the news II.** Answers will vary.

3. **Summaries.**

   a) The mean price of the electric smoothtop ranges is $1001.50.

   b) In order to find the median and the quartiles, the list must be ordered.
      565   750   850   900   1000   1050   1050   1200   1250   1400
      The median price of the electric ranges is $1025.
      Quartile 1 = $850 and Quartile 3 = $1200.

   c) The range of the distribution of prices is Max – Min = $1400 – $565 = $835.
      The IQR = Q3 – Q1 = $1200 - $850 = $350.

4. **More summaries.**

   a) The mean annual number of deaths from tornadoes is 56.27.

   b) In order to find the median and the quartiles, the list must be ordered.
      25   30   33   39   39   40   53   67   69   94   130
      The median annual number of deaths from tornadoes is 40.
      Quartile 1 = 33 deaths, and Quartile 3 = 69 deaths.
      (Some statisticians consider the median to be part of the both the lower and upper halves
      of the ordered list. This changes the position of the quartiles slightly. If median is
      included, Q1 = 36, Q3 = 68. In practice, it rarely matters, since these measures of position
      are best for large data sets.)

   c) The range of the distribution of deaths is Max – Min = 130 – 25 = 105 deaths.
      The IQR = Q3 – Q1 = 69 – 33 = 36 deaths. (Or, the IQR = 68 – 38 = 32 deaths, if the median is
      included in both halves of the ordered list.)

5. **Mistake.**

   a) As long as the boss's true salary of $200,000 is still above the median, the median will be
      correct. The mean will be too large, since the total of all the salaries will decrease by
      $2,000,000 - $200,000 = $1,800,000, once the mistake is corrected.

   b) The range will likely be too large. The boss's salary is probably the maximum, and a lower
      maximum would lead to a smaller range. The IQR will likely be unaffected, since the new
      maximum has no effect on the quartiles. The standard deviation will be too large, because
      the $2,000,000 salary will have a large squared deviation from the mean.

6. **Sick days.**

   The company probably uses the mean, while the union uses the median number of sick
   days. The mean will likely be higher, since it is affected by probable right skew. Some
   employees may have many sick days, while most have relatively few.

7. **Payroll.**

   a) The mean salary is $\dfrac{(1200 + 700 + 6(400) + 4(500))}{12} = \$525$.

   The median salary is the middle of the ordered list:
   400   400   400   400   400   400   500   500   500   500   700   1200
   The median is $450.

   b) Only two employees, the supervisor and the inventory manager, earn more than the mean wage.

   c) The median better describes the wage of the typical worker. The mean is affected by the two higher salaries.

   d) The IQR is the better measure of spread for the payroll distribution. The standard deviation and the range are both affected by the two higher salaries.

8. **Singers.**

   a) 5-number summary: 60, 65, 66, 70, 76

   b) The boxplot of heights of the choir members is at the right.

   c) The mean height of the singers is 67.12 inches, and the standard deviation of the heights is 3.79 inches.

   d) The histogram of heights of the choir members is below the boxplot.

   e) The distribution of the heights of the choir members is bimodal (probably due to differences in height of men and women) and skewed slightly to the right. The median is 66 inches. The distribution is fairly spread out, with the middle 50% of the heights falling between 65 and 70 inches. There are no gaps or outliers in the distribution.

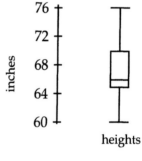

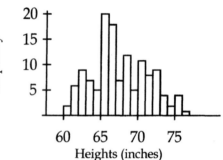

9. **Standard deviation.**

   a) Set 2 has the greater standard deviation. Both sets have the same mean, 6, but set two has values that are generally farther away from the mean.
   SD(Set 1) = 2.24   SD(Set 2) = 3.16

   b) Set 2 has the greater standard deviation. Both sets have the same mean (15), maximum (20), and minimum (10), but 11 and 19 are farther from the mean than 14 and 16.
   SD(Set 1) = 3.61   SD(Set 2) = 4.53

   c) The standard deviations are the same. Set 2 is simply Set 1 + 80. Although the measures of center and position change, the spread is exactly the same.
   SD(Set 1) = 4.24   SD(Set 2) = 4.24

10. Standard deviation.

a) Set 2 has the greater standard deviation. Both sets have the same mean (7), maximum (10), and minimum (4), but 6 and 8 are farther from the mean than 7.
SD(Set 1) = 2.12   SD(Set 2) = 2.24

b) The standard deviations are the same. Set 1 is simply Set 2 + 90. Although the measures of center and position are different, the spread is exactly the same.
SD(Set 1) = 36.06   SD(Set 2) = 36.06

c) Set 2 has the greater standard deviation. The central 4 values of Set 2 are simply the central 4 values of Set 1 +40, but the maximum and minimum of Set 2 are farther away from the mean than the maximum and minimum of Set 1. Range(Set 1) = 18 and Range(Set 2) = 22. Since the Range of Set 2 is greater than the Range of Set 1, the standard deviation is also larger.
SD(Set 1) = 6.03   SD(Set 2) = 7.04

11. Home runs.

5-number summary: 8, 14, 24.5, 33, 61   IQR = Q3 – Q1 = 33 – 14 = 19 home runs.

Using the Outlier Rule (more than 1.5 IQRs beyond the quartiles) to find the fences:

Upper Fence: $Q3 + 1.5(IQR) = 33 + 1.5(19)$
$= 33 + 28.5$
$= 61.5$

Lower Fence: $Q1 - 1.5(IQR) = 14 - 1.5(19)$
$= 14 - 28.5$
$= -14.5$

Home Runs

Any year with more than 61.5 home runs or less than –14.5 home runs (a meaningless fence) is considered an outlier. Technically, the year in which Maris hit 61 home runs is **not** an outlier, but according to a less formal definition (any point that departs from the overall pattern), it seems strange. For this reason, 61 homeruns is displayed as an outlier in the boxplot.

12. Campsites.

a) The distribution of the number of campsites in public parks in Vermont is skewed to the right, so median and IQR are appropriate measures of center and spread.

b) IQR = Q3 – Q1 = 78 – 28 = 50.
Using the Outlier Rule (1.5 IQRs beyond quartiles):

Upper Fence: $Q3 + 1.5(IQR) = 78 + 1.5(50)$
$= 78 + 75$
$= 153$

Lower Fence: Well below 0 campsites.

There are 3 parks with greater than 180 campsites. These are definitely outliers. There are 2 parks with between 150 and 160 campsites each. These may be outliers as well.

c) A boxplot of the distribution of number of campsites is at the right.

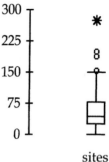

sites

d) The distribution of the number of campsites at public parks in Vermont is unimodal and skewed to the right. The center of the distribution is approximately 44 campsites. The distribution of campsites is quite spread out, with several high outliers. These parks have in excess of 150 campsites each.

13. Marriage age.

The distribution of marriage age of U.S. men is skewed right, with a typical man (as measured by the median) first marrying at around 24 years old. The middle 50% of male marriage ages is between about 23 and 26 years. For U.S. women, the distribution of marriage age is also skewed right, with median of around 21 years. The middle 50% of female marriage age is between about 20 and 23 years. When comparing the two distributions, the most striking feature is that the distributions are nearly identical in spread, but have different centers. Females typically seem to marry earlier than males. In fact, between 50% and 75% of the women marry at a younger age than *any* man.

14. Fuel economy.

Cars with 4 cylinders generally get better gas mileage than cars with 6 cylinders, which generally get better gas mileage than cars with 8 cylinders. Additionally, the greater the number of cylinders, the more consistent the mileage becomes. 4 cylinder cars typically get between 27 – 33 mpg, 6 cylinder cars typically get between 18 – 22 mpg, and large cars typically get between 16 – 19 mpg.

15. Wines.

a) A Seneca Lake winery had the maximum case price of approximately $150.

b) A Seneca Lake winery had the minimum case price of approximately $50.

c) The Keuka Lake wineries were consistently high.

d) Cayuga Lake wineries and Seneca Lake wineries have approximately the same average case price, at about $100, while a typical Keuka Lake winery has a case price of about $130. Keuka Lake wineries have consistently high case prices, between $120 and $140, with one low outlier at about $85 per case. Cayuga Lake wineries have case prices from $70 to $135, and Seneca Lake wineries have highly variable case prices, from $50 to $150.

16. Ozone.

a) April had the highest recorded ozone level, approximately 440.

b) February had the largest IQR of ozone level, approximately 50.

c) August had the smallest range of ozone levels, approximately 50.

d) January had a slightly lower median ozone level than June, 340 and 350, respectively, but June's ozone levels were much more consistent.

e) Generally, ozone levels rose through the winter and were highest in the spring, then fell through the summer and were lowest in the fall. Additionally, ozone levels were very consistent in the summer, became more variable in the fall, were most variable in the winter, and became more consistent through the spring.

17. Unemployment.

The distribution of unemployment rates reported by the U.S. Bureau of Labor Statsitics must have had about 30% of the rates above 4.2% and the rest below. In other words, 4.2% should be slightly below the third quartile (25% of rates above). Other than that, the distribution cannot be accurately determined. An example of one possible distribution is shown in the boxplot at the right.

% Unemployed

18. Wild card Summer Olympics.

a) Since the mean time is larger than the median time, the distribution of qualifying swim times should be skewed to the right.

b) Even if the distribution was generally symmetric, some swimmers with very high times might have been outliers in the distribution.

c) The distribution of qualifying times in the men's 100 meter swim at the 2000 Olympics in Sydney was unimodal and skewed slightly to the right, with all but one swimmer having times between 47 and 63 seconds. One swimmer, Eric Moussambani, had a qualifying time of over 110 seconds.

19. Test scores.

a) Class 3 had the highest mean score, probably somewhere in the 70s. The other two classes had mean scores in the 60s.

b) Class 3 had the highest median score, probably somewhere in the 80s. The other two classes had median scores in the 60s.

c) Class 3 has median higher than the mean, since the distribution of scores is skewed to the left. The mean is pulled toward the tail. The other two classes have roughly symmetric distributions, with mean and median approximately equal.

d) Class 1 has the smallest standard deviation. Most scores are clustered close to the mean.

e) Class 1 probably has the smallest IQR. However, without the actual scores, it is impossible to calculate the exact IQR of these classes. We can estimate them. Class 1 has 24 students, so Q1 is between the 6th and 7th scores, somewhere between 50 and 60. Q3 is between the 18th and 19th scores, somewhere between 70 and 80. The IQR is at least 10 and at most 30. Class 3 also has 24 scores. Q1 is between 60 and 70. Q3 is between the 18th score (80-90) and the 19th score (90-100), meaning it could be between 80 and 100. The IQR is at least 10 and at most 40. Class 2 seems to have the largest IQR, of at least 30 and at most 50.

20. Test scores.

  a) Class 3 did better overall. Class 3 has Q1 at about the same score as the median of either of the other two classes. This means that 75% of the students in Class 3 scored higher than the median of the other two classes.

  b) Classes 1 and 2 had distributions of test scores that were roughly symmetric, while Class 3 had a distribution of scores that was skewed to the left. All classes were fairly spread out, with scores ranging from approximately 30 to 100.

  c) Class A is Class 1   Class B is Class 2   Class C is Class 3

21. Still rockin'.

  a) The histogram and boxplot of the distribution of "crowd crush" victims' ages both show that a typical crowd crush victim was approximately 18 - 20 years of age, that the range of ages is 36 years, that there are two outliers, one victim at age 36 - 38 and another victim at age 46 – 48.

  b) This histogram shows that there may have been two modes in the distribution of ages of "crowd crush" victims, one at 18 - 20 years of age and another at 22 – 24 years of age. Boxplots, in general, can show symmetry and skewness, but not features of shape like bimodality or uniformity.

  c) Median is the better measure of center, since the distribution of ages has outliers. Median is more resistant to outliers than the mean.

  d) IQR is a better measure of spread, since the distribution of ages has outliers. IQR is more resistant to outliers than the standard deviation.

22. Golf courses.

  a) The range of the lengths of the golf courses in Max – Min = 6796 – 5185 = 1611 yards.

  b) In any distribution, 50% of scores lie between Quartile 1 and Quartile 3. In this case, Quartile 1 = 5585.75 yards and Quartile 3 = 6131 yards.

  c) The distribution of golf course lengths appears roughly symmetric, so the mean is the appropriate measure of center. In this case, the mean is 5892.91 yards.

  d) The distribution of the lengths of all the golf courses in Vermont is roughly unimodal and symmetric. The mean length of the golf courses is approximately 5900 yards. Vermont has golf courses anywhere from 5185 yards to 6796 yards long. There are no outliers in the distribution.

23. Graduation?

  a) The distribution of the percent of incoming college freshman who graduate on time is roughly symmetric. The mean and the median are reasonably close to one another and the quartiles are approximately the same distance from the mean.

b) Upper Fence: Q3 + 1.5(IQR) = 74.75 + 1.5(74.75 - 59.15)

$$= 74.75 + 23.4$$

$$= 98.15$$

Lower Fence: Q1 - 1.5(IQR) = 59.15 - 1.5(74.75 - 59.15)

$$= 59.15 - 23.4$$

$$= 35.75$$

Since the maximum value of the distribution of the percent of incoming freshmen who graduate on time is 87.4% and the upper fence is 98.15%, there are no high outliers. Likewise, since the minimum is 43.2% and the lower fence is 35.75%, there are no low outliers. Since the minimum and maximum percentages are within the fences, all percentages must be within the fences.

c) A boxplot of the distribution of the percent of incoming freshmen who graduate on time is at the right.

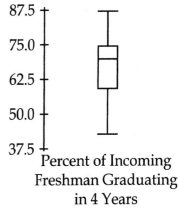

Percent of Incoming
Freshman Graduating
in 4 Years

d) The distribution of the percent of incoming freshmen who graduate on time is roughly symmetric, with mean of approximately 68% of freshmen graduating on time. Universities surveyed had between 43.2% and 87.4% of students graduating on time, with the middle 50% of universities reporting between 59.15% and 74.75% graduating on time.

24. Wineries.

a) The distribution of size of Finger Lakes wineries is skewed heavily to the right. The mean size is a great deal higher than the median size. With a minimum of 6 acres and a range of 244 acres, the maximum must be 6 + 244 = 250 acres. The maximum size is almost 200 acres above Quartile 3.

b) Upper Fence: Q3 + 1.5(IQR) = 55 + 1.5(55 - 18.5)

$$= 55 + 54.75$$

$$= 109.75$$

Lower Fence: Q1 - 1.5(IQR) = 18.5 - 1.5(55 - 18.5)

$$= 18.5 - 54.75$$

$$= -36.25$$

The maximum of 250 acres is well above the upper fence of 109.75 acres. Therefore, there is at least one high outlier, 250 acres. Since the lower fence is negative, there are no low outliers, since it is certainly impossible to have a winery with negative size.

c) The boxplot of the distribution of sizes of Finger Lakes wineries is at the right. There may be additional outliers, but we are sure that there is at least one, the maximium.

d) The distribution of sizes of Finger Lakes wineries is skewed to the right. Many wineries have moderate sizes, with the middle 50% of wineries consisting of 18.5 to 55 acres. The smallest winery is 6 acres. At least one winery is comparatively bigger, at 250 acres.

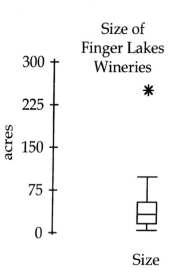

25. Caffeine.

a) *Who* – 45 volunteeers. *What* – Level of caffeine consumption and memory test score. *When* – Not specified. *Where* – Not specified. *Why* – The student researchers want to see the possible effects of caffeine on memory. *How* – It appears that the researchers imposed the treatment of level of caffeine consumption in an experiment. However, this point is not clear. Perhaps they allowed the subjects to choose their own level of caffeine.

b) *Variables* – Caffeine level is a categorical variable with three levels: no caffeine, low caffeine, and high caffeine. Test score is a quantitative variable, measured in number of items recalled correctly.

c)

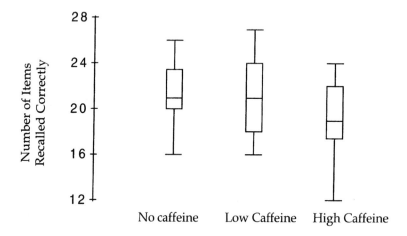

d) The groups consuming no caffeine and low caffeine had comparable memory test scores. A typical score from these groups was around 21. However, the scores of the group consuming no caffeine were more consistent, with a smaller range and smaller interquartile range than the scores of the group consuming low caffeine. The group consuming high caffeine had lower memory scores in general, with a median score of about 19. No one in the high caffeine group scored above 24, but 25% of each of the other groups scored above 24.

26. Rainmakers.

a) Median and IQR, as well as the quartiles are the appropriate summary statistics, since the distribution of amount of rain produced is either skewed to the right or has high outliers. Indication of skewness and outliers can be seen in the comparison of median and mean. The mean amount of rain produced is significantly higher than the median for both seeded and unseeded clouds. Skewness or outliers pulled up the sensitive mean.

b) There is evidence that that the seeded clouds produced more rain. The median and both quartiles are higher than the corresponding statistics for unseeded clouds. In fact, the median amount of rainfall for seeded clouds is 221.60 acre-feet, about 5 times the median amount for unseeded clouds.

27. States.

a) The distribution of state populations is skewed heavily to the right. Therefore, the median and IQR are the appropriate measures of center and spread.

b) The mean population must be larger than the median population. The extreme values on the right affect the mean greatly and have no effect on the median.

c) There are 51 entries in the stemplot, so the 26th entry must be the median. Counting in the ordered stemplot gives median = 4 million people. The middle of the lower 50% of the list (26 state populations) is between the 13th and 14th population, or 1.5 million people. The middle of the upper half of the list (26 state populations) is between the 13th and 14th population from the top, or 6 million people. The IQR = Q3 – Q1 = 6 – 1.5 = 4.5 million people.

d) The distribution of population for the 50 U.S. States and Washington, D.C. is skewed heavily to the right. The median population is 4 million people, with 50% of states having populations between 1 and 6 million people. There is one outlier, a state with 34 million people. The next highest population is only 21 million.

28. Population growth.

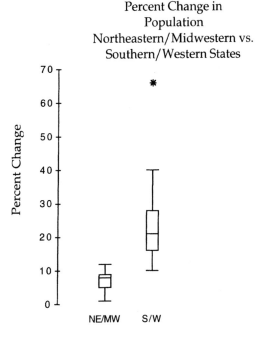

29. Derby speeds.

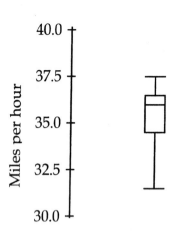

Derby Speeds

a) The median speed is the speed at which 50% of the winning horses ran slower. Find 50% on the left, move straight over to the graph and down to a speed of about 36 mph.

b) Quartile 1 is at 25% on the left, and Quartile 3 is at 75% on the left. Matching these to the ogive, Q1 = 34.5 mph and Q3 = 36.5 mph.

c) Range = Max – Min = 38 – 31 = 7 mph
   IQR = Q3 – Q1 = 36.5 – 34.5 = 2 mph

d) A boxplot of winning Kentucky Derby Speeds is at the right.

e) The distribution of winning speeds in the Kentucky Derby is skewed to the left. The lowest winning speed is just under 31 mph and the fastest speed is about 37.5 mph. The median speed is approximately 36 mph, and 75% of winning speeds are above 34.5 mph. Only a few percent of winners have had speeds below 33 mph.

30. Cholesterol.

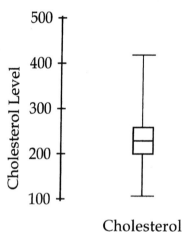

Cholesterol

A boxplot for the distribution of cholesterol levels of 1400 men is at the right. The five number summary is estimated from the ogive to be: 100, 200, 230, 260, 425.

The distribution of cholesterol levels is skewed to the right, and tightly clustered around the median. The median cholesterol level is approximately 230, with the middle 50% of cholesterol levels between 200 and 260. A very small percentage of men had cholesterol below 150 or above 325.

31. Reading scores.

a) The highest score for boys was 6, which is higher than the highest score for girls, 5.9.

b) The range of scores for boys is greater than the range of scores for girls.
   Range = Max – Min   Range(Boys) = 4   Range(Girls) = 3.1

c) The girls had the greater IQR.
   IQR = Q3 – Q1   IQR(Boys) = 4.9 – 3.9 = 1   IQR(Girls) = 5.2 – 3.8 = 1.4

d) The distribution of boys' scores is more skewed. The quartiles are not the same distance from the median. In the distribution of girls' scores, Q1 is 0.7 units below the median, while Q3 is 0.7 units above the median.

e) Overall, the girls did better on the reading test. The median, 4.5, was higher than the median for the boys, 4.3. Additionally, the upper quartile score was higher for girls than boys, 5.2 compared to 4.9. The girls' lower quartile score was slightly lower than the boys' lower quartile score, 3.8 compared to 3.9.

f) The overall mean is calculated by weighting each mean by the number of students.

$$\frac{14(4.2)+11(4.6)}{25} = 4.38$$

32. SAT scores.

a) Parallel boxplots comparing the scores of boys and girls SAT scores are at the right.

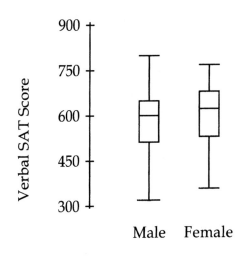

b) Females in this graduating class scored slightly higher on the Verbal SAT, with a median of 625, compared to the median of 600 for the males. Additionally, the females had higher first and third quartiles. The IQR of the males' scores was slightly smaller, than the IQR for the females' scores, indicating a bit more consistency in male scores. However, the overall spread of male scores was greater than that of female scores, with males having both the minimum and maximum score. Both distributions of scores were slightly skewed to the left.

33. Phone calls.

a) A boxplot of the distribution of Net2Phone rates is at the right.

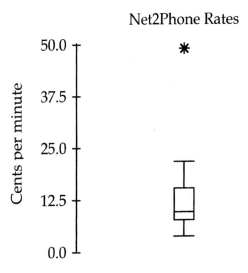

b) The mean rate for Net2Phone is 13.7 cents per minute. The median rate is 9.9 cents per minute. The rates for India and Pakistan are outliers, so the median is the appropriate measure of center.

c) The IQR of the distribution of Net2Phone rates is 7.6 cents per minute and the standard deviation is 11.8 cents per minute. The rates for India and Pakistan are outliers, so the IQR is the appropriate measure of spread.

d) Upper Fence: Q3 + 1.5(IQR) = 15.5 + 1.5(15.5 - 7.9)

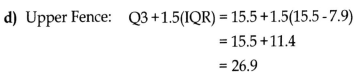

$$= 15.5 + 11.4$$
$$= 26.9$$

Lower Fence: Q1 - 1.5(IQR) = 7.9 - 1.5(15.5 - 7.9)

$$= 7.9 - 11.4$$
$$= -3.5$$

Any country that has a rate higher than 26.9 cents per minute is an outlier. Both India and Pakistan have rates of 49 cents per minute. No country could be a low outlier. It is impossible to have a rate lower than –3.5 cents per minute.

e) The distribution of the long distance rates for Net2Phone is unimodal and symmetric except for two outliers at 49 cents per minute. These are the rates for India and Pakistan. The median rate is 9.9 cents per minute. The middle 50% of rates are between 7.9 and 15.5 cents per minute for an IQR of 7.6 cents per minute.

f) The rates of these 24 countries may not be representative of the rates of all 250 countries that are serviced by Net2Phone. It would be unwise to assume that these rates could tell us anything about Net2Phone rates in general.

34. Job growth.

a) A histogram of the job growth rates predicted by Standard and Poor's DRI is shown at the right. A boxplot, stemplot, or dotplot would also have been an acceptable display.

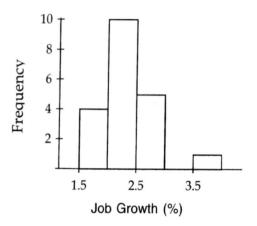

b) The mean predicted growth rate is 2.317% and the median predicted growth rate is 2.235%. The mean is higher because the outlier, the predicted growth rate in Las Vegas, NV, pulls it up.

c) The median would be the appropriate measure of center of the distribution of predicted job growth rates, since the outlier does not affect it.

d) The standard deviation of the distribution of predicted job growth rates is 0.427% and the IQR is 0.515%.

e) The IQR is the appropriate measure of spread, because the outlier influences the standard deviation.

f) If 1.2% were subtracted from each of the predicted job growth rates, the mean and median would each decrease by 1.2%. The standard deviation and the IQR would not change.

g) If we were to omit Las Vegas, an outlier, the mean would decrease. The outlier was pulling it up. The standard deviation would decrease, since the presence of the outlier gave the impression of more spread. The median and IQR would be relatively unaffected, since those measures are resistant to the presence of outliers, although they would change slightly, since they are each based upon relative position. With the outlier removed, there would only be 19 job predicted job growth rates, instead of 20. This would cause the median and the quartiles to shift down slightly.

h) The distribution of job growth rates predicted by Standard and Poor's DRI is unimodal and symmetric except for the one outlier, Las Vegas at 3.72%. The median growth rate for these cities is 2.235%. The middle 50% of the cities had growth rates between 2.045% and 2.560%, for an interquartile range of 0.515%.

35. Math scores.

a) 5-number summary: 275, 448, 499, 531, 604
   IQR: 83
   Mean: 487.2
   Standard deviation: 72.3

b)

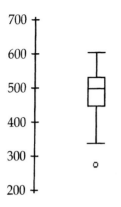

The distribution of average math achievement scores for eighth graders in 38 nations is unimodal, skewed slightly to the left, and centered at about 500. South Africa's score of 275 is the lowest and is a departure from the overall pattern of the data. Not including this score, scores range from 337 to 604, with the middle 50% of the scores falling between 448 and 531. The United States score of 502, although well above the mean of 487.2, is actually near the middle of the distribution. Several low scores pull down the mean, making the median a better measure of center.

36. Prisons.

   a) A histogram of the increase in federal prison populations for northeastern and midwestern states in 1999 is shown at the right.

   b) Since the distribution is roughly unimodal and symmetric, mean and standard deviation are the appropriate measures of center and spread. The five number summary will also be useful.
Mean = 5.08%
Std. Dev. = 1.79%
5-number summary:
1.3%, 3.95%, 5.4%, 6.35%, 8%.

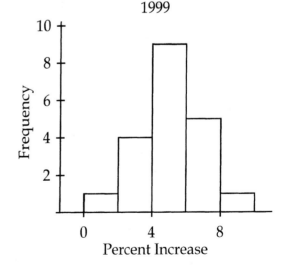

   c) The distribution of percent increases in federal prison populations in 20 northeastern and midwestern states during 1999 is unimodal and roughly symmetric. There are no outliers.
The mean percent increase was 5.08% with a standard deviation of 1.79%. The middle 50% of states had an increase of between 3.95% and 6.35%.

37. Gasoline usage.

In 2000, per capita gasoline usage by state in the United States averaged approximately 500 gallons. The distribution of gasoline usage was slightly skewed to the left, with two low outliers, North Carolina and Hawaii. These states used much less gasoline per person than other states. The IQR of the distribution was 82 gallons per person, with the middle 50% of states having gasoline usage between 453 and 535 gallons per person.

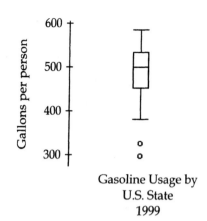

Gasoline Usage by
U.S. State
1999

38. Industrial experiment.

First of all, there is an extreme outlier in the distribution of distances for the slow speed drilling. One hole was drilled almost an inch away from the center of the target! If that distance is correct, the engineers at the computer production plant should investigate the slow speed drilling process closely. It may be plagued by extreme, intermittent inaccuracy. The outlier in the slow speed drilling process is so extreme that no graphical display can display the distribution in a meaningful way while including that outlier. That distance should be removed before looking at a plot of the drilling distances.

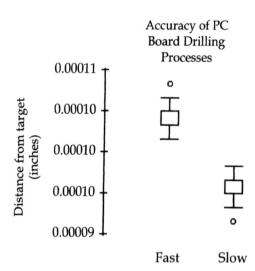

Accuracy of PC
Board Drilling
Processes

With the outlier removed, we can determine that the slow drilling process is more accurate. The greatest distance from the target for the slow drilling process, 0.000098 inches, is still more accurate than the smallest distance for the fast drilling process, 0.000100 inches.

39. Customer Database.

a) The mean of 54.41 is meaningless. The codes assigned to the titles are categories, even though the categories are represented by numbers. Averaging is only appropriate for quantitative data.

b) Typically, the mean and standard deviation are influenced by outliers and skewness.

c) There is no point in discussing the reason for the differences in the summary statistics, because the summary statistics are only appropriate for quantitative data. The title codes are categorical.

40. Zip codes revisited

a) Neither mean nor median is appropriate for these data. Zip codes are categories, and mean and median are only appropriate for quantitative data.

b) Neither standard deviation nor IQR is appropriate for these data. Zip codes are categories, and standard deviation and IQR are only appropriate for quantitative data.

c) The statistics cannot tell us very much since zip codes are categorical. However, there is *some* information in the first digit of zip codes. They indicate a general East (0-1) to West (8-9) direction. So, the distribution shows that a large portion of their sales occurs in the West and another in the 32000 area. But a bar chart of the first digits would be the appropriate display to show this information.

41. Eye and hair color.

The graph is not appropriate. Boxplots are for quantitative data, and these are categorical data, although coded as numbers. The numbers used for hair color and eye color are arbitrary, so the boxplot and any accompanying statistics for eye color make no sense.

42. Stereograms.

a) The two variables discussed in the description are fusion time and treatment group.

b) Fusion time is a quantitative variable, measured in seconds. Treatment group is a categorical variable, with subjects either receiving verbal clues only, or visual and verbal clues.

c) Generally, the Visual/Verbal group had shorter fusion times than the No/Verbal group. The median for the Visual/Verbal group was approximately the same as the lower quartile for the No/Verbal group. The No/Verbal Group also had an extreme outlier, with at least one subject whose fusion time was approximately 50 seconds. There is evidence that visual information may reduce fusion time.

43. Stereograms, revisited.

The re-expression using logarithms has a distribution that is more symmetric than the original distribution of fusion times, and the re-expression has no outliers. This symmetry makes it easier to compare the two groups.

44. Stereograms, yet again.

The reciprocal re-expression of the fusion times has a distribution that shows skewness, as in the original distribution of fusion times, although it has no outliers. Of the three, the logarithmic re-expression seems to be the best.

Chapter 6 – The Standard Deviation as a Ruler and the Normal Model

1. **Payroll.**

   a) The distribution of salaries in the company's weekly payroll is skewed to the right. The mean salary, $700, is higher than the median, $500.

   b) The IQR, $600, measures the spread of the middle 50% of the distribution of salaries.

   Q3 - Q1 = IQR

   Q3 = Q1 + IQR   50% of the salaries are found between $350 and $950.

   Q3 = $350 + $600

   Q3 = $950

   c) If a $50 raise were given to each employee, all measures of center or position would increase by $50. The minimum would change to $350, the mean would change to $750, the median would change to $550, and the first quartile would change to $400. Measures of spread would not change. The entire distribution is simply shifted up $50. The range would remain at $1200, the IQR would remain at $600, and the standard deviation would remain at $400.

   d) If a 10% raise were given to each employee, all measures of center, position, and spread would increase by 10%.

Minimum = $330	Mean = $770	Median = $550	Range = $1320
IQR = $660	First Quartile = $385	St. Dev. = $440	

2. **Hams.**

   a) Range = Maximum – Minimum = 7.45 – 4.15 = 3.30 pounds
   IQR = Q3 – Q1 = 6.55 – 5.6 = 0.95 pounds

   b) The distribution of weights of hams is slightly skewed to the left because the mean is lower than the median and the first quartile is farther from the median than the third quartile.

   c) All of the statistics are multiplied by 16 in the conversion from pounds to ounces.
   Mean = 96 oz.   St. Dev. = 10.4 oz.   First Quartile = 89.6 oz.
   Third Quartile = 104.8 oz.   Median = 99.2 oz.   IQR = 15.2 oz.
   Range = 52.8 oz.

   d) Measures of position increase by 30 ounces. Measures of spread remain the same.
   Mean = 126 oz.   St. Dev. = 10.4 oz.   First Quartile = 119.6 oz.
   Third Quartile = 134.8 oz.   Median = 129.2 oz.   IQR = 15.2 oz.
   Range = 52.8 oz.

   e) If a 10-pound ham were added to the distribution, the mean would change, since the total weight of all the hams would increase. The standard deviation would also increase, since 10 pounds is far away from the mean. The overall spread of the distribution would increase. The range would increase, since 10 pounds would be the new maximum. The median, quartiles, and IQR may not change. These measures are summaries of the middle 50% of the distribution, and are resistant to the presence of outliers, like the 10-pound ham.

3. **SAT or ACT?**

   Measures of center and position (lowest score, top 25% above, mean, and median) will be multiplied by 40 and increased by 150 in the conversion from ACT to SAT by the rule of thumb. Measures of spread (standard deviation and IQR) will only be affected by the multiplication.

   Lowest score = 910   Mean = 1230   Standard deviation = 120
   Top 25% above = 1350   Median = 1270   IQR = 240

4. **Cold U?**

   Measures of center and position (maximum, median, and mean) will be multiplied by $\frac{9}{5}$ and increased by 32 in the conversion from Fahrenheit to Celsius. Measures of spread (range, standard deviation, IQR) will only be affected by the multiplication.

   Maximum temperature = 51.8°F   Range = 59.4°F
   Mean = 33.8°F   Standard deviation = 12.6°F
   Median = 35.6°F   IQR = 28.8°F

5. **Temperatures.**

   In January, with mean temperature 36° and standard deviation in temperature 10°, a high temperature of 55° is almost 2 standard deviations above the mean. In July, with mean temperature 74° and standard deviation 8°, a high temperature of 55° is more than two standard deviations below the mean. A high temperature of 55° is less likely to happen in July, when 55° is farther away from the mean.

6. **Placement Exams.**

   On the French exam, the mean was 72 and the standard deviation was 8. The student's score of 82 was 10 points, or 1.25 standard deviations, above the mean. On the math exam, the mean was 68 and the standard deviation was 12. The student's score of 86 was 18 points or 1.5 standard deviations above the mean. The student did better on the math exam.

7. **Final Exams.**

   a) Anna's average is $\frac{83+83}{2} = 83$.   Megan's average is $\frac{77+95}{2} = 86$.

   Only Megan qualifies for language honors, with an average higher than 85.

   b) On the French exam, the mean was 81 and the standard deviation was 5. Anna's score of 83 was 2 points, or 0.4 standard deviations, above the mean. Megan's score of 77 was 4 points, or 0.8 standard deviations below the mean.

   On the Spanish exam, the mean was 74 and the standard deviation was 15. Anna's score of 83 was 9 points, or 0.6 standard deviations, above the mean. Megan's score of 95 was 21 points, or 1.4 standard deviations, above the mean.

   Measuring their performance in standard deviations is the only fair way in which to compare the performance of the two women on the test.

Anna scored 0.4 standard deviations above the mean in French and 0.6 standard deviations above the mean in Spanish, for a total of 1.0 standard deviation above the mean.

Megan scored 0.8 standard deviations below the mean in French and 1.4 standard deviations above the mean in Spanish, for a total of only 0.6 standard deviations above the mean.

Anna did better overall, but Megan had the higher average. This is because Megan did very well on the test with the higher standard deviation, where it was comparatively easy to do well.

8. MP3s.

a) Standard deviation measures variability, which translates to consistency in everyday use. A type of batteries with a small standard deviation would be more likely to have lifespans close to their mean lifespan than a type of batteries with a larger standard deviation.

b) RockReady batteries have a higher mean lifespan and smaller standard deviation, so they are the better battery. 8 hours is $2\frac{2}{3}$ standard deviations below the mean lifespan of RockReady and $1\frac{1}{2}$ standard deviations below the mean lifespan of DuraTunes. DuraTunes batteries are more likely to *fail* before the 8 hours have passed.

c) 16 hours is $2\frac{1}{2}$ standard deviations higher than the mean lifespan of DuraTunes, and $2\frac{2}{3}$ standard deviations higher than the mean lifespan of RockReady. Neither battery has a good chance of lasting 16 hours, but DuraTunes batteries have a greater chance than RockReady batteries.

9. Professors.

The standard deviation of the distribution of years of teaching experience for college professors must be 6 years. College professors can have between 0 and 40 (or possibly 50) years of experience. A workable standard deviation would cover most of that range of values with ±3 standard deviations around the mean. If the standard deviation were 6 months ($\frac{1}{2}$ year), some professors would have years of experience 10 or 20 standard deviations away from the mean, whatever it is. That isn't possible. If the standard deviation were 16 years, ±2 standard deviations would be a range of 64 years. That's way too high. The only reasonable choice is a standard deviation of 6 years in the distribution of years of experience.

10. Rock concerts.

The standard deviation of the distribution of the number of fans at the rock concerts would most likely be 2000. A standard deviation of 200 fans seems much too consistent. With this standard deviation, the band would be very unlikely to draw more than a 1000 fans (5 standard deviations!) above or below the mean of 21,359 fans. It seems like rock concert attendance could vary by much more than that. If a standard deviation of 200 fans is too small, then so is a standard deviation of 20 fans. 20,000 fans is too large for a likely standard deviation in attendance, unless they played several huge venues. Zero attendance is only a bit more than 1 standard deviation below the mean, although it seems very unlikely. 2000 fans is the most reasonable standard deviation in the distribution of number of fans at the concerts.

11. Guzzlers.

a) The Normal model for auto fuel economy is at the right.

b) Approximately 68% of the cars are expected to have highway fuel economy between 18.6 mpg and 31.0 mpg.

c) Approximately 16% of the cars are expected to have highway fuel economy above 31 mpg.

d) Approximately 13.5% of the cars are expected to have highway fuel economy between 31 mpg and 37 mpg.

e) The worst 2.5% of cars are expected to have fuel economy below approximately 12.4 mpg.

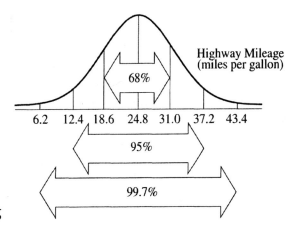

12. IQ.

a) The Normal model for IQ scores is at the right.

b) Approximately 95% of the IQ scores are expected to be within the interval 68 to 132 IQ points.

c) Approximately 16% of IQ scores are expected to be above 116 IQ points.

d) Approximately 13.5% of IQ scores are expected to be between 68 and 84 IQ points.

e) Approximately 2.5% of the IQ scores are expected to be above 132.

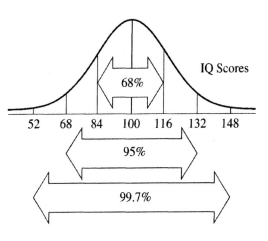

13. Winter Olympics 2002 downhill.

a) The 2002 Winter Olympics downhill times have mean of 102.71 second and standard deviation 3.01 seconds. 99.7 seconds is 1 standard deviation below the mean. If the Normal model is appropriate, 16% of the times should be below 99.7 seconds.

b) Only 3 out of 53 times (5.7%) are below 99.7 seconds.

c) The percentages in parts a and b do not agree because the Normal model is not appropriate in this situation.

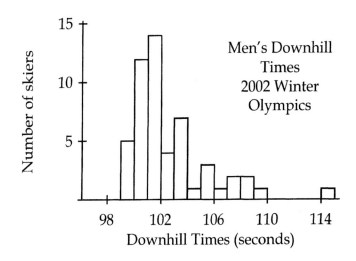

d) The histogram of 2002 Winter Olympic Downhill times is skewed to the right, and has a high outlier. The Normal model is not appropriate for the distribution of times, because the distribution is not unimodal and symmetric.

14. Rivets.

a) The Normal model for the distribution of shear strength of rivets is at the right.

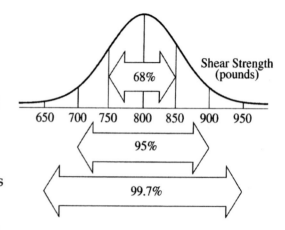

b) 750 pounds is 1 standard deviation below the mean, meaning that the Normal model predicts that approximately 16% of the rivets are expected to have a shear strength of less than 750 pounds. These rivets are a poor choice for a situation that requires a shear strength of 750 pounds, because 16% of the rivets would be expected to fail. That's too high a percentage.

c) Approximately 97.5% of the rivets are expected to have shear strengths below 900 pounds.

d) In order to make the probability of failure very small, these rivets should only be used for applications that require shear strength several standard deviations below the mean, probably farther than 3 standard deviations. (The chance of failure for a required shear strength 3 standard deviations below the mean is still approximately 3 in 2000.) For example, if the required shear strength is 550 pounds (5 standard deviations below the mean), the chance of one of these bolts failing is approximately 1 in 1,000,000.

15. Trees.

a) The Normal model for the distribution of tree diameters is at the right.

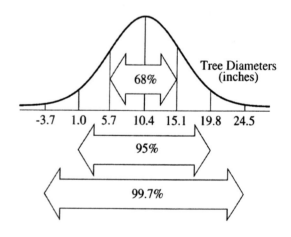

b) Approximately 95% of the trees are expected to have diameters between 1.0 inch and 19.8 inches.

c) Approximately 2.5% of the trees are expected to have diameters less than an inch.

d) Approximately 34% of the trees are expected to have diameters between 5.7 inches and 10.4 inches.

e) Approximately 16% of the trees are expected to have diameters over 15 inches.

16. Trees, part II.

The use of the Normal model requires a distribution that is unimodal and symmetric. The distribution of tree diameters is neither unimodal nor symmetric, so use of the Normal model is not appropriate.

17. TV watching.

a) Approximately 16% of the college students are expected to watch less than 1 standard deviation below the mean number of hours of TV.

b) The distribution of the number of hours of TV watched per week has mean 3.66 hours and standard deviation 4.93 hours. According to the Normal model, students who watch fewer than 1 standard deviation below the mean number of hours of TV are expected to watch less than –1.27 hours of TV per week. Of course, it is impossible to watch less than 0 hours of TV, let alone less than –1.27 hours.

c) The distribution of the number of hours of TV watched per week by college students is skewed heavily to the right. Use of the Normal model is not appropriate for this distribution, since it is not unimodal and symmetric.

18. Customer data base.

a) The median of 93% is the better measure of center for the distribution of the percentage of white residents in the neighborhoods, since the distribution is skewed to the left. Median is a better summary for skewed distributions since the median is resistant to effects of the skewness, while the mean is pulled toward the tail.

b) The IQR of 17% is the better measure of spread for the distribution of the percentage of white residents in the neighborhoods, since the distribution is skewed to the left. IQR is a better summary for skewed distributions since the IQR is resistant to effects of the skewness, and the standard deviation is not.

c) According to the Normal model, approximately 68% of neighborhoods are expected to have a percentage of whites within 1 standard deviation of the mean.

d) The mean percentage of whites in a neighborhood is 83.59%, and the standard deviation is 22.26%. 83.59% ± 22.26% = 61.33% to 105.85%. Estimating from the graph, more than 80% of the neighborhoods have a percentage of whites greater than 61.33%.

e) The distribution of the percentage of whites in the neighborhoods is strongly skewed to the left. The Normal model is not appropriate for this distribution. There is a discrepancy between c) and d) because c) is wrong!

19. Normal models.

a)   b)

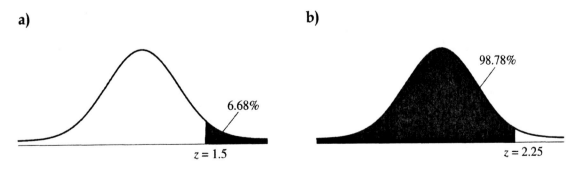

c)

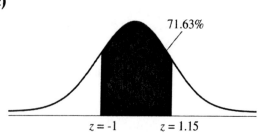

d)

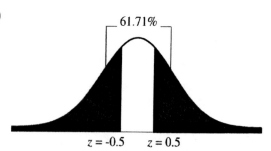

20. Normal models, again.

a)

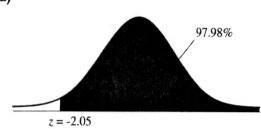

b)

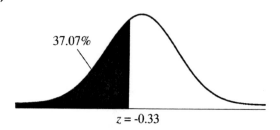

c)

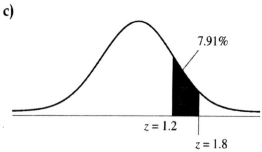

d)

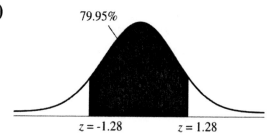

21. More Normal models.

a)

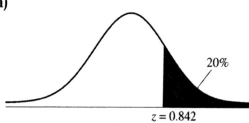

b)

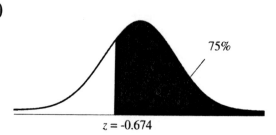

c)

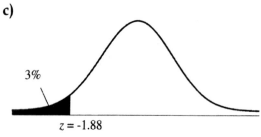

d)

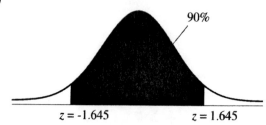

22. Yet another Normal model.

a)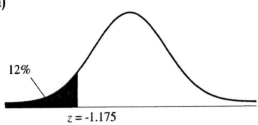

12%

$z = -1.175$

b)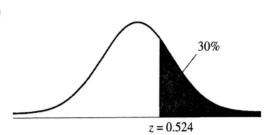

30%

$z = 0.524$

c)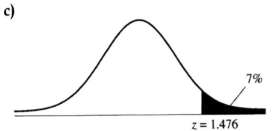

7%

$z = 1.476$

d)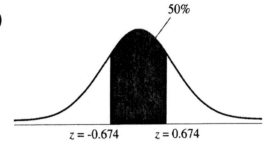

50%

$z = -0.674$ $z = 0.674$

23. Parameters.

a)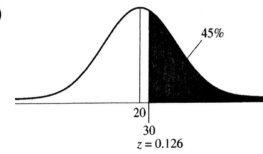

45%

20
30
$z = 0.126$

$$z = \frac{y - \mu}{\sigma}$$

$$0.126 = \frac{30 - 20}{\sigma}$$

$$0.126\sigma = 10$$

$$\sigma = 79.58$$

b)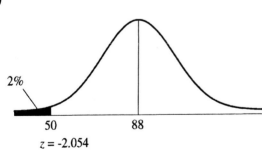

2%

50   88

$z = -2.054$

$$z = \frac{y - \mu}{\sigma}$$

$$-2.054 = \frac{50 - 88}{\sigma}$$

$$-2.054\sigma = -38$$

$$\sigma = 18.50$$

c)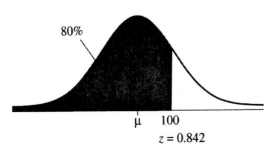

80%

μ 100

$z = 0.842$

$$z = \frac{y - \mu}{\sigma}$$

$$0.842 = \frac{100 - \mu}{5}$$

$$(0.842)(5) = 100 - \mu$$

$$\mu = 95.79$$

d)

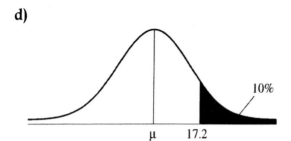

$$z = \frac{y - \mu}{\sigma}$$

$$1.282 = \frac{17.2 - \mu}{15.6}$$

$$(1.282)(15.6) = 17.2 - \mu$$

$$\mu = -2.79$$

24. Parameters.

a)

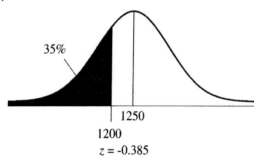

$$z = \frac{y - \mu}{\sigma}$$

$$-0.385 = \frac{1200 - 1250}{\sigma}$$

$$-0.385\sigma = -50$$

$$\sigma = 129.87$$

b)

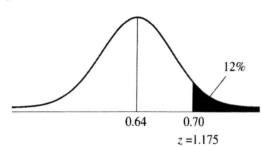

$$z = \frac{y - \mu}{\sigma}$$

$$1.175 = \frac{0.70 - 0.64}{\sigma}$$

$$1.175\sigma = 0.06$$

$$\sigma = 0.05$$

c)

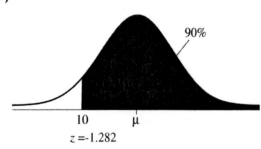

$$z = \frac{y - \mu}{\sigma}$$

$$-1.282 = \frac{10 - \mu}{0.5}$$

$$\mu = 10.64$$

d)

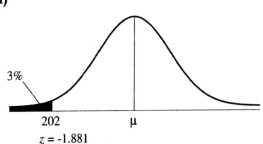

$$z = \frac{y - \mu}{\sigma}$$

$$-1.881 = \frac{202 - \mu}{220}$$

$$\mu = 615.77$$

25. Cholesterol.

a) The Normal model for cholesterol levels of adult American women is at the right.

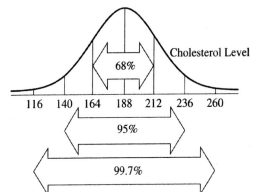

b)

$$z = \frac{y - \mu}{\sigma}$$

$$z = \frac{200 - 188}{24}$$

$$z = 0.5$$

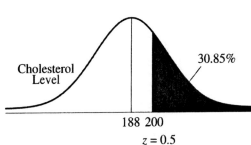

According to the Normal model, 30.85% of American women are expected to have cholesterol levels over 200.

c)

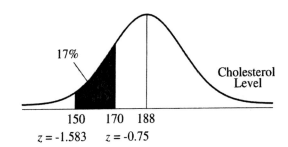

According to the Normal model, 17.00% of American women are expected to have cholesterol levels between 150 and 170.

d)

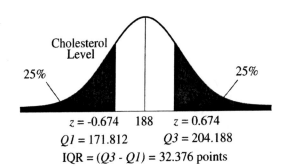

According to the Normal model, the interquartile range of the distribution of cholesterol levels of American women is approximately 32.38 points.

e)

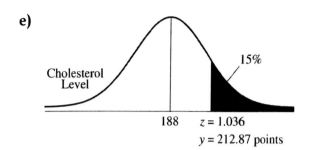

$$z = \frac{y - \mu}{\sigma}$$

$$1.036 = \frac{y - 188}{24}$$

$$y = 212.87$$

According to the Normal model, the highest 15% of women's cholesterol levels are above approximately 212.87 points.

26. Tires.

a) A tread life of 40,000 miles is 3.2 standard deviations above the mean tread life of 32,000. According to the Normal model, only approximately 0.07% of tires are expected to have a tread life greater than 40,000 miles. It would not be reasonable to hope that your tires lasted this long.

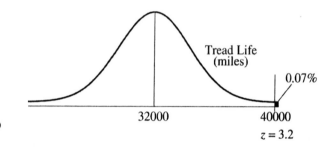

b)

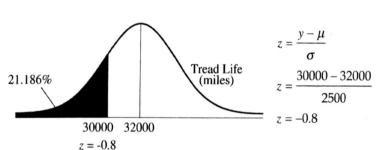

$$z = \frac{y - \mu}{\sigma}$$

$$z = \frac{30000 - 32000}{2500}$$

$$z = -0.8$$

According to the Normal model, approximately 21.19% of tires are expected to have a tread life less than 30,000 miles.

c)

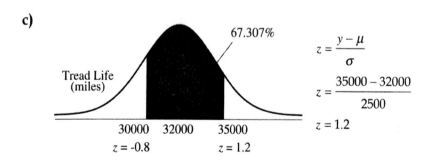

$$z = \frac{y - \mu}{\sigma}$$

$$z = \frac{35000 - 32000}{2500}$$

$$z = 1.2$$

According to the Normal model, approximately 67.31% of tires are expected to last between 30,000 and 35,000 miles.

d)

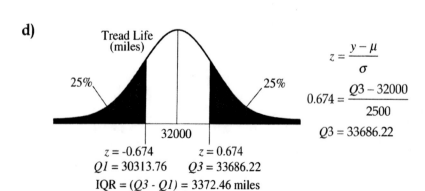

$$z = \frac{y - \mu}{\sigma}$$

$$0.674 = \frac{Q3 - 32000}{2500}$$

$$Q3 = 33686.22$$

According to the Normal model, the interquartile range of the distribution of tire tread life is expected to be 3372.46 miles.

e)

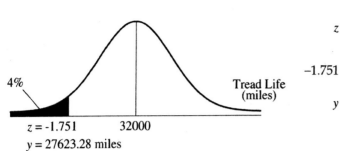

$$z = \frac{y - \mu}{\sigma}$$

$$-1.751 = \frac{y - 32000}{2500}$$

$$y = 27623.28$$

According to the Normal model, 1 of every 25 tires is expected to last less than 27,623.28 miles. If the dealer is looking for a round number for the guarantee, 27,000 miles would be a good tread life to choose.

27. Kindergarten.

a)

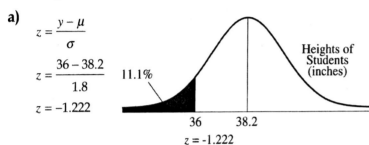

$$z = \frac{y - \mu}{\sigma}$$

$$z = \frac{36 - 38.2}{1.8}$$

$$z = -1.222$$

According to the Normal model, approximately 11.1% of kindergarten kids are expected to be less than three feet (36 inches) tall.

b)

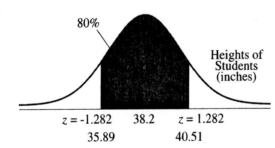

$$z = \frac{y - \mu}{\sigma}$$

$$-1.282 = \frac{y_1 - 38.2}{1.8}$$

$$y_1 = 35.89$$

$$z = \frac{y - \mu}{\sigma}$$

$$1.282 = \frac{y_2 - 38.2}{1.8}$$

$$y_2 = 40.51$$

According to the Normal model, the middle 80% of kindergarten kids are expected to be between 35.89 and 40.51 inches tall. (The appropriate values of $z = \pm1.282$ are found by using right and left tail percentages of 10% of the Normal model.)

c)

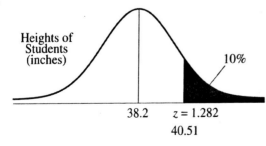

$$z = \frac{y - \mu}{\sigma}$$

$$1.282 = \frac{y - 38.2}{1.8}$$

$$y = 40.51$$

According to the Normal model, the tallest 10% of kindergarteners are expected to be at least 40.51 inches tall.

28. Body temperature.

a) According to the Normal model (and based upon the 68-95-99.7 rule), 95% of people's body temperatures are expected to be between 96.8° and 99.6°. Virtually all people (99.7%) are expected to have body temperatures between 96.1° and 101.3°.

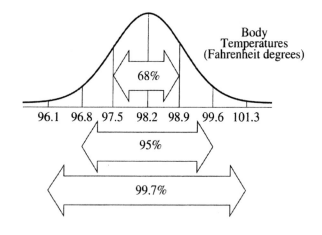

b)

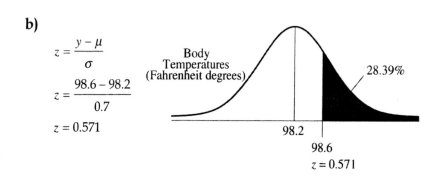

$$z = \frac{y - \mu}{\sigma}$$

$$z = \frac{98.6 - 98.2}{0.7}$$

$$z = 0.571$$

According to the Normal model, approximately 28.39% of people are expected to have body temperatures above 98.6°.

c)

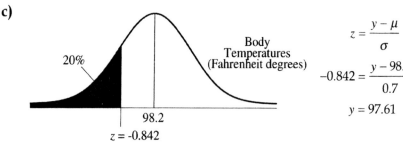

$$z = \frac{y - \mu}{\sigma}$$

$$-0.842 = \frac{y - 98.2}{0.7}$$

$$y = 97.61$$

According to the Normal model, the coolest 20% of all people are expected to have body temperatures below 97.6°.

29. First steps.

A good way to visualize the solution to this problem is to look at the distance between 10 and 13 months in two different scales. First, 10 and 13 months are 3 months apart. When measured in standard deviations, the respective z-scores, -1.645 and 0.674, are 2.319 standard deviations apart. So, 3 months must be the same as 2.319 standard deviations.

According to the Normal model, the mean age at which babies develop the ability to walk is 12.1 months, with a standard deviation of 1.3 months.

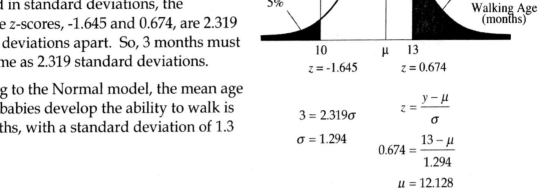

$$3 = 2.319\sigma$$

$$\sigma = 1.294$$

$$z = \frac{y - \mu}{\sigma}$$

$$0.674 = \frac{13 - \mu}{1.294}$$

$$\mu = 12.128$$

30. Trout.

A good way to visualize the solution to this problem is to look at the distance between 2 and 5 pounds in two different scales. First, 2 and 5 pounds are 3 pounds apart. When measured in standard deviations, the respective z-scores, -0.772 and 1.555, are 2.327 standard deviations apart. So, 3 pounds must be the same as 2.327 standard deviations.

According to the Normal model, the mean weight of adult trout is expected to be 3.0 pounds, with a standard deviation of 1.3 pounds.

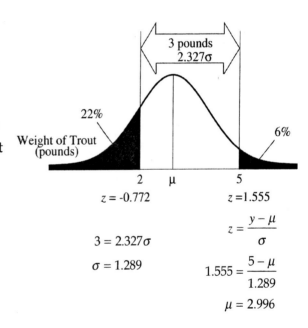

$$3 = 2.327\sigma$$

$$\sigma = 1.289$$

$$z = \frac{y - \mu}{\sigma}$$

$$1.555 = \frac{5 - \mu}{1.289}$$

$$\mu = 2.996$$

31. Eggs.

a)

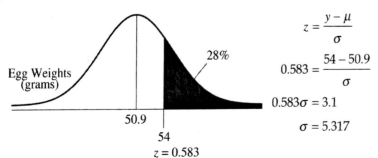

$$z = \frac{y - \mu}{\sigma}$$

$$0.583 = \frac{54 - 50.9}{\sigma}$$

$$0.583\sigma = 3.1$$

$$\sigma = 5.317$$

According to the Normal model, the standard deviation of the egg weights for young hens is expected to be 5.3 grams.

b)

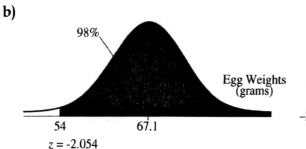

$$z = \frac{y - \mu}{\sigma}$$

$$-2.054 = \frac{54 - 67.1}{\sigma}$$

$$-2.054\sigma = -13.1$$

$$\sigma = 6.377$$

According to the Normal model, the standard deviation of the egg weights for older hens is expected to be 6.4 grams.

c) The younger hens lay eggs that have more consistent weights than the eggs laid by the older hens. The standard deviation of the weights of eggs laid by the younger hens is lower than the standard deviation of the weights of eggs laid by the older hens.

d) A good way to visualize the solution to this problem is to look at the distance between 54 and 70 grams in two different scales. First, 54 and 70 grams are 16 grams apart. When measured in standard deviations, the respective z-scores, -1.405 and 1.175, are 2.580 standard deviations apart. So, 16 grams must be the same as 2.580 standard deviations.

According to the Normal model, the mean weight of the eggs is 62.7 grams, with a standard deviation of 6.2 grams.

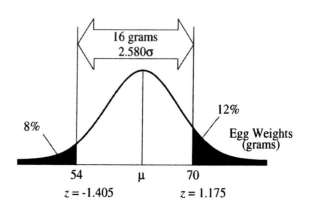

$$16 = 2.58\sigma$$

$$\sigma = 6.202$$

$$z = \frac{y - \mu}{\sigma}$$

$$1.175 = \frac{70 - \mu}{6.202}$$

$$\mu = 62.713$$

32. Tomatoes.

a)

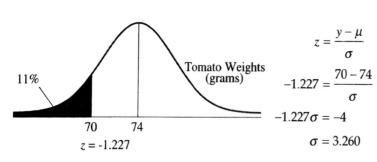

$$z = \frac{y - \mu}{\sigma}$$

$$-1.227 = \frac{70 - 74}{\sigma}$$

$$-1.227\sigma = -4$$

$$\sigma = 3.260$$

According to the Normal model, the standard deviation of the weights of Roma tomatoes now being grown is 3.26 grams.

b)

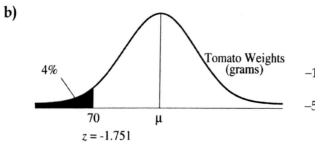

$$z = \frac{y - \mu}{\sigma}$$

$$-1.751 = \frac{70 - \mu}{3.260}$$

$$-5.708 = 70 - \mu$$

$$\mu = 75.71$$

According to the Normal model, the target mean weight for the tomatoes should be 75.71 grams.

c)

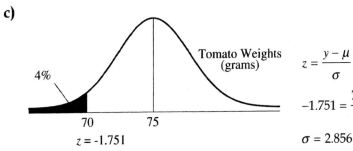

Tomato Weights (grams)

$$z = \frac{y - \mu}{\sigma}$$

$$-1.751 = \frac{70 - 75}{\sigma}$$

$$\sigma = 2.856$$

According to the Normal model, the standard deviation of these new Roma tomatoes is expected to be 2.86 grams.

$z = -1.751$

d) The weights of the new tomatoes have a lower standard deviation than the weights of the current variety. The new tomatoes have more consistent weights.

Review of Part I – Exploring and Understanding Data

1. **Bananas.**

   a) A histogram of the prices of bananas from 15 markets, as reported by the USDA, appears at the right.

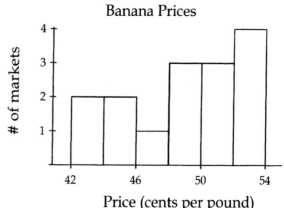

   b) The distribution of banana prices is skewed to the left, so median and IQR are appropriate measures of center and spread.
   Median = 49 cents per pound
   IQR = 7 cents per pound

   c) The distribution of the prices of bananas from 15 markets, as reported by the USDA, is unimodal and skewed to the left. The center of the distribution is approximately 50 cents, with the lowest price 42 cents per pound and the highest price 53 cents per pound.

2. **Prenatal care.**

   a) $\frac{5.4+3.9+6.1}{3} = 5.1\overline{3}$, so the overall rate of 5.1 deaths per thousand live births is equal to the average of the rates for Intensive, Adequate, and Inadequate prenatal care, when rounded to the nearest tenth. There is no reason this should be the case unless the number of women receiving each type of prenatal care is approximately the same.

   b) Yes, the results indicate (but do not prove) that adequate prenatal care is important for pregnant women. The mortality rate is quite a bit lower for women with adequate care than for other women.

   c) No, the results do not suggest that a woman pregnant with twins should be wary of seeking too much medical care. Intensive care is given for emergency conditions. The data do not suggest that the level of care is the cause of the higher mortality.

3. **Singers.**

   The distribution of heights of each voice part is roughly symmetric. The basses and tenors are generally taller than the altos and sopranos, with the basses being slightly taller than the tenors. The sopranos and altos have about the same median height. Heights of basses and sopranos are more consistent than altos and tenors.

4. **Dialysis.**

   There are only three patients currently on dialysis. With so few patients, no display is needed. We know that one patient has had his or her toes amputated and that two patients have developed blindness. What we don't know is whether or not the patient that has had his or her toes amputated has also developed blindness. Even if we wanted to, we do not have enough information to make an appropriate display.

5. **Beanstalks.**

   a) The greater standard deviation for the distribution of women's heights means that their heights are more variable than the heights of men.

   b) The z-score for women to qualify is 3.25 compared with 4.16 for men, so it is harder for men to qualify. (There appears to be some error in the statistics provided from the National Health Survey. Mean heights of 63.6" (almost 5'4") for men and 60.9" (almost 5'1") for women seem low. The z-scores confirm this skepticism. If the statistics were correct, Beanstalk Clubs would be expected to have virtually no members. According to the Normal model, only 0.0016% of men and 0.058% of women would be expected to qualify for membership!)

6. **Bread.**

   a) The distribution of the number of loaves sold each day in the last 100 days at the Clarksburg Bakery is unimodal and skewed to the right. The mode is near 100, with the majority of days recording fewer than 120 loaves sold. The number of loaves sold ranges from 95 to 140.

   b) The mean number of loaves sold will be higher than the median number of loaves sold, since the distribution of sales is skewed to the right. The mean is sensitive to this skewness, while the median is resistant.

   c) Create a boxplot with quartiles at 97 and 105.5, median at 100. The IQR is 8.5 so the upper fence is at 105.5 + 1.5(8.5) = 118.25. There are several high outliers. There are no low outliers because the min at 95 lies well within the lower fence at 97 – 1.5(8.5) = 84.25. One possible boxplot is at the right.

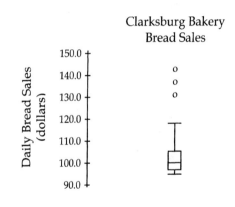

   d) The distribution of daily bread sales is not symmetric, but rather skewed to the right. The Normal model is not appropriate for this distribution. No conclusions can be drawn.

7. **Watsamatta University.**

   a) *Who* – Local residents near Watsamatta University. *What* – Age, whether or not the respondent attended college, and whether or not the respondent had a favorable opinion of Watsamatta University. *When* – Not specified. *Where* – Region around Watsamatta University. *Why* – The information will be included in a report to the University's directors. *How* – 850 local residents were surveyed by phone.

   b) There is one quantitative variable, age, probably measured in years. There are two categorical variables, college attendance (yes or no), and opinion of Watsamatta University (favorable or unfavorable).

   c) There are several problems with the design of the survey. No mention is made of a random selection of residents. Furthermore, there may be a non-response bias present. People with an unfavorable opinion of the university may hang up as soon as the staff member identifies himself or herself. Also, response bias may be introduced by the interviewer. The responses of the residents may be influenced by the fact that employees of the university are asking the questions. There may be greater percentage of favorable responses to the survey than truly exist.

8. Acid Rain.

a) The Normal model for pH level of rainfall in the Shenandoah Mountains is at the right.

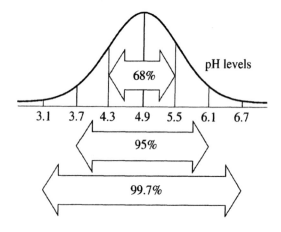

b)

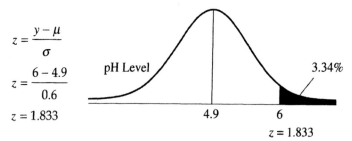

$$z = \frac{y - \mu}{\sigma}$$

$$z = \frac{6 - 4.9}{0.6}$$

$$z = 1.833$$

According to the Normal model, 3.34% of the rainstorms are expected to produce rainfall with pH levels above 6.

c)

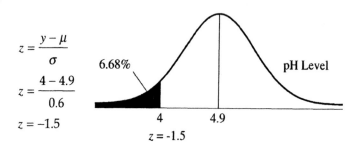

$$z = \frac{y - \mu}{\sigma}$$

$$z = \frac{4 - 4.9}{0.6}$$

$$z = -1.5$$

According to the Normal model, 6.68% of rainstorms are expected to produce rainfall with pH levels below 4.

d)

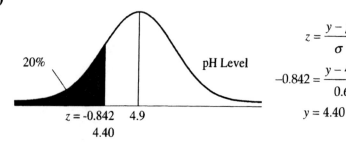

$$z = \frac{y - \mu}{\sigma}$$

$$-0.842 = \frac{y - 4.9}{0.6}$$

$$y = 4.40$$

According to the Normal model, the most acidic 20% of storms have pH below 4.40.

e)

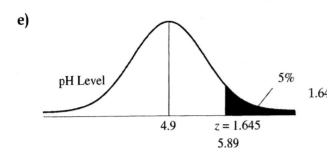

$$z = \frac{y - \mu}{\sigma}$$

$$1.645 = \frac{y - 4.9}{0.6}$$

$$y = 5.89$$

According to the Normal model, the least acidic 5% of storms have pH above 5.89.

f)

$$z = \frac{y - \mu}{\sigma}$$

$$-0.674 = \frac{Q1 - 4.9}{0.6}$$

$$Q1 = 4.50$$

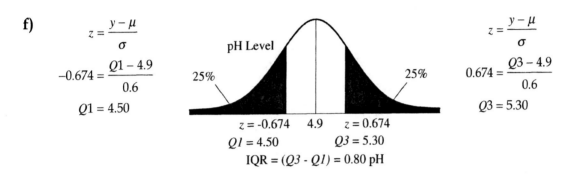

$$z = \frac{y - \mu}{\sigma}$$

$$0.674 = \frac{Q3 - 4.9}{0.6}$$

$$Q3 = 5.30$$

According to the Normal model, the IQR of the pH levels of the rainstorms is 0.80.

9. Fraud detection.

   a) Even though they are numbers, the SIC code is a categorical variable. A histogram is a quantitative display, so it is not appropriate.

   b) The Normal model will not work at all. The Normal model is for modeling distributions of unimodal and symmetric quantitative variables. SIC code is a categorical variable.

10. Streams.

   a) Stream Name – categorical; Substrate – categorical; pH – quantitative; Temperature – quantitative (ºC); BCI – quantitative.

   b) Substrate is a categorical variable, so a pie chart or a bar chart would be a useful display.

11. Cramming.

   a) Comparitive boxplots of the distributions of Friday and Monday scores are at the right.

   b) The distribution of scores on Friday was generally higher by about 5 points. Students fared worse on Monday after preparing for the test on Friday. The spreads are about the same, but the scores on Monday are slightly skewed to the right.

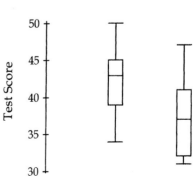

c) A histogram of the distribution of change in test score is at the right.

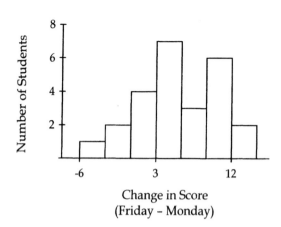

d) The distribution of changes in score is roughly unimodal and symmetric, and is centered near 4 points. Changes ranged from a student who scored 5 points higher on Monday, to two students who each scored 14 points higher on Friday. Only three students did better on Monday.

12. Computers and the Internet.

The conclusion is not sound. Many homes may have both a personal computer and access to the Internet. (In Chapter 14, we will say that these percentages may not be added because they are not disjoint.)

13. Let's play cards.

a) Suit is a categorical variable.

b) In the game of Go Fish, the denomination is not ordered. Numbers are merely matched with one another. You may have seen children's Go Fish decks that have symbols or pictures on the cards instead of numbers. These work just fine.

c) In the game of Gin Rummy, the order of the cards is important. During the game, ordered "runs" of cards are assembled (with Jack = 11, Queen = 12, King = 13), and at the end of the hand, points are totaled from the denomination of the card (face cards = 10 points). However, even in Gin Rummy, the denomination of the card sometimes behaves like a categorical variable. When you are collecting 3s, for example, order doesn't matter.

14. Accidents.

a) The distances from home are organized in categories, so a bar chart is provided at the right. A pie chart would also be useful, since the percentages represent parts of a whole.

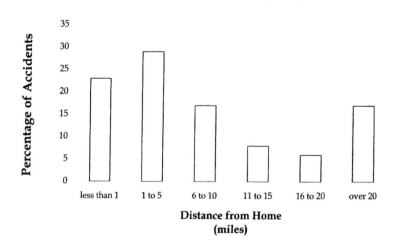

b) A greater percentage of accidents happen close to home than further away. But it is likely that people drive more miles close to home as well. These data do not indicate that driving near home is dangerous.

15. Hard water.

a) The variables in this study are both quantitative. Annual mortality rate for males is measured in deaths per 100,000. Calcium concentration is measured in parts per million.

b) The distribution of calcium concentration is skewed right, with many towns having concentrations below 25 ppm. The rest of the towns have calcium concentrations which are distributed in a fairly uniform pattern from 25 ppm to 100 ppm, tapering off to a maximum concentration around 150 ppm.
The distribution of mortality rates is unimodal and symmetric, with center approximately 1500 deaths per 100,000. The distribution has a range of 1000 deaths per 100,000, from 1000 to 2000 deaths per 100,000.

16. Hard water II.

a) The overall mean mortality rate is $\dfrac{34(1631.59)+27(1388.85)}{34+27}=1524.15$ deaths per 100,000.

b) The distribution of mortality rates for the towns north of Derby is generally higher than the distribution of mortality rates for the towns south of Derby. Fully half of the towns south of Derby have mortality rates lower than any of the towns north of Derby. A quarter of the northern towns have rates higher than any of the Southern towns.

17. Seasons.

a) The two histograms have different horizontal and vertical scales. This makes a quick comparison impossible.

b) The center of the distribution of average temperatures in January in is the low 30s, compared to a center of the distribution of July temperatures in the low 70s. The January distribution is also much more spread out than the July distribution. The range is over 50 degrees in January, compared to a range of over 20 in July. The distribution of average temperature in January is skewed slightly to the right, while the distribution of average temperature in July is roughly symmetric.

c) The distribution of difference in average temperature (July – January) for 60 large U.S. cities is slightly skewed to the left, with median at approximately 44 degrees. There are several low outliers, cities with very little difference between their average July and January temperatures. The single high outlier is a city with a large difference in average temperature between July and January. The middle 50% of differences are between approximately 38 and 46 degrees.

18. Old Faithful.

The distribution of time gaps between eruptions of Old Faithful is bimodal. A large cluster of time gaps has a mode at approximately 80 minutes and a slightly smaller cluster of time gaps has a mode at approximately 50 minutes. The distribution around each mode is fairly symmetric.

19. Old Faithful?

a) The distribution of duration of the 222 eruptions is bimodal, with modes at approximately 2 minutes and 4.5 minutes. The distribution is fairly symmetric around each mode.

b) The bimodal shape of the distribution of duration of the 222 eruptions suggests that there may be two distinct groups of eruption durations. Summary statistics would try to summarize these two groups as a single group, which wouldn't make sense.

c) The intervals between eruptions are generally longer for long eruptions than the intervals for short eruptions. Over 75% of the short eruptions had intervals of approximately 60 minutes or less, while almost all of the long eruptions had intervals of more than 60 minutes.

20. Teen drivers.

Involvement in fatal crashes is not independent of age. If the variables were independent, we would expect the percentage of fatal crashes involving teen drivers to be the same as the overall percentage of teen drivers.

21. Liberty's nose.

a) The distribution of the ratio of arm length to nose length of 18 girls in a statistics class is unimodal and roughly symmetric, with center around 15. There is one low outlier, a ratio of 11.8. A boxplot is provided at the right. A histogram or stemplot is also an appropriate display.

b) In the presence of an outlier, the 5-number summary is the appropriate choice for summary statistics. The 5-number summary is 11.8, 14.4, 15.25, 15.7, 16.9. The IQR is 1.3.

c) The ratio of 9.3 for the Statue of Liberty is very low, well below the lowest ratio in the statistics class, 11.8, which is already a low outlier. Compared to the girls in the statistics class, the Statue of Liberty's nose is very long in relation to her arm.

22. Winter Olympics 2002 speed skating.

a)

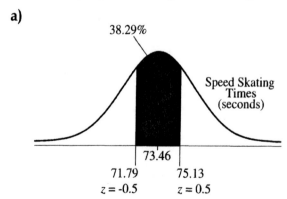

Being within 1.67 seconds of the mean when the standard deviation is 3.33 seconds is the same as being within half a standard deviation of the mean. If the Normal model is appropriate, we expect approximately 38.29% of the speed skating times to be within this interval.

b) Only three speed skating times, 71.96, 74.75, and 74.94 seconds were within the interval 71.79 – 75.13 seconds. This is only 6% (3 of 50) of the times.

c) A histogram of the distribution of speed skating times is at the right. The distribution is bimodal, reflecting the male and female times clustered around each mode. There were two distinct groups of times within this distribution, and it probably should have been displayed as two distributions. At any rate, the Normal model is not appropriate, since the distribution of the speed skating times is not unimodal and symmetric.

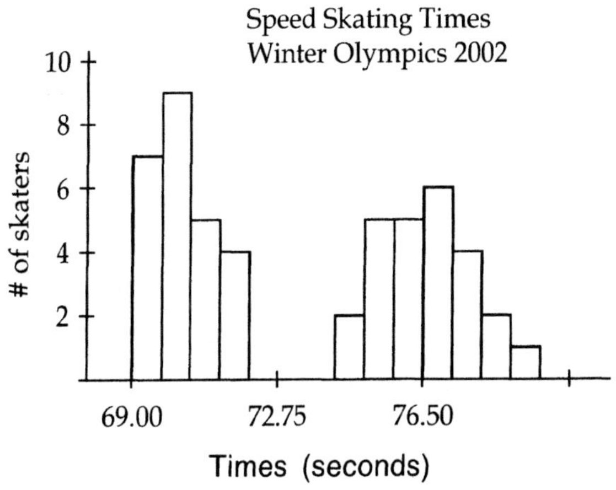

23. Sample.

Overall, the follow-up group was insured only 11.1% of the time as compared to 16.6% for the not traced group. At first, it appears that group is associated with presence of health insurance. But for blacks, the follow-up group was quite close (actually slightly higher) in terms of being insured: 8.9% to 8.7%. The same is true for whites. The follow-up group was insured 83.3% of the time, compared to 82.5% of the not traced group. When broken down by race, we see that group is not associated with presence of health insurance for either race. This demonstrates Simpson's paradox, because the overall percentages lead us to believe that there is an association between health insurance and group, but we see the truth when we examine the situation more carefully.

24. Sluggers.

a) The 5-number summary for McGwire's career is 3, 25.5, 36, 50.5, 70. The IQR is 25.

b) By the outlier test, 1.5(IQR) = 37.5. There are no homerun totals less than Q1 – 37.5 or greater than Q3 + 37.5. Technically, there are no outliers. However, the seasons in which McGwire hit fewer than 22 homeruns stand out as a separate group.

c) Parallel boxplots comparing the homerun careers of Mark McGwire and Babe Ruth are at the right.

d) Without the injured seasons, McGwire and Ruth's home run production distributions look similar. (Note: Ruth's seasons as a pitcher were not included.) Ruth's median is a little higher, and he was a little more consistent (less spread), but McGwire had the two highest season totals.

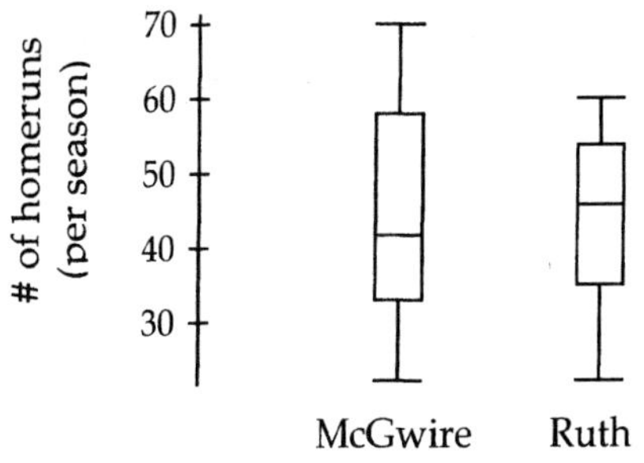

e) A back-to-back stem-and-leaf display of the homerun careers of McGwire and Ruth is at the right.

f) From the stem-and-leaf display, we can see that Ruth was much more consistent. During most of his seasons, Ruth had homerun totals in the 40s and 50s. The shape of McGwire's distribution of homeruns is revealed to be skewed to the right.

Ruth		McGwire
	7	0
0	6	5
944	5	28
9766611	4	29
54	3	2399
52	2	2

7|0 = 70 homeruns per season

25. Be quick!

a) The Normal model for the distribution of reaction times is at the right.

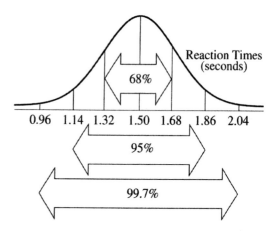

b) The distribution of reaction times is unimodal and symmetric, with mean 1.5 seconds, and standard deviation 0.18 seconds. According to the Normal model, 95% of drivers are expected to have reaction times between 1.14 seconds and 1.86 seconds.

c)

$$z = \frac{y - \mu}{\sigma}$$

$$z = \frac{1.25 - 1.50}{0.18}$$

$$z = -1.389$$

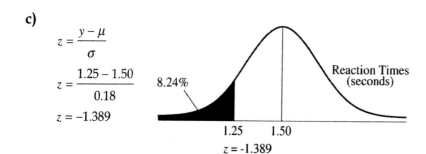

According to the Normal model, 8.24% of drivers are expected to have reaction times below 1.25 seconds.

d)

$$z = \frac{y - \mu}{\sigma}$$

$$z = \frac{1.6 - 1.5}{0.18}$$

$$z = 0.556$$

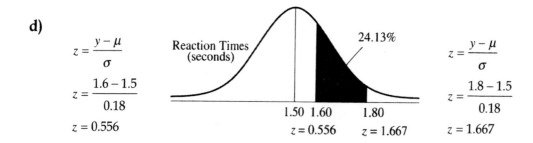

$$z = \frac{y - \mu}{\sigma}$$

$$z = \frac{1.8 - 1.5}{0.18}$$

$$z = 1.667$$

According to the Normal model, 24.13% of drivers are expected to have reaction times between 1.6 seconds and 1.8 seconds.

e)

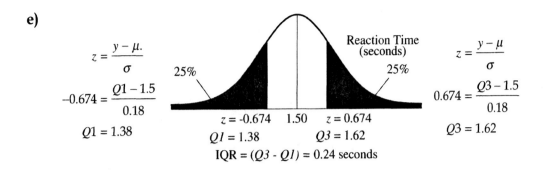

According to the Normal model, the interquartile range of the distribution of reaction times is expected to be 0.24 seconds.

f)

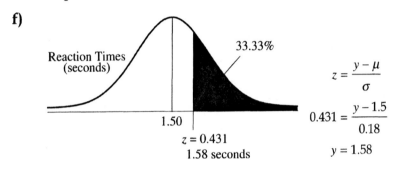

According to the Normal model, the slowest 1/3 of all drivers are expected to have reaction times of 1.58 seconds or more. (Remember that a high reaction time is a SLOW reaction time!)

26. Music and memory.

a) *Who* – 62 people. *What* – Type of music and number of objects remembered correctly. *When* – Not specified. *Where* – Not specified. *Why* – Researchers hoped to determine whether or not music affects memorization ability. *How* – Data were gathered in a completely randomized experiment.

b) Type of music (Rap, Mozart, or None) is a categorical variable. Number of items remembered is a quantitative variable.

c) Accurate boxplots cannot be constructed, because we do not have all the data. By performing outlier tests, we can determine that there are no low outliers (the minimums are all within the fences), but the Rap group and the Mozart group each have at least one high outlier (the maximum in each group is above the fence). Some possible boxplots are at the right.

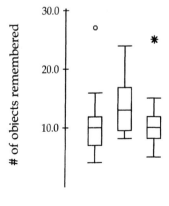

d) Mozart and Rap had very similar distributions of the number if objects remembered. The scores for None are, if anything, slightly higher than the other two groups. It is clear that groups listening to music (Rap or Mozart) did **not** score higher than those who listened to None.

27. Wines.

a) A histogram of the case prices for 36 Finger Lakes region wines is at the right.

Case Prices of Finger Lakes Wines

b) Since the distribution of case prices is unimodal and symmetric, the mean and standard deviation are appropriate measures of center and spread. The mean case price is $100.25, and the standard deviation is $25.54.

c) The distribution of case prices of Finger Lakes wines is unimodal and symmetric, with mean $100.25 and standard deviation $25.54. The cheapest case was $52 and the most expensive was $151.

d) 23 of the 36 wines (64%) had case prices within one standard deviation of the mean, or within the interval $74.71 - $125.79. This is fairly close to the 68% predicted by the Normal model. The Normal model may be useful for modeling the distribution of Finger Lakes wine case prices.

28. Pay.

The distribution of hourly wages for Chief Executives has a mean less than the median, indicating a distribution that is skewed to the left. Many Chief Executives are likely to have high hourly wages, and a few have low hourly wages, pulling the mean down. The distribution of hourly wages for General and Operations Managers has a mean higher than the median, indicating a distribution that is skewed to the right. Many General and Operations Managers have comparatively low hourly wages, and a few have high hourly wages, pulling the mean up.

29. Engines.

a) The count of cars is 38.

b) The mean displacement is higher than the median displacement, indicating a distribution of displacements that is skewed to the right. There are likely to be several very large engines in a group that consists of mainly smaller engines.

c) Since the distribution is skewed, the median and IQR are useful measures of center and spread. The median displacement is 148.5 cubic inches and the IQR is 126 cubic inches.

d) Your neighbor's car has an engine that is bigger than the median engine, but 227 cubic inches is smaller than the third quartile of 231, meaning that at least 25% of cars have a bigger engine than your neighbor's car. Don't be impressed!

e) Using the Outlier Rule (more than 1.5 IQRs beyond the quartiles) to find the fences:

Upper Fence: Q3 + 1.5(IQR) = 231 + 1.5(126) = 420 cubic inches.

Lower Fence: Q1 – 1.5(IQR) = 105 – 1.5(126) = - 84 cubic inches.

Since there are certainly no engines with negative displacements, there are no low outliers. Q1 + Range = 105 + 275 = 380 cubic inches. This means that the maximum must be less than 380 cubic inches. Therefore, there are no high outliers (engines over 420 cubic inches).

f) It is not reasonable to expect 68% of the car engines to measure within one standard deviation of the mean. The distribution engine displacements is skewed to the right, so the Normal model is not appropriate.

g) Multiplying each of the engine displacements by 16.4 to convert cubic inches to cubic centimeters would affect measures of position and spread. All of the summary statistics (except the count!) could be converted to cubic centimeters by multiplying each by 16.4.

30. Engines, again.

a) The distribution of horsepower is roughly uniform, with a bit of skew to the right, as the number of cars begins to taper off after about 125 horsepower. The center of the distribution is about 100 horsepower. The lowest horsepower is around 60 and the highest is around 160.

b) The interquartile range is Q3 – Q1 = 125 – 78 = 47 horsepower.

c) Using the Outlier Rule (more than 1.5 IQRs beyond the quartiles) to find the fences:

Upper Fence: Q3 + 1.5(IQR) = 125 + 1.5(47) = 195.5 horsepower

Lower Fence: Q1 – 1.5(IQR) = 78 – 1.5(47) = 7.5 horsepower

From the histogram, we can see that there are no cars with horsepower ratings anywhere near these fences, so there are no outliers.

d) The distribution of horsepower is uniform, not unimodal, and not very symmetric, so the Normal model is probably not a very good model of the distribution of horsepower.

e) Within one standard deviation of the mean is roughly the interval 75 – 125 horsepower. By dividing the bars of the histogram up into boxes representing one car, and taking half of the boxes in the bars representing 70-79 and 120-129, I counted 22 (of the 38) cars in the interval. Approximately 58% of the cars are within one standard deviation of the mean.

f) Adding 10 horsepower to each car would increase the measures of position by 10 horsepower and leave the measures of spread unchanged. Mean, median, 25th percentile and 75th percentile would each increase by 10. The standard deviation, interquartile range, and range would remain the same.

31. Age and party.

a) 1101 of 4002, or approximately 27.5%, of all voters surveyed were Republicans.

b) This was a representative telephone survey conducted by Gallup, a reputable polling firm. It is likely to be a reasonable estimate of the percentage of all voters who are Republicans.

c) 1001 + 1004 = 2005 of 4002, or approximately 50.1%, of all voters surveyed were under 30 or over 65 years old.

d) 409 of 4002, or approximately 10.2%, of all voters surveyed were Independents under the age of 30.

e) 409 of the 1497 Independents surveyed, or approximately 27.3%, were under the age of 30.

f) 409 of the 1001 respondents under 30, or approximately 40.9%, were Independents.

32. Age and party II.

a) The marginal distribution of party affiliation is:
   Republican – 27.5% Democrat – 35.1%   Independent – 37.4%
   (As counts: Republican – 1101   Democrat – 1404   Independent – 1497)

b)

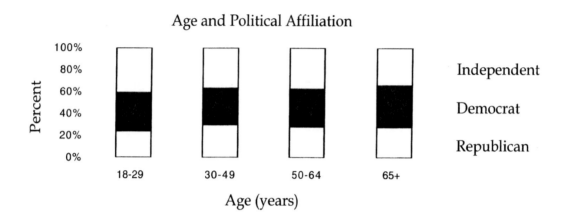

c) Political affiliation appears to be largely unrelated to age. According to The Gallup Poll, the percentages of Independents, Democrats and Republicans within four age categories are roughly the same, with approximately 35-40% Independent, 33-38% Democrat, and 25-30% Republican. However, there is some evidence that younger voters are more likely to be Independent than older voters.

d) The percentages of Independents, Democrats, and Republicans are roughly the same within each age category. Age and political affiliation appear to be independent. At the very least, there is no evidence of a strong association between the two.

33. Herbal medicine.

a) *Who* – 100 customers. *What* – Researchers asked whether or not the customer had taken the cold remedy and had customers rate the effectiveness of the remedy on a scale from 1 to 10. *When* – Not specified. *Where* – Store where natural health products are sold. *Why* – The researchers were from the Herbal Medicine Council, which sounds suspiciously like a group that might be promoting the use of herbal remedies. *How* – Researchers conducted personal interviews with 100 customers. No mention was made of any type of random selection.

b) "Have you taken the cold remedy?" is a categorical variable. Effectiveness on a scale of 1 to 10 is a categorical variable, as well, with respondents rating the remedy by placing it into one of 10 categories.

c) Very little confidence can be placed in the Council's conclusions. Respondents were people who already shopped in a store that sold natural remedies. They may be pre-disposed to thinking that the remedy was effective. Furthermore, no attempt was made to randomly select respondents in a representative manner. Finally, the Herbal Medicine Council has an interest in the success of the remedy.

34. Public opinion.

a) The percentage of respondents unaccounted for is 3% or more. This is too large to be rounding error. Perhaps some people responded that they knew of her, but their opinion was neither favorable nor unfavorable. At any rate, this discrepancy needs to be explained before we can put much faith in these data.

b) The people surveyed in June 2002 are different people than the people surveyed in July 2002. The difference in the percentages is due to sampling error, the natural variability inherent in sampling.

c) Segmented bar charts for the distribution of opinion of Martha Stewart by date are below.

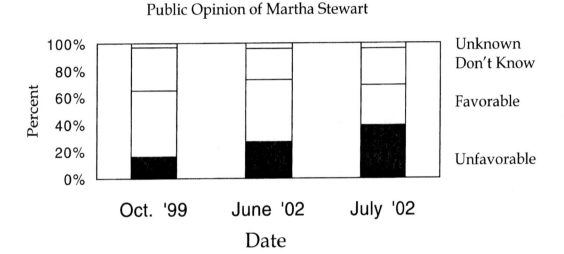

d) Pie charts for the distribution of opinion of Martha Stewart by date are below.

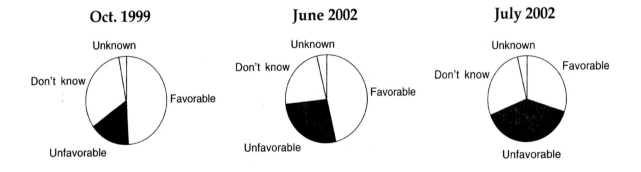

e) A timeplot of public opinion of Martha Stewart over time appears below.

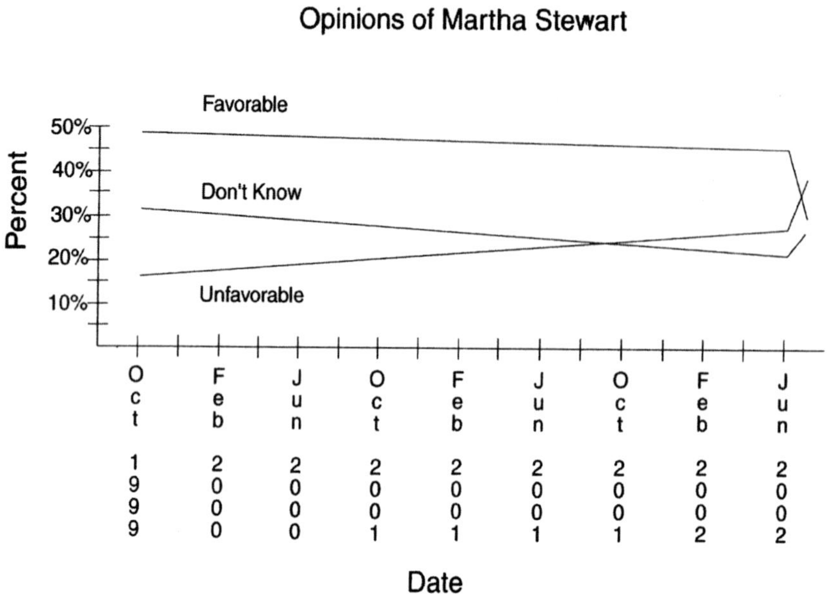

Opinions of Martha Stewart

f) Each plot shows that public opinion of Martha Stewart became less favorable over time. The pie charts clearly show the proportions of the population with each opinion during each time period, but are difficult to compare. The timeplot shows the sharp rise in the percentage of unfavorable ratings between June 2002 and July 2002, but gives the undesirable impression that we could estimate the public opinion for the months in between data points, when we really have no information about what happened in the time gaps. The segmented bar chart shows the increase in the percentage of people who rated Martha Stewart unfavorably, but gives a false impression of evenness in time scale. Each of the three plots has its strengths and weaknesses as a display.

g) In October 1999, public opinion of Martha Stewart was largely favorable. By June 2002, Martha Stewart became more well-known, with the percentage of people who stated that they did not know of her dropping from 32% to 23%. During the same time period, the percentage of people who rated her unfavorably increased from 16% to 27%, while the percentage of people who rated her favorably remained fairly steady. In July 2002, after Martha Stewart came under attack amidst rumors of insider trading, the percentage of people rating her favorably decreased sharply, while the percentage of people rating her unfavorably increased.

35. Bike safety.

a) *Who* – Years from 1991 to 2000. *What* – Number of bicycle injuries reported. *When* – 1991 to 2000. *Where* – Massachusetts. *Why* – The information was collected for a report by the Governor's Highway Safety Bureau. *How* – Although not specifically stated, the information was probably collected from government agency or hospital records.

b)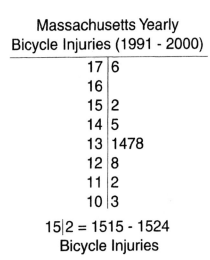

Massachusetts Yearly
Bicycle Injuries (1991 - 2000)

```
17 | 6
16 |
15 | 2
14 | 5
13 | 1478
12 | 8
11 | 2
10 | 3
```
15|2 = 1515 - 1524
Bicycle Injuries

c)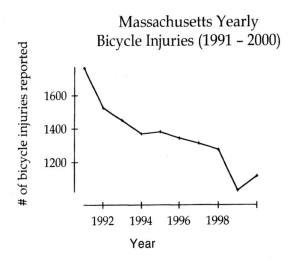

Massachusetts Yearly
Bicycle Injuries (1991 – 2000)

d) The stem-and-leaf display of the number of yearly bicycle injuries reported in Massachusetts shows that the center of the distribution is approximately 1350 bicycle injuries per year. Additionally, we can see that the year in which 1763 injuries were reported had an unusually high number of injuries. This is not visible on the timeplot.

e) The timeplot of the number of yearly bicycle injuries reported in Massachusetts shows that the number of injuries per year has declined over time.

f) In the years 1991 – 2000, the number of reported bicycle injuries per year has declined from approximately 1800 injuries in 1991 to approximately 1100 injuries in 2000. The number of injuries was unusually high in 1991. In the middle of the decade, the number of bicycle injuries reported was fairly steady at approximately 1300 injuries.

36. Profits.

a) The 5-number summary of the profits as a percent of sales of 29 of the *Forbes* 500 largest US corporations is: -9, 1, 4, 9.5, 25

(If you got –9, 1, 4, 9, 25, don't worry. Some statisticians figure quartiles of small sets differently than others. No one seems to care much which you use, since quartiles are much more useful in large data sets, anyway, where this doesn't matter.)

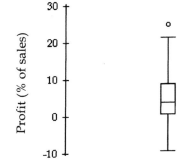

b) The boxplot of the distribution of the profits as a percent of sales of 29 of the *Forbes* 500 largest US corporations is at the right.

c) The mean profit is 4.72%, and the standard deviation of the distribution of profits is 7.55%.

d) The distribution of profits is unimodal and symmetric, centered around 4% of sales. The middle 50% of companies report profit between 1% and 9.5%. There are two companies with unusually high profits, 22% and 25%, although only 25% is technically an outlier.

37. Some assembly required.

a) A good way to visualize the solution to this problem is to look at the distance between 1 and 2 hours in two different scales. First, 1 and 2 hours are 1 hour apart. When measured in standard deviations, the respective z-scores, -0.674 and 1.645, are 2.319 standard deviations apart. So, 1 hour must be the same as 2.319 standard deviations.

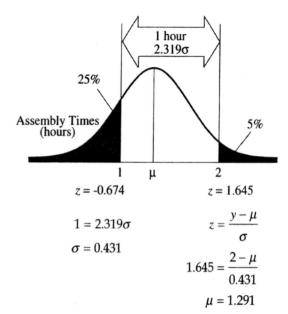

According to the Normal model, the mean assembly time is 1.29 hours and the standard deviation is 0.43 hours.

$$1 = 2.319\sigma$$

$$\sigma = 0.431$$

$$z = \frac{y - \mu}{\sigma}$$

$$1.645 = \frac{2 - \mu}{0.431}$$

$$\mu = 1.291$$

b)

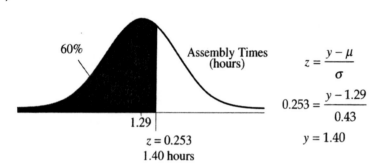

$$z = \frac{y - \mu}{\sigma}$$

$$0.253 = \frac{y - 1.29}{0.43}$$

$$y = 1.40$$

According to the Normal model, the company would need to claim that the desk takes "less than 1.40 hours to assemble", not the catchiest of slogans!

c)

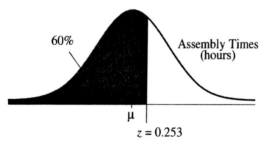

$$z = \frac{y - \mu}{\sigma}$$

$$0.253 = \frac{1 - \mu}{0.43}$$

$$\mu = 0.89$$

According to the Normal model, the company would have to lower the mean assembly time to 0.89 hour (53.4 minutes).

d) The new instructions and part-labeling may have helped lower the mean, but it also may have changed the standard deviation, making the assembly times more consistent as well as lower.

38. Crime and punishment.

a) The homicide rate rose sharply from 1900 to 1930, then declined dramatically until the early 40's. It started increasing quickly in the early 60's to around 1980 where it remained fairly stable until 1990, when it started declining again.

b) Executions rose sharply from 1900 to 1940, then fell as quickly, reaching near 0 in the decade 1970-1979. The last two decades have shown increase, although the number of executions remains comparatively low.

c) Through the last century, homicide rates and executions in the US have risen and fallen roughly together. This does not appear to support the idea of threat of execution as a deterrent to homicide. If capital punishment were a deterrent, we would expect a rise in the number of executions to correspond with a decline in the homicide rate. Certainly, no cause-and-effect relationship can be implied, because there may be lurking variables.

Chapter 7 – Scatterplots, Association, and Correlation

1. **Association.**

   a) Either weight in grams or weight in ounces could be the explanatory or response variable. Greater weights in grams correspond with greater weights in ounces. The association between weight of apples in grams and weight of apples in ounces would be positive, linear, and perfect. Each apple's weight would simply be measured in two different scales. The points would line up perfectly.

   b) Circumference is the explanatory variable, and weight is the response variable, since one-dimensional circumference explains three-dimensional volume (and therefore weight). For apples of roughly the same size, the association would be positive, linear, and strong. If the sample of apples contained very small and very large apples, the association's true curved form would become apparent.

   c) There would be no association between shoe size and GPA of college freshmen.

   d) Number of miles driven is the explanatory variable, and gallons remaining in the tank is the response variable. The greater the number of miles driven, the less gasoline there is in the tank. If a sample of different cars is used, the association is negative, linear, and moderate. If the data is gathered on different trips with the same car, the association would be strong.

2. **Association.**

   a) Price for each T-Shirt is the explanatory variable, and number of T-Shirts sold is the response variable. The association would be negative, curved, and moderate. A very low price would likely lead to very high sales, and a very high price would lead to low sales, eventually zero T-Shirts sold.

   b) Depth of the water is the explanatory variable, and water pressure is the response variable. The deeper you dive, the greater the water pressure. The association is positive, linear, and strong. For every 33 feet of depth, the pressure increases by one atmosphere (14.7 psi).

   c) Depth of the water is the explanatory variable, and visibility is the response variable. The deeper you dive, the lower the visibility. The association is negative, linear, and moderate if a sample of different bodies of water is used. If the same body of water has visibility measured at different depths, the association would be strong.

   d) At first, it appears that there should be no association between weight of elementary school students and score on a reading test. However, with weight as the explanatory variable and score as the response variable, the association is positive, linear, and moderate. Students who weigh more are likely to do better on reading tests because of the lurking variable of age. Certainly, older students generally weigh more and generally are better readers. Therefore, students who weigh more are likely to be better readers. This does not mean that weight causes higher reading scores.

3. `**Association.**

   a) Altitude is the explanatory variable, and temperature is the response variable. As you climb higher, the temperature drops. The association is negative, linear, and strong.

b) At first, it appears that there should be no association between ice cream sales and air conditioner sales. When the lurking variable of temperature is considered, the association becomes more apparent. When the temperature is high, ice cream sales tend to increase. Also, when the temperature is high, air conditioner sales tend to increase. Therefore, there is likely to be an increase in the sales of air conditioners whenever there is an increase in the sales of ice cream. The association is positive, linear, and moderate. Either one of the variables could be used as the explanatory variable.

c) Age is the explanatory variable, and grip strength is the response variable. The association is neither negative nor positive, but is curved, and moderate in scatter, due to the variability in grip strength among people in general. The very young would have low grip strength, and grip strength would increase as age increased. After reaching a maximum (at whatever age physical prowess peaks), grip strength would decline again, with the elderly having low grip strengths.

d) Blood alcohol content is the explanatory variable, and reaction time is the response variable. As blood alcohol level increase, so does the time it takes to react to a stimulus. The association is positive, probably curved, and strong. The scatterplot would probably be almost linear for low concentrations of alcohol in the blood, and then begin to rise dramatically, with longer and longer reaction times for each incremental increase in blood alcohol content.

4. Association.

a) Time spent talking on the phone is the explanatory variable, and cost of the call is the response variable. The longer you spend talking, the more the call costs. The association is positive, linear, and moderate, since some long distance companies charge more than others.

b) Distance from lightning is the explanatory variable, and time delay of the thunder is the response variable. The farther away you are from the strike, the longer it takes the thunder to reach your ears. The association is positive, linear, and fairly strong, since the speed of sound is not a constant. Sound travels at a rate of around 770 miles per hour, depending on the temperature.

c) Distance from the streetlight is the explanatory variable, and brightness is the response variable. The further away from the light you are, the less bright it appears. The association is negative, curved, and strong. Distance and light intensity follow an inverse square relationship. Doubling the distance to the light source reduces the intensity by a factor of four.

d) There is likely very little association between the weight of the car and the age of the owner. However, some might say that older drivers tend to drive larger cars. (Anyone who has seen my grandfather's car can attest to this!) If that is the case, there may be a positive, linear, and very weak association between weight of a car and the age of its owner.

5. a) None of the scatterplots show little or no association, although # 4 is very weak.

b) #3 and #4 show negative association. Increases in one variable are generally related to decreases in the other variable.

c) #2, #3, and #4 each show a linear association.

d) #2 shows a moderately strong association.

e) #1 and #3 each show a very strong association. #1 shows a curved association and #3 shows a linear association.

6. **Association.**

   a) #1 shows little or no association.

   b) #4 shows a negative association.

   c) #2 and #4 each show a linear association.

   d) #3 shows a moderately strong, curved association.

   e) #2 and #4 each show a very strong association, although some might classify the association as merely "strong".

7. **Performance IQ scores _vs._ brain size.**

   The scatterplot of IQ scores _vs._ Brain Sizes is scattered, with no apparent pattern. There appears to be little or no association between the IQ scores and brain sizes displayed in this scatterplot.

8. **Kentucky derby.**

   Winning speeds in the Kentucky Derby have generally increased over time. The association between year and speed seems slightly curved, with a greater rate of increase in winning speed before 1950 and a smaller rate of increase after 1950, suggesting that winning speeds have leveled off over time.

9. **Firing pottery.**

   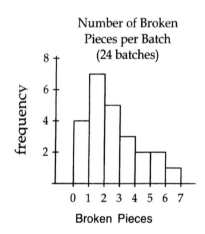

   a) A histogram of the number of broken pieces is at the right.

   b) The distribution of the number broken pieces per batch of pottery is skewed right, centered around 1 broken piece per batch. Batches had from 0 and 6 broken pieces. The scatterplot does not show the center or skewness of the distribution.

   c) The scatterplot shows that the number of broken pieces increases as the batch number increases. If the 8 daily batches are numbered sequentially, this indicates that batches fired later in the day generally have more broken pieces. This information is not visible in the histogram.

10. Coffee sales.

a) A histogram of daily sales is at the right.

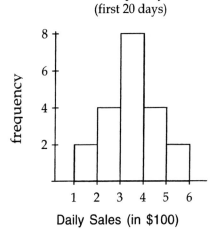

Coffee Shop Daily Sales
(first 20 days)

Daily Sales (in $100)

b) The scatterplot shows that, in general, the sales have been increasing over time. The histogram does not show this.

c) The histogram shows that the mean of the daily sales for the coffee shop was between $300 and $400, and that this happened on 8 days. The scatterplot does not show this.

11. Matching.

a) 0.006   **b)** 0.777   **c)** - 0.923   **d)** - 0.487

12. Matching.

a) - 0.977   **b)** 0.736   **c)** 0.951   **d)** - 0.021

13. Lunchtime.

a) The correlation between time toddlers spent at the table and the number of calories consumed by the toddlers is $r = -0.649$.

b) If time spent at the table were recorded in hours instead of minutes, the correlation would not change at all. Correlation is calculated from z-scores, and since these unitless measures of position relative to the mean are unaffected by changes in scale, correlation is likewise unaffected.

c) The correlation between time spent at the table and calorie consumption for toddlers is –0.649, a moderate, negative correlation. Toddlers who spent more time at the table tended to consume fewer calories.

d) The analyst's remark is speculative. There are many possible explanations for the behavior of the toddlers. The data merely show us an association between time spent at the table and calories consumed by toddlers, and association is not the same thing as a cause-and-effect relationship.

14. Vehicle weights.

a) A scatterplot of the Static Weight vs. Weight-in-Motion of the test truck is at the right.

b) The association between static weight and weight-in-motion is positive, strong, and roughly linear. There may be a hint of a curve in the scatterplot.

c) As the static weight of the test truck increased, so did the weight-in-motion, but the relationship appears weaker for heavier trucks.

d) The correlation between static weight and weight-in-motion is $r = 0.965$.

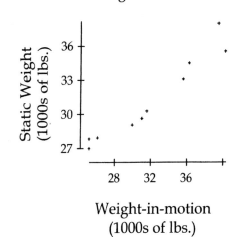

Weight of a Test Truck

Static Weight (1000s of lbs.)

Weight-in-motion (1000s of lbs.)

e) Weighing the trucks in kilograms instead of pounds would not change the correlation. Correlation, like z-score, has no units. It is a numerical measure of the degree of linear association between two variables.

f) When the test truck weighed approximately 35,500 pounds, it weighed higher in motion. The scale may need to be recalibrated. If the scale were calibrated exactly, we would expect the points to line up perfectly, with no curve, and no deviations from the pattern.

15. Fuel economy.

a) A scatterplot of average fuel economy vs. horsepower ratings is at the right.

b) There is a strong, negative, linear association between horsepower and mileage of the selected vehicles.

c) The correlation between horsepower and mileage of the selected vehicles is $r = -0.878$.

d) Generally, vehicles in the selected group with more horsepower have lower mileage. There are two vehicles with extremely high horsepower ratings and low mileage. If these two vehicles were removed, the correlation would not be as high.

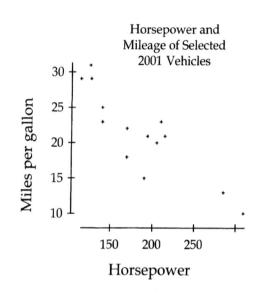

16. Drug abuse.

a) A scatterplot of percentage of teens who have used other drugs vs. percentage who have used marijuana in the U.S. and 10 Western European countries is at the right.

b) The correlation between the percent of teens who have used marijuana and the percent of teens who have used other drugs is $r = 0.934$.

c) The association between the percent of teens who have used marijuana and the percent of teens who have used other drugs is positive, strong, and linear. Countries with higher percentages of teens who have used marijuana tend to have higher percentages of teens who have used other drugs.

d) These results do not confirm that marijuana is a "gateway drug". An association exists between the percent of teens who have used marijuana and the percent of teens who have used other drugs. This does not mean that one caused the other.

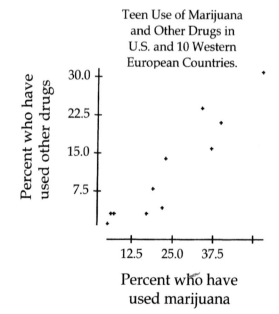

17. Burgers.

There is no apparent association between the number of grams of fat and the number of milligrams of sodium in several brands of fast food burgers. The correlation is only $r = 0.199$, which is close to zero, an indication of no association. One burger had a much lower fat content than the other burgers, at 19 grams of fat, with 920 milligrams of sodium. Without this (comparatively) low fat burger, the correlation would have been even lower.

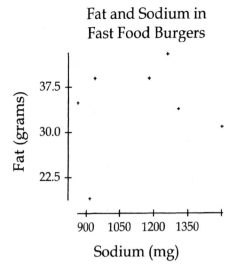

18. Burgers.

a) A scatterplot of fat content vs. calorie content of several fast food burgers is at the right.

b) The correlation between the number of calories and the number of grams of fat in several fast food burgers is $r = 0.961$.

c) The association between the number of calories and the number of grams of fat in several fast food burgers is positive, linear, and strong. Typically, burgers with higher fat content have more calories.

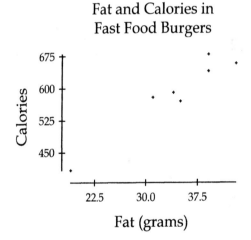

19. Attendance.

a) Number of runs scored and attendance are quantitative variables, the relationship between them appears to be linear, and there are no outliers, so calculating a correlation is appropriate.

b) The association between attendance and runs scored is positive, linear, and moderate in strength. Generally, as the number of runs scored increases, so does attendance.

c) There is evidence of an association between attendance and runs scored, but a cause-and-effect relationship between the two is not implied. There may be lurking variables that can account for the increases in each. For example, perhaps winning teams score more runs and also have higher attendance. We don't have any basis to make a claim of causation.

20. Second inning.

a) Winning teams generally enjoy greater attendance at their home games. The association between attendance and number of wins is positive, somewhat linear, but not very strong.

b) The association between scoring runs and attendance has the strongest correlation, with $r = 0.740$.

c) The correlation between number of runs scored and number of wins is $r = 0.680$, indicating a possible moderate association. However, since there is no scatterplot of wins vs. runs provided, we can't be sure the relationship is linear. Correlation may not be an appropriate measure of the strength of the association.

21. Politics.

The candidate might mean that there is an **association** between television watching and crime. The term correlation is reserved for describing linear associations between quantitative variables. We don't know what type of variables "television watching" and "crime" are, but they seem categorical. Even if the variables are quantitative (hours of TV watched per week, and number of crimes committed, for example), we aren't sure that the relationship is linear. The politician also seems to be implying a cause-and-effect relationship between television watching and crime. Association of any kind does not imply causation.

22. Association.

The researcher should have plotted the data first. A strong, curved relationship may have a very low correlation. In fact, correlation is only a useful measure of the strength of a linear relationship.

23. Correlation errors.

a) If the association between GDP and infant mortality is linear, a correlation of −0.772 shows a moderate, negative association. Generally, as GDP increases, infant mortality rate decreases.

b) Continent is a categorical variable. Correlation measures the strength of linear associations between quantitative variables.

c) Correlation must be between −1 and 1, inclusive. Correlation can never be 1.22.

d) A correlation, no matter how strong, cannot prove a cause-and-effect relationship.

24. Sample survey.

Even though zipcodes are numbers, they are categorical variables representing different geographic areas. Likewise, even thought the variable *datasource* has numerical values, it is also categorical, representing the source from which the data were acquired. Correlation is only an appropriate measure of the strength of linear association between quantitative variables.

25. Baldness and heart disease.

Even though the variables baldness and heart disease were assigned numerical values, they are categorical. Correlation is only an appropriate measure of the strength of linear association between quantitative variables. Their conclusion is meaningless.

26. Oil production.

a) The correlation between oil production and year is $r = 0.117$.

b) No. The reporter should have looked at a scatterplot of the relationship between oil production and year. The association between oil production and year is very strong. It just isn't linear. Correlation is not an appropriate measure of the strength of a curved association.

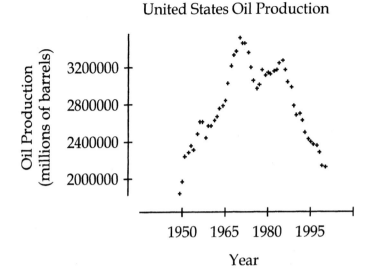

27. Planets.

a) The association between Position Number of each planet and its distance from the sun (in millions of miles) is very strong, positive and curved. The scatterplot is at the right.

b) The relationship between Position Number and distance from the sun is not linear. Correlation is a measure of the degree of *linear* association between two variables.

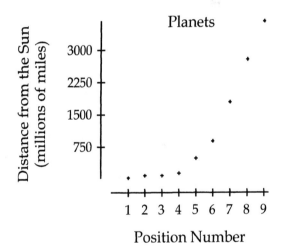

c) The scatterplot of the logarithm of distance versus Position Number (shown at the right) still shows a strong, positive relationship, but it is straighter than the previous scatterplot. It still shows a curve in the scatterplot, but it is straight enough that correlation may now be used as an appropriate measure of the strength of the relationship between logarithm of distance and Position Number, which will in turn give an indication of the strength of the association.

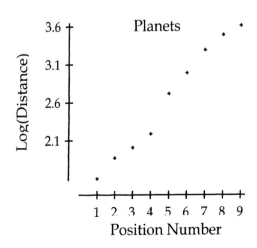

28. Internet journals.

a) The association between the number of Internet journals and the year is positive, strong, and curved. As the years passed, the number of Internet journals increased. A scatterplot of number of Internet journals vs. year is at the right.

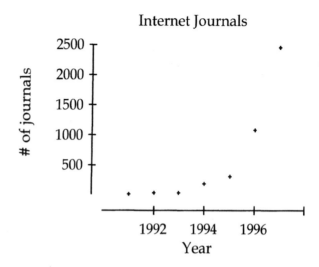

b) A scatterplot of logarithm of the number of journals vs. year is at the right. This re-expression makes the association more linear. Correlation may now be used as a measure of the strength of the association between the logarithm of the number of Internet journals and the year, which can give us and idea of the strength of the association.

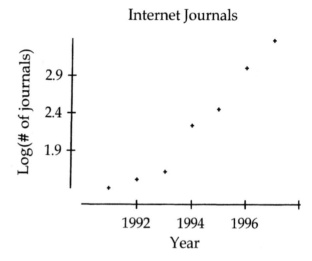

Chapter 8 – Linear Regression

1. Regression equations.

$\bar{x}$	s_x	$\bar{y}$	s_y	r	$\hat{y} = b_0 + b_1 x$
a) 10	2	20	3	0.5	$\hat{y} = 12.5 + .75x$
b) 2	0.06	7.2	1.2	–0.4	$\hat{y} = 23.2 - 8x$
c) 12	6	152	30	–0.8	$\hat{y} = 200 - 4x$
d) 2.5	1.2	25	100	0.6	$\hat{y} = -100 + 50x$

a)

$$b_1 = r\left(\frac{s_y}{s_x}\right)$$
$$b_1 = 0.5\left(\frac{3}{2}\right)$$
$$b_1 = 0.75$$

$$\hat{y} = b_0 + b_1 x$$
$$\bar{y} = b_0 + b_1\bar{x}$$
$$20 = b_0 + 0.75(10)$$
$$b_0 = 12.5$$

b)

$$b_1 = r\left(\frac{s_y}{s_x}\right)$$
$$b_1 = -0.4\left(\frac{1.2}{0.06}\right)$$
$$b_1 = -8$$

$$\hat{y} = b_0 + b_1 x$$
$$\bar{y} = b_0 + b_1\bar{x}$$
$$7.2 = b_0 - 8(2)$$
$$b_0 = 23.2$$

c)

$$\hat{y} = b_0 + b_1 x$$
$$\bar{y} = b_0 + b_1\bar{x}$$
$$\bar{y} = 200 - 4(12)$$
$$\bar{y} = 152$$

$$b_1 = r\left(\frac{s_y}{s_x}\right)$$
$$-4 = -0.8\left(\frac{s_y}{6}\right)$$
$$s_y = 30$$

d)

$$\hat{y} = b_0 + b_1 x$$
$$\bar{y} = b_0 + b_1\bar{x}$$
$$\bar{y} = -100 + 50(2.5)$$
$$\bar{y} = 25$$

$$b_1 = r\left(\frac{s_y}{s_x}\right)$$
$$50 = r\left(\frac{100}{1.2}\right)$$
$$r = 0.6$$

2. More regression equations.

$\bar{x}$	s_x	$\bar{y}$	s_y	r	$\hat{y} = b_0 + b_1 x$
a) 30	4	18	6	–0.2	$\hat{y} = 27 - 0.3x$
b) 100	18	60	10	0.9	$\hat{y} = 10 + 0.5x$
c) 4	0.8	50	15	0.8	$\hat{y} = -10 + 15x$
d) 6	1.2	18	4	–0.6	$\hat{y} = 30 - 2x$

a)

$$b_1 = r\left(\frac{s_y}{s_x}\right)$$
$$b_1 = -0.2\left(\frac{6}{4}\right)$$
$$b_1 = -0.3$$

$$\hat{y} = b_0 + b_1 x$$
$$\bar{y} = b_0 + b_1\bar{x}$$
$$18 = b_0 - 0.3(30)$$
$$b_0 = 27$$

b)

$$b_1 = r\left(\frac{s_y}{s_x}\right)$$
$$b_1 = 0.9\left(\frac{10}{18}\right)$$
$$b_1 = 0.5$$

$$\hat{y} = b_0 + b_1 x$$
$$\bar{y} = b_0 + b_1\bar{x}$$
$$60 = b_0 + 0.5(100)$$
$$b_0 = 10$$

c)

$$\hat{y} = b_0 + b_1 x$$
$$\bar{y} = b_0 + b_1 \bar{x}$$
$$50 = -10 + 15(\bar{x})$$
$$\bar{x} = 4$$

$$b_1 = r\left(\frac{s_y}{s_x}\right)$$
$$15 = r\left(\frac{15}{0.8}\right)$$
$$r = 0.8$$

d)

$$\hat{y} = b_0 + b_1 x$$
$$\bar{y} = b_0 + b_1 \bar{x}$$
$$18 = 30 - 2(\bar{x})$$
$$\bar{x} = 6$$

$$b_1 = r\left(\frac{s_y}{s_x}\right)$$
$$-2 = -0.6\left(\frac{4}{s_x}\right)$$
$$s_x = 1.2$$

3. **Residuals.**

   a) The scattered residuals plot indicates an appropriate linear model.

   b) The curved pattern in the residuals plot indicates that the linear model is not appropriate. The relationship is not linear.

   c) The fanned pattern indicates heteroscedastic data. The models predicting power decreases as the values of the explanatory variable increase.

4. **Residuals.**

   a) The curved pattern in the residuals plot indicates that the linear model is not appropriate. The relationship is not linear.

   b) The fanned pattern indicates heteroscedastic data. The models predicting power increases as the values of the explanatory variable increase.

   c) The scattered residuals plot indicates an appropriate linear model.

5. **Least squares.**

   If the 4 x-values are plugged into $\hat{y} = 7 + 1.1x$, the 4 predicted values are $\hat{y} = 18, 29, 51$ and 62, respectively. The 4 residuals are $-8, 21, -31$, and 18. The squared residuals are $64, 441, 961$, and 324, respectively. The sum of the squared residuals is 1790. Least squares means that no other line has a sum lower than 1790. In other words, it's the best fit.

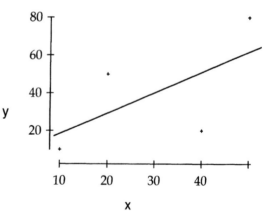

6. **Least squares.**

   If the 4 x-values are plugged into $\hat{y} = 1975 - 0.45x$, the 4 predicted values are $\hat{y} = 1885, 1795, 1705$, and 1615, respectively. The 4 residuals are $65, -145, 95$, and -15. The squared residuals are $4225, 21025, 9025$, and 225, respectively. The sum of the squared residuals is $34,500$. Least squares means that no other line has a sum lower than $34,500$. In other words, it's the best fit.

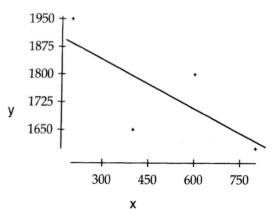

7. **Real estate.**

   a) The explanatory variable (x) is size, measured in square feet, and the response variable (y) is price measured in thousands of dollars. Therefore, the units of the slope are thousands of dollars per square foot.

   b) 71.4% of the variability in price can be explained by variability in size. (In other words, 71.4% of the variability in price can be explained by the linear model.)

   c) The slope of the regression line predicting price from size should be positive. Bigger homes are expected to cost more.

   d) The correlation between size and price is $r = \sqrt{R^2} = \sqrt{0.714} = 0.845$. The positive value of the square root is used, since the relationship is believed to be positive.

   e) The price of a home that is one standard deviation above the mean size would be predicted to be 0.845 standard deviations (in other words r standard deviations) above the mean price.

   f) The price of a home that is two standard deviations below the mean size would be predicted to be 1.69 (or 2×0.845) standard deviations below the mean price.

8. **Baseball.**

   a) The linear model is appropriate. Although the relationship is not strong, it is reasonably straight, and the residuals plot shows no pattern.

   b) 33.3% of the variability in attendance can be explained by variability in the number of wins. (In other words, 33.3% of the variability can be explained by the model.)

   c) The correlation between attendance and number of wins is $r = \sqrt{R^2} = \sqrt{0.333} = 0.577$. The positive value of the square root is used, since the relationship is positive.

   d) A team that is two standard deviations above the mean in number of wins would be expected to have attendance that is 1.154 (or 2×0.577) standard deviations above the mean attendance.

   e) A team that is one standard deviation below the mean in attendance would be expected to have a number of wins that is 0.577 standard deviations (in other words, r standard deviations) below the mean number of wins. The correlation between two variables is the same, regardless of the direction in which predictions are made. Be careful, though, since the same is NOT true for predictions made using the slope of the regression equation. Slopes are valid only for predictions in the direction for which they were intended.

9. **More real estate.**

   a) According to the linear model, the price of a home is expected to increase $61 (0.061 thousand dollars) for each additional square-foot in size.

   b)
   $$\widehat{price} = 47.82 + 0.061x$$
   $$\widehat{price} = 47.82 + 0.061(3000)$$
   $$\widehat{price} = 230.82$$

   According to the linear model, a 3000 square-foot home is expected to have a price of $230,820.

c)

$$\hat{price} = 47.82 + 0.061x$$

$$\hat{price} = 47.82 + 0.061(1200)$$

$$\hat{price} = 121.02$$

According to the linear model, a 1200 square-foot home is expected to have a price of $121,020. The asking price is $121,020 - $6000 = $115,020. $6000 is the (negative) residual.

10. Second inning

a) *Attendance* $= 5773.27 + 517.609(Wins)$ is the equation of the regression line that predicts attendance from the number of games won by American League baseball teams.

b)

$$\hat{Attendance} = 5773.27 + 517.609(Wins)$$

$$\hat{Attendance} = 5773.27 + 517.609(50)$$

$$\hat{Attendance} = 31,653.72$$

The model predicts that a team with 50 wins will have attendance of 31,653.72 people.

c) For each additional win, the model predicts an increase in attendance of 517.609 people.

d) A negative residual means that the teams actual attendance is lower than the attendance model predicts for a team with as many wins.

e)

$$\hat{Attendance} = 5773.27 + 517.609(Wins)$$

$$\hat{Attendance} = 5773.27 + 517.609(43)$$

$$\hat{Attendance} = 28,030.457$$

The predicted attendance for the Cardinals was 28,030.457. The actual attendance of 38,988 gives a residual of 38,988 – 28,030.457 = 10,957.543. The Cardinals had almost 11,000 more people attending than the model predicted.

11. What slope?

The only slope that makes sense is 300 pounds per foot. 30 pounds per foot is too small. For example, a Honda Civic is about 14 feet long, and a Cadillac DeVille is about 17 feet long. If the slope of the regression line was 30 pounds per foot, the Cadillac would be predicted to outweigh the Civic by only 90 pounds! (The real difference is about 1500 pounds.) Similarly, 3 pounds per foot is too small. A slope of 3000 pounds per foot would predict a weight difference of 9000 pounds (4.5 tons) between Civic and DeVille. The only answer that is even reasonable is 300 pounds per foot, which predicts a difference of 900 pounds. This isn't very close to the actual difference of 1500 pounds, but at least it is in the right ballpark.

12. What slope?

The only slope that makes sense is 1 foot in height per inch in circumference. 0.1 feet per inch is too small. A trunk would have to increase in diameter by 10 inches for every foot in height. If that were true, pine trees would be all trunk! 10 feet per inch (and, similarly 100 feet per inch) is too large. If pine trees reach a maximum height of 60 feet, for instance, then the variation in circumference of the trunk would only be 6 inches. Pine tree trunks certainly come in more sizes than that. The only slope that is reasonable is 1 foot in height per inch in circumference.

13. Misinterpretations.

   a) R^2 is an indication of the strength of the model, not the appropriateness of the model. A scattered residuals plot is the indicator of an appropriate model.

   b) Regression models give predictions, not actual values. The student should have said, "The model predicts that a bird 10 inches tall is expected to have a wingspan of 17 inches."

14. More misinterpretations.

   a) R^2 measures the amount of variation explained by the model. Literacy rate determines 64% of *the variability* in life expectancy.

   b) Regression models give predictions, not actual values. The student should have said, "The slope of the line shows that an increase of 5% in literacy rate *is associated with an expected* 2-year improvement in life expectancy."

15. ESP.

   a) First, since no one has ESP, you must have scored 2 standard deviations above the mean by chance. On your next attempt, you are unlikely to duplicate the extraordinary event of scoring 2 standard deviations above the mean. You will likely "regress" towards the mean on your second try, getting a lower score. If you want to impress your friend, don't take the test again. Let your friend think you can read his mind!

   b) Your friend doesn't have ESP, either. No one does. Your friend will likely "regress" towards the mean score on his second attempt, as well, meaning his score will probably go up. If the goal is to get a higher score, your friend should try again.

16. SI jinx.

   Athletes, especially rookies, usually end up on the cover of Sports Illustrated for extraordinary performances. If these performances represent the upper end of the distribution of performance for this athlete, future performance is likely to regress toward the average performance of that athlete. An athlete's average performance usually isn't notable enough to land the cover of SI. Of course, there are always exceptions, like Michael Jordan, Tiger Woods, Serena Williams, and others.

17. SAT scores.

   a) The association between SAT Math scores and SAT Verbal Scores is linear, moderate in strength, and positive. Students with high SAT Math scores typically have high SAT Verbal scores.

   b) One student got a 500 Verbal and 800 Math. That set of scores doesn't seem to fit the pattern.

   c) $r = 0.685$ indicates a moderate, positive association between SAT Math and SAT Verbal, but only because the scatterplot shows a linear relationship. Students who scored one standard deviation above the mean in SAT MAth are expected to score 0.685 standard deviations above the mean in SAT Verbal. Additionally, $R^2 = (0.685)^2 = 0.469225$, so 46.9% of the variability in math score is explained by variability in verbal score.

d) The scatterplot of verbal and math scores shows a relationship that is straight enough, so a linear model is appropriate.

$$b_1 = r\left(\frac{s_{Math}}{s_{Verbal}}\right)$$

$$\hat{y} = b_0 + b_1 x$$

$$\bar{y} = b_0 + b_1 \bar{x}$$

$$b_1 = 0.685\left(\frac{96.1}{99.5}\right)$$

$$612.2 = b_0 + 0.661593(596.3)$$

$$b_1 = 0.661593$$

$$b_0 = 217.692$$

The equation of the least squares regression line for predicting SAT Math score from SAT Verbal score is $\widehat{Math} = 217.692 + 0.662(Verbal)$.

e) For each additional point in verbal score, the model predicts an increase of 0.662 points in math score. A more meaningful interpretation might be scaled up. For each additional 10 points in verbal score, the model predicts an increase of 6.62 points in math score.

f)

$$\widehat{Math} = 217.692 + 0.662(Verbal)$$

$$\widehat{Math} = 217.692 + 0.662(500)$$

$$\widehat{Math} = 548.692$$

According to the model, a student with a verbal score of 500 is expected to have a math score of 548.692.

g)

$$\widehat{Math} = 217.692 + 0.662(Verbal)$$

$$\widehat{Math} = 217.692 + 0.662(800)$$

$$\widehat{Math} = 747.292$$

According to the model, a student with a verbal score of 800 is expected to have a math score of 747.292. Since she actually scored 800 on math, her residual is 800 – 747.292 = 52.708 points

18. Success in college

a) A scatterplot showed the relationship between combined SAT score and GPA to be reasonably linear, so a linear model is appropriate.

$$b_1 = r\left(\frac{s_{GPA}}{s_{SAT}}\right)$$

$$\hat{y} = b_0 + b_1 x$$

$$\bar{y} = b_0 + b_1 \bar{x}$$

$$b_1 = 0.47\left(\frac{0.56}{83}\right)$$

$$2.66 = b_0 + 0.003171(1222)$$

$$b_1 = 0.003171$$

$$b_0 = -1.215$$

The regression equation predicting GPA from SAT score is:
$$\widehat{GPA} = -1.215 + 0.0032(SAT)$$

b) The model predicts that a student with an SAT score of 0 would have a GPA of –1.215. The *y*-intercept is not meaningful in this context, since both scores are impossible.

c) The model predicts that an additional 100 points on the SAT is associated with an increase of 0.32 points in GPA.

d)

$$G\hat{P}A = -1.215 + 0.0032(SAT)$$

$$G\hat{P}A = -1.215 + 0.0032(1400)$$

$$G\hat{P}A = 3.265$$

According to the model, a student with an SAT score of 1400 is expected to have a GPA of 3.265.

e) According to the model, SAT score is not a very good predictor of college GPA. $R^2 = (0.47)^2 = 0.2209$, which means that only 22.09% of the variability in GPA can be predicted by the model. The rest of the variability is determined by other factors.

f) A student would prefer to have a positive residual. A positive residual means that the student's actual GPA is higher than the model predicts for someone with the same SAT score.

19. SAT, take 2.

a) $r = 0.685$. The correlation between SAT Math and SAT Verbal is a unitless measure of the degree of linear association between the two variables. It doesn't depend on the order in which you are making predictions.

b) The scatterplot of verbal and math scores shows a relationship that is straight enough, so a linear model is appropriate.

$$b_1 = r\left(\frac{s_{Verbal}}{s_{Math}}\right)$$

$$b_1 = 0.685\left(\frac{99.5}{96.1}\right)$$

$$b_1 = 0.709235$$

$$\hat{y} = b_0 + b_1 x$$

$$\bar{y} = b_0 + b_1 \bar{x}$$

$$596.3 = b_0 + 0.709235(612.2)$$

$$b_0 = 162.106$$

The equation of the least squares regression line for predicting SAT Verbal score from SAT Math score is: $Ve\hat{r}bal = 162.106 + 0.709(Math)$

c) A positive residual means that the student's actual verbal score is higher than the score the model predicts for someone with the same math score.

d)

$$Ve\hat{r}bal = 162.106 + 0.709(Math)$$

$$Ve\hat{r}bal = 162.106 + 0.709(500)$$

$$Ve\hat{r}bal = 516.606$$

According to the model, a person with a math score of 500 is expected to have a verbal score of 516.606 points.

e)

$$M\hat{a}th = 217.692 + 0.662(Verbal)$$

$$M\hat{a}th = 217.692 + 0.662(516.606)$$

$$M\hat{a}th = 559.685$$

According to the model, a person with a verbal score of 516.606 is expected to have a math score of 559.685 points.

f) The prediction in part e) does not cycle back to 500 points because the regression equation used to predict math from verbal is a different equation than the regression equation used to predict verbal from math. One was generated by minimizing squared residuals in the verbal direction, the other was generated by minimizing squared residuals in the math direction. If a math score is one standard deviation above the mean, its predicted verbal score regresses toward the mean. The same is true for a verbal score used to predict a math score.

20. Success, part 2.

$$b_1 = r\left(\frac{s_{SAT}}{s_{GPA}}\right)$$

$$b_1 = 0.47\left(\frac{83}{0.56}\right)$$

$$b_1 = 69.660714$$

$$\hat{y} = b_0 + b_1 x$$

$$\bar{y} = b_0 + b_1 \bar{x}$$

$$1222 = b_0 + 69.660714(2.66)$$

$$b_0 = 1036.703$$

The regression equation to predict SAT score from GPA is:

$$\hat{SAT} = 1036.703 + 69.661(GPA)$$

$$\hat{SAT} = 1036.703 + 69.661(3)$$

$$\hat{SAT} = 1245.686$$

The model predicts that a student with a GPA of 3.0 is expected to have an SAT score of 1245.686.

21. Used cars.

a) We are attempting to predict the price in dollars of used Toyota Corollas from their age in years. A scatterplot of the relationship is at the right.

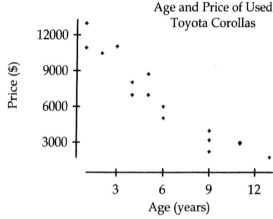

Age and Price of Used Toyota Corollas

b) There is a strong, negative, linear association between price and age of used Toyota Corollas.

c) The scatterplot provides evidence that the relationship is linear. A linear model will likely be an appropriate model.

d) Since $R^2 = 0.894$, simply take the square root to find r. $\sqrt{0.894} = 0.946$. Since association between age and price is negative, $r = -0.946$.

e) 89.4% of the variability in price of a used Toyota Corolla can be explained by variability in the age of the car.

f) The relationship is not perfect. Other factors, such as options, condition, and mileage explain the rest of the variability in price.

22. Drug abuse.

a) The scatterplot shows a positive, strong, linear relationship. It is straight enough to make the linear model the appropriate model.

b) 87.3% of the variability in percentage of other drug usage can be explained by percentage of marijuana use.

c) $R^2 = 0.873$, so $r = \sqrt{0.873} = 0.934$ (since relationship is positive).

$$b_1 = r\left(\frac{s_O}{s_M}\right)$$

$$b_1 = 0.934\left(\frac{10.2}{11.6}\right)$$

$$b_1 = 0.82128$$

$$\hat{y} = b_0 + b_1 x$$

$$\bar{y} = b_0 + b_1\bar{x}$$

$$15.6 = b_0 + 0.82128(23.9)$$

$$b_0 = -4.029$$

The regression equation used to predict the percentage of teens who use other drugs from the percentage who have used marijuana is:

$$\hat{Other} = -4.029 + 0.821(Marijuana)$$

d) According to the model, each additional percent of teens using marijuana is expected to add 0.821 percent to the percentage of teens using other drugs.

e) The results do not *confirm* marijuana as a gateway drug. They do indicate an *association* between marijuana and other drug usage, but association does not imply causation.

23. More used cars.

a) The scatterplot from Exercise 21 shows that the relationship is straight, so the linear model is appropriate.
The regression equation to predict the price of a used Toyota Corolla from its age is $\hat{Price} = 12319.6 - 924(Years)$.
The computer regression output used is at the right.

Dependent variable is: **Price**
No Selector
R squared = 89.4%   R squared (adjusted) = 88.7%
s = 1221 with 17 - 2 = 15 degrees of freedom

Source	Sum of Squares	df	Mean Square	F-ratio
Regression	187830720	1	187830720	126
Residual	22346074	15	1489738	

Variable	Coefficient	s.e. of Coeff	t-ratio	prob
Constant	12319.6	575.7	21.4	≤ 0.0001
Age	-924.000	82.29	-11.2	≤ 0.0001

b) According to the model, for each additional year in age, the car is expected to drop $924 in price.

c) The model predicts that a new Toyota Corolla (0 years old) will cost $12,319.60.

d)

$$\hat{Price} = 12319.6 - 924(Years)$$

$$\hat{Price} = 12319.6 - 924(7)$$

$$\hat{Price} = 5851.60$$

According to the model, an appropriate price for a 7-year old Toyota Corolla is $5,851.60.

e) Buy the car with the negative residual. Its actual price is lower than predicted.

f)

$$\hat{Price} = 12319.6 - 924(Years)$$

$$\hat{Price} = 12319.6 - 924(10)$$

$$\hat{Price} = 3079.60$$

According to the model, a 10-year-old Corolla is expected to cost $3,079.60. The car has an actual price of $1500, so its residual is $1500 − $3079.60 = − $1579.60 The car costs $1579.60 less than predicted.

g) The model would not be useful for predicting the price of a 20-year-old Corolla. The oldest car in the list is 13 years old. Predicting a price after 20 years would be an extrapolation.

24. Veggie burgers.

a)

$$\hat{fat} = 6.8 + 0.97(protein)$$

$$\hat{fat} = 6.8 + 0.97(14)$$

$$\hat{fat} = 20.38$$

According to the model, a burger with 14 grams of protein is expected to have 20.38 grams of fat.

b) From the package, the actual fat content of the veggie burger is 10 grams. The residual is $10 - 20.38 = -10.38$ grams of fat. The veggie burgers have about 10.4 fewer grams of fat than is predicted by the model for a regular burger with a similar protein content.

c) The new veggie burger has 14 grams of protein and 10 grams of fat. The veggie burger has about 10.4 fewer grams of fat than a typical regular burger with a similar protein content.

25. Burgers.

a) The scatterplot of calories vs. fat content in fast food hamburgers is at the right. The relationship appears linear, so a linear model is appropriate.

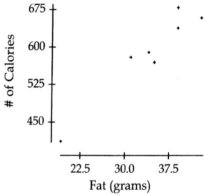

Fat and Calories of Fast Food Burgers

```
Dependent variable is:   Calories
No Selector
R squared = 92.3%     R squared (adjusted) = 90.7%
s = 27.33  with  7 - 2 = 5  degrees of freedom
```

Source	Sum of Squares	df	Mean Square	F-ratio
Regression	44664.3	1	44664.3	59.8
Residual	3735.73	5	747.146	

Variable	Coefficient	s.e. of Coeff	t-ratio	prob
Constant	210.954	50.10	4.21	0.0084
Fat	11.0555	1.430	7.73	0.0006

b) From the computer regression output, $R^2 = 92.3\%$.
92.3% of the variability in the number of calories can be explained by the variability in the number of grams of fat in a fast food burger.

c) From the computer regression output, the regression equation that predicts the number of calories in a fast food burger from its fat content is: $\hat{Calories} = 210.954 + 11.0555(Fat)$

d) The residuals plot at the right shows no pattern. The linear model appears to be appropriate.

e) The model predicts that a fat free burger would have 210.954 calories. Since there are no data values close to 0, this is an extrapolation outside the data and isn't of much use.

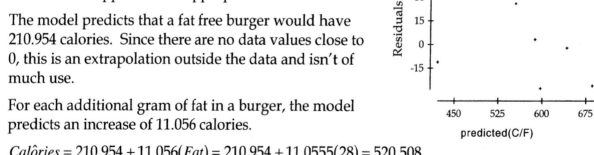

f) For each additional gram of fat in a burger, the model predicts an increase of 11.056 calories.

g) $\hat{Calories} = 210.954 + 11.056(Fat) = 210.954 + 11.0555(28) = 520.508$
The model predicts a burger with 28 grams of fat will have 520.508 calories. If the residual is +33, the actual number of calories is $520.508 + 33 \approx 553.5$ calories.

26. Chicken.

a) The scatterplot is fairly straight, so the linear model is appropriate.

b) The correlation of 0.947 indicates a strong, linear, positive relationship between fat and calories for chicken sandwiches.

c)

$$b_1 = r\left(\frac{s_{Cal}}{s_{Fat}}\right)$$

$$\hat{y} = b_0 + b_1 x$$

$$b_1 = 0.947\left(\frac{144.2}{9.8}\right)$$

$$\bar{y} = b_0 + b_1\bar{x}$$

$$472.7 = b_0 + 13.934429(20.6)$$

$$b_1 = 13.934429$$

$$b_0 = 185.651$$

The linear model for predicting calories from fat in chicken sandwiches is:

$$Cal\hat{o}ries = 185.651 + 13.934(Fat)$$

d) For each additional gram of fat, the model predicts an increase in 13.934 calories.

e) According to the model, a fat-free chicken sandwich would have 185.651 calories. This is probably an extrapolation, although without the actual data, we can't be sure.

f) In this context, a negative residual means that a chicken sandwich has fewer calories than the model predicts.

g) For the chicken sandwich:

$$Cal\hat{o}ries = 185.651 + 13.934(Fat)$$
$$Cal\hat{o}ries = 185.651 + 13.934(35)$$
$$Cal\hat{o}ries = 673.341$$

For the burger:

$$Cal\hat{o}ries = 210.954 + 11.056(Fat)$$
$$Cal\hat{o}ries = 210.954 + 11.056(35)$$
$$Cal\hat{o}ries = 597.914$$

A chicken sandwich with 35 grams of fat is predicted to have more calories than a burger with 35 grams of fat.

h) Using the chicken sandwich model:

$$Cal\hat{o}ries = 185.651 + 13.934(Fat)$$
$$Cal\hat{o}ries = 185.651 + 13.934(26)$$
$$Cal\hat{o}ries = 547.935$$

Using the burger model:

$$Cal\hat{o}ries = 210.954 + 11.056(Fat)$$
$$Cal\hat{o}ries = 210.954 + 11.056(26)$$
$$Cal\hat{o}ries = 498.41$$

A Filet-O-Fish sandwich, at 470 calories, has fewer calories than expected for a burger and many fewer calories than expected for a chicken sandwich. The fish sandwich has a relationship between fat and calories that is similar to the burgers.

27. A second helping of burgers.

a) The model from Exercise 25 was for predicting number of calories from number of grams of fat. In order to predict grams of fat from the number of calories, a new linear model needs to be generated.

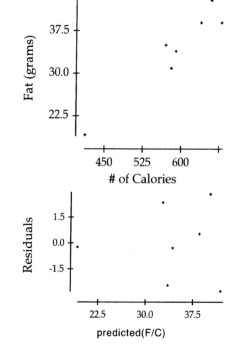

Calories and Fat in Fast Food Burgers

b) The scatterplot at the right shows the relationship between number fat grams and number of calories in a set of fast food burgers. The association is strong, positive, and linear. Burgers with higher numbers of calories typically have higher fat contents. The relationship is straight enough to apply a linear model.

Dependent variable is: **Fat**
No Selector
R squared = 92.3%   R squared (adjusted) = 90.7%
s = 2.375 with 7 - 2 = 5 degrees of freedom

Source	Sum of Squares	df	Mean Square	F-ratio
Regression	337.223	1	337.223	59.8
Residual	28.2054	5	5.64109	

Variable	Coefficient	s.e. of Coeff	t-ratio	prob
Constant	-14.9622	6.433	-2.33	0.0675
Calories	0.083471	0.0108	7.73	0.0006

The linear model for predicting fat from calories is:
$\hat{Fat} = -14.9622 + 0.083471(Calories)$
The model predicts that for every additional 100 calories, the fat content is expected to increase by about 8.3 grams.

The residuals plot shows no pattern, so the model is appropriate. $R^2 = 92.3\%$, so 92.3% of the variability in fat content can be explained by the model.

$\hat{Fat} = -14.9622 + 0.083471(Calories)$

$\hat{Fat} = -14.9622 + 0.083471(600)$

$\hat{Fat} \approx 35.1$

According to the model, a burger with 600 calories is expected to have 35.1 grams of fat.

28. Cost of living.

a) The association between cost of living in 2000 and 2001 is linear, positive and strong. The linearity of the scatterplot indicates that the linear model is appropriate.

b) $R^2 = (0.957)^2 = 0.9158$. This means that 91.6% of the variability in cost of living in 2001 can be explained by variability in cost of living in 2000.

c) Moscow had a cost of living of 136.1% of New York's in 2000.

$\hat{cost01} = 25.41 + 0.69(cost00)$

$\hat{cost01} = 25.41 + 0.69(136.1)$

$\hat{cost01} = 119.319$

According to the model, Moscow is predicted to have a cost of living in 2001 that is about 119.3% of New York's. Moscow actually had a cost of living in 2001 that was 132.4% of New York's. Moscow's cost of living in 2001 was about 13.1% more than predicted. Moscow's residual was about +13.1%.

29. Cell phones.

a) $A\hat{n}l = -5.1 + 0.44(200) = 82.9$ minutes of analog talk time is predicted by the model.

b) Since $r = 0.94$, $R^2 = 0.8836$. In other words, 88.36% of the variability in analog talk time can be explained by variability in digital talk time, and the scatterplot is linear. The model should be fairly accurate.

c) The model predicts that for each additional minute of digital talk time, we can expect 0.44 minutes of additional analog talk time.

d) A battery with a positive residual provides more actual talk time than the linear model would predict.

30. Candy.

The scatterplot of Halloween candy sales by year, at the right, shows an association that is positive, linear, and strong. In general, sales of Halloween candy have increased as the years have increased. The scatterplot is straight enough to justify the use of the linear model. The linear regression output from a computer program is below.

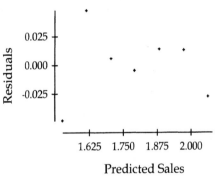

Yearly Sales of Halloween Candy

```
Dependent variable is:   Halloween Candy Sales(millions of dollars)
No Selector
R squared = 97.5%     R squared (adjusted) = 97.0%
s = 0.0339  with  7 - 2 = 5  degrees of freedom
```

Source	Sum of Squares	df	Mean Square	F-ratio
Regression	0.226980	1	0.226980	197
Residual	0.005763	5	0.001153	

Variable	Coefficient	s.e. of Coeff	t-ratio	prob
Constant	-178.099	12.82	-13.9	≤ 0.0001
Year	0.090036	0.0064	14.0	≤ 0.0001

The linear model for predicting candy sales from the year is :
$\hat{Sal}es = -178.099 + 0.090036(Year)$.

This model is appropriate, since the residuals plot (at the right) shows no apparent pattern. The model explains 97.5% of the variability in candy sales, so the model should be quite accurate. The model predicts that Halloween candy sales will be 2.15 million dollars.

$\hat{Sal}es = -178.099 + 0.090036(Year)$
$\hat{Sal}es = -178.099 + 0.090036(2002)$
$\hat{Sal}es \approx 2.15$

Since the model is appropriate and accurate, the prediction should be a good one. However, the year 2002 is a slight extrapolation outside of the range of the data, so we must assume the linear model continues according to the same pattern.

31. El Niño.

a) The correlation between CO_2 level and mean temperature is $r = \sqrt{R^2} = \sqrt{0.334} = 0.5779$.

b) 33.4% of the variability in mean temperature can be explained by variability in CO_2 level.

c) Since the scatterplot of CO_2 level and mean temperature shows a relationship that is straight enough, use of the linear model is appropriate. The linear regression model that predicts mean temperature from CO_2 level is: $\widehat{MeanTemp} = 15.3066 + 0.004(CO_2)$

d) The model predicts that an increase in CO_2 level of 1 ppm is associated with an increase of 0.004 °C in mean temperature.

e) According to the model, the mean temperature is predicted to be 15.3066 °C when there is no CO_2 in the atmosphere. This is an extrapolation outside of the range of data, and isn't very meaningful in context, since there is always CO_2 in the atmosphere. We want to use this model to study the change in CO_2 level and how it relates to the change in temperature.

f) The residuals plot shows no apparent patterns. The linear model appears to be an appropriate one.

g)

$$\widehat{MeanTemp} = 15.3066 + 0.004(CO_2)$$

$$\widehat{MeanTemp} = 15.3066 + 0.004(364)$$

$$\widehat{MeanTemp} = 16.7626$$

According to the model, the temperature is predicted to be 16.7626 °C when the CO_2 level is 364 ppm.

32. Birth rates.

a) A scatterplot of the live birth rates in the US over time is at the right. The association is negative, strong, and appears to be curved, with one low outlier, the rate of 14.8 live births per 1000 women age 15 – 44 in 1975. Generally, as time passes, the birth rate is getting lower.

US Birth Rate

b) Although the association is slightly curved, it is straight enough to try a linear model. The linear regression output from a computer program is shown below:

```
Dependent variable is:   Rate
No Selector
R squared = 61.0%    R squared (adjusted) = 54.5%
s = 1.206  with  8 - 2 = 6  degrees of freedom
```

Source	Sum of Squares	df	Mean Square	F-ratio
Regression	13.6542	1	13.6542	9.39
Residual	8.72081	6	1.45347	

Variable	Coefficient	s.e. of Coeff	t-ratio	prob
Constant	246.051	74.99	3.28	0.0168
Year	-0.115935	0.0378	-3.06	0.0221

The linear regression model for predicting birth rate from year is:
$$\widehat{Birthrate} = 246.051 - 0.115935(Year)$$

c) The residuals plot, at the right, shows a slight curve. The model yields positive residuals near the ends of the range of data, and negative residuals in the middle. The linear model may not be appropriate. At the very least, be cautious when using this model.

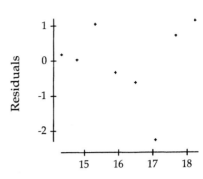

Predicted # of live births per 1000 women

d) The model predicts that each passing year is associated with a decline in birth rate of 0.116 births per 100 women.

e)

$$\widehat{Birthrate} = 246.051 - 0.116(Year)$$

$$\widehat{Birthrate} = 246.051 - 0.115935(1978)$$

$$\widehat{Birthrate} \approx 16.73$$

The model predicts about 16.73 births per 1000 women in 1978.

f) If the actual birth rate in 1978 was 15.0 births per 1000 women, the model has a residual of 15.0 − 16.73 = −1.73 births per 1000 women. This means that the model predicted 1.73 births higher than the actual rate.

g)

$$\widehat{Birthrate} = 246.051 - 0.115935(Year)$$

$$\widehat{Birthrate} = 246.051 - 0.115935(2005)$$

$$\widehat{Birthrate} \approx 13.60$$

According to the model, the birth rate in 2005 is predicted to be 13.60 births per 1000 women. This prediction seems a bit low. It is an extrapolation outside the range of the data, and furthermore, the model only explains 61% of the variability in birth rate. Don't place too much faith in this prediction.

h)

$$\widehat{Birthrate} = 246.051 - 0.116(Year)$$

$$\widehat{Birthrate} = 246.051 - 0.115935(2020)$$

$$\widehat{Birthrate} \approx 11.86$$

According to the model, the birth rate in 2020 is predicted to be 11.86 births per 1000 women. This prediction is an extreme extrapolation outside the range of the data, which is dangerous. No faith should be placed in this prediction.

33. Body fat.

a) The scatterplot of % body fat and weight of 20 male subjects, at the right, shows a strong, positive, linear association. Generally, as a subject's weight increases, so does % body fat. The association is straight enough to justify the use of the linear model.

The linear model that predicts % body fat from weight is:
$$\%\widehat{Fat} = -27.3763 + 0.249874(Weight)$$

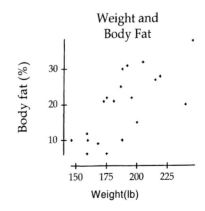

Weight and Body Fat

b) The residuals plot, at the right, shows no apparent pattern. The linear model is appropriate.

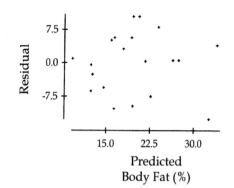

c) According to the model, for each additional pound of weight, body fat is expected to increase by about 0.25%.

d) Only 48.5% of the variability in % body fat can be explained by the model. The model is not expected to make predictions that are accurate.

e)

$\%F\hat{a}t = -27.3763 + 0.249874(Weight)$

$\%F\hat{a}t = -27.3763 + 0.249874(190)$

$\%F\hat{a}t \approx 20.09976$

According to the model, the predicted body fat for a 190-pound man is 20.09976%. The residual is $21 - 20.09976 \approx 0.9\%$.

34. Body fat, again.

The scatterplot of % body fat and waist size is at the right. The association is strong, linear, and positive. As waist size increases, % body fat has a tendency to increase, as well. The scatterplot is straight enough to justify the use of the linear model.

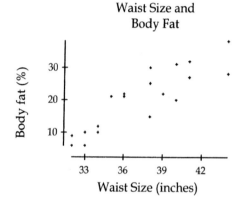

The linear model for predicting % body fat from waist size is : $\%F\hat{a}t = -62.557 + 2.222(Waist)$.

For each additional inch in waist size, the model predicts an increase of 2.222% body fat.

78.7% of the variability in % body fat can be determined by waist size. The residuals plot, at right, shows no apparent pattern. The residuals plot and the relatively high value of R^2 indicate an appropriate model with more predicting power than the model based on weight.

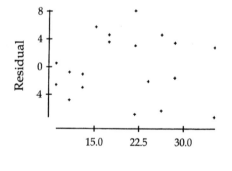

35. Hard water.

a) There is a fairly strong, negative, linear relationship between calcium concentration (in ppm) in the water and mortality rate (in deaths per 100,000). Towns with higher calcium concentrations tended to have lower mortality rates.

b) The linear regression model that predicts mortality rate from calcium concentration is $Mort\hat{a}lity = 1676 - 3.23(Calcium)$.

c) The model predicts a decrease of 3.23 deaths per 100,000 for each additional ppm of calcium in the water. For towns with no calcium in the water, the model predicts a mortality rate of 1676 deaths per 100,000 people.

d) Exeter had 348.6 fewer deaths per 100,000 people than the model predicts.

e)

$$Mort\hat{a}lity = 1676 - 3.23(Calcium)$$

$$Mort\hat{a}lity = 1676 - 3.23(100)$$

$$Mort\hat{a}lity = 1353$$

The town of Derby is predicted to have a mortality rate of 1353 deaths per 100,000 people.

f) 43% of the variability in mortality rate can be explained by variability in calcium concentration.

36. Gators.

a) Weight is the proper dependent variable. The researchers can estimate length from the air, and use length to predict weight, as desired.

b) The correlation between an alligator's length and weight is $r = \sqrt{R^2} = \sqrt{0.836} = 0.914$.

c) The linear regression model that predicts and alligator's weight from its length is :
   $$Wei\hat{g}ht = -393 + 5.9(Length).$$

d) For each additional inch in length, the model predicts an increase of 5.9 pounds in weight.

e) The estimates made using this model should be fairly accurate. The model explains 83.6% of the variability in weight. However, care should be taken. With no scatterplot, and no residuals plot, we cannot verify the regression condition of linearity. The association between length and weight may be curved, in which case, the linear model is not appropriate.

37. Internet.

The association between the year and the number of Internet journals is very strong, positive, but curved. Later years generally have higher numbers of journals, but the increase per year is not constant. The "straight enough" condition is not satisfied, so the linear model is not appropriate.

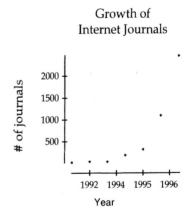

Growth of Internet Journals

38. Oil production.

   a) The linear regression model that predicts oil production (millions of barrels) from year is:
$\hat{Oil} = -3,705,119 + 3289.53(Year)$.

   b) $\hat{Oil} = -3,705,119 + 3289.53(Year) = -3,705,119 + 3289.53(2001) = 2,877,230.53$ million barrels is the prediction for 2001.

   c) The last time oil production was that high was 1988, and production has been going down since 1972. This prediction does not appear to be accurate.

   d) When modeling data, always start with a picture. Had we done this, we would have seen the strong, curved association between oil production and year. The "straight enough" condition is not satisfied, so a linear model is not appropriate.

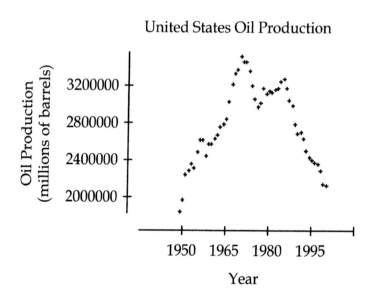

United States Oil Production

Chapter 9 – Regression Wisdom

1. **Marriage age.**

   a) The trend in age at first marriage for American women is very strong over the entire time period recorded on the graph, but the direction and form are different for different time periods. The trend appears to be somewhat linear, and consistent at around 22 years, up until about 1940, when the age seemed to drop dramatically, to under 21. From 1940 to about 1970, the trend appears non-linear and slightly positive. From 1975 to the present, the trend again appears linear and positive. The marriage age rose rapidly during this time period.

   b) The association between age at first marriage for American women and year is strong over the entire time period recorded on the graph, but some time periods have stronger trends than others.

   c) The correlation, or the measure of the degree of linear association is not high for this trend. The graph, as a whole, is non-linear. However, certain time periods, like 1975 to present, have a high correlation.

   d) Overall, the linear model is not appropriate. The scatterplot is not "straight enough" to satisfy the condition. You could fit a linear model to the time period from 1975 to 1995, but this seems unnecessary. The ages for each year are reported, and, given the fluctuations in the past, extrapolation seems risky.

2. **Ages of couples.**

   a) The correlation between age difference and year is $r = \sqrt{R^2} = \sqrt{0.716} \approx -0.846$. The negative value is used since the scatterplot shows that the association is negative, strong, and linear.

   b) The linear regression model that predicts age difference from year is:
   $(Men - \hat{W}omen) = 33.483 - 0.015756(Year)$. This model predicts that each passing year is associated with a decrease of approximately 0.016 years in the difference between male and female marriage age. A more meaningful comparison might be to say that the model predicts a decrease of approximately 0.16 years in the age difference for every 10 years that pass.

   c)

   $(Men - \hat{W}omen) = 33.483 - 0.015756(Year)$

   $(Men - \hat{W}omen) = 33.483 - 0.015756(2010)$

   $(Men - \hat{W}omen) \approx 1.81344$

   According to the model, the age difference between men and women at first marriage is expected to be approximately 1.81 years. (This figure is very sensitive to the number of decimal places used in the model.)

   d) The latest data point is before the year 2000. Extrapolating for 2010 is risky because it depends on the assumption that the trend in age at first marriage will continue in the same manner.

3. Marriage age revisited.

a) Modeling decisions may vary, but the important idea is using a subset of the data that allows us to make an accurate prediction for the year in which we are interested. We might model a subset to predict the marriage age in 2005, and model another subset to predict the marriage age in 1911.

Female Marriage Age

In order to predict the average marriage age of American women in 2005, use the data points from the most recent trend only. The data points from 1975 – 1995 look straight enough to apply the linear regression model.

Regression output from a computer program is given below, as well as a residuals plot.

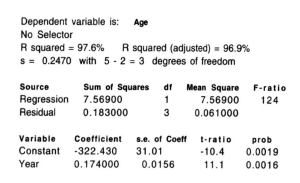

Dependent variable is: **Age**
No Selector
R squared = 97.6%   R squared (adjusted) = 96.9%
s = 0.2470 with 5 - 2 = 3 degrees of freedom

Source	Sum of Squares	df	Mean Square	F-ratio
Regression	7.56900	1	7.56900	124
Residual	0.183000	3	0.061000	

Variable	Coefficient	s.e. of Coeff	t-ratio	prob
Constant	-322.430	31.01	-10.4	0.0019
Year	0.174000	0.0156	11.1	0.0016

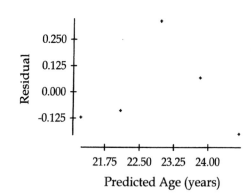

The linear model used to predict average female marriage age from year is:
$A\hat{g}e = -322.430 + 0.174(Year)$. The residuals plot shows no apparent pattern (although the number of points used is small), and the value of R^2 is high. 97.6% of the variability in average female age at first marriage is determined by variability in the year. The model predicts that each year that passes is associated with an increase of 0.174 years in the average female age at first marriage.

$$A\hat{g}e = -322.430 + 0.174(Year)$$
$$A\hat{g}e = -322.430 + 0.174(2005)$$
$$A\hat{g}e \approx 26.44$$

According to the model, the average age at first marriage for women in 2005 will be 26.44 years old. Care should be taken with this prediction, however. It represents an extrapolation of 10 years beyond the highest year.

b) This prediction is for a year that is 10 years higher than the highest year for which we have an average female marriage age. Don't place too much faith in this extrapolation.

c) An extrapolation of more than 50 years into the future would be absurd. There is no reason to believe the trend would continue. In fact, given the situation, it is very unlikely that the pattern would continue in this fashion. The model given in part a) predicts that the marriage age will be 34.27 years. Realistically, that seems quite high.

4. Ages of couples, again.

a) The data from the late 1800s to 1950 are high leverage points. Since they generally follow the same linear trend as the 1975 – 1998 data, those data points increase the correlation and the R^2 value.

b) The residuals plot shows no apparent pattern, so the linear model is appropriate.

c) For every 10 years that pass, the model predicts a decrease of approximately 0.24 years in average age difference at first marriage.

d) The *y*-intercept is the prediction of the model in year 0, over 2000 years ago. An extrapolation that far into the past is not meaningful. The earliest year for which we have data is 1975.

5. Good model?

a) The student's reasoning is not correct. A scattered residuals plot, not high R^2, is the indicator of an appropriate model. Once the model is deemed appropriate, R^2 is used as a measure of the strength of the model.

b) The model may not allow the student to make accurate predictions. The data may be curved, in which case the linear model would not fit well.

6. Bad model?

a) The student's model may, in fact, be appropriate. Low R^2 simply means that the model is not accurate. The model explains only 13% of the variability in the response variable. If the residuals plot shows no pattern, this model may be appropriate.

b) The predictions are not likely to be very accurate, but they may be the best that the student can get. $R^2 = 13\%$ indicates a great deal of scatter around the regression line, but if the residuals plot is not patterned, there probably isn't a better model. The two variables that are being studied by the student have a weak association.

7. Reading.

a) The principal's description of a strong, positive trend is misleading. First of all, "trend" implies a change over time. These data were gathered during one year, at different grade levels. To observe a trend, one class's reading scores would have to be followed through several years. Second, the strong, positive relationship only indicates the yearly improvement that would be expected, as children get older. For example, the 4th graders are reading at approximately a 4th grade level, on average. This means that the school's students are progressing adequately in their reading, not extraordinarily. Finally, the use of average reading scores instead of individual scores increases the strength of the association.

b) The plot appears very straight. The correlation between grade and reading level is very high, probably between 0.9 and 1.0.

c) If the principal had made a scatterplot of all students' scores, the correlation would have likely been lower. Averaging reduced the scatter, since each grade level has only one point instead of many, which inflates the correlation.

d) If a student is reading at grade level, then that student's reading score should equal his or her grade level. The slope of that relationship is 1. That would be "acceptable", according to the measurement scale of reading level. Any slope greater than 1 would indicate above grade level reading scores, which would certainly be acceptable as well. A slope less than 1 would indicate below grade level average scores, which would be unacceptable.

8. Grades.

Perhaps the best way to start is to discuss the type of graph that would have been useful. The admissions officer should have made a scatterplot with a coordinate for each freshman, matching each individual's SAT score with his or her respective GPA. Then, if the cloud of points was straight enough, the officer could have attempted to fit a linear model, and assessed its appropriateness and strength.

As is, the graph of combined SAT score versus mean Freshman GPA indicates, very generally, that higher SAT achievement is associated with higher mean Freshman GPA, but that's about it.

The first concern is the SAT scores. They have been grouped into categories. We cannot perform any type of regression analysis, because this variable is not quantitative. We don't even know how many students are in each category. There may be one student with an SAT score in the 1500s, and 300 students in the 1200s. On this graph, these possibilities are given equal weight!

Even if the SAT scores were at all useful to us, the GPAs given are averages, which would make the association appear stronger than it actually is.

Finally, a connected line graph isn't a useful model. It doesn't simplify the situation at all, and may, in fact, give the false impression that we could interpolate between the data points.

9. Heating.

a) The model predicts a decrease in $2.13 in heating cost for an increase in temperature of 1° Fahrenheit. Generally, warmer months are associated with lower heating costs.

b) When the temperature is 0° Fahrenheit, the model predicts a monthly heating cost of $133.

c) When the temperature is around 32° Fahrenheit, the predictions are generally too high. The residuals are negative, indicating that the actual values are lower than the predicted values.

d)

$\hat{C} = 133 - 2.13(temp)$

$\hat{C} = 133 - 2.13(10)$

$\hat{C} = \$111.70$

According to the model, the heating cost in a month with average daily temperature 10° Fahrenheit is expected to be $111.70.

e) The residual for a 10° day is approximately –$6, meaning that the actual cost was $6 less than predicted, or $111.70 – $6 = $105.70.

f) The model is not appropriate. The residuals plot shows a definite curved pattern. The association between monthly heating cost and average daily temperature is not linear.

g) A change of scale from Fahrenheit to Celsius would not affect the relationship. Associations between quantitative variables are the same, no matter what the units.

10. Speed.

a) The model predicts that as speed increases by 1 mile per hour, the fuel economy is expected to decrease by 0.1 miles per gallon.

b) For this model, the *y*-intercept is the predicted mileage at a speed of 0 miles per hour. It's not possible to get 32 miles per gallon if you aren't moving.

c) The residuals are negative for the higher gas mileages. This means that the model is predicting higher than the actual mileage.

d)

$$\hat{mpg} = 32 - 0.1(mph)$$
$$\hat{mpg} = 32 - 0.1(50)$$
$$\hat{mpg} = 27$$

When a car is driven at 50 miles per hour, the model predicts mileage of 27 miles per gallon.

e)

$$\hat{mpg} = 32 - 0.1(mph)$$
$$\hat{mpg} = 32 - 0.1(45)$$
$$\hat{mpg} = 27.5$$

When a car is driven at 45 miles per hour, the model predicts mileage of 27.5 miles per gallon. From the graph, the residual at 27.5 mpg is +1. The actual gas mileage is 27.5 + 1 = 28.5 mpg.

f) The association between fuel economy and speed is probably quite strong, but not linear.

g) The linear model is not the appropriate model for the association between fuel economy and speed. The residuals plot has a clear pattern. If the linear model were appropriate, we would expect scatter in the residuals plot.

11. Unusual points.

a) **1)** The point is an outlier, because it is not in the cloud of data points.
2) The point is not influential. It has the *potential* to be influential, because its position far from the mean of the explanatory variable gives it high leverage. However, the point is not *exerting* much influence, because it reinforces the association.
3) If the point were removed, the correlation would become weaker. The point heavily reinforces the positive association. Removing it would weaken the association.
4) The slope would remain roughly the same, since the point is not influential.

b) **1)** The point is an outlier. It departs from the overall pattern.
2) The point is influential. The point alone gives the scatterplot the appearance of an overall negative direction, when the points are actually fairly scattered.
3) If the point were removed, the correlation would become weaker. Without the point, there would be very little evidence of linear association.

4) The slope would increase, from a negative slope to a slope near 0. Without the point, the slope of the regression line would be nearly flat.

c) **1)** Since it departs from the overall pattern, the point is an outlier.
2) The point is somewhat influential. It is well away from the mean of the explanatory variable, and has enough leverage to change the slope of the regression line, but only slightly.
3) If the point were removed, the correlation would become stronger. Without the point, the positive association would be reinforced.
4) The slope would increase slightly, becoming steeper after the removal of the point. The regression line would follow the general cloud of points more closely.

d) **1)** The point is an outlier. It departs from the overall pattern.
2) The point is not influential. It is very close to the mean of the explanatory variable, and the regression line is anchored at the point $(\bar{x}, \bar{y})$, and would only pivot if it were possible to minimize the sum of the squared residuals. No amount of pivoting will reduce the residual for the stray point, so the slope would not change.
3) If the point were removed, the correlation would become slightly stronger, decreasing to become more negative. The point detracts from the overall pattern, and its removal would reinforce the association.
4) The slope would remain roughly the same. Since the point is not influential, its removal would not affect the slope.

12. More unusual points.

a) **1)** The point is an outlier. It is a departure from the pattern.
2) The point is influential. It is well away from the mean of the explanatory variable, and has enough leverage to change the slope of the regression line.
3) If the point were removed, the correlation would become stronger. Without the point, the positive association would be reinforced.
4) The slope would increase, becoming steeper after the removal of the point. The regression line would follow the general cloud of points more closely.

b) **1)** The point is an outlier. It is a departure from the pattern.
2) The point is influential. The point alone gives the scatterplot the appearance of an overall positive direction, when the points are actually fairly scattered.
3) If the point were removed, the correlation would become weaker. Without the point, there would be very little evidence of linear association.
4) The slope would decrease, from a positive slope to a slope near 0. Without the point, the slope of the regression line would be nearly flat.

c) **1)** The point is an outlier. It is a departure from the pattern.
2) The point is not influential. It is very close to the mean of the explanatory variable, and the regression line is anchored at the point $(\bar{x}, \bar{y})$, and would only pivot if it were possible to minimize the sum of the squared residuals. No amount of pivoting will reduce the residual for the stray point, so the slope would not change.
3) If the point were removed, the correlation would become slightly stronger. The point detracts from the overall pattern, and its removal would reinforce the association.
4) The slope would remain roughly the same. Since the point is not influential, its removal would not affect the slope.

d) 1) The point is an outlier, because it not in the cloud of data points.
2) The point is not influential. It has the *potential* to be influential, because its position far from the mean of the explanatory variable gives it high leverage. However, the point is not *exerting* much influence, because it reinforces the association.
3) If the point were removed, the correlation would become weaker. The point heavily reinforces the association. Removing it would weaken the association.
4) The slope would remain roughly the same, since the point is not influential.

13. The extra point.

(1) Point e is very influential. Its addition will give the appearance of a strong, negative correlation like $r = -0.90$.

(2) Point d is influential (but not as influential as point e). Its addition will give the appearance of a weaker, negative correlation like $r = -0.40$.

(3) Point c is directly below the middle of the group of points. Its position is directly below the mean of the explanatory variable. It has no influence. Its addition will leave the correlation the same, $r = 0.00$.

(4) Point b is almost in the center of the group of points, but not quite. Its addition will give the appearance of a very slight positive correlation like $r = 0.05$.

(5) Point a is very influential. Its addition will give the appearance of a strong, positive correlation like $r = 0.75$.

14. The extra point revisited.

(1) Point d is influential. Its addition will pull the slope of the regression line toward point d, resulting in the steepest negative slope, a slope of -0.45.

(2) Point e is very influential, but since it is far away from the group of points, its addition will only pull the slope down slightly. The slope is -0.30.

(3) Point c is directly below the middle of the group of points. Its position is directly below the mean of the explanatory variable. It has no influence. Its addition will leave the slope the same, 0.

(4) Point b is almost in the center of the group of points, but not quite. It has very little influence, but what influence it has is positive. The slope will increase very slightly with its addition, to 0.05.

(5) Point a is very influential. Its addition will pull the regression line up to its steepest positive slope, 0.85.

15. Gestation.

a) The association would be stronger if humans were removed. The point on the scatterplot representing human gestation and life expectancy is an outlier from the overall pattern and detracts from the association. Humans also represent an influential point. Removing the humans would cause the slope of the linear regression model to increase, following the pattern of the non-human animals much more closely.

b) The study could be restricted to non-human animals. This appears justifiable, since one could point to a number of environmental factors that could influence human life expectancy and gestation period, making them incomparable to those of non-human animals.

c) The correlation is moderately strong. The model explains 72.2% of the variability in gestation period of non-human animals.

d) For every year increase in life expectancy, the model predicts an increase of approximately 15.5 days in gestation period.

e)

$$\hat{Gest} = -39.5172 + 15.4980(LifEx)$$

$$\hat{Gest} = -39.5172 + 15.4980(20)$$

$$\hat{Gest} \approx 270.4428$$

According to the linear model, monkeys with a life expectancy of 20 years are expected to have gestation periods of about 270.5 days. Care should be taken when assessing the accuracy of this prediction. First of all, the residuals plot has not been examined, so the appropriateness of the model is questionable. Second, it is unknown whether or not monkeys were included in the original 17 non-human species studied. Since monkeys and humans are both primates, the monkeys may depart from the overall pattern as well.

16. Elephants and hippos.

a) Hippos are more of a departure from the pattern. Removing that point would make the association appear to be stronger.

b) The slope of the regression line would increase, pivoting away from the hippos point.

c) Anytime data points are removed, there must be a justifiable reason for doing so, and saying, "I removed the point because the correlation was higher without it" is not a justifiable reason.

d) Elephants are an influential point. With the elephants included, the slope of the linear model is 15.4980 days gestation per year of life expectancy. When they are removed, the slope is 11.6 days per year. The decrease is significant.

17. Law enforcement.

A scatterplot (with regression line) of assaults (per 1000) and kills/injuries (per 1000) for law enforcement officers authorized to carry firearms is at the right. There is one influential point, the National Park Service (highlighted).

The model we have is highly influenced by the National Park Service, and is of no use as a predictor of kills/injury rates from assault rates.

Below this scatterplot is another, with the influential point removed. Notice how removing the National Park Service changes the slope of the regression line.

There doesn't seem to be a good reason to remove the National Park Service. Although the National Park Service seems like a unique branch of law enforcement, so do all of the others. Each type of officer faces different situations on the job than the others. This is no reason to exclude a particular branch.

Even if we could somehow justify removing the National Park Service, the slope of the regression line is nearly 0, indicating a model with extremely weak predicting ability. In essence, the regression line always predicts the mean killed-injured rate.

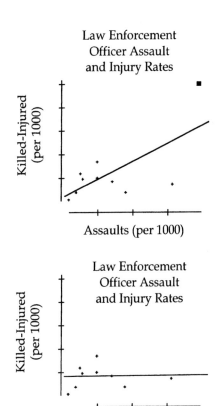

Law Enforcement Officer Assault and Injury Rates

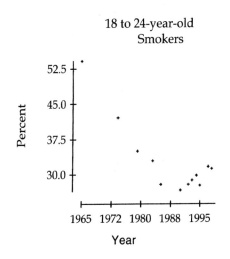

Law Enforcement Officer Assault and Injury Rates

18. Smoking.

The analysis that follows is one of several good models that may be used to predict the percentage of smokers among males ages 18 to 24. The important feature to recognize is that these data consist of two distinct trends. Your modeling decisions may vary slightly from these, but that is fine as long as those decisions are justified.

A scatterplot (at the right) of year vs. percent of males ages 18 to 24 who smoke shows two distinct trends. From 1965 to 1985, there is a strong, negative linear association between year and percent smokers. As time passed, the percentage of smokers decreased. For the years 1990 to 1998, there is a reasonably strong, positive, linear association between year and percent smokers. The percentage of smokers increased during this time period. Two linear models, used together, will fit the relationship well.

Model I (1965 – 1985)

Dependent variable is: **Percentage**
No Selector
R squared = 98.8%   R squared (adjusted) = 98.3%
s = 1.302 with 5 - 2 = 3 degrees of freedom

Source	Sum of Squares	df	Mean Square	F-ratio
Regression	405.059	1	405.059	239
Residual	5.08900	3	1.69633	

Variable	Coefficient	s.e. of Coeff	t-ratio	prob
Constant	2521.62	160.7	15.7	0.0006
Year	-1.25592	0.0813	-15.5	0.0006

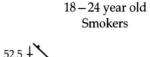
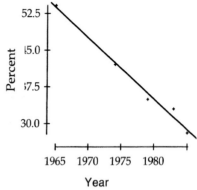

$\hat{\%} = 2521.62 - 1.25592(Year)$ is a good model for the years 1965 to 1985. A scatterplot of the relationship, with regression line, is shown at the right. $R^2 = 98.8\%$, so the model explains 98.8% of the variability in the percentage of males 18 – 24 years old who smoke. The residuals plot is scattered indicating an appropriate model.

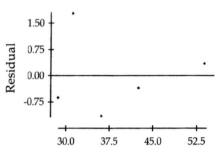

Predicted Percentage

Model II (1990 – 1998)

Dependent variable is: **Percentage**
No Selector
R squared = 77.1%   R squared (adjusted) = 72.6%
s = 0.9874 with 7 - 2 = 5 degrees of freedom

Source	Sum of Squares	df	Mean Square	F-ratio
Regression	16.4427	1	16.4427	16.9
Residual	4.87442	5	0.974884	

Variable	Coefficient	s.e. of Coeff	t-ratio	prob
Constant	-1152.14	287.6	-4.01	0.0103
Year	0.592378	0.1442	4.11	0.0093

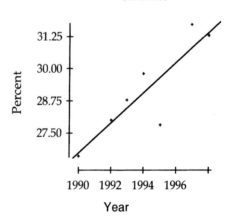

$\hat{\%} = -1152.14 + 0.592378(year)$ is a decent model for the years 1990 – 1998. $R^2 = 77.1\%$, so only 77.1% of the variability in percent smokers is explained by the model. The residuals plot is acceptable, but might have a bit of a pattern. Removing points for 1995 and 1998 would increase R^2 and leave scattered residuals, but removing these points is impossible to justify. Removing 2 of 7 data points for no reason (other than the fact that they don't seem to fit the pattern we *think* is present) just isn't a good idea. The model given for these years seems to be the best we can do.

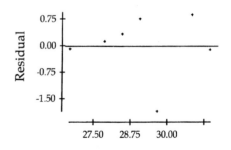

Predicted Percentage

19. Illegitimate births.

The analysis that follows is one of several good models that may be used to predict the percentage of illegitimate births. The important feature to recognize is that these data consist of two distinct trends. Your modeling decisions may vary slightly from these, but that is fine as long as those decisions are justified.

A scatterplot (at the right) of year vs. percent of unmarried births shows two distinct trends. From 1980 to 1994, there is a strong, positive linear association between year and percent of unmarried births. For the years 1995 to 1998, there is also a strong, positive, linear association, but the percent of unmarried births increases much more slowly from year to year, almost to the point of being flat. Two linear models will fit the relationship well.

Model I (1980—1994)

```
Dependent variable is:    Births
No Selector
R squared = 99.0%     R squared (adjusted) = 98.8%
s =  0.5645  with  7 - 2 = 5  degrees of freedom
```

Source	Sum of Squares	df	Mean Square	F-ratio
Regression	160.621	1	160.621	504
Residual	1.59314	5	0.318628	

Variable	Coefficient	s.e. of Coeff	t-ratio	prob
Constant	-2021.41	91.25	-22.2	≤ 0.0001
Year	1.02991	0.0459	22.5	≤ 0.0001

$\hat{\%} = -2021.41 + 1.0299(year)$ is a good model for the years 1980 – 1994. A scatterplot of the relationship, with regression line, is shown above and to the right. $R^2 = 99\%$, so the model explains 99% of the variability in percent of unmarried births. The residuals plot (at the right) is scattered, indicating an appropriate model.

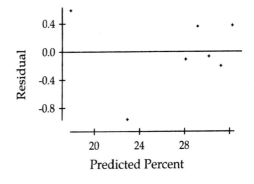

Model II (1995–1998)

Dependent variable is: **Births**
No Selector
R squared = 85.3%   R squared (adjusted) = 77.9%
s = 0.1183 with 4 - 2 = 2 degrees of freedom

Source	Sum of Squares	df	Mean Square	F-ratio
Regression	0.162000	1	0.162000	11.6
Residual	0.028000	2	0.014000	

Variable	Coefficient	s.e. of Coeff	t-ratio	prob
Constant	-326.920	105.6	-3.09	0.0905
Year	0.180000	0.0529	3.40	0.0766

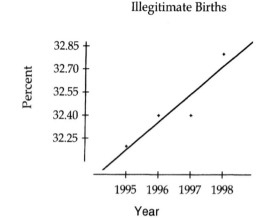

Illegitimate Births

$\hat{\%} = -326.92 + 0.18(year)$ is a good model for the years 1995 – 1998. Although not as accurate as the first model, R^2 = 85.3%, which means that the model accounts for 85.3% of the variability in percent of unmarried births. The residuals plot is scattered, indicating an appropriate model. Great care should be taken in using this model for predictions, since it was developed from only four data points. The slope of the regression line indicates that for each year that passes, the model predicts an increase of only 0.18% unmarried births. The rate may have actually leveled out.

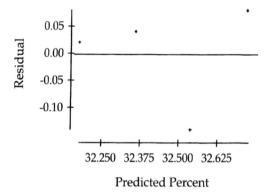

20. Life expectancy.

a) The scatterplot of births per woman and life expectancy is at the right. The association is strong, linear, and negative. Countries with higher life expectancies tend to have a lower number of births per woman. There is one outlier, Costa Rica, with an unreasonable 25 births per woman, and a life expectancy of 79 years.

b) Costa Rica has a birth rate of 25 births per woman, which common sense would indicate is impossible. There is ample justification to leave that point out of all further calculations. Probably, a decimal point was omitted from 2.5 births per woman, but there is no way to be sure. Omit this strange data point!

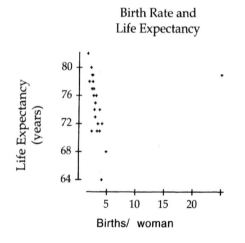

Birth Rate and
Life Expectancy

c)

Dependent variable is:		Life Exp.			
No Selector					
26 total cases of which 1 is missing					
R squared = 66.4% R squared (adjusted) = 65.0%					
s = 2.444 with 25 - 2 = 23 degrees of freedom					

Source	Sum of Squares	df	Mean Square	F-ratio
Regression	272.021	1	272.021	45.5
Residual	137.419	23	5.97475	

Variable	Coefficient	s.e. of Coeff	t-ratio	prob
Constant	86.8137	1.915	45.3	≤ 0.0001
Births/ w...	-4.35623	0.6456	-6.75	≤ 0.0001

Without Costa Rica, $R^2 = 66.4\%$, so $r = \sqrt{R^2} = \sqrt{0.664} = -0.815$.

66.4% of the variability in life expectancy is explained by variability in the number of births per year.

d) The linear model that predicts a country's life expectancy from the number of births per woman is $Lif\hat{e}Exp = 86.8137 - 4.35623(Births)$.

e) The linear model, without Costa Rica, is an appropriate model. The residuals plot shows no pattern.

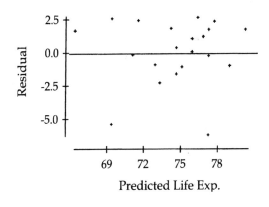

f) According to the model, each additional birth per woman is expected to correspond with a decrease of approximately 4.36 years. When the number of births per woman is 0, the model predicts that the life expectancy will be 86.8 years. This figure is an extrapolation below the range of the data, and doesn't hold any meaning.

g) The government leaders should not suggest that women have fewer children in order to raise the life expectancy. Although there is evidence of an association between the birth rate and life expectancy, this does not mean that one causes the other. There may be lurking variables involved, such as economic conditions, social factors, or level of health care.

21. Inflation.

a) The trend in Consumer Price Index is strong, non-linear, and positive. Generally, CPI has increased over the years, but the rate of increase has become much greater since approximately 1972. Other characteristics include fluctuations in CPI in the years prior to 1950.

b) In order to effectively predict the CPI in 2010, use only the most recent trend. The trend since 1972 is straight enough to apply the linear model. Prior to 1972, the trend is radically different from that of recent years, and is of no use in predicting CPI for 2010.

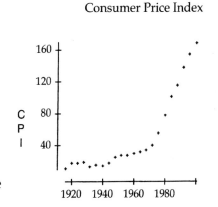

Dependent variable is: **CPI2**
No Selector
R squared = 99.6%   R squared (adjusted) = 99.5%
s = 3.241 with 8 - 2 = 6 degrees of freedom

Source	Sum of Squares	df	Mean Square	F-ratio
Regression	14908.5	1	14908.5	1420
Residual	63.0062	6	10.5010	

Variable	Coefficient	s.e. of Coeff	t-ratio	prob
Constant	-9247.62	248.3	-37.2	≤ 0.0001
Year2	4.71012	0.1250	37.7	≤ 0.0001

Consumer Price Index

The linear model that predicts CPI from year is $\hat{CPI} = -9247.62 + 4.71012(year)$. $R^2 = 99.6\%$, meaning that the model predicts 99.6% of the variability in CPI. The residuals plot shows no pattern, so the linear model is appropriate. According to the model, the CPI is expected to increase by \$4.70 each year, for 1972 − 2000.

$\hat{CPI} = -9247.62 + 4.71012(year)$

$\hat{CPI} = -9247.62 + 4.71012(2010)$

$\hat{CPI} \approx 219.72$

In 2010, CPI is predicted to be \$219.72

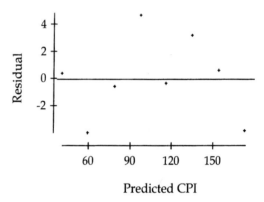

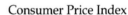

c) In the scatterplot of year versus log(CPI), differences in CPI are accentuated. We can clearly see the impact on CPI of historical events such as the stock market crash of 1929, the Great Depression in the years following, and World War II.

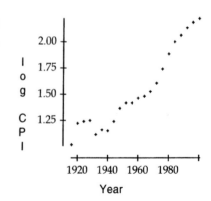

Consumer Price Index

Chapter 10 – Re-expressing Data:
It's Easier Than You Think

1. Models.

a) $\hat{y} = 1.2 + 0.8x$
$\hat{y} = 1.2 + 0.8(2)$
$\hat{y} = 2.8$

b) $\ln \hat{y} = 1.2 + 0.8x$
$\ln \hat{y} = 1.2 + 0.8(2)$
$\ln \hat{y} = 2.8$
$\hat{y} = e^{2.8} = 16.44$

c) $\sqrt{\hat{y}} = 1.2 + 0.8x$
$\sqrt{\hat{y}} = 1.2 + 0.8(2)$
$\sqrt{\hat{y}} = 2.8$
$\hat{y} = 2.8^2 = 7.84$

d) $\dfrac{1}{\hat{y}} = 1.2 + 0.8x$
$\dfrac{1}{\hat{y}} = 1.2 + 0.8(2)$
$\dfrac{1}{\hat{y}} = 2.8$
$\hat{y} = \dfrac{1}{2.8} = 0.36$

e) $\hat{y} = 1.2(2)^{0.8}$
$\hat{y} = 2.09$

2. More models.

a) $\hat{y} = 1.2 + 0.8 \log x$
$\hat{y} = 1.2 + 0.8 \log(2)$
$\hat{y} = 1.44$

b) $\log \hat{y} = 1.2 + 0.8x$
$\log \hat{y} = 1.2 + 0.8(2)$
$\log \hat{y} = 2.8$
$\hat{y} = 10^{2.8} = 630.96$

c) $\hat{y} = 1.2 + 0.8\sqrt{x}$
$\hat{y} = 1.2 + 0.8\sqrt{2}$
$\hat{y} = 2.33$

d) $\hat{y} = 1.2(0.8^x)$
$\hat{y} = 1.2(0.8^2)$
$\hat{y} = 0.77$

e) $\hat{y} = 0.8x^2 + 1.2x + 1$
$\hat{y} = 0.8(2)^2 + 1.2(2) + 1$
$\hat{y} = 6.6$

3. Gas mileage.

a) The association between weight and gas mileage of cars is fairly linear, strong, and negative. Heavier cars tend to have lower gas mileage.

b) For each additional thousand pounds of weight, the linear model predicts a decrease of 7.652 miles per gallon in gas mileage.

c) The linear model is not appropriate. There is a curved pattern in the residuals plot. The model tends to underestimate gas mileage for cars with relatively low and high gas mileages, and overestimates the gas mileage of cars with average gas mileage.

d) The residuals plot for the re-expressed relationship is much more scattered. This is an indication of an appropriate model.

e) The linear model that predicts the number of gallons per 100 miles in gas mileage from the weight of a car is: $\widehat{Gal} / 100 = 0.625 + 1.178(Weight)$.

f) For each additional 1000 pounds of weight, the model predicts that the car will require an additional 1.178 gallons to drive 100 miles.

g)

$$\hat{Gal}/100 = 0.625 + 1.178(Weight)$$
$$\hat{Gal}/100 = 0.625 + 1.178(3.5)$$
$$\hat{Gal}/100 = 4.748$$

According to the model, a car that weighs 3500 pounds (3.5 thousand pounds) is expected to require approximately 4.748 gallons to drive 100 miles, or 0.04748 gallons per mile.

This is $\dfrac{1}{0.04748} \approx 21.06$ miles per gallon.

4. Pressure.

The scatterplot at the right shows a strong, curved, negative association between the height of the cylinder and the pressure inside. Because of the curved form of the association, a linear model is not appropriate.

Re-expressing the pressure as the reciprocal of the pressure produces a scatterplot (below the first) that is much straighter. Computer regression output for the height versus the reciprocal of pressure is below.

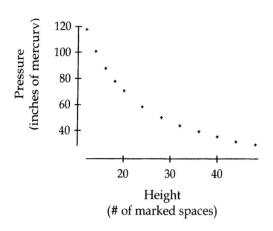

Boyle's Pressure and Volume

Dependent variable is: **recip pressure**
No Selector
R squared = 100.0%   R squared (adjusted) = 100.0%
s = 0.0001 with 12 - 2 = 10 degrees of freedom

Source	Sum of Squares	df	Mean Square	F-ratio
Regression	0.000841	1	0.000841	75241
Residual	0.000000	10	0.000000	

Variable	Coefficient	s.e. of Coeff	t-ratio	prob
Constant	-7.66970e-5	0.0001	-0.982	0.3494
Height	7.13072e-4	0.0000	274	≤ 0.0001

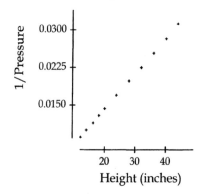

Height vs. Reciprocal of Pressure

The reciprocal re-expression is very straight (perfectly straight, as far as the statistical software is concerned!). $R^2 = 100\%$, meaning that 100% of the variability in the reciprocal of pressure is explained by the model. The equation of the model is:

$$\frac{1}{\hat{Pressure}} = -0.000077 + 0.000713(Height).$$

5. Brakes.

a) The association between speed and stopping distance is strong, positive, and appears straight. Higher speeds are generally associated with greater stopping distances. The linear regression model, with equation $\hat{Distance} = -65.9 + 5.98(Speed)$, has $R^2 = 96.9\%$, meaning that the model explains 96.9% of the variability in stopping distance. However, the residuals plot has a curved pattern. The linear model is not appropriate. A model using re-expressed variables should be used.

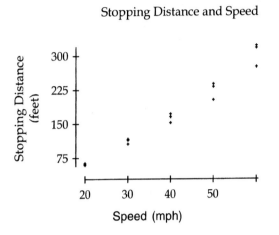

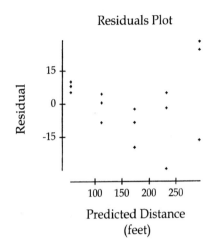

b) Stopping distances appear to be relatively higher for higher speeds. This increase in the rate of change might be able to be straightened by taking the square root of the response variable, stopping distance. The scatterplot of Speed versus $\sqrt{Distance}$ seems like it might be a bit straighter.

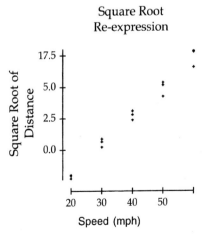

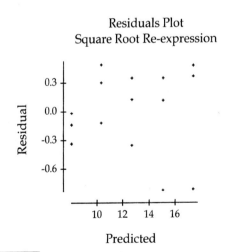

c) The model for the re-expressed data is $\sqrt{\hat{Distance}} = 3.303 + 0.235(Speed)$. The residuals plot shows no pattern, and $R^2 = 98.4\%$, so 98.4% of the variability in the square root of the stopping distance can be explained by the model.

d)

$$\sqrt{Dis\hat{t}ance} = 3.303 + 0.235(Speed)$$

$$\sqrt{Dis\hat{t}ance} = 3.303 + 0.235(55)$$

$$\sqrt{Dis\hat{t}ance} = 16.228$$

$$Dis\hat{t}ance = 16.228^2 \approx 263.4$$

According to the model, a car traveling 55 mph is expected to require approximately 263.4 feet to come to a stop.

e)

$$\sqrt{Dis\hat{t}ance} = 3.303 + 0.235(Speed)$$

$$\sqrt{Dis\hat{t}ance} = 3.303 + 0.235(70)$$

$$\sqrt{Dis\hat{t}ance} = 19.753$$

$$Dis\hat{t}ance = 19.753^2 \approx 390.2$$

According to the model, a car traveling 70 mph is expected to require approximately 390.2 feet to come to a stop.

f) The level of confidence in the predictions should be quite high. R^2 is high, and the residuals plot is scattered. The prediction for 70 mph is a bit of an extrapolation, but should still be reasonably close.

6. Pendulum.

a) The scatterplot shows the association between the length of string and the number of swings a pendulum took every 20 seconds to be strong, negative, and curved. A pendulum with a longer string tended to take fewer swings in 20 seconds. The linear model is not appropriate, because the association is curved.

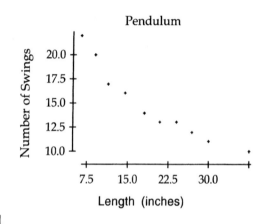

b) Curvature in a negative relationship sometimes is an indication of a reciprocal relationship. Try re-expressing the response variable with the reciprocal.

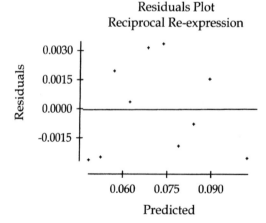

c) The reciprocal re-expression yields the model $\dfrac{1}{\widehat{Swings}} = 0.0367 + 0.00176(Length)$. The residuals plot is scattered, and $R^2 = 98.1\%$, indicating that the model explains 98.1% of the variability in the reciprocal of the number of swings. The model is both appropriate and accurate.

d) $\dfrac{1}{\widehat{Swings}} = 0.0367 + 0.00176(Length) = 0.0367 + 0.00176(4) = 0.04374$

$\widehat{Swings} = \dfrac{1}{0.04374} \approx 22.9$

According to the model, a pendulum with a 4″ string is expected to swing approximately 22.9 times in 20 seconds.

e) $\dfrac{1}{\widehat{Swings}} = 0.0367 + 0.00176(Length) = 0.0367 + 0.00176(48) = 0.12118$

$\widehat{Swings} = \dfrac{1}{0.12118} \approx 8.3$. The model predicts 8.3 swings in 20 seconds for a 48″ string.

f) Confidence in the predictions is fairly high. The model is appropriate, as indicated by the scattered residuals plot, and accurate, indicated by the high value of R^2. The only concern is the fact that these predictions are slight extrapolations. The lengths of the strings aren't too far outside the range of the data, so the predictions should be reasonably accurate.

7. Baseball salaries.

a) The scatterplot of year versus highest salary is moderately strong, positive and curved. The highest salary has generally increased over the years. Since the scatterplot shows a curved relationship, the linear model is not appropriate. Using the logarithm re-expression on the salaries seems to straighten the scatterplot considerably.

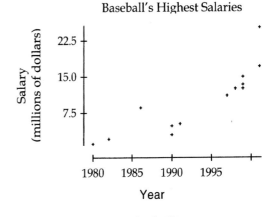

Baseball's Highest Salaries

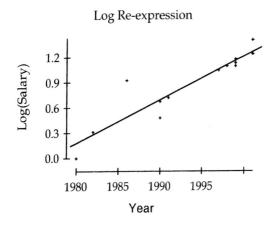

Log Re-expression

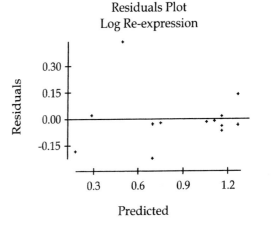

Residuals Plot
Log Re-expression

b) $\log(Sal\hat{a}ry) = -113.473 + 0.05734(Year)$ is a good mode for predicting the highest salary from the year. The residuals plot is scattered and $R^2 = 84.9\%$, indicating that 84.9% of the variability in highest salary can be explained by the model.

c) $\log(Sal\hat{a}ry) = -113.473 + 0.05734(Year) = -113.473 + 0.05734(2005) = 1.4937$

$Sal\hat{a}ry = 10^{1.4937} \approx 31.2$

According to the model, the highest salary in baseball is expected to be approximately 31.2 million dollars. (Be careful with decimal places here. When the slope is small and the value of the explanatory variable is large, like a year, too much rounding of the slope can be detrimental to the accuracy of predictions. When using logs, this is an even bigger problem!)

8. Planet distances and years.

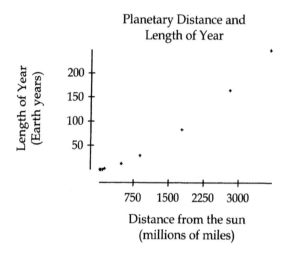

Planetary Distance and Length of Year

a) The association between distance from the sun and planet year is strong, positive, and curved concave upward. Generally, planets farther from the sun have longer years than closer planets.

b) The rate of change in length of year per unit distance appears to be increasing, but not exponentially. Re-expressing with the logarithm of each variable may straighten a plot such as this. The scatterplot and residuals plot for the linear model relating log(Distance) and log(Length of Year) appear below.

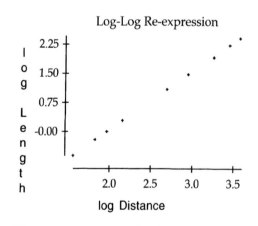

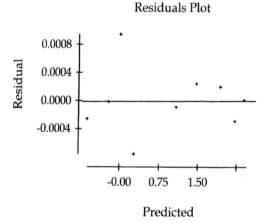

The regression model for the log-log re-expression is :
$\log(Len\hat{g}th) = -2.955 + 1.501(\log(Distance))$.

c) $R^2 = 100\%$, so the model explains 100% of the variability in the log of the length of the planetary year, at least according to the accuracy of the statistical software. The residuals plot is scattered, and the residuals are all extremely small. This is a very accurate model.

9. Planet distances and order.

a) The association between planetary position and distance from the sun is strong, positive, and curved (below left). A good re-expression of the data is position versus Log(Distance). The scatterplot with regression line (below center) shows the straightened association. The equation of the model is $\log(Dist\hat{a}nce) = 1.245 + 0.271(Position)$. The residuals plot (below right) may have some pattern, but after trying several re-expressions, this is the best that can be done. $R^2 = 98.2\%$, so the model explains 98.2% of the variability in the log of the planets distance from the sun.

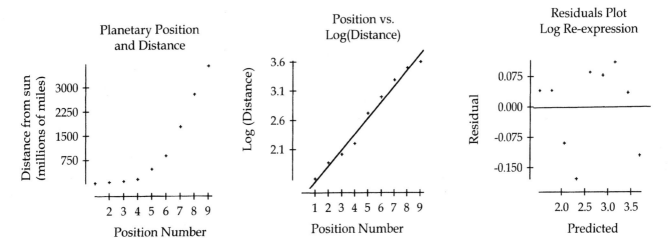

b) At first glance, this model appears to provide little evidence to support the contention of the astronomers. Pluto appears to fit the pattern, although Pluto's distance from the sun is a bit less than expected. A model generated without Pluto does not have a dramatically improved residuals plot, does not have a significantly higher R^2, nor a different slope. Pluto does not appear to be influential.

But don't forget that a logarithmic scale is being used for the vertical axis. The higher up the vertical axis you go, the greater the effect of a small change.

$$\log(Dist\hat{a}nce) = 1.24521 + 0.270922(Position)$$
$$\log(Dist\hat{a}nce) = 1.24521 + 0.270922(9)$$
$$\log(Dist\hat{a}nce) = 3.683508$$
$$Dist\hat{a}nce = 10^{3.683508} \approx 4825$$

According to the model, the 9th planet in the solar system is predicted to be approximately 4825 million miles away from the sun. Pluto is actually 3666 million miles away.

Pluto doesn't fit the pattern for position and distance in the solar system. In fact, the model made with Pluto included isn't a good one, because Pluto influences those predictions. The model without Pluto, $\log(Dist\hat{a}nce) = 1.20259 + 0.283706(Position)$, works much better. It has a high R^2, and scattered residuals plot. This new model predicts that the 9th planet should be a whopping 5701 million miles away from the sun! There is evidence that the astronomers are correct. Pluto doesn't behave like planet in its relation to position and distance.

10. Planets, part 3.

Using the revised planetary numbering system, and straightening the scatterplot using the same methods as in Exercise 9, the new model, $\log(Distance) = 1.3058 + 0.232923(Position)$, is a slightly better fit. The residuals plot is more scattered, and R^2 is slightly higher, with the improved model explaining 99.4% of the variability in the log of distance from the sun.

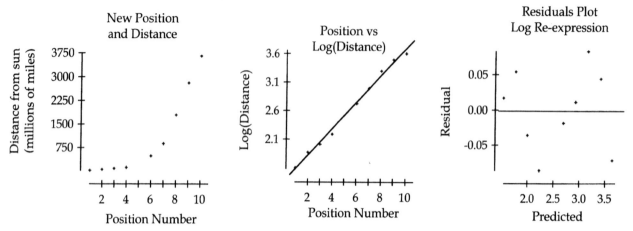

Pluto still doesn't fit very well. The new model predicts that Pluto, as 10th planet, should be about 4315 million miles away. That's about 650 million miles farther away. A better model yet is $\log(Distance) = 1.28514 + 0.238826(Position)$, a model made with the new numbering system and with Pluto omitted.

11. Quaoar, planets, part 4.

a)

$$\log(Distance) = 1.20259 + 0.283706(Position)$$
$$\log(Distance) = 1.20259 + 0.283706(9)$$
$$\log(Distance) = 3.755944$$
$$Distance = 10^{3.755944} \approx 5701$$

Using the model predicting distance from the sun from position, the 9th planet is expected to be about 5701 million miles from the sun. At 4000 million miles, Quaoar is a better fit than Pluto, but still not a good fit.

b) With Quaoar as the 9th planet, the model that relates planetary position to distance is $\log(Distance) = 1.23679 + 0.273447(Position)$.

$$\log(Distance) = 1.23679 + 0.273447(Position)$$
$$\log(Distance) = 1.23679 + 0.273447(9)$$
$$\log(Distance) = 3.697813$$
$$Distance = 10^{3.697813} \approx 4987$$

According to the new model, the 9th planet is predicted to be 4987 million miles from the sun. Quaoar, at approximately 4000 million miles, is a better fit than Pluto, but doesn't appear to be an additional planet.

12. Models and laws, planets part 5

The re-expressed data relating distance and year length (Exercise 8), are better described by their model than the re-expressed data relating position and distance (Exercises 9-11). The model relating distance and year length has $R^2 = 100\%$, and a very scattered residuals plot (with miniscule residuals), possibly a natural "law". If planets in another solar system followed the Titius-Bode pattern, this belief would be reinforced. Similarly, if data were acquired from planets in another solar system that did not follow this pattern, we would be unlikely to think that this relationship was a universal law.

13. Logs (not logarithms).

a) The association between the diameter of a log and the number of board feet of lumber is strong, positive, and curved. As the diameter of the log increases, so does the number of board feet of lumber contained in the log.

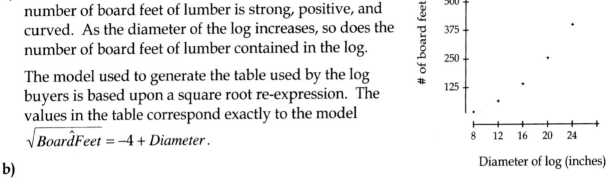

Doyle Log Scale

The model used to generate the table used by the log buyers is based upon a square root re-expression. The values in the table correspond exactly to the model

$$\sqrt{\widehat{BoardFeet}} = -4 + Diameter.$$

b)

$$\sqrt{\widehat{BoardFeet}} = -4 + Diameter$$
$$\sqrt{\widehat{BoardFeet}} = -4 + (10)$$
$$\sqrt{\widehat{BoardFeet}} = 6$$
$$\widehat{BoardFeet} = 36$$

According to the model, a log 10" in diameter is expected to contain 36 board feet of lumber.

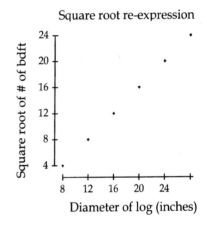

Square root re-expression

c)

$$\sqrt{\widehat{BoardFeet}} = -4 + Diameter$$
$$\sqrt{\widehat{BoardFeet}} = -4 + (36)$$
$$\sqrt{\widehat{BoardFeet}} = 32$$
$$\widehat{BoardFeet} = 1024$$

According to the model, a log 36" in diameter is expected to contain 1024 board feet of lumber.

Normally, we would be cautious of this prediction, because it is an extrapolation beyond the given data, but since this is a prediction made from an exact model based on the volume of the log, the prediction will be accurate.

14. Weight lifting.

a) The association between weight class and weight lifted for gold medal winners in weightlifting at the 2000 Olympics is strong, positive, and curved. The linear model that best fits the data is
$$\widehat{Lift} = 179.918 + 2.401(WeightClass).$$

Although this model explains 96.9% of the variability in weight lifted, it does not fit the data well.

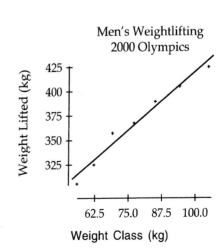

Men's Weightlifting 2000 Olympics

b) The residuals plot for the linear model shows a curved pattern, indicating that the linear model has failed to model the association well. A re-expressed model might fit the association between weight class and weight lifted better than the linear model.

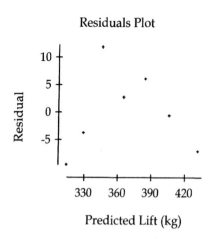

Residuals Plot

c) A first attempt at re-expressing the data might be one of the logarithmic re-expressions. These models do not decrease the amount of curvature in the residuals plot, and seem to fit no better than the linear model. Squaring the weight lifted seems to straighten the scatterplot somewhat, but the residuals plot still has a curved pattern similar to the residuals plot of the linear model.

Cubing the weight lifted gives the model $\left(\hat{Lift}\right)^3 = -25{,}010{,}077 + 977{,}390.41(WeightClass)$

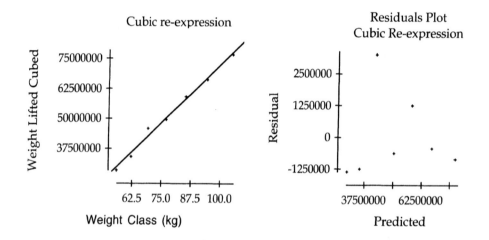

d) The model generate with the cubic re-expression is a better model, because the model has a residuals plot that shows less pattern than the residuals plot for the linear model. Also, the re-expressed model explains 99.1% of the variability in the cube of weight lifted. Care must be taken in using this model, however. The presence of one unusual point, the lift of 357.5 kg in the 69 kg weight class seems to be affecting the model. The residuals plot still has the same curved pattern present in all other residuals plots, but cubing the lift of that one outlier has stretched the vertical scale, making the pattern less apparent. This model is better than the linear model, but still may not be appropriate.

e) Boevski, from Bulgaria, lifted 357.5 kg. His lift does not fit the pattern established by the other lifts. The lift in this weight class is predicted to be 348.8 kg. According to the model, Boevski lifted 8.7 kg (over 19 pounds) more than expected.

f) Without Boevski, the cubic re-expression yields the following model:

$$\left(\hat{Lift}\right)^3 = -27{,}160{,}661 + 997{,}521(WeightClass)$$

g)

$$\left(\hat{Lift}\right)^3 = -27,160,661 + 997,521(WeightClass)$$

$$\left(\hat{Lift}\right)^3 = -27,160,661 + 997,521(69)$$

$$\left(\hat{Lift}\right)^3 = 41,668,288$$

$$\hat{Lift} = \sqrt[3]{41,668,288} = 346.7$$

According to the model, the winner of the 69 kg weight class was predicted to lift 346.7 kg

h) Since Boevski actually lifted 357.5 kg, his residual was 357.5 – 346.7 = 10.8. He lifted 10.8 kg (almost 24 pounds) more than predicted!

15. Life expectancy.

The association between year and life expectancy is strong, curved and positive. As the years passed, life expectancy for white males has increased. The bend (in mathematical language, the change in curvature) in the scatterplot makes it impossible to straighten the association by re-expression.
If all of the data points are to be used, the linear model, $Li\hat{f}eExp = -454.938 + 0.265697(Year)$ is probably the best model we can make. It doesn't fit the data well, but accounts for 93.6% of the variability in life expectancy.

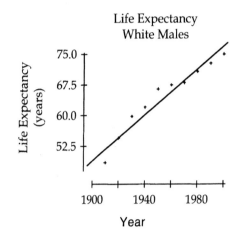

In order to make better future predictions, we might use only a subset of the data points. The scatterplot is very linear from 1970 to 2000, and these recent years are likely to be more indicative of the current trend in life expectancy. Using only these four data points, we might be able to make more accurate future predictions.
The model $Li\hat{f}eExp = -379.020 + 0.227(Year)$ explains 99.6% of the variability in life expectancy, and has a scattered residuals plot. Great care should be taken in using this model for predictions, since it is based upon only 4 points, and, as always, prediction into the

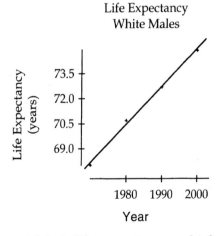

future is risky. This is especially true since the response variable is life expectancy, which cannot be expected to increase at the same rate indefinitely.

16. Tree growth.

a) The association between age and average diameter of grapefruit trees is strong, curved, and positive. Generally, older trees have larger average diameters. The linear model for this association, $Average\widehat{Diameter} = 1.973 + 0.463(Age)$ is not appropriate. The residuals plot shows a clear pattern.

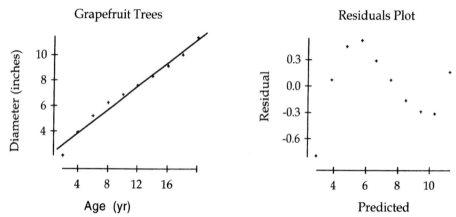

Because of the change in curvature in the association, these data cannot be straightened by re-expression.

b) If diameters from individual trees were given, instead of averages, the association would have been weaker. Individual observations are more variable than averages.

17. Slower is cheaper.

The association between speed and mileage is strong and curved. For speeds under 50 miles per hour, higher speeds are associated with higher mileage. For speeds in excess of 50 mph, higher speeds are associated with lower mileage. For this reason, the scatterplot cannot be straightened by the methods used in this chapter.

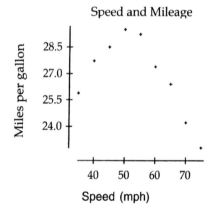

The most logical modeling decision at this point is to divide the curve up into two linear models. The association seems fairly linear from 35−50 mph, and also fairly linear from 55−75 mph.

For lower speeds, the model is $Mil\hat{e}age = 18.040 + 0.232(Speed)$. The model has some pattern in the residuals plot, but no re-expression seems to change it. The model explains 96.7% of the variability in mileage. For low speeds, the model predicts an increase of 0.232 miles per gallon for each additional mile per hour of speed.

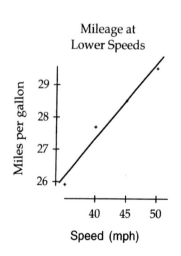

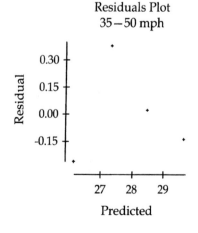

For high speed, the model is
Mileage = 46.800 − 0.320(*Speed*).
The residuals plot is
scattered and the model
explains 99.1% of the
variability in mileage. For
higher speeds, the model
predicts a decrease in
mileage of 0.320 miles per
gallon for each additional
mile per hour of speed

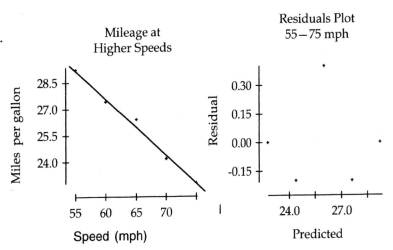

18. Orange production.

The association between the
number of oranges per tree and
the average weight of the
oranges is strong, negative, and
appears linear at first look.
Generally, trees that contain
larger numbers of oranges have
lower average weight per
orange. The linear model may
look appropriate, until the
residuals plot is examined.
Even though the linear model is
a close fit, the residuals plot
shows a strong curved pattern.
The data should be re-expressed.

Plotting the number of oranges per
tree and the reciprocal of the
average weight per orange
straightens the relationship
considerably. The residuals plot
shows little pattern and the value
of R^2 indicates that the model
explains 99.8% of the variability in
the reciprocal of the average
weight per orange. The more
appropriate model is:

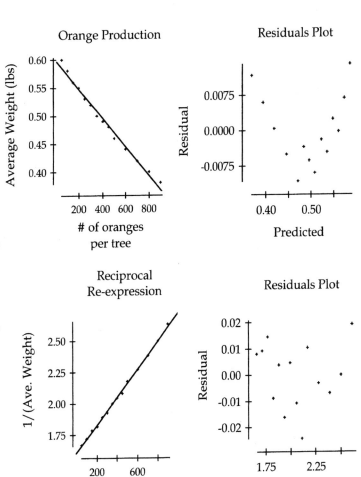

$$\frac{1}{\widehat{Ave.wt}} = 1.603 + 0.00112(\#\,Oranges\,/\,Tree).$$

19. Years to live.

a) The association between the age and estimated additional years of life for black males is strong, curved, and negative. Older men generally have fewer estimated years of life remaining. The square root re-expression of the data, $\sqrt{Year\hat{s}Left} = 8.39253 - 0.069553(Age)$, straightens the data considerably, but has an extremely patterned residuals plot. The model is not a mathematically appropriate model, but fits so closely that it should be fine for predictions within the range of data. The model explains 99.8% of the variability in the estimated number of additional years.

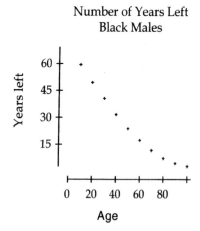

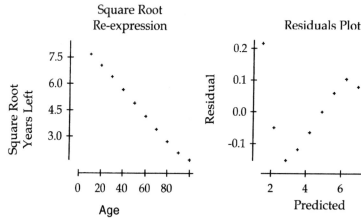

b)

$$\sqrt{Year\hat{s}Left} = 8.39253 - 0.069553(Age)$$

$$\sqrt{Year\hat{s}Left} = 8.39253 - 0.069553(18)$$

$$\sqrt{Year\hat{s}Left} = 7.140936$$

$$Year\hat{s}Left = 7.140936^2 \approx 50.99$$

According to the model, an 18-year-old black male is expected to live and additional 50.99 years, for a total age of 68.99 years.

20. Oil production.

The association between the year and the number of barrels of oil is strong and curved. Oil production generally increased until the mid-1980s, and then decreased until the year 2000. Because of the association reverses direction, no re-expression of the *y*-variable will straighten this scatterplot.

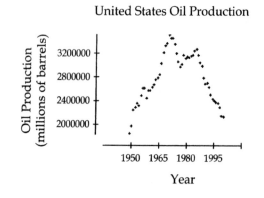

Since no re-expressed model can straighten this association, no faith would be replaced in any prediction.

21. Internet.

a) The association between the year and the number of electronic journals is strong, positive, and curved. The number of electronic journals rapidly increased in the last several years. Clearly, the linear model would not be appropriate for modeling these data.

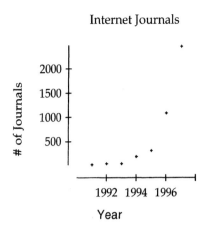

A good model is the log re-expression $\text{Log}(\widehat{Journals}) = -686.763 + 0.345547(Year)$. The residuals plot is scattered, and the value of R^2 indicates that the model explains 95.9% of the variability in the number of journals.

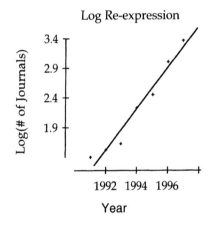

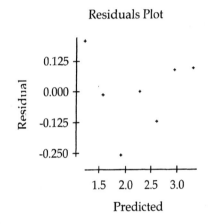

b)

$\text{Log}(\widehat{Journals}) = -686.763 + 0.345547(Year)$

$\text{Log}(\widehat{Journals}) = -686.763 + 0.345547(2000)$

$\text{Log}(\widehat{Journals}) = 4.334$

$\widehat{Journals} = 10^{4.334} \approx 21,429$

According to the model, there are expected to be approximately 21,429 electronic journals in the year 2000.

c) The estimate of 21,429 journals in 2000 seems very high, and seems unlikely to be accurate. The year 2000 is an extrapolation 3 years beyond the most recent year included in the data set. The model is increasing exponentially, but it is quite unlikely that the number of journals could keep increasing exponentially.

Review of Part II – Exploring Relationships Between Variables

1. **College.**

    % over 50: $r = 0.69$ The only moderate, positive correlation in the list.
    % under 20: $r = -0.71$ Moderate, negative correlation (-0.98 is too strong)
    % Gr. on time: $r = -0.51$ Moderate, negative correlation (not as strong as %under 20)
    % Full-time Fac.: $r = 0.09$ No correlation.

2. **Togetherness.**

    a) If no meals are eaten together, the model predicts a GPA of 2.73.

    b) For an increase of one meal per week eaten together, the model predicts an increase of 0.11 in GPA.

    c) The model will predict the mean GPA for the mean number of meals, 3.78.

    $\hat{gpa} = 2.73 + 0.11meals$ The mean GPA is 3.15.
    $\hat{gpa} = 2.73 + 0.11(3.78)$
    $\hat{gpa} = 3.15$

    d) A negative residual means that the student's actual GPA was lower than the GPA predicted by the model. The model over-predicted the student's GPA.

    e) Although there is evidence of an association between GPA and number of meals eaten together per week, this is not necessarily a cause-and-effect relationship. There may be other variables that are related to GPA and meals, such as parental involvement and family income.

3. **Wines.**

    a) There does not appear to be an association between ages of vineyards and the cost of wines. $r = \sqrt{R^2} = \sqrt{0.027} = 0.164$, indicating a very weak association, at best. The model only explains 2.7% of the variability in case price. Furthermore, the regression equation appears to be influenced by two outliers, wines from vineyards over 30 years old, with relatively high case prices.

    b) This analysis tells us nothing about wines worldwide. There is no reason to believe that the results for the Finger Lakes region are representative of the wines of the world.

    c) The linear equation used to predict case price from age of the vineyard is:
    $\hat{CasePrice} = 92.765 + 0.567284(Years)$

    d) This model is not useful because only 2.7% of the variability in case price is accounted for by the ages of the vineyards. Furthermore, the slope of the regression line seems influenced by the presence of two outliers, wines from vineyards over 30 years old, with relatively high case prices.

4. **More wine.**

    a) There is no evidence of an association between size of the winery and case price.

b) One vineyard is approximately 250 acres, with a relatively low case price. This point has high leverage.

c) If the point were removed, the correlation would be expected to increase, from a slightly negative correlation, to a correlation that is slightly positive. The point is an outlier in the *x*-direction and low in the *y*-direction. It is giving the association the artificial appearance of a slightly negative relationship.

d) If the point were removed, the slope would be expected to increase, from slightly negative to slightly positive. The point is "pulling" the regression line down.

5. **More twins?**

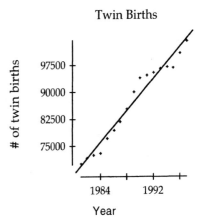

a) The association between year and the number of twin births is strong, positive, and appears non-linear. Generally, the number of twin births has increased over the years. The linear model that predicts the number of twin births from the year is:
$\hat{Twins} = -4316980 + 2214.19(Year)$

b) For each year that passes, the model predicts that the number of twins born will increase by approximately 2214 twin births.

c)
$$\hat{Twins} = -4,316,980 + 2214.19(Year)$$
$$\hat{Twins} = -4,316,980 + 2214.19(2002)$$
$$\hat{Twins} \approx 115,828.4$$

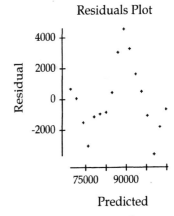

According to the model, there are expected to be 115,828.4 twin births in the US in 2002. However, the scatterplot appears non-linear, and there is no reason to believe the number of twin births will keep increasing at the same rate for 5 years beyond the last recorded year. Faith in this prediction is very low.

d) The residuals plot shows a definite curved pattern. The association is not linear, so the linear model is not appropriate.

6. **Dow Jones.**

a) $r = \sqrt{R^2} = \sqrt{0.658} = 0.811$. Since the slope of the regression equation is positive, we know that the correlation is also positive.

b) The linear model that predicts the Dow from the year is $\hat{Dow} = -603,335 + 305.471(Year)$.

c) This model predicts that the Dow was expected to be –603,335 in the year 0, which doesn't have contextual meaning, since this model is only for the Dow since 1972. Furthermore, the model predicts that the Dow is expected to increase by approximately 305.5 each year.

d) The residuals plot shows a definite pattern. A single linear model is not appropriate. Before attempting to fit a linear model, look at the scatterplot. If it is not straight enough, the linear model cannot be used.

7. Acid rain.

a) $r = \sqrt{R^2} = \sqrt{0.27} = -0.5196$. The association between pH and BCI appears negative in the scatterplot, so use the negative value of the square root.

b) The association between pH and BCI is negative, moderate, and linear. Generally, higher pH is associated with lower BCI. Additionally, BCI appears more variable for higher values of pH.

c) In a stream with average pH, the BCI would be expected to be average, as well.

d) In a stream where the pH is 3 standard deviations above average, the BCI is expected to be 1.56 standard deviations below the mean level of BCI. ($r(3) = -0.5196(3) = -1.56$)

8. Manatees.

a) The explanatory variable is the to be the number of powerboat registrations. This is the relationship about which the biologists are concerned. They believe that the high number of manatees killed is related to the increase in powerboat registrations.

b) The association between the number of powerboat registrations and the number of manatees killed in Florida is fairly strong, linear, and positive. Higher numbers of powerboat registrations are associated with higher numbers of manatees killed.

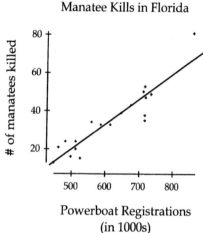
Manatee Kills in Florida

c) The correlation between the number of powerboat registrations and the number of manatees killed in Florida is $r = 0.924$.

d) $R^2 = 85.4\%$. Variability in the number of powerboat registrations explains 85.4% of the variability in the number of manatees killed.

e) There is an association between the number of powerboat registrations and the number of manatees killed, but that is no reason to assume a cause-and-effect relationship. There may be lurking variables that affect one or the other of the variables.

9. A Manatee model.

a) The association between the number of powerboat registrations and the number of manatees killed is straight enough to try a linear model.
$\widehat{Kills} = -45.8933 + 0.131759(ThousandBoats)$ is the best fitting model. The residuals plot is scattered, so the linear model is appropriate.

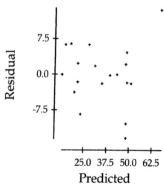
Residuals Plot

b) For every additional 10,000 powerboats registered, the model predicts that an additional 1.32 manatees will be killed.

c) The model predicts that if no powerboats were registered, the number of manatee deaths would be approximately −46. This is an extrapolation beyond the scope of the data, and doesn't have much contextual meaning.

d)

$$\hat{Kills} = -45.8933 + 0.131759(ThousandBoats)$$

$$\hat{Kills} = -45.8933 + 0.131759(860)$$

$$\hat{Kills} \approx 67.4$$

The model predicted 67.4 manatee kills in 2001, when the number of powerboat registrations was 860,000. The actual number of kills was 81. The model under-predicted the number of kills by 13.6.

e) Negative residuals are better for the manatees. A negative residual suggests that the actual number of kills was below the number of kills predicted by the model.

f) Over time, the number of powerboat registrations has increased and the number of manatees killed has increased. The trend may continue, resulting in a greater number of manatee deaths in the future.

10. Grades.

a)

$$\hat{fin} = 10 + 0.9mid$$

$$\hat{fin} = 10 + 0.9(70)$$

$$\hat{fin} = 73$$

According to the model, Susan is predicted to earn a score of 73 on the final exam.

b) Susan's residual is 80 – 73 = 7 points. She scored 7 points higher than predicted.

c)

$$b_1 = r\left(\frac{s_y}{s_x}\right)$$

$$0.9 = r\left(\frac{12}{10}\right)$$

$$r = 0.75$$

The correlation between midterm exam score and final exam score is 0.75.

d)

$$\hat{fin} = 10 + 0.9mid$$

$$100 = 10 + 0.9mid$$

$$mid = 100$$

In order to have a predicted final exam score of 100, a student would need to have a midterm exam score of 100, as well.

e) This linear model is designed to predict final exam scores based upon midterm exam scores. It does not predict midterm scores from final exam scores. In order to predict in this direction, a linear model would have to be generated with final exam score as the explanatory variable and midterm exam score as the response variable. (Notice that part d is NOT predicting midterm from final, but rather asking what *actual* midterm score is required to result in a *prediction* of 100 for the final exam score.)

f) From part d, a student with a midterm score of 100 is predicted have a final exam score of 100. The students residual is 15 – 100 = –85.

g) The R^2 value of the regression will increase. This student's large negative residual would detract from the overall pattern of the data, allowing the model to explain less of the variability in final exam score. Removing it would increase the strength of the association.

h) The slope of the linear model would increase. This student's large negative residual would "pull" the regression line down, perhaps even making the association appear negative. The removal of this point would allow the line to snap back up to the true positive association.

11. Traffic.

a)

$$b_1 = r\left(\frac{s_y}{s_x}\right)$$

$$-0.352 = r\left(\frac{9.68}{27.07}\right)$$

$$r = -0.984$$

The correlation between traffic density and speed is $r = -0.984$

b) $R^2 = (-0.984)^2 = 0.969$.
The variation in the traffic density explains 96.9% of the variation in speed.

c)

$$spêed = 50.55 - 0.352cars$$
$$spêed = 50.55 - 0.352(50)$$
$$spêed = 32.95$$

According to the linear model, when traffic density is 50 cars per mile, the average speed of traffic on a moderately large city thoroughfare is expected to be 32.95 miles per hour.

d)

$$spêed = 50.55 - 0.352cars$$
$$spêed = 50.55 - 0.352(56)$$
$$spêed = 30.84$$

According to the linear model, when traffic density is 56 cars per mile, the average speed of traffic on a moderately large city thoroughfare is expected to be 30.84 miles per hour. If traffic is actually moving at 32.5 mph, the residual is $32.5 - 30.84 = 1.66$ miles per hour.

e)

$$spêed = 50.55 - 0.352cars$$
$$spêed = 50.55 - 0.352(125)$$
$$spêed = 6.55$$

According to the linear model, when traffic density is 125 cars per mile, the average speed of traffic on a moderately large city thoroughfare is expected to be 6.55 miles per hour. The point with traffic density 125 cars per minute and average speed 55 miles per hour is considerably higher than the model would predict. If this point were included in the analysis, the slope would increase.

f) The correlation between traffic density and average speed would become weaker. The influential point (125, 55) is a departure from the pattern established by the other data points.

g) The correlation would not change if kilometers were used instead of miles in the calculations. Correlation is a "unitless" measure of the degree of linear association based on z-scores, and is not affected by changes in scale. The correlation would remain the same, $r = -0.984$.

12. Cramming.

a) The correlation between the Friday scores and the Monday scores on the Spanish Test is $r = 0.473$.

b) The scatterplot shows a weak, linear, positive association between Friday score and Monday score. Generally, students who scored high on Friday also tended to score high on Monday.

Spanish Test Scores

c) A student with a positive residual scored higher on Monday's test than the model predicted.

d) A student with a Friday score that is one standard deviation below average is expected to have a Monday score that is 0.473 standard deviations below Monday's average score. The distribution of scores for Monday had mean 37.24 points and standard deviation 5.02 points, so the student's score is predicted to be approximately $37.24 - (0.473)(5.02) = 34.87$.

e) The regression equation for the linear model that predicts Monday score from Friday score is: $\widehat{Monday} = 14.59 + 0.536(Friday)$.

f)

$\widehat{Monday} = 14.5921 + 0.535666(Friday)$

$\widehat{Monday} = 14.5921 + 0.535666(40)$

$\widehat{Monday} \approx 36.0$

According to the model, a student with a Friday score of 40 is expected to have a Monday score of about 36.0.

13. Correlations.

a) Weight, with a correlation of -0.903, seems to be most strongly associated with fuel economy, since the correlation has the largest magnitude (distance from zero). However, without looking at a scatterplot, we can't be sure that the relationship is linear. Correlation might not be an appropriate measure of the strength of the association if the association is non-linear.

b) The negative correlation between weight and fuel economy indicates that, generally, cars with higher weights tend to have lower mileages than cars with lower weights. Once again, this is only correct if the association between weight and fuel economy is linear.

c) $R^2 = (-0.903)^2 = 0.815$. The variation in weight accounts for 81.5% of the variation in mileage. Once again, this is only correct if the association between weight and fuel economy is linear.

14. Autos revisited.

a) Displacement and weight show the strongest association, with a correlation of 0.951. Generally, cars with larger engines are heavier than cars with smaller engines. However, without looking at a scatterplot, we can't be sure that the relationship is linear. Correlation might not be an appropriate measure of the strength of the association if the association is non-linear.

b) The strong correlation between displacement and weight is not necessarily a sign of a cause-and-effect relationship. Price of the car might be confounded with weight and displacement. More expensive luxury cars may have extra features that result in higher weights. One of these features might be a larger engine. Another difficulty with assigning a cause and an effect is that we can't be sure of the direction of the relationship. Certainly, the larger engine adds to the weight of the car, but maybe the larger engine is needed to power heavier cars.

c) The correlation would not change if cubic centimeters or liters were used instead of cubic inches in the calculations. Correlation is a "unitless" measure of the degree of linear association based on z-scores, and is not affected by changes in scale.

d) As long as the association between fuel economy and engine displacement was linear, a car whose engine displacement is one standard deviation above the mean would be predicted to have a fuel economy that is 0.786 standard deviations below the mean. (The correlation between the variables is –0.786, so a change in direction is indicated.) If the relationship were non-linear, the relative fuel economy could not be determined.

15. Cars, one more time.

a) The linear model that predicts the horsepower of an engine from the weight of the car is:
$\widehat{Horsepower} = 3.49834 + 34.3144(Weight)$.

b) The weight is measured in thousands of pounds. The slope of the model predicts an increase of about 34.3 horsepower for each additional unit of weight. 34.3 horsepower for each additional thousand pounds makes more sense than 34.3 horsepower for each additional pound.

c) Since the residuals plot shows no pattern, the linear model is appropriate for predicting horsepower from weight.

d)

$\widehat{Horsepower} = 3.49843 + 34.3144(Weight)$

$\widehat{Horsepower} = 3.49843 + 34.3144(2.595)$

$\widehat{Horsepower} \approx 92.544$

According to the model, a car weighing 2595 pounds is expected to have 92.543 horsepower. The actual horsepower of the car is: $92.544 + 22.5 \approx 115.0$ horsepower.

16. Colorblind.

Gender and colorblindness are both categorical variables. Correlation is a measure of the strength of a linear relationship between quantitative variables. The proper terminology is to say gender is associated with colorblindness.

17. Old Faithful.

a) The association between the duration of eruption and the interval between eruptions of Old Faithful is fairly strong, linear, and positive. Long eruptions are generally associated with long intervals between eruptions. There are also two distinct clusters of data, one with many short eruptions followed by short intervals, the other with many long eruptions followed by long intervals, with only a few medium eruptions and intervals in between.

b) The linear model used to predict the interval between eruptions is:
$\widehat{Interval} = 33.9668 + 10.3582(Duration)$.

c) As the duration of the previous eruption increases by one minute, the model predicts an increase of about 10.4 minutes in the interval between eruptions.

d) $R^2 = 77.0\%$, so the model accounts for 77% of the variability in the interval between eruptions. The predictions should be fairly accurate, but not precise. Also, the association appears linear, but we should look at the residuals plot to be sure that the model is appropriate before placing too much faith in any prediction.

e)

$\widehat{Interval} = 33.9668 + 10.3582(Duration)$

$\widehat{Interval} = 33.9668 + 10.3582(4)$

$\widehat{Interval} \approx 75.4$

According to the model, if an eruption lasts 4 minutes, the next eruption is expected to occur in approximately 75.4 minutes.

f) The actual eruption at 79 minutes is 3.6 minutes later than predicted by the model. The residual is 79 – 75.4 = 3.6 minutes. In other words, the model under-predicted the interval.

18. Which croc?

a) The associations between the head sizes and body sizes for the crocodiles appear to be strong. 97.2% of the variability in Indian Crocodile length and 98% of the variability in Australian Crocodile length is explained by the variability in head size. (This assertion is only valid if the association between head and body length is linear for each crocodile.)

b) The slopes of the two models are similar. Indian Crocodiles are predicted to increase in length 7.4 centimeters for each centimeter increase in head length, and Australian Crocodiles are predicted to increase in length 7.72 centimeters for each centimeter increase in head length. (These predictions are only valid if the association between head and body length is linear for each crocodile.) The values of R^2 are also similar.

c) The two models have different values for the y-intercept. According to the models, the Indian Crocodile is smaller.

d) Indian Crocodile Model

$\widehat{IBody} = -69.3693 + 7.40004(IHead)$

$\widehat{IBody} = -69.3693 + 7.40004(62)$

$\widehat{IBody} = 389.43318$

Australian Crocodile Model

$\widehat{ABody} = -20.2245 + 7.71726(IHead)$

$\widehat{ABody} = -20.2245 + 7.71726(62)$

$\widehat{ABody} = 458.24562$

The appropriate models predict body lengths of 389.4 centimeters and 458.2 centimeters for Indian Crocodiles and Australian Crocodiles, respectively. The actual length of 380 centimeters indicates that this is probably an Indian Crocodile. The prediction is closer to the actual length for that model.

19. How old is that tree?

a) The correlation between tree diameter and tree age is $r = 0.888$. Although the correlation is moderately high, this does not suggest that the linear model is appropriate. We must look at a scatterplot in order to verify that the relationship is straight enough to try the linear model. After finding the linear model, the residuals plot must be checked. If the residuals plot shows no pattern, the linear model can be deemed appropriate.

b) The association between diameter and age of these trees is fairly strong, somewhat linear, and positive. Trees with larger diameters are generally older.

c) The linear model that predicts age from diameter of trees is: $A\hat{g}e = -0.974424 + 2.20552(Diameter)$. This model explains 78.9% of the variability in age of the trees.

d) The residuals plot shows a curved pattern, so the linear model is not appropriate. Additionally, there are several trees with large residuals.

e) The largest trees are generally above the regression line, indicating a positive residual. The model is likely to underestimate these values.

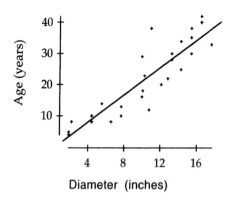

Tree Age and Diameter

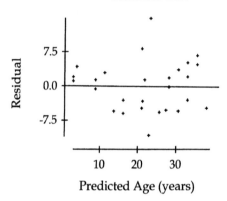

Residuals Plot

20. Improving trees.

a) The re-expressed data, diameter squared versus age, is straighter than the original. This model appears to fit much better.

b) The linear model that predicts age from diameter squared is: $A\hat{g}e = 7.23961 + 0.113011(Diameter^2)$. This model explains 78.7% of the variability in the age of the trees.

c) The residuals plot shows random scatter. This model appears to be appropriate, although there are still some trees with large residuals.

d)

$$A\hat{g}e = 7.23961 + 0.113011(Diameter^2)$$

$$A\hat{g}e = 7.23961 + 0.113011(18^2)$$

$$A\hat{g}e \approx 43.855$$

According to the model, a tree with a diameter of 18" is expected to be approximately 43.9 years old.

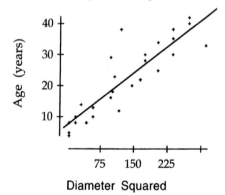

Square Re-expression

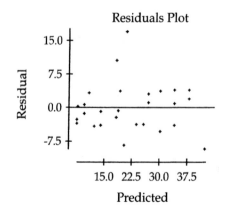

Residuals Plot

21. New homes.

New homes might have areas somewhere between 1000 and 5000 square feet. Doubling the area of a house from 1000 square feet to 2000 square feet would be predicted to increase the price by either 0.008(1000) = $8 thousand, 0.08(1000) = $80 thousand, 0.8(1000) = $800 thousand, or 8(1000) = $8000 thousand. The only reasonable answer is $80 thousand, so the slope must be 0.08.

22. Smoking and pregnancy

a) The association between year and the percent of expectant mothers who smoked cigarettes during their pregnancies is strong, roughly linear, and negative. The percentage has decreased steadily since 1990.

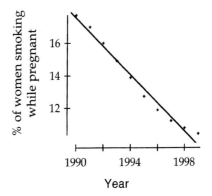

b) The correlation between year and percent of expectant mothers who smoked cigarettes during their pregnancies is $r = -0.991$. This may not be an appropriate measure of strength, since the scatterplot shows a slight bend.

c) The use of averages instead of individual percentages for each of the 50 cities results in a correlation that is artificially strong. The correlation of the averaged data is "more negative" than the correlation of the individual percentages would have been.

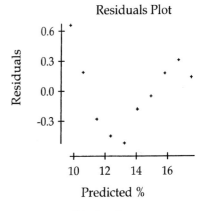

d) The linear model that predicts the percent of expectant mothers who smoked during their pregnancies from the year is: $\hat{\%} = 1745.84 - 0.868485(Year)$. This model explains 98.1% of the variability in the percent. According to this model, for each year that passes, the average percent of women who smoked while pregnant decreases by 0.868%. This model does not appear to be appropriate, since the residuals plot shows a pattern. However, it is unlikely that we can provide a better model, since the scatterplot shows a bend that cannot be straightened by re-expression.

23. No smoking?

a) The model from Exercise 22 is for predicting the percent of expectant mothers who smoked during their pregnancies from the year, not the year from the percent.

b) The model that predicts the year from the percent of expectant mothers who smoked during pregnancy is: $\hat{Year} = 2009.92 - 1.12968(\%)$. This model predicts that 0% of mothers will smoke during pregnancy in $2009.92 - 1.12968(0) = 2009.92 \approx 2010$.

c) The lowest data point corresponds to 10.4% of expectant mothers smoking during pregnancy in 1998. The prediction for 0% is an extrapolation outside the scope of the data. There is no reason to believe that the model will be accurate at that point.

24. Tips.

a) There is a very weak, linear, positive relationship between tip size and rating of service. Generally, better tips are associated with better ratings of service, but the relationship is so weak that it may simply be due to sampling error.

b) $R^2 = 1.21\%$. Only about 1% of the variability in tip size is explained by variability in the rating of service.

25. Move south?

There is a strong, roughly linear, negative association between mean January temperature and latitude. U.S. cities with higher latitudes generally have lower mean January temperatures. There are two outliers, cities with higher mean January temperatures than the pattern would suggest.

26. Correlations.

a) Latitude appears to be the better predictor of average January temperature, so long as the relationship between latitude and temperature is linear. The correlation, –0.848, is stronger than the correlation between altitude and temperature, –0.369.

b) The correlation would be the same, –0.848. Correlation is a measure of the degree of linear association between two quantitative variables and is unaffected by changes in units.

c) The correlation would be the same, –0.369. Correlation is a measure of the degree of linear association between two quantitative variables and is unaffected by changes in units.

d) $(-0.369)(2) = -0.738$. If a city has an altitude 2 standard deviations above the mean, its average January temperature is expected to be 0.738 standard deviations below the mean average January temperature.

27. Winter in the city.

a) $R^2 = (-0.848)^2 \approx 0.719$. The variation in latitude explains 71.9% of the variability in average January temperature.

b) The negative correlation indicates that the as latitude increases, the average January temperature generally decreases.

c)

$$b_1 = r\left(\frac{s_y}{s_x}\right)$$

$$\hat{y} = b_0 + b_1 x$$

$$b_1 = -0.848\left(\frac{13.49}{5.42}\right)$$

$$\bar{y} = b_0 + b_1\bar{x}$$

$$26.44 = b_0 - 2.1106125(39.02)$$

$$b_1 = -2.1106125$$

$$b_0 = 108.79610$$

The equation of the linear model for predicting January temperature from latitude is:

$$Jan\hat{T}emp = 108.796 - 2.111(Latitude)$$

d) For each additional degree of latitude, the model predicts a decrease of approximately 2.1°F in average January temperature.

e) The model predicts that the mean January temperature will be approximately 108.8°F when the latitude is 0°. This is an extrapolation, and may not be meaningful.

f)

$$Jan\hat{T}emp = 108.796 - 2.111(Latitude)$$

$$Jan\hat{T}emp = 108.796 - 2.111(40)$$

$$Jan\hat{T}emp \approx 24.4$$

According to the model, the mean January temperature in Denver is expected to be 24.4°F.

g) In this context, a positive residual means that the actual average temperature in the city was higher than the temperature predicted by the model. In other words, the model underestimated the average January temperature.

28. Depression.

First of all, no association between variables can imply a cause-and-effect relationship. There may be lurking variables that explain the increase in both Internet use and depression. Additionally, provided the association is linear, only 4.6% of the variability in depression level can be explained by variability in Internet use. This is a very weak linear association at best.

29. Jumps.

a) The association between Olympic long jump distances and high jump heights is strong, linear, and positive. Years with longer long jumps tended to have higher high jumps.

b) There is an association between long jump and high jump performance, but it is likely that training and technique have improved over time and affected both jump performances.

c) The correlation would be the same, 0.92. Correlation is a measure of the degree of linear association between two quantitative variables and is unaffected by changes in units.

d) In a year when the high jumper jumped one standard deviation better than the average jump, the long jumper would be predicted to jump $r = 0.92$ standard deviations above the average long jump.

30. Modeling jumps.

a)

$$b_1 = r\left(\frac{s_y}{s_x}\right)$$

$$b_1 = 0.917\left(\frac{7.26}{20.71}\right)$$

$$b_1 = 0.321459$$

$$\hat{y} = b_0 + b_1 x$$

$$\bar{y} = b_0 + b_1\bar{x}$$

$$83.04 = b_0 + 0.321459(314.10)$$

$$b_0 = -17.930272$$

The linear model that predicts high jumps heights from long jump distances is:
$$Hi\hat{g}h = -17.930 + 0.321(Long)$$

b) For each additional inch jumped in the long jump, the model predicts an increase of approximately 0.321 inches in the high jump.

c)

$$Hi\hat{g}h = -17.930 + 0.321(Long)$$

$$Hi\hat{g}h = -17.930 + 0.321(340)$$

$$Hi\hat{g}h \approx 91.21$$

According to the model, the high jump height is expected to be approximately 91.21 inches in a year when the long jump distance is 340 inches.

d) This equation cannot be used to predict long jump distance from high jump height, because it was specifically designed to predict high jump height from long jump distance.

e)

$$b_1 = r\left(\frac{s_y}{s_x}\right)$$

$$b_1 = 0.917\left(\frac{20.71}{7.26}\right)$$

$$b_1 = 2.615850$$

$$\hat{y} = b_0 + b_1 x$$

$$\bar{y} = b_0 + b_1\bar{x}$$

$$314.10 = b_0 + 2.615850(83.04)$$

$$b_0 = 96.879816$$

The linear model that predicts long jump distances from high jump distances is:

$$L\hat{o}ng = 96.880 + 2.616(High)$$

31. French.

a) Most of the students would have similar weights. Regardless of their individual French vocabularies, the correlation would be near 0.

b) There are two possibilities. If the school offers French at all grade levels, then the correlation would be positive and strong. Older students, who typically weigh more, would have higher scores on the test, since they would have learned more French vocabulary. If French is not offered, the correlation between weight and test score would be near 0. Regardless of weight, most students would have horrible scores.

c) The correlation would be near 0. Most of the students would have similar weights and vocabulary test scores. Weight would not be a predictor of score.

d) The correlation would be positive and strong. Older students, who typically weigh more, would have higher test scores, since they would have learned more French vocabulary.

32. Twins.

a) There is a strong, fairly linear, positive trend in pre-term twin birth rates. As the year of birth increases, the pre-term twin birth rate increases.

b) The highest pre-term twin birth rate is for mothers receiving "adequate" prenatal care, and the lowest pre-term twin birth rate is for mothers receiving "inadequate" prenatal care. The slope is about the same for these relations. Mothers receiving "intensive" prenatal care had a pre-term twin birth rate that was higher than mothers receiving "inadequate" care and lower than mothers receiving "adequate" prenatal care. However, the rate of increase in pre-term twin birth rate is greater for mothers receiving "intensive" prenatal care than the rate of increase for the other groups.

c) Avoiding medical care would not be a good idea. There are likely lurking variables explaining the differences in pre-term twin birth rate. For instance, the level of pre-natal care may actually be determined by complications early in the pregnancy that may result in a pre-term birth.

33. Lunchtime.

The association between time spent at the table and number of calories consumed by toddlers is moderate, roughly linear, and negative. Generally, toddlers who spent a longer time at the table consumed fewer calories than toddlers who left the table quickly. The scatterplot between time at the table and calories consumed is straight enough to justify the use of the linear model. The linear model that predicts the time number of calories consumed by a toddler from the time spent at the table is $Calôries = 560.7 - 3.08(Time)$. For each additional minute spent at the table, the model predicts that the number of calories consumed will be approximately 3.08 fewer. Only 42.1% of the variability in the number of calories consumed can be explained by the variability in time spent at the table. The residuals plot shows no pattern, so the linear model is appropriate, if not terribly useful for prediction.

34. Gasoline.

a) The association between the year and the number of gallons of leaded gasoline available is linear, very strong, and negative. As the years have passed, the number of gallons of leaded gasoline has decreased steadily. The linear model that predicts the number of gallons available based upon the year is:

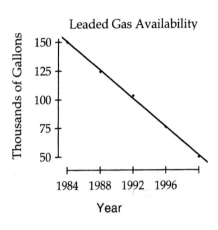

$ThousandG\hat{a}llons = 12,451.2 - 6.2(Year)$

The residuals plot shows no pattern, so the linear model is appropriate. $R^2 = 99.8\%$, so the variability in year explains 99.8% of the variability in the number of gallons available. According to the model, there will be approximately 20,200 gallons available in 2005.

b) The model is designed to predict the number of gallons available based on the year. The question asks the students to predict the year based on the number of gallons available, and models only predict in one direction.

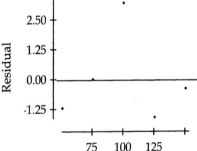

c) The linear regression model that predicts the year from the number of gallons available is:

$Y\hat{e}ar = 2008.22 - 0.161(ThousandGallons)$

The model predicts that 0 gallons of leaded gasoline will be available in about 2008. (In this case, the prediction is the y-intercept of the model.)

d) The association between year and the number of gallons of leaded gasoline available is very strong. In fact, it is so strong that the models actually do a decent job of predicting in the wrong direction! The model designed to minimize the sum of squared residuals in the response direction is actually pretty good at minimizing the sum of the squared residuals in the explanatory direction.

35. Tobacco and alcohol.

The first concern about these data is that they consist of averages for regions in Great Britain, not individual households. Any conclusions reached can only be about the regions, not the individual households living there. The second concern is the data point for Northern Ireland. This point has high leverage, since it has the highest household tobacco spending and the lowest household alcohol spending. With this point included, there appears to be only a weak positive association between tobacco and alcohol spending. Without the point, the association is much stronger. In Great Britain, with the exception of Northern Ireland, higher levels of household spending on tobacco are associated with higher levels of household spending on tobacco. It is not necessary to make the linear model, since we have the household averages for the regions in Great Britain, and the model wouldn't be useful for predicting in other countries or for individual households in Great Britain.

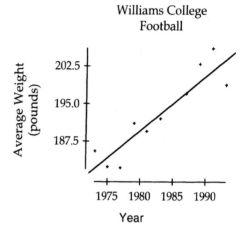

36. Football weights

a) The association between year and average weight of the members of the Williams College football team is strong, linear, and positive. Over the years, there has been a general increase in the average weight of the team. The linear model that predicts average weight from year is: $Ave.\widehat{Weight} = -1971.26 + 1.09137(Year)$. According to the model, average weight has increased by approximately one pound per year since 1973. The model explains 83.8% of the variability in average weight.

b) The residuals plot shows no pattern, so the linear model is appropriate. However, there are a couple of things to remember. First, the model is for predicting the average weight of the players on the team. Individual weights would be much more variable. Second, since we are dealing with weights, it is not reasonable to use the model for extrapolations. There is no reason to believe that average weights before 1973 or after 1993 would follow the model.

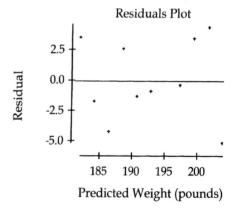

c)

$Ave.\widehat{Weight} = -1971.26 + 1.09137(Year)$

$Ave.\widehat{Weight} = -1971.26 + 1.09137(2003)$

$Ave.\widehat{Weight} \approx 214.75$

The model predicts that the average weight of the Williams College football team will be approximately 214.75 pounds in 2003. This prediction might be pretty close, but we shouldn't place to much faith in it. 2003 is ten years later than the last year for which we have data.

d) The model predicts that the average weight of the Williams College football team will be approximately 323.88 pounds in 2103. This is not reasonable. The prediction is based upon an extrapolation of 110 years.

e) The model predicts that the average weight of the Williams College football team will be approximately 1306.11 pounds in 3003. This is absurd. The prediction is based upon an extrapolation of 1010 years.

37. Models.

a)

$\hat{y} = 2 + 0.8 \ln x$

$\hat{y} = 2 + 0.8 \ln(10)$

$\hat{y} \approx 3.842$

b)

$\log \hat{y} = 5 - 0.23x$

$\log \hat{y} = 5 - 0.23(10)$

$\log \hat{y} = 2.7$

$\hat{y} = 10^{2.7} \approx 501.187$

c)

$\dfrac{1}{\sqrt{\hat{y}}} = 17.1 - 1.66x$

$\dfrac{1}{\sqrt{\hat{y}}} = 17.1 - 1.66(10) = 0.5$

$\hat{y} = \dfrac{1}{0.5^2} = 4$

38. Williams vs. Texas.

a) The association between year and average weight for the University of Texas football team is strong, roughly linear, and positive. The average weight has generally gone up over time. The linear model is: $Ave.\widehat{Weight} = -1121.66 + 0.67326(Year)$. This model explains 92.8% of the variability in average weight, but the residuals plot shows a possible pattern. The linear model may not be appropriate.

b)

$-1971.26 + 1.09137(Year) = -1121.66 + 0.67326(Year)$

$0.41811(Year) = 849.60$

$Year \approx 2032$

According to these models, the predicted weights will be the same some time during the year 2032. The average weight of the Williams College team is predicted to be more than the average weight of the University of Texas team any time after 2032.

c) This information is not likely to be accurate. The year 2032 is an extrapolation for both of the models, each of which has been shown to be of little use for even small extrapolations.

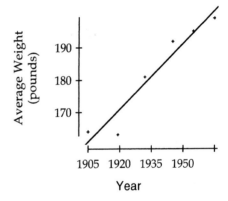

University of Texas

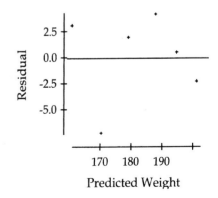

Residuals Plot

39. Vehicle weights.

a)

$\hat{Wt} = 10.85 + 0.64\,scale$

$\hat{Wt} = 10.85 + 0.64(31.2)$

$\hat{Wt} = 30.818$

According to the model, a truck with a scale weight of 31,200 pounds is expected to weigh 30,818 pounds.

b) If the actual weight of the truck is 32,120 pounds, the residual is 32,120 – 30,818 = 1302 pounds. The model underestimated the weight.

c)

$\hat{Wt} = 10.85 + 0.64\,scale$

$\hat{Wt} = 10.85 + 0.64(35.590)$

$\hat{Wt} = 33.6276$ thousand pounds

The predicted weight of the truck is 33,627.6 pounds. If the residual is –2440 pounds, the actual weight of the truck is 33,627.6 – 2440 = 31,187.6 pounds.

d) $R^2 = 93\%$, so the model explains 93% of the variability in weight, but some of the residuals are 1000 pounds or more. If we need to be more accurate than that, then this model will not work well.

e) Negative residuals will be more of a problem. Police would be issuing tickets to trucks whose weights had been overestimated by the model. The U.S. justice system is based upon the principle of innocence until guilt is proven. These truckers would be unfairly ticketed, and that is worse than allowing overweight trucks to pass.

40. Profit

a) The re-expressed data are more symmetric, with no outliers. That's good for regression because there is less of a chance for influential points. (Additionally, symmetric distributions of the explanatory and response variables will help ensure that the residuals around a line through the data are more unimodal and symmetric. We will learn more about why this is important in a later chapter.)

b) The association between log(sales) and log(profit) is linear, positive, and strong. The residuals plot shows no pattern. This model appears to be appropriate, and is surely better than the model generated with the original data.

c) The linear model is: $\log(\hat{Profit}) = -0.106259 + 0.647798 \log(Sales)$

d)

$\log(\hat{Profit}) = -0.106259 + 0.647798 \log(Sales)$

$\log(\hat{Profit}) = -0.106259 + 0.647798 \log(2500)$

$\log(\hat{Profit}) = 2.0949197$

$10^{2.0949197} \approx 124.43$

According to the model, a company with sales of 2.5 billion dollars is expected to have profits of about 124.43 million dollars.

41. Down the drain.

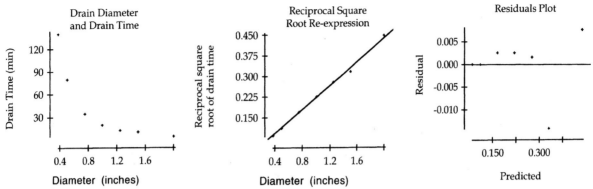

The association between diameter of the drain plug and drain time of this water tank is strong, curved, and negative. Tanks with larger drain plugs have lower drain times. The linear model is not appropriate for the curved association, so several re-expressions of the data were tried. The best one was the reciprocal square root re-expression, resulting in the

equation $\dfrac{1}{\sqrt{\widehat{DrainTime}}} = 0.00243 + 0.219(Diameter)$.

The re-expressed data is nearly linear, and although the residuals plot might still indicate some pattern and has one large residual, this is the best of the models examined. The model explains 99.7% of the variability in drain time.

42. Chips.

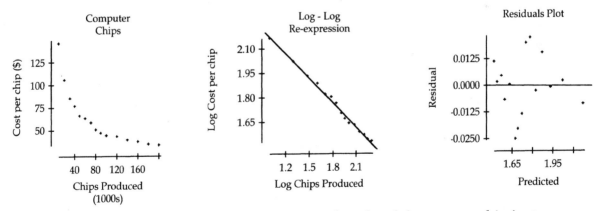

The association between the number of chips produced and the cost per chip is strong, negative, and curved. As the number of chips produced increases, the cost per chip decreases. The linear model is not appropriate for this curved association, so several re-expressions of the data were tried. Re-expressing each variable using logarithms results in a scatterplot that is nearly linear.

The model, $\log(\widehat{CostperChip}) = 2.67492 - 0.501621(\log(ChipsProduced))$, has a residuals plot that shows no pattern and $R^2 = 99.5\%$. The model explains 99.5% of the variability in the cost per chip.

43. Least Squares.

If the 4 x-values are plugged into $\hat{y} = 810 - 3x$, the 4 predicted values are $\hat{y} = 774, 738, 702$ and 666, respectively. The 4 residuals are $26, -58, 38, -6$. The squared residuals are $676, 3364, 1444,$ and 36, respectively. The sum of the squared residuals is 5520. Least squares means that no other line has a sum lower than 5520. In other words, it's the best fit.

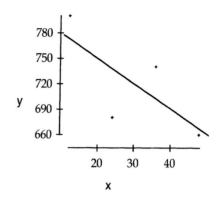

Chapter 11 – Understanding Randomness

1. **Coin toss.**

   A coin flip is random, because the outcome cannot be predicted beforehand.

2. **Casino.**

   If the outcome of the video roulette game cannot be predicted ahead of time, then it is truly random. Each of the individual outcomes (numbers 01-36, plus 0 and 00 for American roulette or 01-36, plus 0, for European roulette) should be equally likely. Probably, this is not technically the case, since the video roulette machine would use a random number generator. These generators are only pseudorandom.

3. **The lottery.**

   In state lotteries, a machine pops up numbered balls. If the lottery were truly random, the outcome could not be predicted and the outcomes would be equally likely. It is random only if the balls generate numbers in equal frequencies.

4. **Games**

   Answers may vary.

   Rolling one or two dice: If the dice are fair, then each outcome, 1 through 6 should be equally likely.

   Spinning a spinner: Each outcome should be equally likely, but the spinner might be more likely to land on one outcome than another due to friction or design.

   Shuffling cards and dealing a hand: If the cards are shuffled adequately (7 times for riffle shuffling), the cards will be approximately equally likely to be in any given hand.

5. **Bad simulations**

   a) The outcomes are not equally likely. For example, the probability of getting 5 heads in 9 tosses is not the same as the probability of getting 0 heads, but the simulation assumes they are equally likely.

   b) The even-odd assignment assumes that the player is equally likely to score or miss the shot. In reality, the likelihood of making the shot depends on the player's skill.

   c) Suppose a hand has four aces. This might be represented by 1,1,1,1, and any other number. The likelihood of the first ace in the hand is not the same as for the second or third or fourth. But with this simulation, the likelihood is the same for each.

6. **More bad simulations.**

   a) The numbers would represent the sums, but the sums are not all equally likely. For example, the probability of rolling a 7 is 6/36, but the probability of getting a 2 is only 1/36. The simulation assumes they are equally likely.

   b) The number of boys in a family of 5 children is not equally likely. For example, having a total of 5 boys is less likely than having 3 boys out of 5 children. The simulation assigns the same likelihood to each event.

c) The likelihood for out, single, double, triple, and home run are not the same. The outcome of an at bat depends on the player's skill. The simulation assumes that these outcomes are equally likely.

7. **Wrong conclusion.**

The conclusion should indicate that the simulation **suggests** that the average length of the line would be 3.2 people. Future results might not match the simulated results exactly.

8. **Another wrong conclusion.**

The simulation **suggests** that 24% of the people might contract the disease. The simulation does not represent what happened, but what might have happened.

9. **Cereal.**

Answers will vary. A component is the simulation of the picture in one box of cereal. One possible way to model this component is to generate random digits 0-9. Let 0 and 1 represent Tiger Woods, 2-4 represent Lance Armstrong, and 5-9 represent Serena Williams. Each run will consist of 5 random digits, and the response variable will be whether or not a complete set of pictures is simulated. Runs in which at least one of each picture is simulated will be a success. The total number of successes divided by the total number of runs will be the simulated probability of ending up with a complete set of pictures. According to the simulation, the probability of getting a complete set of pictures is expected to be about 51.5%.

10. **Cereal, again.**

Answers will vary. A component is the simulation of the picture in one box of cereal. One possible way to model this component is to generate random digits 0-9. Let 0 and 1 represent Tiger Woods, 2-4 represent Lance Armstrong, and 5-9 represent Serena Williams. Each run will consist of generating random numbers until a 0 or 1 is generated. The response variable will be the number of digits generated until the first 0 or 1. The total number of digits generated divided by the total number of runs will be the simulated average number of boxes required to get a Tiger Woods picture. According to the simulation, in order to be reasonably assured of getting a Tiger Woods picture, expect to buy about 5 boxes.

11. **Multiple choice**

Answers will vary. A component is one multiple-choice question. One possible way to model this component is to generate random digits 0-9. Let digits 0-7 represent a correct answer, and let digits 8 and 9 represent an incorrect answer. Each run will consist of 6 random digits. The response variable is whether or not all 6 simulated questions are answered correctly (all 6 digits are 0-7). The total number of successes divided by the total number of runs will be the simulated probability of getting all 6 questions right. According to the simulation, the probability of getting all 6 multiple-choice questions correct is expected to be about 26%.

12. **Lucky guessing.**

Answers will vary. A component is one multiple-choice question. One possible way to model this component is to generate random digits 0-9. Let the digit 0 represent a correct answer, and let digits 1, 2, and 3 represent an incorrect answer. Ignore digits 4-9. Each run will consist of 6 usable random digits. The response variable is whether or not all 6 simulated questions are answered correctly. The total number of successes divided by the total number of runs will be the simulated probability of getting all 6 questions right. Few simulations will have any runs getting all 6 correct, leading us to conclude that the probability of getting all 6 questions correct is very small. (The true probability is 0.00024). It isn't likely that your friend is telling the truth.

13. Driving test.

Answers will vary. A component is one drivers test, but this component will be modeled differently, depending on whether or not it is the first test taken. One possible way to model this component is to generate pairs of random digits 00-99. Let 01-34 represent passing the first test and let 35-99 and 00 represent failing the first test. Let 01-72 represent passing a retest, and let 73-99 and 00 represent failing a retest. To simulate one run, generate pairs of random numbers until a pair is generated that represents passing a test. Begin each run using the "first test" representation, and switch to the "retest" representation if failure is indicated on the first simulated test. The response variable is the number of simulated tests required to achieve the first passing test. The total number of simulated tests taken divided by the total number of runs is the simulated average number of tests required to pass. According to the simulation, the number of driving tests required to pass is expected to be about 1.9.

14. Still learning?

Answers will vary. A component is one drivers test, but this component will be modeled differently, depending on whether or not it is the first test taken. One possible way to model this component would be to generate pairs of random digits 00-99. Let 01-34 represent passing the first test and let 35-99 and 00 represent failing the first test. Let 01-72 represent passing a retest, and let 73-99 and 00 represent failing a retest. To simulate one run, generate pairs of random numbers until a pair is generated that represents passing a test. Begin each run using the "first test" representation, and switch to the "retest" representation if failure is indicated on the first simulated test. The response variable is whether or not the drivers test is passed within two attempts. The total number of simulated *failed* tests divided by the total number of runs is the simulated percentage of those tested who do not have a driver's license after two attempts. According to the simulation, the percentage of those tested that still do not pass within 2 tests is expected to be about 17%.

15. Basketball strategy.

Answers will vary. A component is one foul shot. One possible way to model this component would be to generate pairs of random digits 00-99. Let 01-72 represent a made shot, and let 73-99 and 00 represent a missed shot. The response variable is the number of shots made in a "one and one" situation. If the first shot simulated represents a made shot, simulate a second shot. If the first shot simulated represents a miss, the run is over. The simulated average number of points is the total number of simulated points divided by the total number of runs. According to the simulation, the player is expected to score about 1.24 points.

16. Blood donors.

Answers will vary. A component is one donor. One possible way to model this component is to generate pairs of random digits 00-99. Let 01-44 represent a type O donor, and let 45-99 and 00 represent a donor who is not type O. The response variable is the number of pairs of digits generated until 3 type O donors are simulated. Once 3 type O donors are simulated, the run is over. The simulated average number of donors required is the total number of pairs of digits generated divided by the total number of runs. According to the simulation, about 6.8 donors are required to be reasonably assured of getting 3 type O donors.

17. Free groceries.

Answers will vary. A component is the selection of one card with the prize indicated. One possible way to model the prize is to generate pairs of random digits 00-99. Let 01-10 represent $200, let 11-20 represent $100, let 21-40 represent $50, and let 41-99 and 00 represent $20. Repeated pairs of digits must be ignored. (For this reason, a simulation in which random digits 0-9 are generated with 0 representing $200, 1 representing $100, etc., is NOT acceptable. Each card must be individually represented.) A run continues until the total simulated prize is greater than $500. The response variable is the number of simulated customers until the payoff is greater than $500. The simulated average number of customers is the total number of simulated customers divided by the number of runs. According to the simulation, about 10.2 winners are expected each week.

18. Find the ace.

Answers will vary. A component is turning over one card. One way to model the cards turned over is to generate random digits 0-9. Let the digit 0 represent the ace, and let digits 1, 2, 3, and 4 each represent one of the other four cards. Ignore digits 5-9. A run consists of simulating turning over the cards until the ace is drawn. Each card must be represented individually; repeated digits must be ignored. The response variable is the number of simulated cards drawn until the ace is drawn, with $100 being awarded if the ace is drawn first, and $50, $20, $10, or $5 if the ace is drawn second, third, fourth, or fifth, respectively. The simulated average dollar amount of music the store is expected to give away is the total dollar amount of music given away divided by the number of runs. According to the simulation, the dollar amount given away is expected to be about $37.

19. The family.

Answers will vary. Each child is a component. One way to model the component is to generate random digits 0-9. Let 0-4 represent a boy and let 5-9 represent a girl. A run consists of generating random digits until a child of each gender is simulated. The response variable is the number of children simulated until this happens. The simulated average family size is the number of digits generated in each run divided by the total number of runs. According to the simulation, the expected number of children in the family is about 3.

20. A bigger family.

Answers will vary. Each child is a component. One way to model the component is to generate random digits 0-9. Let 0-4 represent a boy and let 5-9 represent a girl. A run consists of generating random digits until two children of each gender are simulated. The response variable is the number of children simulated until this happens. The simulated average family size is the number of digits generated in each run divided by the total number of runs. According to the simulation, the expected number of children in the family is slightly less than 6.

21. Dice game.

Answers will very. Each roll of the die is a component. One way of modeling this component is to generate random digits 0-9. The digits 1-6 correspond to the numbers on the faces of the die, and digits 7-9 and 0 are ignored. A run consists of generating random numbers until the sum of the numbers is exactly 10. If the sum exceeds 10, the last roll must be ignored and simulated again, but still counted as a roll. The response variable is the number of rolls until the sum is exactly 10. The simulated average number of rolls until this happens is the total number of rolls simulated divided by the number of runs. According to the simulation, expect to roll the die about 7.5 times.

22. Parcheesi.

Answers will vary. Each roll of two dice is a component. One way of modeling this component is to generate random digits 0-9. The digits 1-6 correspond to the numbers on the faces of the die, and digits 7-9 and 0 are ignored. For this simulation, look at the digits in usable pairs of digits, and consider the sum, as well as the numbers themselves. A run consists of generating usable pairs of digits until the sum is 3, or until at least one of the dice shows a 3. The response variable is the number of pairs of usable numbers generated until this happens. The simulated average number of rolls is the total number of rolls divided by the number of runs. According to the simulation, expect to roll the dice about 2.7 times.

23. The hot hand.

Answers may vary. Each shot is a component. One way of modeling this component is to generate pairs of random digits 00-99. Let 01-65 represent a made shot, and let 66-99 and 00 represent a missed shot. A run consists of 20 simulated shots. The response variable is whether or not the 20 simulated shots contained a run of 6 or more made shots. To find the simulated percentage of games in which the player is expected to have a run of 6 or more made shots, divide the total number of successes by the total number of runs. According to the simulation, the player is expected to make 6 or more shots in a row in about 40% of games. This isn't unusual. The announcer was wrong to characterize her performance as extraordinary.

24. The World Series.

Answers may vary. Each game is a component. One way of modeling this component is to generate pairs of random digits 00-99. Let 01-55 represent a win by the favored team, and let 56-99 and 00 represent a win by the underdog. A run consists of generating pairs until one team has 4 simulated wins. The response variable is whether or not the underdog wins. The simulated percentage of World Series wins is the total number of successes divided by the total number of runs. According to simulation, the underdog is expected to win the World Series about 39% of the time.

25. Job Discrimination

Answers may vary. Each person hired is a component. One way of modeling this component is to generate pairs of random digits 00-99. Let 01-10 represent each of the 10 women, and let 11-22 represent each of the 12 men. Ignore 23-99 and 00. A run consists of 3 usable pairs of numbers. Ignore repeated pairs of digits, since the same man or woman cannot be hired more than once. The response variable is whether or not all 3 simulated hires are women. The simulated percentage of the time that 3 women are expected to be hired is the number of successes divided by the number of runs. According to the simulation, the 3 people hired will all be women about 7.8% of time. This seems a bit strange, but not quite strange enough to be evidence of job discrimination.

26. Teammates.

Answers will vary. Each player chosen is a component. One way to model this component is to generate random numbers 0-9. Let 1 and 2 represent the first couple, 3 and 4 the second couple, 5 and 6 the third couple, and 7 and 8 the fourth couple. Ignore 9 and 0. A run consists of generating random digits, ignoring repeats, and organizing them into pairs, until pairs representing the first three teams are generated. (The final team is assigned by default.) The response variable is whether or not each of the simulated teams is a pairing other than 1-2, 3-4, 5-6, or 7-8. The simulated percentage of the time this is expected to happen is the total number of successes (times that the pairings are *different* than the couples) divided by the total number of trials. According to the simulation, all players are expected to be paired with someone other than the person with whom he or she came to the party about 37.5% of the time.

27. Second team.

Answers will vary. Each player chosen is a component. One way to model this component is to generate random numbers 0-9. Let digits 1-4 represent the four players who are to be chosen. Ignore digits 5-9 and 0. A run consists of generating a sequence of random numbers that represents the order in which the cards where chosen. Since each number represents a person, and people cannot be chosen more than once, ignore repeated numbers. The response variable is whether or not any digit in the generated sequence matches the corresponding digit in the sequence 1234 (or any other sequence of the four numbers, as long as it is determined ahead of time). The simulated percentage of the time this is expected to happen is the total number of successes (times that the sequences have no matching corresponding digits) divided by the number of runs. According to the simulation, all players are expected to be paired with someone other than the person with whom he or she came to the party about 37.5% of the time.

28. Tires.

Answers may vary. Each tire is a component. According to the Normal model, the probability that one tire will last at least 30,000 miles is about 78.8%. To model this component, generate triples of random digits 000-999. Let 001-788 represent a tire that lasts at least 30,000 miles, and let 789-999 and 000 represent a tire that does not last at least 30,000 miles. (The simulation can be adjusted with fewer or more digits, depending on the accuracy desired and the rounding of the probability of a tire lasting 30,000 miles.) A run consists of four simulated tires. The response variable is whether or not all four simulated tires lasted at least 30,000 miles. The percentage of the time all four tires are expected to last more than 30,000 miles is the total number of successes divided by the number of trials. According to the simulation, all four tires are expected to last this long about 38.5% of the time.

29. Freshmen.

Answers may vary. Each envelope is a component. According to the Normal model, the probability that one student will score over 1200 on the SAT is approximately 10.6%. One way to model this component is to generate triples of random digits 000-999. Let 001-106 represent a student with score over 1200, and let 107-999 and 000 represent a student with score below 1200. A run consists of simulating envelopes until three are found which represent scores above 1200. The response variable is the number of envelopes simulated until three are found that represent scores above 1200. The simulated average number of envelopes required is the total number of envelopes simulated divided by the number of runs. According to the simulation, the number of envelopes required would be about 25.

Chapter 12 – Sample Surveys

1. **Population** – human resources directors of Fortune 500 companies
   Parameter – proportion of directors who don't feel that surveys intruded on their workday
   Sampling Frame – list of HR directors at Fortune 500 companies
   Sample – the 23% of the HR directors who responded
   Method – questionnaire mailed to all (non-random)
   Bias – nonresponse. It is impossible to generalize about the opinions all HR directors because the nonrespondents would likely have a different opinion about the question. HR directors who feel that surveys intrude on their workday would be less likely to respond.

2. **Population** – all U.S. adults
   Parameter – proportion that feels marijuana should be legalized for medicinal purposes
   Sampling Frame – none given –potentially all people with access to web site
   Sample – those visiting the web site who responded
   Method – voluntary response (no randomization employed)
   Bias – voluntary response sample. Those who visit the website and respond may be predisposed to a particular answer.

3. **Population** – all U.S. adults
   Parameter – proportion who have used and benefited from alternative medical treatments.
   Sampling Frame – all Consumers Union subscribers
   Sample – those subscribers who responded
   Method – not specified, but probably a questionnaire mailed to all subscribers
   Bias – nonresponse bias, specifically voluntary response bias. Those who respond may have strong feelings one way or another.

4. **Population** – all U.S. adults
   Parameter – type of content that bothers most on TV
   Sampling Frame – all U.S. adults
   Sample – 1423 randomly selected U.S. citizens
   Method – not specified, but at least "random"
   Bias – no apparent bias

5. **Population** – adults
   Parameter – proportion who think drinking and driving is a serious problem
   Sampling Frame – bar patrons
   Sample – every 10th person leaving the bar
   Method – systematic sampling
   Bias – undercoverage. Those interviewed had just left a bar, and may have opinions about drinking and driving that differ from the opinions of the population in general.

6. **Population** – city voters
   Parameter – not clear. They might be interested in the percentage of voters favoring various issues.
   Sampling Frame – all city residents
   Sample – as many residents as they can find in one block from each district. No

randomization is specified, but hopefully a block is selected at random within each district.
Method – multistage sampling; stratified by district and clustered by block.
Bias – convenience sampling. Once the block is randomly chosen as the cluster, every resident living in that block should be surveyed, not just those that were conveniently available. A random sample of each block could be also be taken, but we wouldn't refer to that as "cluster" sampling, but rather multi-stage, with stratification by district, a simple random sample of one block within each district, and another simple random sample of residents within the block.

7. **Population** – soil around a former waste dump
   Parameter – concentrations of toxic chemicals in the soil
   Sampling Frame – accessible soil around the dump
   Sample – 16 soil samples
   Method – not clear. There is no indication that the samples were selected randomly.
   Bias – possible convenience sample. Since there is no indication of randomization, the samples may have been taken from easily accessible areas. Soil in these areas may be more or less polluted than the soil in general.

8. **Population** – all cars
   Parameter – proportion of cars with up-to-date (or out-of-date) registrations, insurance, or safety inspections.
   Sampling Frame – cars on that road
   Sample – cars stopped by the roadblock
   Method – cluster sample of an area, stopping all cars within the cluster
   Bias – undercoverage. The cars stopped might not be representative of all cars because of time of day and location. The locations are probably not chosen randomly, so might represent areas in which it is easy to set up a roadblock, resulting in a convenience sample.

9. **Population** – snack food bags
   Parameter – weight of bags, proportion passing inspection
   Sampling Frame – all bags produced each day
   Sample – 10 bags, one from each of 10 randomly selected cases
   Method – multistage sampling. Presumably, they take a simple random sample of 10 cases, followed by a simple random sample of one bag from each case.
   Bias – no indication of bias

10. **Population** – milk produced by a dairy farm
    Parameter – whether or not the milk contains dirt, antibiotics, or other foreign matter
    Sampling Frame – milk produced by the farm on any given day
    Sample – milk produced by the farm on the day of inspection
    Method – not specified
    Bias – unbiased, as long as the day of inspection is randomly chosen. This might not be the case, however, since the farms might be spread out over a wide geographic area. Inspectors might tend to visit farms that are near one another on the same day, resulting in a convenience sample.

11. Parent opinion, part 1.

a) This is a voluntary response sample. Only those who see the ad, feel strongly about the issue, and have web access will respond.

b) This is cluster sampling, but probably not a good idea. The opinions of parents in one school may not be typical of the opinions of all parents.

c) This is an attempt at a census, and will probably suffer from nonresponse bias.

d) This is stratified sampling. If the follow-up is carried out carefully, the sample should be unbiased.

12. Parent opinion, part 2.

a) This sampling method suffers from voluntary response bias. Only those who see the show and feel strongly will call.

b) Although this method may result in a more representative sample than the method in part **a)**, this is still a voluntary response sample. Only strongly motivated parents attend PTA meetings.

c) This is multistage sampling, stratified by elementary school and then clustered by grade. This is a good design, as long as the parents in the class respond. There should be follow-up to get the opinions of parents who do not respond.

d) This is systematic sampling. As long as a starting point is randomized, this method should produce reliable data.

13. Wording the survey.

a) Responses to these questions will differ. Question 1 will probably get "no" answers, and Question 2 will probably get "yes" answers. This is response bias, based on the wording of the questions.

b) A question with neutral wording might be: "Do you think standardized tests are appropriate for deciding whether a student should be promoted to the next grade?"

14. Survey questions.

a) The question is biased toward "yes" answers because of the word "pollute". A better question might be: "Should companies be responsible for any costs of environmental clean up?"

b) The question is biased toward "no" because of the preamble "18-year-olds are old enough to serve in the military. A better question might be: "Do you think the drinking age should be lowered from 21?"

c) The question seems unbiased.

d) The question is biased toward "yes" because of the phrase "great tradition". A better question might be: "Do you favor continued funding for the space program?"

15. Phone surveys.

a) A simple random sample is difficult in this case because there is a problem with undercoverage. People with unlisted phone numbers and those without phones are not in the sampling frame. People who are at work, or otherwise away from home, are included in the sampling frame. These people could never be in the sample itself.

b) One possibility is to generate random phone numbers and call at random times, although obviously not in the middle of the night! This would take care of the undercoverage of people at work during the day, as well as people with unlisted phone numbers, although there is still a problem avoiding undercoverage of people without phones.

c) Under the original plan, those families in which one person stays home are more likely to be included. Under the second plan, many more are included. People without phones are still excluded.

d) Follow-up of this type greatly improves the chance that a selected household is included, increasing the reliability of the survey.

e) Random dialers allow people with unlisted phone numbers to be selected, although they may not be the most willing participants. There is a reason that the phone number is unlisted. Time of day will still be an issue, as will people without phones.

16. Cell phone survey.

Cell phones are more likely to be used by middle class and upper class individuals. This will result in an undercoverage bias. As cell phone use grows, this will be less of a problem. Also, many cell phone plans require the users to pay airtime for incoming calls. That seems like a sure way to irritate the respondent, and result in response bias toward negative responses.

17. Arm length.

a) Answers will vary. My arm length is 3 hand widths and 2 finger widths.

b) The parameter estimated by 10 measurements is the true length of your arm. The population is all possible measurements of your arm length.

c) The population is now the arm lengths of your friends. The average now estimates the mean of the arm lengths of your friends.

d) These 10 arm lengths are unlikely to be representative of the community, or the country. Your friends are likely to be of the same age, and not very diverse.

18. Fuel economy.

a) The statistic calculated is the mean mileage for the last six fill-ups.

b) The parameter of interest is the mean mileage for the vehicle.

c) The driving conditions for the last six fill-ups might not be typical of the overall driving conditions. For instance, the last six fill-ups might all be in winter, when mileage might be lower than expected.

d) The EPA is trying to estimate the mean gas mileage for all cars of this make, model, and year.

19. Accounting.

a) Assign numbers 001-120 to each order. Generate 10 random numbers 001-120, and select those orders to recheck.

b) The supervisor should perform a stratified sample, randomly checking a certain percentage of each type of sales, retail and wholesale.

20. Happy workers?

a) A small sample will probably consist mostly laborers, with few foremen, and maybe no project managers. Also, there is a potential for response bias based on the interviewer if a member of management asks directly about discontent. Workers who want to keep their jobs will likely tell the management that everything is fine!

b) Assign a number from 001 to 439 to each employee. Use a random number table or software to select the sample.

c) The simple random sample might not give a good cross section of the different types of employees. There are relatively few foremen and project managers, and we want to make sure their opinions are noted, as well as the opinions of the laborers.

d) A better strategy would be to stratify the sample by job type. Sample a certain percentage of each job type.

e) Answers will vary. Assign each person a number from 01-14, and generate 2 usable random numbers from a random number table or software.

21. Quality control.

a) Select three cases at random, then select one jar randomly from each case.

b) Generate three random numbers between 61-80, with no repeats, to select three cases. Then assign each of the jars in the case a number 01-12, and generate one random number for each case to select the three jars, one from each case.

c) This is not a simple random sample, since there are groups of three jars that cannot be the sample. For example, it is impossible for three jars in the same case to be the sample. This would be possible if the sample were a simple random sample.

22. A fish story.

What conclusions they may be able to make will depend on whether fish with discolored scales are equally likely to be caught as those without. It also depends on the level of compliance by fisherman. If fish are not equally likely to be caught, or fishermen more disposed to bring discolored fish, the results will be biased.

23. Sampling methods.

a) This method would probably result in undercoverage of those doctors that are not listed in the Yellow Pages. Using the "line listings" seems fair, as long as all doctors are listed, but using the advertisements would not be a typical list of doctors.

b) This method is not appropriate. This cluster sample will probably contain listings for only one or two types of businesses, not a representative cross-section of businesses.

24. More sampling methods.

a) A petition may pressure people into support. Additionally, some people may not be home on a Saturday, especially those who have taken their kids out to play in a distant park! We are undercovering a group made up of people who probably have a specific opinion.

b) If the food at the largest cafeteria is representative, this should be OK. However, those who really don't like the food won't be eating there. That group is undercovered.

Chapter 13 – Experiments

1. **a)** This is an experiment, since treatments were imposed.
    b) The subjects studied were 30 patients with bipolar disorder.
    c) The experiment has 1 factor (omega-3 fats from fish oil), at 2 levels (high dose of omega-3 fats from fish oil and no omega-3 fats from fish oil).
    d) 1 factor, at 2 levels gives a total of 2 treatments.
    e) The response variable is "improvement", but there is no indication of how the response variable was measured.
    f) There is no information about the design of the experiment.
    g) The experiment is blinded, since the use of a placebo keeps the patients from knowing whether or not they received the omega-3 fats from fish oils. It is not stated whether or not the evaluators of the "improvement" were blind to the treatment, which would make the experiment double-blind.
    h) Although it needs to be replicated, the experiment can determine whether or not omega-3 fats from fish oils cause improvements in patients with bipolar disorder, at least over the short term. The experiment design would be stronger is it were double-blind.

2. **a)** This is an experiment, since treatments were imposed.
    b) The subjects were men aged 60 to 75.
    c) The experiment has 1 factor (an exercise program), at 1 level.
    d) There is only 1 treatment. There is no control group.
    e) The response variable is change in strength, measured by a pre-test and post-test.
    f) The design is matched, with each man's strength level after the program compared to his strength level before the program.
    g) No blinding is indicated, and seems unlikely. The participants certainly know that they have been exercising. Perhaps the evaluators are blinded and do not know whether the subject is taking his pre-test or post-test for strength evaluation.
    h) Conclusions from this experiment only apply to men 60 to 75 who participate in similar exercise programs. However, since there is no random assignment and no control group, other factors might explain the difference in strength.

3. **a)** This is an observational study. The researchers are simply studying traits that already exist in the subjects, not imposing new treatments.
    b) This is a prospective study. The subjects were identified first, then traits were observed.
    c) The subjects are roughly 200 men and women with moderately high blood pressure and normal blood pressure. There is no information about the selection method.
    d) The parameters of interest are memory and reaction time.
    e) An observational study has no random assignment, so there is no way to know that high blood pressure caused subjects to do worse on memory and reaction time tests. A lurking variable, such as age or overall health, might have been the cause. The most we can say is that there was an association between blood pressure and scores on memory and reaction time tests in this group, and recommend a controlled experiment to attempt to determine whether or not there is a cause-and-effect relationship.

4. a) This is an observational study. The researchers are simply studying traits that already exist in the subjects, not imposing new treatments.

   b) This is a prospective study. The subjects were identified first, then traits were observed.

   c) The subjects were disabled women aged 65 and older, with and without a vitamin B-12 deficiency. The selection process is not stated.

   d) The parameter of interest is the percentage of women in each group who suffered severe depression.

   e) There is no random assignment, so a cause-and-effect relationship between B-12 deficiency and depression cannot be established. The most that can be determined is an association, if this is supported by the data.

5. a) This is an observational study. The researchers are simply studying traits that already exist in the subjects, not imposing new treatments.

   b) This is a retrospective study. Researchers studied medical records that already existed.

   c) The subjects were 360,000 Swedish men. The selection process is not stated.

   d) The parameter of interest is the percentage of men with kidney cancer among different groups.

   e) There is no random assignment, so a cause-and-effect relationship between weight or high blood pressure and kidney cancer cannot be established.

6. a) This is an experiment, since treatments were imposed on randomly assigned groups.

   b) There were 459 subjects in this experiment.

   c) There is 1 factor (diet), at 3 levels (DASH diet, the control, and an unspecified diet).

   d) 1 factor, with 3 levels, results in 3 treatments.

   e) The response variable is systolic blood pressure.

   f) The experiment is probably completely randomized. There is no mention of any blocking.

   g) There is no mention of any blinding.

   h) This experiment indicates that the DASH diet lowers systolic blood pressure by an average of 6.7 points.

7. a) This is an experiment, since treatments were imposed on randomly assigned groups.

   b) 24 post-menopausal women were the subjects in this experiment.

   c) There is 1 factor (type of drink), at 2 levels (alcoholic and non-alcoholic). (Supplemental estrogen is not a factor in the experiment, but rather a blocking variable. The subjects were not given estrogen supplements as part of the experiment.)

   d) 1 factor, with 2 levels, is 2 treatments.

   e) The response variable is an increase in estrogen level.

   f) This experiment utilizes a blocked design. The subjects were blocked by whether or not they used supplemental estrogen. This design reduces variability in the response variable of estrogen level that may be associated with the use of supplemental estrogen.

   g) This experiment does not use blinding.

   h) This experiment indicates that drinking alcohol leads to increased estrogen level among those taking estrogen supplements.

8. a) This is an experiment, since treatments were imposed on randomly assigned groups.

   b) The subjects were 40 volunteers suffering from insomnia.

c) There are 2 factors in this experiment (dessert and exercise). The dessert factor has 2 levels (no dessert and normal dessert). The exercise factor has 2 levels (no exercise and an exercise program).

d) 2 factors, with 2 levels each, results in 4 treatments.

e) The response variable is improvement in ability to sleep.

f) This experiment is probably completely randomized.

g) This experiment does not use blinding.

h) This experiment indicates that insomniacs who refrain from desserts and exercise will experience improved ability to sleep.

9. a) This is an experiment, since treatments were imposed on randomly assigned groups.

b) The experimental units were 10 randomly selected garden locations.

c) There is 1 factor (type of trap), at 2 levels (the different types of trap).

d) 1 factor, at 2 levels, results in 2 treatments.

e) The response variable is the number of bugs in each trap.

f) This experiment incorporates a blocked design. Placing one of each type of trap at each location reduces variability due to location in the garden.

g) There is no mention of blinding.

h) This experiment can determine which trap is more effective at catching bugs.

10. a) This is an observational study.

b) The study is retrospective. Results were obtained from pre-existing medical records.

c) The subjects in this study were 981 women who lived near the site of dioxin release.

d) The parameter of interest is the incidence of breast cancer.

e) As there is no random assignment, there is no way to know that the dioxin levels caused the increase in breast cancer. There may have been lurking variables that were not identified.

11. a) This is an observational study.

b) The study is retrospective. Results were obtained from pre-existing church records.

c) The subjects of the study are women in Finland. The data were collected from church records dating 1640 to 1870, but the selection process is unknown.

d) The parameter of interest is lifespan.

e) For this group, having sons was associated with a decrease in lifespan of an average of 34 weeks per son, while having daughters was associated with an unspecified increase in lifespan. As there is no random assignment, there is no way to know that having sons caused a decrease in lifespan.

12. a) This is an observational study.

b) The study is retrospective. Findings were based upon women's work histories.

c) The subjects in the study were 763 women with breast cancer and 741 women without the disease. The selection process is not stated.

d) The parameter of interest is the incidence of breast cancer.

e) As there is no random assignment, there is no way to know that working nights caused the increased risk of breast cancer. There may have been lurking variables that were not identified.

13. a) This is an observational study. (Although some might say that the sad movie was "imposed" on the subjects, this was merely a stimulus used to trigger a reaction, not a treatment designed to attempt to influence some response variable. Researchers merely wanted to observe the behavior of two different groups when each was presented with the stimulus.)

   b) The study is prospective. Researchers identified subjects, and then observed them after the sad movie.

   c) The subjects in this study were people with and without depression. The selection process is not stated.

   d) The parameter of interest is crying response to sad situations.

   e) There is no apparent difference in crying response to sad movies for the depressed and nondepressed groups.

14. a) This is an experiment, since treatments were imposed on randomly assigned groups.

   b) The subjects were volunteers exposed to a cold virus.

   c) There is 1 factor (herbal compound), at 2 levels (herbal compound and sugar solution).

   d) 1 factor, at 2 levels, results in 2 treatments.

   e) The response variable is the severity of cold symptoms.

   f) There is no mention of any randomness is the design. Hopefully, subjects were randomly assigned to treatment groups.

   g) The experiment uses blinding. The use of a sugar solution as a placebo kept the subjects from knowing whether or not they had received the herbal compound. If the doctors responsible for assessing the severity of the patients' colds were also unaware of the treatment group assignments, then the experiment incorporates double blinding.

   h) There is no evidence to suggest that the herbal treatment is effective.

15. a) This is an experiment.

   b) The subjects were rats.

   c) There is 1 factor (sleep deprivation), at 4 levels (no deprivation, 6, 12, and 24 hours).

   d) 1 factor, at 4 levels, results in 4 treatments.

   e) The response variable is glycogen level in the brain.

   f) There is no mention of randomness in the design. Hopefully, the rats were randomly assigned to treatment groups.

   g) There is no mention of blinding.

   h) As long as proper randomization was employed, the conclusion could be that rats deprived of sleep have significantly lower glycogen levels and may need sleep to restore that brain energy fuel. Extrapolating to humans would be speculative.

16. a) This is an experiment.

   b) The subjects were racing greyhounds.

   c) There is 1 factor (level of vitamin C in diet). The 3 levels of diet were not specified.

   d) One factor, at 3 levels, results in 3 treatments.

   e) The response variable is speed.

   f) The experiment uses a matched design. Each greyhound was given each of the 3 levels of diet, in random order. The matched design reduces variation due to the racing ability of each greyhound.

g) There is no mention of blinding.

h) Greyhounds that eat diets high in vitamin C run more slowly than greyhounds with diets lower in vitamin C.

17. a) This is an experiment. Subjects were randomly assigned to treatments.

b) The subjects were people experiencing migraines.

c) There are 2 factors (pain reliever and water temperature). The pain reliever factor has 2 levels (pain reliever or placebo), and the water temperature factor has 2 levels (ice water and regular water).

d) 2 factors, at 2 levels each, results in 4 treatments.

e) The response variable is the level of pain relief.

f) The experiment is completely randomized.

g) The subjects are blinded to the pain reliever factor through the use of a placebo. The subjects are not blinded to the water factor. They will know whether they are drinking ice water or regular water.

h) The experiment may indicate whether pain reliever alone or in combination with ice water give pain relief, but patients are not blinded to ice water, so the placebo effect may also be the cause of any relief seen due to ice water.

18. a) This is an experiment. Hopefully, dogs are randomly assigned to different treatment groups.

b) The subjects are inactive dogs.

c) There is 1 factor (type of dog food), at 2 levels (low-calorie and standard). One possible difficulty with this experiment is that some owners might feed their dogs more food than others. We will assume that the dog food company has given the owners specific instructions about the quantity of food required, based on the size of each dog.

d) 1 factor, at 2 levels, results in 2 treatments.

e) The response variable is the weight of the dogs.

f) The experiment uses blocking by size of breed. Blocking by size reduces variation in weight that may be due to overall size of the dog.

g) Assuming that the dog owners do not know which type of dog food their dog is receiving, the experiment is blinded.

h) Assuming the dog owners followed the prescribed feeding levels, there could be a conclusion as to whether or not the dog food helped the dogs maintain a healthy weight.

19. Tomatoes.

Answers may vary. Number the tomatoes plants 1 to 24. Use a random number generator to randomly select 24 numbers from 1 to 24 without replication. Assign the tomato plants matching the first 8 numbers to the first group, the second 8 numbers to the second group, and the third group of 8 numbers to the third group.

20. Tomatoes II.

Answers may vary. Number the tomato plants 1 to 24. Use a random number generator to randomly select 24 numbers from 1 to 24 without replication. Assign the tomato plants matching the first group of 4 numbers to the first treatment (no fertilizer, natural watering), the second group of 4 numbers to the second treatment (no fertilizer, daily water), the third group of 4 numbers to the third treatment (half fertilizer, natural watering), and so on to the sixth treatment.

21. Mozart.

a) The differences in spatial reasoning scores between the students listening to Mozart and the students sitting quietly were more than would have been expected from ordinary sampling variation.

b)

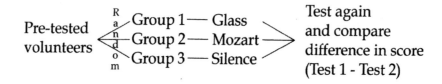

c) The Mozart group seems to have the smallest median difference in spatial reasoning test score and thus the *least* improvement, but there does not appear to be a significant difference.

d) No, the results do not prove that listening to Mozart is beneficial. If anything, there was generally less improvement. The difference does not seem significant compared with the usual variation one would expect between the three groups. Even if type of music has no effect on test score, we would expect some variation between the groups.

22. More Mozart.

Answers may vary. Suppose you select the next 12 batches of sake for the experiment, in order to replicate the experiment. For each batch, divide the yeast into two parts. Randomly select half of the yeast and play Mozart for that half and no Mozart for the other half. When the sake is done, have trained sake testers taste samples from each group and rate the sake. In order to eliminate potential bias on the part of the testers, blinding should be used. Do not tell the tasters whether or not the sake has been brewed with Mozart playing next to the yeast. Compare the ratings for the Mozart and the non-Mozart sake. Try to keep other variables (type of yeast, size of batch etc.) under control.

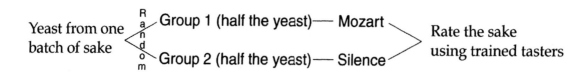

23. Frumpies.

a) They should perform a survey. Randomly select a group of children, ages 10 to 13, and have them taste the cereal. Ask the children a question with as little bias as possible, such as, "Do you like the cereal?"

b) Answers may vary. Get volunteers age 10 to 13. Each volunteer will taste both cereals, randomizing the order in which they taste them. Compare the percentage of favorable ratings for each cereal.

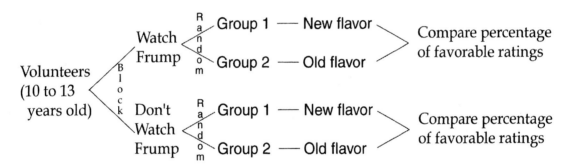

c) Answers may vary. From the volunteers, identify the children who watch Frump, and identify the children who do not watch Frump. Use a blocked design to reduce variation in cereal preference that may be associated with watching the Frump cartoon.

24. Full moon.

Answers may vary. Use a retrospective observational study. For example, collect records from a random selection of police and emergency room logs for the past 3 years. Find the number of cases for the days when there is a full moon, when there is a waxing moon, a waning moon, and when the moon is nearly dark. Compare the numbers for each group.

25. Wine.

a) This is a prospective observational study. The researchers followed a group of children born at a Copenhagen hospital between 1959 and 1961.

b) The results of the Danish study report a link between high socioeconomic status, education, and wine drinking. Since people with high levels of education and higher socioeconomic status are also more likely to be healthy, the relation between health and wine consumption might be explained by the confounding variables of socioeconomic status and education.

c) These studies prove none of these. While the variables have a relation, there is no indication of a cause-and-effect relationship. The only way to determine causation is through a controlled, randomized, and replicated experiment.

26. Swimming.

a) The swimmers showed a rate of depression that was lower than would be expected from a sample of that size drawn at random from the population. This rate was so low that it was unlikely to be due to natural sampling variation.

b) This is a retrospective observational study. There was no imposition of treatments. The researchers simply identified a group and evaluated them for depression.

c) The news reports made a claim of a cause-and-effect relationship. Causation can only be determined through the use of a controlled, randomized, and replicated experiment, not an observational study. The difference in depression rates might be explained by lurking variables. For example, swimmers might tend to have higher incomes than the general population. Swimmers need to have access to a pool, either by having their own, or paying for a membership to a health club. Perhaps it is their financial situation that makes them happier, not the swimming. Another possible explanation is a reversal of the direction of the relationship implied by the news reports. Perhaps depression makes people not want to swim.

d) Answers may vary. Give the subjects a test to measure depression. Then randomly assign the 120 subjects to one of three groups: the control group (no exercise program), the anaerobic exercise group, and the aerobic exercise group. Monitor subjects' exercise (have them report to a particular gym or pool). At the end of 12 weeks, administer the depression test again. Compare the post-exercise and pre-exercise depression scores.

```
                    R ⟋ Group 1(60) — Aerobic         Post-test and
120 volunteers      a                            ⟍
(pre-test for      ⟨n ⟩  Group 2(60) — Anaerobic  ⟩  compare difference
depression)         d ⟍ Group 3(60) — No exercise  ⟋ in depression
                    o
                    m
```

27. Dowsing.

a) Arrange the 20 containers in 20 separate locations. Number the containers 01 – 20, and use a random number generator to identify the 10 containers that should be filled with water.

b) We would expect the dowser to be correct about 50% of the time, just by guessing. A record of 60% (12 out of 20) does not appear to be significantly different than the 10 out of 20 expected.

c) Answers may vary. A high level of success would need to be observed. 90% to 100% success (18 to 20 correct identifications) would be convincing.

28. Healing.

```
                    R ⟋ Group 1 — vitamin E pill
Post-surgical       a                          ⟍     Compare time
volunteers         ⟨n ⟩                         ⟩    until recovery
                    d ⟍ Group 2 — placebo pill  ⟋
                    o
                    m
```

Answers will vary. This double-blind experiment has 1 factor (vitamin E), at 2 levels (vitamin E and no vitamin E), resulting in 2 treatments. The response variable measured is the time it takes the patient to recover from the surgery. Randomly select half of the patients who agree to the study to get large doses of vitamin E after surgery. Give the other patients in the study a similar looking placebo pill. Monitor their progress, recording the time until they have reached an easily agreed upon level of healing. Have the evaluating doctor blinded to whether the patient received the vitamin E or the placebo. Compare the number of days until recovery of the two groups.

29. Reading.

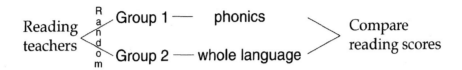

Answers may vary. This experiment has 1 factor (reading program), at 2 levels (phonics and whole language), resulting in 2 treatments. The response variable is reading score on an appropriate reading test after a year in the program. After randomly assigning students to teachers, randomly assign half the reading teachers in the district to use each method. There may be variation in reading score based on school within the district, as well as by grade. Blocking by both school and grade will reduce this variation.

30. Gas mileage.

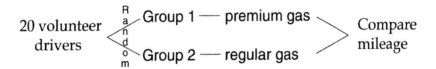

Answers may vary. This experiment has 1 factor (type of gasoline), at 2 levels (premium and regular), resulting in two treatments. The response variable is gas mileage. An experiment diagram for a simple design appears above. Randomly assign each of the 20 volunteers to the premium or regular groups. Ask them to keep driving logs (the number of miles driven and the gallons of gasoline) for one month. Compare the differences in the fuel economy for the two groups.

Stronger designs would control for several variables that may have an effect on fuel economy, such as size of engine, type of driving (for example, city or highway), and driving style (for example, if the is driver aggressive, or if the driver exceeds the speed limit). With only 20 volunteers, it would be difficult to block for all of these variables, but a matched design would work well. Have each volunteer use regular gasoline for a specified time period and record the mileage, and also use premium for a specified time period. Randomize which type of gasoline is used first.

31. Weekend deaths.

a) The difference between death rate on the weekend and death rate during the week is greater than would be expected due to natural sampling variation.

b) This was a prospective observational study. The researchers identified hospitals in Ontario, Canada, and tracked admissions to the emergency rooms. This certainly cannot be an experiment. People can't be assigned to become injured on a specific day of the week!

c) Waiting until Monday, if you were ill on Saturday, would be foolish. There are likely to be confounding variables that account for the higher death rate on the weekends. For example, people might be more likely to engage in risky behavior on the weekend.

d) Alcohol use might have something to do with the higher death rate on the weekends. Perhaps more people drink alcohol on weekends, which may lead to more traffic accidents, and higher rates of violence during these days of the week.

32. Shingles.

a) Answers may vary. This experiment has 1 factor (ointment), at 2 levels (current and new), resulting in 2 treatments. The response variables are the improvements in severity of the case of shingles and the improvements in the pain levels of the patients. Randomly assign the eight patients to either the current ointment or to the new ointment. Before beginning treatment, have doctors assess the severity of the case of shingles for each patient, and ask patients to rate their pain levels. Administer the ointments for a prescribed time, and then have doctors reassess the severity of the case of shingles, and ask patients to once again rate their pain levels. If the neither the patients nor the doctors are told which treatment is given to each patient, the experiment will be double-blind. Compare the improvement levels for each group.

b) Answers may vary. Let numbers 1 through 8 correspond to letter A through H, respectively. Ignore digits 0 and 9, and ignore repeats. The first row contains the random digits, the second row shows the corresponding patient (X indicates an ignored or repeated digit), and the third row shows the resulting group assignment, alternating between Group 1 and Group 2.

```
41098 18329 78458 31685 55259
DAXXH XXCBX GXXEX XXF
11  1   12  2 2    2
```

Group 1 (current ointment): D, A, H, C
Group 2 (new ointment): B, G, E, F

c) Assuming that the ointments looked alike, it would be possible to blind the experiment for the patient and the evaluating doctor. If both the subject and the evaluator are blinded, the experiment is double-blind.

d) Before randomly assigning patients to treatments, identify them as male or female. Having blocks for males and females will eliminate variation in improvement due to gender.

33. Beetles.

Answers may vary. This experiment has 1 factor (pesticide), at 3 levels (pesticide A, pesticide B, no pesticide), resulting in 3 treatments. The response variable is the number of beetle larvae found on each plant. Randomly select a third of the plots to be sprayed with pesticide A, a third with pesticide B, and a third to be sprayed with no pesticide (since the researcher also wants to know whether the pesticides even work at all). To control the experiment, the plots of land should be as similar as possible, with regard to amount of sunlight, water, proximity to other plants, etc. If not, plots with similar characteristics should be blocked together. If possible, use some inert substance as a placebo pesticide on the control group, and do not tell the counters of the beetle larvae which plants have been treated with pesticides. After a given period of time, count the number of beetle larvae on each plant and compare the results.

Plots of corn — Random — Group 1 — pesticide A / Group 2 — pesticide B / Group 3 — no pesticide → Count the number of beetle larvae on each plant and compare

34. SAT Prep.

a) The students were not randomly assigned to the special study course. Those who signed up for the course may be a special group whose scores would have improved anyway, due to motivation, intelligence, parental involvement, or other reasons.

b) Answers may vary. This experiment has 1 factor (study course), at 2 levels (study course, no study course), resulting in 2 treatments. The response variable is improvement in SAT score on the second test. Find a group of volunteers who are willing to participate. Have all volunteers take the SAT exam. Randomly assign the subjects to the study course or no study course groups. After giving the study course to the appropriate group, have both groups take the SAT again. Check to see if the group given the study course had a significant improvement in scores when compared with the group receiving no study course

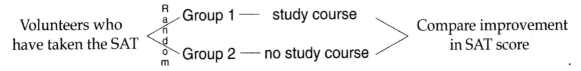

Volunteers who have taken the SAT — Random — Group 1 — study course / Group 2 — no study course → Compare improvement in SAT score

c) After the volunteers have taken the first SAT exam, block the volunteers by Low, Average, and High SAT exam score performance. For each block, replicate the experiment design described in part b.

35. Safety switch.

Answers may vary. This experiment has 1 factor (hand), at 2 levels (right, left), resulting in 2 treatments. The response variable is the difference in deactivation time between left and right hand. Find a group of volunteers. Using a matched design, we will require each volunteer to deactivate the machine with his or her left hand, as well as with his or her right hand. Randomly assign the left or right hand to be used first. Hopefully, this will equalize any variability in time that may result from experience gained after deactivating the machine the first time. Complete the first attempt for the whole group. Now repeat the experiment with the alternate hand. Check the differences in time for the left and right hands. Since the response variable is difference in times for each hand, workers should be blocked into groups based on their dominant hand. Another way to account for this difference would be to use the absolute value of the difference as the response variable. We are interested in whether or not the difference is significantly different from the zero difference we would expect if the machine were just as easy to operate with either hand.

36. Washing clothes.

Answers may vary. This experiment has two factors (water temperature, wash cycle). The factor water temperature has 2 levels (cold, hot), and the factor wash cycle has 2 levels (regular, delicates). 2 factors, at 2 levels each, results in 4 treatments (hot-regular, hot-delicates, cold-regular, cold-delicates). The response variable is the level of cleaning of the grass stains. It would be nice to have 32 shirts with which to experiment, so that we could randomly assign 8 shirts to each treatment group, but equal numbers of shirts in each group are not necessary. After washing, have "laundry experts" rate the cleanliness of each shirt. Compare the level of cleanliness in each group.

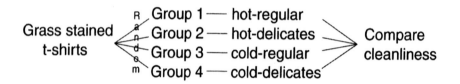

Review of Part III – Gathering Data

1. The researchers performed a prospective observational study, since the children were identified at birth and examined at ages 8 and 20. There were indications of behavioral differences between the group of "preemies", and the group of full-term babies. The "preemies" were less likely to engage in risky behaviors, like use of drugs and alcohol, teen pregnancy, and conviction of crimes. This may point to a link between premature birth and behavior, but there may be lurking variables involved. Without a controlled, randomized, and replicated experiment, a cause-and-effect relationship cannot be determined.

2. A retrospective observational study was performed. There may be a link between tea drinking and survival after a heart attack. Other variables, like overall health and education might also be involved. Since lurking variables may be involved, a controlled, randomized, and replicated experiment must be performed to determine whether or not a cause-and-effect relationship exists between tea drinking and post heart attack survival.

3. The researchers at the Purina Pet Institute performed an experiment, matched by gender and weight. The experiment had one factor (diet), at two levels (allowing the dogs to eat as much as they want, or restricted diet), resulting in two treatments. One of each pair of similar puppies was randomly assigned to each treatment. The response variable was length of life. The researchers were able to conclude that, on average, dogs with a lower-calorie diet live longer.

4. The officials used a random sample. The population is all homes on the property tax list. The parameter of interest is level of radon contamination. The officials' procedure is not clear, but if they make an effort to get some houses from each area in the city, the sample is stratified by area. If the procedure is followed carefully, the officials can use the results of the sample to make inferences about the radon levels in other houses in the county.

5. It is not apparent whether or not the high folate intake was a randomly imposed treatment. It is likely that the high folate intake was simply an observed trait in the women, so this is a prospective observational study. This study indicates that folate may help in reducing colon cancer for those with family histories of the disease. There may be other variables in the diet and lifestyle of these women that accounted for the decreased risk of colon cancer, so a cause-and-effect relationship cannot be inferred.

6. The research team performed a retrospective observational study. There is evidence that the date of first flowering has generally advanced over the last 47 years, but there may be other variables besides climate change that can account for this. The assertion of the researchers is speculative.

7. The fireworks manufacturers are sampling. No information is given about the sampling procedure, so hopefully the tested fireworks are selected randomly. It would probably be a good idea to test a few of each type of firework, so stratification by type seems likely. The population is all fireworks produced each day, and the parameter of interest is the proportion of duds. With a random sample, the manufacturers can make inferences about

the proportion of duds in the entire day's production, and use this information to decide whether or not the day's production is suitable for sale.

8. This appears to be an experiment, although there are no details in the exercise to confirm this assertion. The statement "exposed some laboratory animals" might imply that there were others not exposed, serving as a control. Also, the statement "heightened incidence of damage" seems to imply that there was a comparison of some sort. There is one factor (exposure to phthalates), at two levels (exposure and none), resulting in two treatments. The response variable is male reproductive damage. There is no mention of random allocation to treatment groups. This experiment shows that exposure to phthalates leads to a higher incidence of male reproductive damage in laboratory animals, but speculation that the same may be true in humans is very risky.

9. The researchers performed a retrospective observational study. The data were gathered from pre-existing medical records. Living near strong electromagnetic fields may be associated with an increase in leukemia rates. Since this is not a controlled, randomized, and replicated experiment, lurking variables may be involved in this increased risk. For example, the neighborhood around the antennas may consist of people of the same socioeconomic status. Perhaps some variable linked with socioeconomic status may be responsible for the higher leukemia rates.

10. The medical researchers performed a retrospective observational study. The data were gathered from pre-existing medical records. The study does not *prove* that there is no long-term risk of prostate cancer associated with having a vasectomy, but it does provide evidence to that effect.

11. This is an experiment, blocked by sex of the rat. There is one factor (type of hormone), with at least two levels (leptin, insulin). There is a possibility that an additional level of no hormone (or placebo injection) was used as a control, although this is not specifically mentioned. This results in at least two treatments, possibly three, if a control was used. There is no specific mention of random allocation. There are two response variables: amount of food consumed and weight lost. Hormones can help suppress appetite, and lead to weight loss, in laboratory rats. The male rats responded best to insulin, and the female rats responded best to leptin.

12. The artisan is performing an experiment. There are 2 factors (glaze type and temperature). The glaze type has 4 levels, and the temperature has 3 levels, resulting in 12 treatments (the different combinations of glazes and temperatures). There is no mention of randomization. The response variable is apparent age of the pottery. Assuming that the evaluator is unbiased, the artisan can make a conclusion about the best combination of glaze and temperature.

13. The researchers performed an experiment. There is one factor (gene therapy), at two levels (gene therapy and no gene therapy), resulting in two treatments. The experiment is completely randomized. The response variable is heart muscle condition. The researchers can conclude that gene therapy is responsible for stabilizing heart muscle in laboratory rats.

14. The researchers performed an observational study that was neither prospective nor retrospective. There appears to be a relationship between eye diameter and time of singing.

15. The orange juice plant depends on sampling to ensure the oranges are suitable for juice. The population is all of the oranges on the truck, and the parameter of interest is the proportion of unsuitable oranges. The procedure used is a random sample, stratified by location in the truck. Using this well-chosen sample, the workers at the plant can estimate the proportion of unsuitable oranges on the truck, and decide whether or not to accept the load.

16. The soft drink manufacture is sampling in order to determine whether or not the machine that caps the bottles is working properly. The population is all of the bottle cap seals. The parameter of interest is the whether or not the bottles are sealing properly. They are using a systematic sample, checking bottles at fixed intervals. If any bottles in the sample are not sealed properly, they can tell that the machine may need adjustment or repair.

17. The researchers performed a prospective observational study, since the subjects were identified ahead of time. Physically fit men may have a lower risk of death from cancer than other men.

18. This statistics professor is performing an experiment, blocked by whether or not the students have taken calculus. However, there is probably no randomization, since students usually select their own courses. Hopefully, the two sections contain similar groups of students. There is one factor (use of software), at two levels (software and no software), resulting in two treatments. The response variable is the final exam score. The experiment incorporates blinding, since the graders do not know which students used software and which did not. The professor can decide if computer software is beneficial, and if so, determine whether or not calculus students perform differently than those who have not had calculus.

19. Point spread.

Answers may vary. Perform a simulation to determine the gambler's expected winnings. A component is one game. To model that component, generate random digits 0 to 9. Since the outcome after the point spread is a tossup, let digits 0-4 represent a loss, and let digits 5-9 represent a win. A run consists of 5 games, so generate 5 random digits at a time. The response variable is the profit the gambler makes, after accounting for the $10 bet. If the outcome of the run is 0, 1, or 2 simulated wins, the profit is −$10. If the outcome is 3, 4, or 5 simulated wins, the profit is $0, $10, or $40, respectively. The total profit divided by the number of runs is the average weekly profit. According to the simulation (80 runs were performed), the gambler is expected to break even. His simulated losses equaled his simulated winnings.

20. The lottery.

a) Answers may vary. Perform a simulation to determine the number of plays required to win. Pick 3 numbers from 1 to 20. (These don't need to be randomly generated. Players of the lottery aren't required to pick randomly, so there is no reason we should!) Let's use 1, 2, and 3 to keep it simple. A component is the selection of 1 winning number. Simulate the winning number by generating a random pair of digits from 01 to 20. Depending on the type of lottery you simulate, repeated numbers may have to be ignored. Some lotteries choose the numbers from 5 different sets of numbers, while others choose 5 numbers from a single set of numbers. A run consists of 5 winning numbers, so generate 5 such pairs per run. The response variable is whether or not the numbers 1, 2, and 3 appear in the run. Simply count the number of runs it takes to simulate a win. Just for fun, I performed a couple hundred runs of this simulation, and never got a match for my 3 numbers. You are very unlikely to get a match.

b) With more numbers from which to choose, and more matches required to win, the odds of winning go down dramatically. Winning a state lottery is highly improbable.

21. Everyday randomness.

Answers will vary. Most of the time, events described as "random" are anything but truly random.

22. Cell phone risks.

a) This is an experiment, since treatments were imposed on randomly assigned groups. There is one factor (radio waves), at three levels (digital cell phone radio waves, analog cell phone radio waves, and no radio waves), resulting in three treatments. The response variable is the incidence of brain tumors.

b) The differences in the incidence of tumors between the groups of rats were not great enough to convince the researchers that the differences were due to anything other than sampling variability.

c) Since the research was funded by Motorola, there may have been bias present. The researchers may have felt pressure to declare cell phones safe, and introduced bias, intended or not, into the experiment.

23. Tips.

a) The waiters performed an experiment, since treatments were imposed on randomly assigned groups. This experiment has one factor (candy), at two levels (candy or no candy), resulting in two treatments. The response variable is the percentage of the bill given as a tip.

b) If the decision whether to give candy or not was made before the people were served, the server may have subconsciously introduced bias by treating the customers better. If the decision was made just before the check was delivered, then it is reasonable to conclude that the candy was the cause of the increase in the percentage of the bill given as a tip.

c) "Statistically significant" means that the difference in the percentage of tips between the candy and no candy groups was more than expected due to sampling variability.

24. Tips, take 2.

a) A diagram of the tipping experiment appears below.

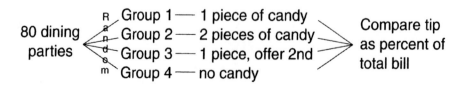

b) This experiment has 1 factor (candy), at 4 levels (1 piece, 2 pieces, 1 piece with an additional piece offered, and no candy).

c) 1 factor at 4 levels results in 4 treatments.

d) The response variable is the percent of the total bill left as a tip.

e) The diners were not aware that they were part of an experiment, so the experiment was blinded. This experiment did not use double-blinding, but there is probably no way to double blind this experiment, since there is no need to blind the evaluator of the response variable. Biased evaluation of the amount of tip left doesn't seem possible.

f) If the waitress knew which tables were going to receive certain treatments, then she might have treated some tables better than others. The waitress should be unaware of the treatment until after the meal, to avoid the introduction of bias. (Note: This is still only single-blinding. Blinding does not refer to the number of people blinded, but rather the type of blinding employed. If the diners and the waitress are unaware of the assignment of treatments, it is for the same purpose, namely to keep the diners from being systematically influenced.)

25. Cloning.

a) The *USA Weekend* survey suffers from voluntary response bias. Only those who feel strongly will pay for the 900 number phone call.

b) Answers may vary. A strong positive response might be created with the question, "If it would help future generations live longer, healthier lives, would you be in favor of human cloning?"

26. Laundry.

a) Answers may vary. Water (quality and temperature) and material can vary. These confounding variables may influence results. The treatments in the experiment must be in environments that are identical, with the exception of the factor being studied.

b) These conditions are unrealistic. This will not help the experimenters determine how well *Sparklekleen* will work in the conditions for which it was intended.

c) If the swatches are stained at the same time, the stains on the swatches washed later will have more time to "set in", causing bias towards *Sparklekleen*. Also, unforeseen variables, like changes in water temperature or pressure won't be equalized through randomization.

d) The conditions under which the *Sparklekleen* was tested are unknown. There is no way to keep the conditions comparable. Furthermore, the company that produced *Sparklekleen* may not produce reliable data. They have a vested interest in the success of their product.

27. When to stop?

a) Answers may vary. A component in this simulation is rolling 1 die. To simulate this component, generate a random digit 1 to 6. To simulate a run, simulate 4 rolls, stopping if a 6 is rolled. The response variable is the sum of the 4 rolls, or 0 if a 6 is rolled. The average number of points scored is the sum of all rolls divided by the total number of runs. According to the simulation, the average number of points scored will be about 5.8.

b) Answers may vary. A component in this simulation is rolling 1 die. To simulate this component, generate a random digit 1 to 6. To simulate a run, generate random digits until the sum of the digits is at least 12, or until a 6 is rolled. The response variable is the sum of the digits, or 0 if a 6 is rolled. The average number of points scored is the sum of all rolls divided by the total number of runs. According to the simulation, the average number of points scored will be about 5.8, similar to the outcome of the method described in part a).

c) Answers may vary. Be careful when making your decision about the effectiveness of your strategy. If you develop a strategy with a higher simulated average number of points than the other two methods, this is only an indication that you may win in the long run. If the game is played round by round, with the winner of a particular round being declared as the player with the highest roll made during that round, the game is much more variable. For example, if Player B rolls a 12 in a particular game, Player A will always lose that game, provided he or she sticks to the strategy. A better way to get a feel for your chances of winning this type of game might be to simulate several rounds, recording whether each player won or lost the round. Then estimate the percentage of the time that each player is expected to win, according to the simulation.

28. Rivets

a)

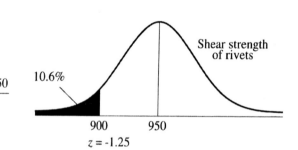

$$z = \frac{x - \mu}{\sigma}$$

$$z = \frac{900 - 950}{40}$$

$$z = -1.25$$

According to the Normal model, approximately 10.6% of rivets are expected to break when tested under a 900-pound load.

b) Answers may vary. The component being simulated is whether or not a rivet will break under a 900-pound load. To model this component, generate triples of random digits, from 000 to 999. Let digits 001 to 106 represent a broken rivet under a 900-pound load. (001 to 106 represents the 10.6% probability of breaking.) Let other triples of digits represent unbroken rivets under a 900-pound load. To simulate a run, generate a triple of random digits. The response variable is whether or not the rivet will break under a 900-pound load. Count the number of simulated rivets until 3 broken rivets are simulated. According to the simulation, you might need to test about 28 rivets before finding 3 that fail at 900-pounds or below.

29. Homecoming.

a) Since telephone numbers were generated randomly, every number that could possibly occur in that community had an equal chance of being selected. This method is "better" than using the phone book, because unlisted numbers are also possible. Those community members who deliberately do not list their phone numbers might not consider this method "better"!

b) Although this method results in a simple random sample of phone numbers, it does not result in a simple random sample of residences. Residences without a phone are excluded, and residences with more than one phone have a greater chance of being included.

c) No, this is not a SRS of local voters. People who respond to the survey may be of the desired age, but not registered to vote. Additionally, some voters who are contacted may choose not to participate.

d) This method does not guarantee an unbiased sample of households. Households in which someone answered the phone may be more likely to have someone at home when the phone call was generated. The attitude about homecoming of these households might not be the same as the attitudes of the community at large.

30. Youthful appearance.

a) The differences is guessed age were greater than differences that could be explained by natural sampling variability.

b) Dr. Weeks is implying that having sex caused the people to have a more youthful appearance. It seems more plausible that younger-looking people are more sexually active than older-looking people, because of their age.

31. Smoking and Alzheimer's.

a) The studies do not prove that smoking offers any protection from Alzheimer's. The studies merely indicate an association. There may be other variables that can account for this association.

b) Alzheimer's usually shows up late in life. Since smoking is known to be harmful, perhaps smokers have died of other causes before Alzheimer's can be seen.

c) The only way to establish a cause-and-effect relationship between smoking and Alzheimer's is to perform a controlled, randomized, and replicated experiment. This is unlikely to ever happen, since the factor being studied, smoking, has already been proven harmful. It would be unethical to impose this treatment on people for the purposes of this experiment. A prospective observational study could be designed in which groups of smokers and nonsmokers are followed for many years and the incidence of Alzheimer's disease is tracked.

32. Antacids.

a) This is a randomized experiment, blocked by gender.

b) Experiments always use volunteers. This is not a problem, since experiments are testing response to a treatment, not attempting to determine an unknown population parameter. The randomization in an experiment is random assignment to treatment groups, not random selection from a population.

c) Since the experiment is studying the effects of an antacid, the placebo may actually confound the experiment, since the introduction of _any_ substance, even a sugar pill, into the digestive system may have an effect on acid reflux. (The use of some sort of placebo is always recommended, but in some cases it may be difficult to find a placebo that truly has no effect, beyond the expected "placebo effect", of course!)

33. Sex and violence.

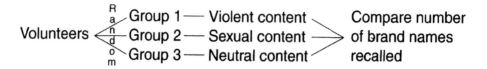

This experiment has one factor (program content), at three levels (violent, sexual, and neutral), resulting in three treatments. The response variable is the number of brand names recalled after watching the program. Numerous subjects will be randomly assigned to see shows with violent, sexual, or neutral content. They will see the same commercials. After the show, they will be interviewed for their recall of brand names in the commercials.

34. Pubs.

a) *Who* – 900 Englishmen. *What* – The researcher was interested in their reasons for going to the pub. *When* – not stated. *Where* – England. *Why* – The producers of Kaliber alcohol-free beer hoped to show that men went to the pub for reasons other than the alcohol. *How* – Researchers surveyed men regarding their reasons for going to the pub.

b) The researcher surveyed believes that the population is all Englishmen.

c) The most important omitted detail is the selection process. How did the researcher acquire the sample of men? Was randomness used? Is the sample representative of the population of Englishmen?

d) Although not stated, it appears that the researcher simply took convenience samples of those in the pubs.

e) The results may be biased. First of all, an alcohol-free beer producer funded the survey. Respondents may have felt subconscious pressure to indicate that alcohol was not the primary reason for going to the pub. Additionally, admitting that you go to the pub merely for the alcohol is a potentially embarrassing admission. The percentage of pub patrons who go for the alcohol may be significantly higher than 10%.

35. Age and party.

a) The number of respondents is roughly the same for each age category. This may indicate a sample stratified by age category, although it may be a simple random sample.

b) 1404 Democrats were surveyed. $\frac{1404}{4002} \approx 35.1\%$ of the people surveyed were Democrats.

c) If this poll is truly representative, this is probably a good estimate of the percentage of voters who are Democrats. However, telephone surveys are likely to systematically exclude those who do not have phones, a group that is likely to be composed of people who cannot afford phones. If socioeconomic status is linked to political affiliation, this might be a problem.

d) The pollsters were probably attempting to determine whether or not political party is associated with age.

36. Bias?

a) Barone claims that nonresponse bias exists in polls, since conservatives are more likely to refuse to participate than other groups. Nonresponse is less of an issue if it is believed that all groups fail to respond at the same rate.

b) The population of interest is all adults in the United States.

c) The column totals do not add up to 100%. There is information missing, and the discrepancies are too large to be attributed to rounding.

d) The differences observed are similar to differences that may have been observed simply due to natural sampling variability. Differences of this size would be probable, even if no bias exists.

37. Save the grapes.

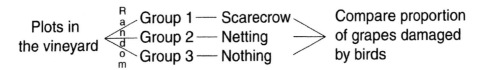

This experiment has one factor (bird control device), at three levels (scarecrow, netting, and no device), resulting in three treatments. Randomly assign different plots in the vineyard to the different treatments, making sure to ensure adequate separation of plots, so that the possible effect of the scarecrow will not be confounded with the other treatments. The response variable to be measured at the end of the season is the proportion of bird-damaged grapes in each plot.

38. Bats.

Answers may vary. This experiment has one factor (type of bat), at two levels (wooden and aluminum), resulting in two treatments. The response variable is the difference is distance the ball is hit with each type of bat. Since players vary in their ability to hit the ball, a matched design should be used, with each batter hitting with both types of bats several times in a randomly chosen order. For each batter, calculate the average difference in distance between the aluminum and wooden bats.

39. Knees.

a) In an experiment, all subjects must be treated as alike as possible. If there were no "placebo surgery", subjects would know that they had not received the operation, and might react differently. Of course, all volunteers for the experiment must be aware of the possibility of being randomly assigned to the placebo group.

b) Experiments always use volunteers. This is not a problem, since experiments are testing response to a treatment, not attempting to determine an unknown population parameter. The randomization in an experiment is random assignment to treatment groups, not random selection from a population. Voluntary response is a problem when sampling, but is not an issue in experimentation..

c) There were differences in the amount of pain relief experienced by the two groups, but these differences were small enough that they could be explained by natural sampling variation, even if surgery were not a factor.

40. NBA draft lottery.

Answers will vary. A component in the simulation is drawing one lottery card. To simulate this component, generate random numbers 01 to 66.

Let numbers 01 to 11 represent the team with the worst record, numbers 12 to 21 represent the team with the second worst record, 22 to 30 represent the team with the third worst record (your team), numbers 31 to 38 represent the fourth worst team, numbers 39 to 45 represent the fifth worst team, numbers 46 to 51 representing the sixth worst team, 52 to 56 represent the seventh worst team, 57 to 60 represent the eighth team, 61 to 63 represent the ninth worst team, 64 and 65 represent the tenth worst team, and 66 represent the team with the best record of the teams not making the playoffs.

A run consists of the assignment of one draft pick position to each of the 11 teams. (You can stop after 2 assignments. We are only concerned with whether or not our team gets to pick first or second.) The response variable is whether or not your team receives first or second pick. To simulate a run, generate a random number. If that number is 22 to 30, stop and record a success. If that number is not 22 to 30, note which team received the first pick and generate another random number, ignoring any number that corresponds to the team receiving first pick. If the second number is 22 to 30, record a success. Otherwise record a failure.

The probability that your team gets first or second pick is the total number of successes divided by the number of runs. According to the simulation, your team should get first or second pick approximately 30% of the time.

41. Security.

a) To ensure that passengers from first-class, as well as coach, get searched, select 2 passengers from first-class and 12 from coach. Using this stratified random sample, 10% of the first-class passengers are searched, as are 10% of the coach passengers.

b) Answers will vary. Number the passengers alphabetically, with 2-digit numbers. Bergman = 01, Bowman = 02, and so on, ending with Testut = 20. Read the random digits in pairs, ignoring pairs 21 to 99 and 00, and ignoring repeated pairs.

```
65|43|67|11|          27|04|
XX XX XX Fontana      XX Castillo
```
The passengers selected for search from first-class are Fontana and Castillo.

c) Number the passengers alphabetically, with 3 digit numbers, 001 to 120. Use the random number table to generate 3-digit numbers, ignoring numbers 121 to 999 and 000, and ignoring repeated numbers. Search the passengers corresponding to the first 12 valid numbers generated.

42. Profiling?

Answers will vary. A component in this simulation is the selection of a passenger to be searched. To simulate this component, generate pairs of random digits 01 to 20. Let 01 to 04 represent the businessmen from the Middle East, and let numbers 05 to 20 represent other passengers. A run consists of the selection of 2 passengers for search. To simulate a run, generate 2 pairs of random digits, ignoring repeats. The response variable is whether of not both passengers selected are Middle Eastern businessmen. If both pairs generated are from 01 to 04, record a "success". Otherwise, record a "failure". The total number of successes divided by the total number of runs is the probability that both passengers selected are Middle Eastern businessmen. According to the simulation, this should happen about 15% of the time. Although relatively small, this percentage does not indicate an event that is extremely unlikely. We can't be certain that the Middle Eastern businessmen were not "profiled", but there is little evidence suggesting that they were.

43. Par 4.

Answers may vary. A component in this simulation is a shot. Use pairs of random digits 00 to 99 to represent a shot. The way in which this component is simulated depends on the type of shot.

For the first shot, let pairs of digits 01 to 70 represent hitting the fairway, and let pairs of digits 71 to 99, and 00, represent not hitting the fairway.

If the first simulated shot hits the fairway, let 01 to 80 represent landing on the green on the second shot, and let 81 to 99, and 00, represent not landing on the green on the second shot. If the first simulated shot does not hit the fairway, let 01 to 40 represent landing on the green on the second shot, and let 41 to 99, and 00, represent not landing on the green on the second shot.

If the second simulated shot does not land on the green, let 01 to 90 represent landing on the green, and 91 to 99, and 00, represent not landing on the green. Keep simulating shots until the shot lands on the green.

Once on the green, let 01 to 20 represent sinking the putt on the first putt, and let 21 to 99, and 00, represent not sinking the putt on the first putt. If second putts are required, continue simulating putts until a putt goes in, with 01 to 90 representing making the putt, and 91 to 99, and 00, representing not making the putt.

A run consists of following the guidelines above until the final putt is made. The response variable is the number of shots required until the final putt is made.

The simulated average score on the hole is the total number of shots required divided by the total number of runs. According to 40 runs of this simulation, a pretty good golfer can be expected to average about 4.2 strokes per hole. Your simulation results may vary.

44. The back nine.

a) Answers may vary. A component in this simulation is a shot. Use pairs of random digits 00 to 99 to represent a shot. The way in which this component is simulated depends on the type of shot.

For the first shot, let pairs of digits 01 to 80 represent hitting the fairway, and let pairs of digits 81 to 99, and 00, represent not hitting the fairway.

If the first simulated shot hits the fairway, let 01 to 80 represent landing on the green on the second shot, and let 81 to 99, and 00, represent not landing on the green on the second shot. If the first simulated shot does not hit the fairway, let 01 to 40 represent landing on the green on the second shot, and let 41 to 99, and 00, represent not landing on the green on the second shot.

If the second simulated shot does not land on the green, let 01 to 90 represent landing on the green, and 91 to 99, and 00, represent not landing on the green. Keep simulating shots until the shot lands on the green.

Once on the green, let 01 to 20 represent sinking the putt on the first putt, and let 21 to 99, and 00, represent not sinking the putt on the first putt. If second putts are required, continues simulating putts until a putt goes in, with 01 to 90 representing making the putt, and 91 to 99, and 00, representing not making the putt.

A run consists of following the guidelines above until the final putt is made. The response variable is the number of shots required until the final putt is made.

The simulated average score on the hole is the total number of shots required divided by the total number of runs. According to 20 runs of this simulation, a pretty good golfer can be expected to average about 3.7 strokes per hole. Your simulation results may vary.

b) Answers may vary. The simulation is set up identically to part a), with the exception of the second shot. Now, let 01 to 10 represent hitting the green, and let 11 to 99, and 00, represent not hitting the green.

According to 20 runs of this simulation, a pretty good golfer can be expected to average about 5.3 strokes per hole. Your simulation results may vary.

c) Answers may vary.

Chapter 14 – From Randomness to Probability

1. Roulette.

If a roulette wheel is to be considered truly random, then each outcome is equally likely to occur, and knowing one outcome will not affect the probability of the next. Additionally, there is an implication that the outcome is not determined through the use of an electronic random number generator.

2. Rain.

When a weather forecaster makes a prediction such as a 25% chance of rain, this means that when weather conditions are like they are now, rain happens 25% of the time in the long run.

3. Winter.

Although acknowledging that there is no law of averages, Knox attempts to use the law of averages to predict the severity of the winter. Some winters are harsh and some are mild over the long run, and knowledge of this can help us to develop a long-term probability of having a harsh winter. However, probability does not compensate for odd occurrences in the short term. Suppose that the probability of having a harsh winter is 30%. Even if there are several mild winters in a row, the probability of having a harsh winter is still 30%.

4. Rain.

The radio announcer is referring to the "law of averages", which is not true. Probability does not compensate for deviations from the expected outcome in the recent past. The weather is not more likely to be bad later on in the winter because of a few sunny days in autumn. The weather makes no conscious effort to even things out, which is what the announcer's statement implies.

5. Cold streak.

There is no such thing as being "due for a hit". This statement is based on the so-called law of averages, which is a mistaken belief that probability will compensate in the short term for odd occurrences in the past. The batter's chance for a hit does not change based on recent successes or failures.

6. Crash.

a) There is no such thing as the "law of averages". The overall probability of an airplane crash does not change due to recent crashes.

b) Again, there is no such thing as the "law of averages". The overall probability of an airplane crash does not change due to a period in which there were no crashes. It makes no sense to say a crash is "due". If you say this, you are expecting probability to compensate for strange events in the past.

7. Fire insurance.

a) It would be foolish to insure your neighbor's house for $300. Although you would probably simply collect $300, there is a chance you could end up paying much more than $300. That risk probably is not worth the $300.

b) The insurance company insures many people. The overwhelming majority of customers pay the insurance and never have a claim. The few customers who do have a claim are offset by the many who simply send their premiums without a claim. The relative risk to the insurance company is low.

8. Jackpot.

a) The Desert Inn can afford to give away millions of dollars on a $3 bet because almost all of the people who bet do not win the jackpot.

b) The press release generates publicity, which entices more people to come and gamble. Of course, the casino wants people to play, because the overall odds are always in favor of the casino. The more people who gamble, the more the casino makes in the long run. Even if that particular slot machine has paid out more than it ever took in, the publicity it gives to the casino more than makes up for it.

9. Spinner.

a) This is a legitimate probability assignment. Each outcome has probability between 0 and 1, inclusive, and the sum of the probabilities is 1.

b) This is a legitimate probability assignment. Each outcome has probability between 0 and 1, inclusive, and the sum of the probabilities is 1.

c) This is not a legitimate probability assignment. Although each outcome has probability between 0 and 1, inclusive, the sum of the probabilities is greater than 1.

d) This is a legitimate probability assignment. Each outcome has probability between 0 and 1, inclusive, and the sum of the probabilities is 1. However, this game is not very exciting!

e) This probability assignment is not legitimate. The sum of the probabilities is 0, and there is one probability, –1.5 , that is not between 0 and 1, inclusive.

10. Scratch off.

a) This is not a legitimate assignment. Although each outcome has probability between 0 and 1, inclusive, the sum of the probabilities is less than 1.

b) This is not a legitimate probability assignment. Although each outcome has probability between 0 and 1, inclusive, the sum of the probabilities is greater than 1.

c) This is a legitimate probability assignment. Each outcome has probability between 0 and 1, inclusive, and the sum of the probabilities is 1.

d) This probability assignment is not legitimate. Although the sum of the probabilities is 1, there is one probability, –0.25 , that is not between 0 and 1, inclusive.

e) This is a legitimate probability assignment. Each outcome has probability between 0 and 1, inclusive, and the sum of the probabilities is 1. This is also known as a 10% off sale!

11. Car repairs.

  a) Since all of the events listed are disjoint, the addition rule can be used.

   1. P(no repairs) = 1 − P(some repairs) = 1 − (0.17 + 0.07 + 0.04) = 1 − (0.28) = 0.72

   2. P(no more than one repair) = P(no repairs or one repair) = 0.72 + 0.17 = 0.89

   3. P(some repairs) = P(one or two or three or more repairs) = 0.17 + 0.07 + 0.04 = 0.28

  b) Since repairs on the two cars are independent from one another, the multiplication rule can be used. Use the probabilities of events from part a) in the calculations.

   1. P(neither will need repair) = (0.72)(0.72) = 0.5184

   2. P(both will need repair) = (0.28)(0.28) = 0.0784

12. Stats projects.

  a) Since all of the events listed are disjoint, the addition rule can be used.

   1. P(some Calculus) = P(one semester or two or more semesters) = 0.32 + 0.13 = 0.45

   2. P(no more than one semester) = P(no Calculus or one semester) = 0.55 + 0.32 = 0.87

  b) Since students with Calculus backgrounds are independent from one another, use the multiplication rule. Use the probabilities of events from part a) in the calculations.

   1. P(neither has studied Calculus) = (0.55)(0.55) = 0.3025

   2. P(both have studied at least one semester of Calculus) = (0.45)(0.45) = 0.2025

   3. P(at least one has had more than one semester of Calculus)
      = 1 − P(neither has studied more than one semester of Calculus)
      = 1 − (0.87)(0.87) = 0.2431

13. M&M's

  a) Since all of the events are disjoint (an M&M can't be two colors at once!), use the addition rule where applicable.

   1. P(brown) = 1 − P(not brown) = 1 − P(yellow or red or orange or blue or green)
      = 1 − (0.20 + 0.20 + 0.10 + 0.10 +0.10) = 0.30

   2. P(yellow or orange) = 0.20 + 0.10 = 0.30

   3. P(not green) = 1 − P(green) = 1 − 0.10 = 0.90

   4. P(striped) = 0

  b) Since the events are independent (picking out one M&M doesn't affect the outcome of the next pick), the multiplication rule may be used.

   1. P(all three are brown) = (0.30)(0.30)(0.30) = 0.027

   2. P(the third one is the first one that is red) = P(not red, not red, red)
      = (0.80)(0.80)(0.20) = 0.128

   3. P(none are yellow) = P(not yellow, not yellow, not yellow) = (0.80)(0.80)(0.80) = 0.512

   4. P(at least one is green) = 1 − P(none are green) = 1 − (0.90)(0.90)(0.90) = 0.271

14. Blood.

a) Since all of the events are disjoint (a person cannot have more than one blood type!), use the addition rule where applicable.

   1. $P(\text{Type AB}) = 1 - P(\text{not Type AB}) = 1 - P(\text{Type O or Type A or Type B})$
      $= 1 - (0.45 + 0.40 + 0.11) = 0.04$

   2. $P(\text{Type A or Type B}) = 0.40 + 0.11 = 0.51$

   3. $P(\text{not Type O}) = 1 - P(\text{Type O}) = 1 - 0.45 = 0.55$

b) Since the events are independent (one person's blood type doesn't affect the blood type of the next), the multiplication rule may be used.

   1. $P(\text{all four are Type O}) = (0.45)(0.45)(0.45)(0.45) \approx 0.041$

   2. $P(\text{no one is Type AB}) = P(\text{not AB, not AB, not AB, not AB})$
      $= (0.96)(0.96)(0.96)(0.96) \approx 0.849$

   3. $P(\text{they are not all Type A}) = 1 - P(\text{all Type A}) = 1 - (0.40)(0.40)(0.40)(0.40) = 0.9744$

   4. $P(\text{at least one person is Type B}) = 1 - P(\text{no one is Type B})$
      $= 1 - (0.89)(0.89)(0.89)(0.89) \approx 0.373$

15. Disjoint or independent?

a) For one draw, the events of getting a red M&M and getting an orange M&M are disjoint events. Your single draw cannot be both red and orange.

b) For two draws, the events of getting a red M&M on the first draw and a red M&M on the second draw are independent events. Knowing that the first draw is red does not influence the probability of getting a red M&M on the second draw.

c) Disjoint events can never be independent. Once you know that one of a pair of disjoint events has occurred, the other one cannot occur, so its probability has become zero. For example, consider drawing one M&M. If it is red, it cannot possible be orange. Knowing that the M&M is red influences the probability that the M&M is orange. It's zero. The events are not independent.

16. Disjoint or independent.

a) For one person, the events of having Type A blood and having Type B blood are disjoint events. One person cannot be have both Type A and Type B blood.

b) For two people, the events of the first having Type A blood and the second having Type B blood are independent events. Knowing that the first person has Type A blood does not influence the probability of the second person having Type B blood.

c) Disjoint events can never be independent. Once you know that one of a pair of disjoint events has occurred, the other one cannot occur, so its probability has become zero. For example, consider selecting one person, and checking his or her blood type. If the person's blood type is Type A, it cannot possibly be Type B. Knowing that the person's blood type is Type A influences the probability that the person's blood type is Type B. It's zero. The events are not independent.

17. Dice.

a) $P(6) = \dfrac{1}{6}$, so $P(\text{all 6's}) = \left(\dfrac{1}{6}\right)\left(\dfrac{1}{6}\right)\left(\dfrac{1}{6}\right) \approx 0.005$

b) $P(\text{odd}) = P(1 \text{ or } 3 \text{ or } 5) = \dfrac{3}{6}$, so $P(\text{all odd}) = \left(\dfrac{3}{6}\right)\left(\dfrac{3}{6}\right)\left(\dfrac{3}{6}\right) = 0.125$

c) $P(\text{not divisible by 3}) = P(1 \text{ or } 2 \text{ or } 4 \text{ or } 5) = \dfrac{4}{6}$

   $P(\text{none divisible by 3}) = \left(\dfrac{4}{6}\right)\left(\dfrac{4}{6}\right)\left(\dfrac{4}{6}\right) \approx 0.296$

d) $P(\text{at least one 5}) = 1 - P(\text{no 5's}) = 1 - \left(\dfrac{5}{6}\right)\left(\dfrac{5}{6}\right)\left(\dfrac{5}{6}\right) \approx 0.421$

e) $P(\text{not all 5's}) = 1 - P(\text{all 5's}) = 1 - \left(\dfrac{1}{6}\right)\left(\dfrac{1}{6}\right)\left(\dfrac{1}{6}\right) \approx 0.995$

18. Slot Machine.

Each wheel runs independently of the others, so the multiplication rule may be used.

a) $P(\text{lemon on 1 wheel}) = 0.30$, so $P(\text{3 lemons}) = (0.30)(0.30)(0.30) = 0.027$

b) $P(\text{bar or bell on 1 wheel}) = 0.50$, so $P(\text{no fruit symbols}) = (0.50)(0.50)(0.50) = 0.125$

c) $P(\text{bell on 1 wheel}) = 0.10$, so $P(\text{3 bells}) = (0.10)(0.10)(0.10) = 0.001$

d) $P(\text{no bell on 1 wheel}) = 0.90$, so $P(\text{no bells on 3 wheels}) = (0.90)(0.90)(0.90) = 0.729$

e) $P(\text{no bar on 1 wheel}) = 0.60$.
    $P(\text{at least one bar on 3 wheels}) = 1 - P(\text{no bars}) = 1 - (0.60)(0.60)(0.60) = 0.784$

19. Champion bowler.

Assuming each frame is independent of others, so the multiplication rule may be used.

a) $P(\text{no strikes in 3 frames}) = (0.30)(0.30)(0.30) = 0.027$

b) $P(\text{makes first strike in the third frame}) = (0.30)(0.30)(0.70) = 0.063$

c) $P(\text{at least one strike in the first three frames}) = 1 - P(\text{no strikes}) = 1 - (0.30)^3 = 0.973$

d) $P(\text{perfect game}) = (0.70)^{12} \approx 0.014$

20. The train.

Assuming the arrival time is independent from one day to the next, the multiplication rule may be used.

a) $P(\text{gets stopped Monday and again on Tuesday}) = (0.15)(0.15) = 0.0225$

b) $P(\text{gets stopped for the first time on Thursday}) = (0.85)(0.85)(0.85)(0.15) \approx 0.092$

c) $P(\text{gets stopped every day}) = (0.15)^5 \approx 0.00008$

d) $P(\text{gets stopped at least once}) = 1 - P(\text{never gets stopped}) = 1 - (0.85)^5 \approx 0.556$

21. Voters.

Since you are calling at random, one person's political affiliation is independent of another's. The multiplication rule may be used.

a) P(all Republicans) = $(0.29)(0.29)(0.29) \approx 0.024$

b) P(no Democrats) = $(1 - 0.37)(1 - 0.37)(1 - 0.37) \approx 0.25$

c) P(at least one Independent) = $1 - P$(no Independents) = $1 - (0.77)(0.77)(0.77) \approx 0.543$

22. Religion.

Since you are calling at random, one person's religion is independent of another's. The multiplication rule may be used.

a) P(all Christian) = $(0.62)(0.62)(0.62)(0.62) \approx 0.148$

b) P(no Jews) = $(1 - 0.12)(1 - 0.12)(1 - 0.12)(1 - 0.12) \approx 0.600$

c) P(at least one person who is nonreligious) = $1 - P$(no nonreligious people)

  = $1 - (0.90)(0.90)(0.90)(0.90) = 0.3439$

23. Tires.

Assume that the defective tires are distributed randomly to all tire distributors so that the events can be considered independent. The multiplication rule may be used.

P(at least one of four tires is defective) = $1 - P$(none are defective)

= $1 - (0.98)(0.98)(0.98)(0.98) \approx 0.078$

24. Pepsi.

Assume that the winning caps are distributed randomly, so that the events can be considered independent. The multiplication rule may be used.

P(you win something) = $1 - P$(you win nothing) = $1 - (0.90)^6 \approx 0.469$

25. 9/11?

a) For any date with a valid three-digit date, the chance is 0.001, or 1 in 1000. For many dates in October through December, the probability is 0. For example, there is no way three digits will make 1015, to match October 15.

b) There are 65 days when the chance to match is 0. (October 10 through October 31, November 10 through November 30, and December 10 through December 31.) That leaves 300 days in a year (that is not a leap year) in which a match might occur. P(no matches in 300 days) = $(0.999)^{300} \approx 0.741$.

c) P(at least one match in a year) = $1 - P$(no matches in a year) = $1 - 0.741 \approx 0.259$

d) P(at least one match on 9/11 in one of the 50 states)
   = $1 - P$(no matches in 50 states) = $1 - (0.999)^{50} \approx 0.049$

Chapter 15 – Probability Rules!

1. **Sample spaces.**

   a) S = { HH, HT, TH, TT} All of the outcomes are equally likely to occur.

   b) S = { 0, 1, 2, 3} All outcomes are not equally likely. A family of 3 is more likely to have, for example, 2 boys than 3 boys. There are three equally likely outcomes that result in 2 boys (BBG, BGB, and GBB), and only one that results in 3 boys (BBB).

   c) S = { H, TH, TTH, TTT} All outcomes are not equally likely. For example the probability of getting heads on the first try is $\frac{1}{2}$. The probability of getting three tails is $\left(\frac{1}{2}\right)^3 = \frac{1}{8}$.

   d) S = {1, 2, 3, 4, 5, 6} All outcomes are not equally likely. Since you are recording only the larger number of two dice, 6 will be the larger when the other die reads 1, 2, 3, 4, or 5. The outcome 2 will only occur when the other die shows 1 or 2.

2. **Sample spaces.**

   a) S = { 2, 3, 4, 5, 6, 7, 8, 9, 10, 11, 12} All outcomes are not equally likely. For example, there are four equally likely outcomes that result in a sum of 5 (1 + 4, 4 + 1, 2 + 3, and 3 + 2), and only one outcome that results in a sum of 2 (1 + 1).

   b) S = {BBB, BBG, BGB, BGG, GBB, GBG, GGB, GGG} All outcomes are equally likely.

   c) S = { 0, 1, 2, 3, 4} All outcomes are not equally likely. For example, there are 4 equally likely outcomes that produce 1 tail (HHHT, HHTH, HTHH, and THHH), but only one outcome that produces 4 tails (TTTT).

   d) S = { 0, 1, 2, 3, 4, 5, 6, 7, 8, 9, 10} All outcomes are not equally likely. A string of 3 heads is much more likely to occur than a string of 10 heads in a row.

3. **Homes.**

   Construct a Venn diagram of the disjoint outcomes. .

   a) *P*(pool or garage) = *P*(pool) + *P*(garage) – *P*(pool and garage)
      = 0.64 + 0.21 – 0.17 = 0.68

   Or, from the Venn: 0.47 + 0.17 + 0.04 = 0.68

   b) *P*(neither)= 1 – *P*(pool or garage) = 1 – 0.68 = 0.32

   Or, from the Venn: 0.32 (the region outside the circles)

   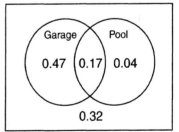

   c) *P*(pool and no garage) = *P*(pool) – *P*(pool and garage) = 0.21 – 0.17 = 0.04

   Or, from the Venn: 0.04 (the region inside pool circle, yet outside garage circle)

4. **Travel.**

    Construct a Venn diagram of the disjoint outcomes.

    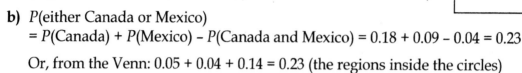

    a) *P*(Canada and not Mexico)
       = *P*(Canada) – *P*(Canada and Mexico) = 0.18 – 0.04 = 0.14

       Or, from the Venn: 0.14
       (region inside the Canada circle, yet outside the Mexico circle)

    b) *P*(either Canada or Mexico)
       = *P*(Canada) + *P*(Mexico) – *P*(Canada and Mexico) = 0.18 + 0.09 – 0.04 = 0.23

       Or, from the Venn: 0.05 + 0.04 + 0.14 = 0.23 (the regions inside the circles)

    c) *P*(neither Canada nor Mexico) = 1 – *P*(either Canada or Mexico) = 1 – 0.23 = 0.77

       Or, from the Venn: 0.77 (the region outside the circles)

5. **Amenities.**

    Construct a Venn diagram of the disjoint outcomes.

    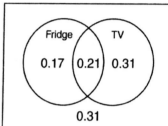

    a) *P*(TV and no refrigerator)   = *P*(TV) – *P*(TV and refrigerator)
                                      = 0.52 – 0.21 = 0.31

       Or, from the Venn: 0.31
       (the region inside the TV circle, yet outside the Fridge circle)

    b) *P*(refrigerator or TV, but not both) =
       = [*P*(refrigerator) – *P*(refrigerator and TV)] +[*P*(TV) – *P*(refrigerator and TV)]
       = [0.38 – 0.21] + [0.52 – 0.21] = 0.48

       This problem is much easier to visualize using the Venn diagram. Simply add the
       probabilities in the two regions for Fridge only and TV only.
       P(refrigerator or TV, but not both) = 0.17 + 0.31 = 0.48

    c) *P*(neither TV nor refrigerator) = 1 – *P*(either TV or refrigerator)
                                         = 1 – [*P*(TV) + *P*(refrigerator) – *P*(TV and refrigerator)]
                                         = 1 – [0.52 + 0.38 – 0.21]
                                         = 0.31

       Or, from the Venn: 0.31 (the region outside the circles)

6. **Workers.**

    Construct a Venn diagram of the disjoint outcomes.

    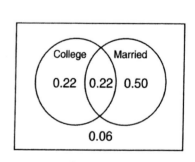

    a) *P*(neither married nor a college graduate)
       = 1 – *P*(either married or college graduate)
       = 1 – [*P*(married) + *P*(college graduate) – *P*(both)]
       = 1 – [0.72 + 0.44 – 0.22]
       = 1 – [0.94]
       = 0.06

       Or, from the Venn: 0.06 (the region outside the circles)

b) P(married, but not a college graduate)   = P(married) – P(married and a college graduate)
$$= 0.72 - 0.22$$
$$= 0.50$$

Or, from the Venn: 0.50 (the region inside the Married circle, yet inside the College circle)

c) P(married or a college graduate) = P(married) + P(college graduate) – P(both)
$$= 0.72 + 0.44 - 0.22$$
$$= 0.94$$

Or, from the Venn diagram: 0.22 + 0.22 + 0.50 = 0.94 (the regions inside the circles)

7. Cards.

a) $P(\text{heart} \mid \text{red}) = \dfrac{P(\text{heart and red})}{P(\text{red})} = \dfrac{13/52}{26/52} = \dfrac{1}{2}$

A more intuitive approach is to think about only the red cards. Half of them are hearts.

b) $P(\text{red} \mid \text{heart}) = \dfrac{P(\text{red and heart})}{P(\text{heart})} = \dfrac{13/52}{13/52} = 1$

Think about only the hearts. They are all red!

c) $P(\text{ace} \mid \text{red}) = \dfrac{P(\text{ace and red})}{P(\text{red})} = \dfrac{2/52}{26/52} = \dfrac{2}{26} \approx 0.077$

Consider only the red cards. Of those 26 cards, 2 of them are aces.

d) $P(\text{queen} \mid \text{face}) = \dfrac{P(\text{queen and face})}{P(\text{face})} = \dfrac{4/52}{12/52} \approx 0.333$

There are 12 faces cards: 4 jacks, 4 queens, and 4 kings. Four of the 12 face cards are queens.

8. Pets.

Organize the counts in a two-way table.

	Cats	Dogs	Total
Male	6	8	14
Female	12	16	28
Total	18	24	42

a) $P(\text{male} \mid \text{cat}) = \dfrac{P(\text{male and cat})}{P(\text{cat})} = \dfrac{6/42}{18/42} = \dfrac{6}{18} \approx 0.333$

Consider only the Cats column. There are 6 male cats, out of a total of 18 cats.

b) $P(\text{cat} \mid \text{female}) = \dfrac{P(\text{cat and female})}{P(\text{female})} = \dfrac{12/42}{28/42} = \dfrac{12}{28} \approx 0.429$

We are interested in the Female row. Of the 28 female animals, 12 are cats.

c) $P(\text{female} \mid \text{dog}) = \dfrac{P(\text{female and dog})}{P(\text{dog})} = \dfrac{^{16}\!/_{42}}{^{24}\!/_{42}} = \dfrac{16}{24} \approx 0.667$

Look at only the Dogs column. There are 24 dogs, and 16 of them are female.

9. Health.

Construct a two-way table of the conditional probabilities, including the marginal probabilities.

	Blood Pressure		
Cholesterol	High	OK	Total
High	0.11	0.21	0.32
OK	0.16	0.52	0.68
Total	0.27	0.73	1.00

a) $P(\text{both conditions}) = 0.11$

b) $P(\text{high blood pressure}) = 0.11 + 0.16 = 0.27$

c) $P(\text{high chol.} \mid \text{high BP}) = \dfrac{P(\text{high chol. and high BP})}{P(\text{high BP})} = \dfrac{0.11}{0.27} \approx 0.407$

Consider only the High Blood Pressure column. Within this column, the probability of having high cholesterol is 0.11 out of a total of 0.27.

d) $P(\text{high BP} \mid \text{high chol.}) = \dfrac{P(\text{high BP and high chol.})}{P(\text{high chol.})} = \dfrac{0.11}{0.32} \approx 0.344$

This time, consider only the high cholesterol row. Within this row, the probability of having high blood pressure is 0.11, out of a total of 0.32.

10. Death penalty.

Construct a two-way table of the conditional probabilities, including the marginal probabilities.

	Death Penalty		
Party	Favor	Oppose	Total
Republican	0.26	0.04	0.30
Democrat	0.12	0.24	0.36
Other	0.24	0.10	0.34
Total	0.62	0.38	1.00

a) $P(\text{favor the death penalty})$
$= 0.26 + 0.12 + 0.24$
$= 0.62$

b) $P(\text{favor death penalty} \mid \text{Republican}) = \dfrac{P(\text{favor death penalty and Republican})}{P(\text{Republican})} = \dfrac{0.26}{0.30} \approx 0.867$

Consider only the Republican row. The probability of favoring the death penalty is 0.26 out of a total of 0.30 for that row.

c) $P(\text{Democrat} \mid \text{favor death penalty}) = \dfrac{P(\text{Democrat and favor death penalty})}{P(\text{favor death penalty})} = \dfrac{0.12}{0.62} \approx 0.194$

Consider only the Favor column. The probability of being a Democrat is 0.12 out of a total of 0.62 for that column.

d) $P(\text{Republican or favor death penalty}) = P(\text{Republican}) + P(\text{favor death pen.}) - P(\text{both})$
$$= 0.30 + 0.62 - 0.26$$
$$= 0.66$$

The overall probabilities of being a Republican and favoring the death penalty are from the marginal distribution of probability (the totals). The candidate can expect 66% of the votes, provided her estimates are correct.

11. Sick kids.

Having a fever and having a sore throat are not independent events, so:

$P(\text{fever and sore throat}) = P(\text{Fever}) P(\text{Sore Throat} \mid \text{Fever}) = (0.70)(0.30) = 0.21$

The probability that a kid with a fever has a sore throat is 0.21.

12. Sick cars.

Needing repairs and paying more than \$400 for the repairs are not independent events. (What happens to the probability of paying more than \$400, if you don't need repairs?!)

$P(\text{needing repairs and paying more than } \$400)$
$= P(\text{needing repairs}) P(\text{paying more than } \$400 \mid \text{repairs are needed})$
$= (0.20)(0.40) = 0.08$

13. Card.

Since cards are not being replaced in the deck, use conditional probabilities throughout.

a)

$P(\text{first heart is drawn on the third card}) = P(\text{no heart}) P(\text{no heart}) P(\text{heart})$
$$= \left(\frac{39}{52}\right)\left(\frac{38}{51}\right)\left(\frac{13}{50}\right) \approx 0.145$$

b)

$P(\text{all three cards are red}) = P(\text{red}) P(\text{red}) P(\text{red})$
$$= \left(\frac{26}{52}\right)\left(\frac{25}{51}\right)\left(\frac{24}{50}\right) \approx 0.118$$

c)

$P(\text{none of the cards are spades}) = P(\text{no spade}) P(\text{no spade}) P(\text{no spade})$
$$= \left(\frac{39}{52}\right)\left(\frac{38}{51}\right)\left(\frac{37}{50}\right) \approx 0.414$$

d)

$P(\text{at least one of the cards in an ace}) = 1 - P(\text{none of the cards are aces})$
$$= 1 - \left[P(\text{no ace}) P(\text{no ace}) P(\text{no ace})\right]$$
$$= 1 - \left(\frac{48}{52}\right)\left(\frac{47}{51}\right)\left(\frac{46}{50}\right) \approx 0.217$$

14. Another hand.

Since cards are not being replaced in the deck, use conditional probabilities throughout.

a)

P(none of the cards are aces) = P(no ace) P(no ace) P(no ace)

$$= \left(\frac{48}{52}\right)\left(\frac{47}{51}\right)\left(\frac{46}{50}\right) \approx 0.783$$

b)

P(all of the cards are hearts) = P(heart) P(heart) P(heart)

$$= \left(\frac{13}{52}\right)\left(\frac{12}{51}\right)\left(\frac{11}{50}\right) \approx 0.013$$

c)

P(the third card is the first red) = P(no red) P(no red) P(red)

$$= \left(\frac{26}{52}\right)\left(\frac{25}{51}\right)\left(\frac{26}{50}\right) \approx 0.127$$

d)

P(at least one of the cards in a diamond) = $1 - P$(none of the cards are diamonds)

$$= 1 - \left[P(\text{no diam.}) \, P(\text{no diam.}) \, P(\text{no diam.})\right]$$

$$= 1 - \left(\frac{39}{52}\right)\left(\frac{38}{51}\right)\left(\frac{37}{50}\right) \approx 0.586$$

15. Batteries.

Since batteries are not being replaced, use conditional probabilities throughout.

a)

P(the first two batteries are good) = P(good) P(good)

$$= \left(\frac{7}{12}\right)\left(\frac{6}{11}\right) \approx 0.318$$

b)

P(at least one of the first three batteries works) = $1 - P$(none of the first three batt. work)

$$= 1 - \left[P(\text{no good}) \, P(\text{no good}) \, P(\text{no good})\right]$$

$$= 1 - \left(\frac{5}{12}\right)\left(\frac{4}{11}\right)\left(\frac{3}{10}\right) \approx 0.955$$

c)

P(the first four batteries are good) = P(good) P(good) P(good) P(good)

$$= \left(\frac{7}{12}\right)\left(\frac{6}{11}\right)\left(\frac{5}{10}\right)\left(\frac{4}{9}\right) \approx 0.071$$

d)

P(pick five to find one good) = P(no good) P(no good) P(no good) P(no good) P(good)

$$= \left(\frac{5}{12}\right)\left(\frac{4}{11}\right)\left(\frac{3}{10}\right)\left(\frac{2}{9}\right)\left(\frac{7}{8}\right) \approx 0.009$$

16. Shirts.

You need two shirts so don't replace them. Use conditional probabilities throughout.

a)

P(the first two are not mediums) = P(not medium) P(not medium)

$$= \left(\frac{16}{20}\right)\left(\frac{15}{19}\right) \approx 0.632$$

b)

P(the first medium is the third shirt) = P(no medium) P(no medium) P(medium)

$$= \left(\frac{16}{20}\right)\left(\frac{15}{19}\right)\left(\frac{4}{18}\right) \approx 0.140$$

c)

P(the first four shirts are extra − large) = P(XL) P(XL) P(XL) P(XL)

$$= \left(\frac{6}{20}\right)\left(\frac{5}{19}\right)\left(\frac{4}{18}\right)\left(\frac{3}{17}\right) \approx 0.003$$

d)

P(at least one of four is a med.) = $1 - P$(none of the first four shirts are mediums)

$$= 1 - \left[P(\text{no med.})\ P(\text{no med.})\ P(\text{no med.})\ P(\text{no med.})\right]$$

$$= 1 - \left(\frac{16}{20}\right)\left(\frac{15}{19}\right)\left(\frac{14}{18}\right)\left(\frac{13}{17}\right) \approx 0.624$$

17. Eligibility.

Construct a Venn diagram of the disjoint outcomes.

a)

P(eligibility) = P(statistics) + P(computer science) − P(both)

$\qquad$ = 0.52 + 0.23 - 0.07

$\qquad$ = 0.68

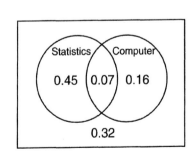

68% of students are eligible for BioResearch, so 100 – 68 = 32% are ineligible.

From the Venn, the region outside the circles represents those students who have taken neither course, and are therefore ineligible for BioResearch.

b)

$$P(\text{computer science} \mid \text{statistics}) = \frac{P(\text{computer science and statistics})}{P(\text{statistics})} = \frac{0.07}{0.52} \approx 0.135$$

From the Venn, consider only the region inside the Statistics circle. The probability of having taken computer science is 0.07 out of a total of 0.52 (the entire Statistics circle).

c) Taking the two courses are not disjoint events, since they have outcomes in common. In fact, 7% of juniors have taken both courses.

d) Taking the two courses are not independent events. The overall probability that a junior has taken a computer science is 0.23. The probability that a junior has taken a computer course given that he or she has taken a statistics course is 0.135. If taking the two courses were independent events, these probabilities would be the same.

18. Benefits.

Construct a Venn diagram of the disjoint outcomes.

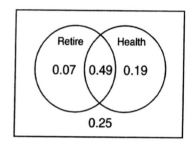

a) $P(\text{neither benefit})$ = 1 − $P(\text{either retirement or health})$
= 1 − [$P(\text{retirement}) + P(\text{health}) - P(\text{both})$]
= 1 − [0.56 + 0.68 − 0.49]
= 0.25

b)

$$P(\text{health insurance} \mid \text{retirement}) = \frac{P(\text{health insurance and retirement})}{P(\text{retirement})} = \frac{0.49}{0.56} = 0.875$$

From the Venn, consider only the region inside the Retirement circle. The probability that a worker has health insurance is 0.49 out of a total of 0.56 (the entire Retirement circle).

c) Having health insurance and a retirement plan are not independent events. 68% of all workers have health insurance, while 87.5% of workers with retirement plans also have health insurance. If having health insurance and a retirement plan were independent events, these percentages would be the same.

d) Having these two benefits are not disjoint events, since they have outcomes in common. 49% of workers have both health insurance and a retirement plan.

19. For sale.

Construct a Venn diagram of the disjoint outcomes.

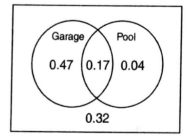

a)

$$P(\text{pool} \mid \text{garage}) = \frac{P(\text{pool and garage})}{P(\text{garage})} = \frac{0.17}{0.64} \approx 0.266$$

From the Venn, consider only the region inside the Garage circle. The probability that the house has a pool is 0.17 out of a total of 0.64 (the entire Garage circle).

b) Having a garage and a pool are not independent events. 26.6% of homes with garages have pools. Overall, 21% of homes have pools. If having a garage and a pool were independent events, these would be the same.

c) No, having a garage and a pool are not disjoint events. 17% of homes have both.

20. On the road again.

Construct a Venn diagram of the disjoint outcomes.

a)

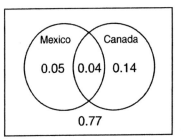

$$P(\text{Canada} \mid \text{Mexico}) = \frac{P(\text{Canada and Mexico})}{P(\text{Mexico})} = \frac{0.04}{0.09} \approx 0.444$$

From the Venn, consider only the region inside the Mexico circle. The probability that an American has traveled to Canada is 0.04 out of a total of 0.09 (the entire Mexico circle).

b) No, travel to Mexico and Canada are not disjoint events. 4% of Americans have been to both countries.

c) No, travel to Mexico and Canada are not independent events. 18% of U.S. residents have been to Canada. 44.4% of the U.S. residents who have been to Mexico have also been to Canada. If travel to the two countries were independent, the percentages would be the same.

21. Cards.

Yes, getting an ace is independent of the suit when drawing one card from a well shuffled deck. The overall probability of getting an ace is 4/52, or 1/13, since there are 4 aces in the deck. If you consider just one suit, there is only 1 ace out of 13 cards, so the probability of getting an ace given that the card is a diamond, for instance, is 1/13. Since the probabilities are the same, getting an ace is independent of the suit.

22. Pets again.

Consider the two-way table from Exercise 8.

Yes, species and gender are independent events. 8 of 24, or 1/3 of the dogs are male, and 6 of 18, or 1/3 of the cats are male. Since these are the same, species and gender are independent events.

	Cats	Dogs	Total
Male	6	8	14
Female	12	16	28
Total	18	24	42

23. Men's health, again.

Consider the two-way table from Exercise 9.

High blood pressure and high cholesterol are not independent events. 28.8% of men with OK blood pressure have high cholesterol, while 40.7% of men with high blood pressure have high cholesterol. If having high blood pressure and high cholesterol were independent, these percentages would be the same.

Blood Pressure

		High	OK	Total
Cholesterol	High	0.11	0.21	0.32
	OK	0.16	0.52	0.68
	Total	0.27	0.73	1.00

24. Politics.

Consider the two-way table from Exercise 10.

Party affiliation and position on the death penalty are not independent events. 86.7% of Republicans favor the death penalty, but only 33.3% of Democrats favor it. If the events were independent, then these percentages would be the same.

<table>
<tr><td rowspan="6" style="writing-mode:vertical-rl">Party</td><td></td><td colspan="3">Death Penalty</td></tr>
<tr><td></td><td>Favor</td><td>Oppose</td><td>Total</td></tr>
<tr><td>Republican</td><td>0.26</td><td>0.04</td><td>0.30</td></tr>
<tr><td>Democrat</td><td>0.12</td><td>0.24</td><td>0.36</td></tr>
<tr><td>Other</td><td>0.24</td><td>0.10</td><td>0.34</td></tr>
<tr><td>Total</td><td>0.62</td><td>0.38</td><td>1.00</td></tr>
</table>

25. Luggage.

Organize the information in a tree diagram.

a) No, the flight leaving on time and the luggage making the connection are not independent events. The probability that the luggage makes the connection is dependent on whether or not the flight is on time. The probability is 0.95 if the flight is on time, and only 0.65 if it is not on time.

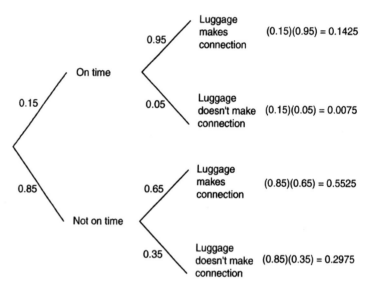

b)

$$P(\text{Luggage}) = P(\text{On time and Luggage}) + P(\text{Not on time and Luggage})$$
$$= (0.15)(0.95) + (0.85)(0.65)$$
$$= 0.695$$

26. Graduation.

Organize the information in a tree diagram.

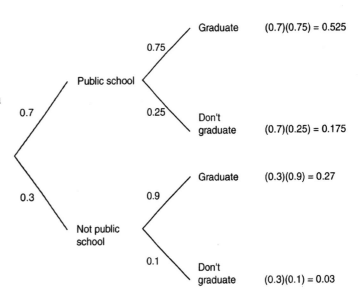

a) Yes, there is evidence to suggest that a freshman's chances to graduate depend upon what kind of high school the student attended. The graduation rate for public school students is 75%, while the graduation rate for others is 90%. If the high school attended was independent of college graduation, these percentages would be the same.

b)

$$P(\text{Graduate}) = P(\text{Public and Graduate}) + P(\text{Not public and Graduate})$$
$$= (0.7)(0.75) + (0.3)(0.9)$$
$$= 0.795$$

Overall, 79.5% of freshmen are expected to eventually graduate.

27. Late luggage.

Refer to the tree diagram constructed for Exercise 25.

$$P(\text{Not on time} \mid \text{No Lug.}) = \frac{P(\text{Not on time and No Lug.})}{P(\text{No Lug.})} = \frac{(0.85)(0.35)}{(0.15)(0.05) + (0.85)(0.35)} \approx 0.975$$

If you pick Leah up at the Denver airport and her luggage is not there, the probability that her first flight was delayed is 0.975.

28. Graduation, part II.

Refer to the tree diagram constructed for Exercise 26.

$$P(\text{Public} \mid \text{Graduate}) = \frac{P(\text{Public and Graduate})}{P(\text{Graduate})} = \frac{(0.7)(0.75)}{(0.7)(0.75) + (0.3)(0.9)} \approx 0.660$$

Overall, 66.0% of the graduates of the private college went to public high schools.

29. Absenteeism.

Organize the information in a tree diagram.

a) No, absenteeism is not independent of shift worked. The rate of absenteeism for the night shift is 2%, while the rate for the day shift is only 1%. If the two were independent, the percentages would be the same.

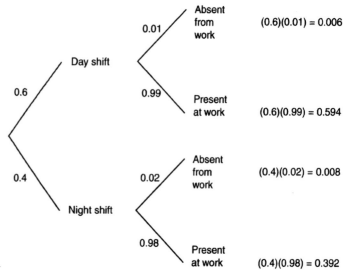

b)

$$P(\text{Absent}) = P(\text{Day and Absent}) + P(\text{Night and Absent}) = (0.6)(0.01) + (0.4)(0.02) = 0.014$$

The overall rate of absenteeism at this company is 1.4%.

30. Lungs and smoke.

Organize the information into a tree diagram.

a) The lung condition and smoking are not independent, since rates of the lung condition are different for smokers and nonsmokers. 57% of smokers have the lung condition by age 60, while only 13% of nonsmokers have the condition by age 60.

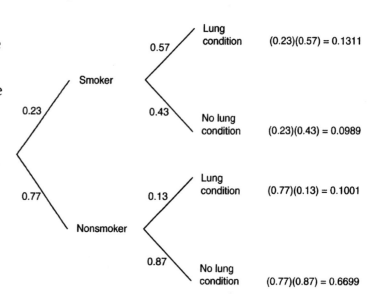

b)

$P(\text{Lung condition})$

$= P(\text{Smoker and Lung Conditon}) + P(\text{Nonsmoker and Lung Condition})$

$= (0.23)(0.57) + (0.77)(0.13)$

≈ 0.231

The probability that a randomly selected 60-year-old has the lung condition is about 0.231.

31. Absenteeism, part II.

Refer to the tree diagram constructed for Exercise 29.

$$P(\text{Night} \mid \text{Absent}) = \frac{P(\text{Night and Absent})}{P(\text{Absent})} = \frac{(0.4)(0.02)}{(0.6)(0.01) + (0.4)(0.02)} \approx 0.571$$

Approximately 57.1% of the company's absenteeism occurs on the night shift.

32. Lungs and Smoke, again.

Refer to the tree diagram constructed for Exercise 30.

$$P(\text{Smoker} \mid \text{Lung cond.}) = \frac{P(\text{Smoker and Lung cond.})}{P(\text{Lung cond.})} = \frac{(0.23)(0.57)}{(0.23)(0.57) + (0.77)(0.13)} \approx 0.567$$

The probability that someone who has the lung condition by age 60 is a smoker is approximately 56.7%.

33. Drunks.

Organize the information into a tree diagram.

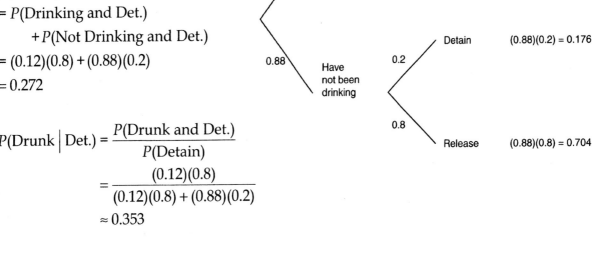

a) $P(\text{Detain} \mid \text{Not Drinking}) = 0.2$

b)
$P(\text{Detain})$
$= P(\text{Drinking and Det.})$
$\quad + P(\text{Not Drinking and Det.})$
$= (0.12)(0.8) + (0.88)(0.2)$
$= 0.272$

c)
$$P(\text{Drunk} \mid \text{Det.}) = \frac{P(\text{Drunk and Det.})}{P(\text{Detain})}$$
$$= \frac{(0.12)(0.8)}{(0.12)(0.8) + (0.88)(0.2)}$$
$$\approx 0.353$$

d)
$$P(\text{Drunk} \mid \text{Release}) = \frac{P(\text{Drunk and Release})}{P(\text{Release})}$$
$$= \frac{(0.12)(0.2)}{(0.12)(0.2) + (0.88)(0.8)}$$
$$\approx 0.033$$

34. Polygraphs.

Organize the information in a tree diagram.

$P(\text{Trustworthy} \mid \text{"Lie" on poly.})$

$= \dfrac{P(\text{Trustworthy} \cap \text{"Lie" on poly.})}{P(\text{"Lie" on poly.})}$

$= \dfrac{(0.95)(0.15)}{(0.95)(0.15)+(0.05)(0.65)}$

≈ 0.814

The probability that a job applicant rejected under suspicion of dishonesty is actually trustworthy is about 0.814.

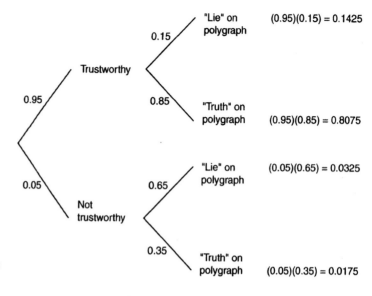

35. Dishwashers.

Organize the information in a tree diagram.

$P(\text{Chuck} \mid \text{Break})$

$= \dfrac{P(\text{Chuck and Break})}{P(\text{Break})}$

$= \dfrac{(0.3)(0.03)}{(0.4)(0.01)+(0.3)(0.01)+(0.3)(0.03)}$

≈ 0.563

If you hear a dish break, the probability that Chuck is on the job is approximately 0.563.

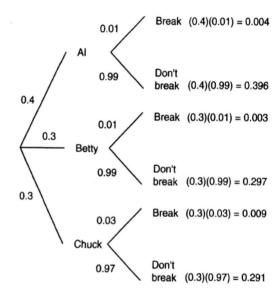

36. Parts.

Organize the information in a tree diagram.

$P(\text{Supplier A} \mid \text{Defective})$

$= \dfrac{P(\text{Supplier A and Defective})}{P(\text{Defective})}$

$= \dfrac{(0.7)(0.01)}{(0.7)(0.01)+(0.2)(0.02)+(0.1)(0.04)}$

≈ 0.467

The probability that a defective component came from supplier A is approximately 0.467.

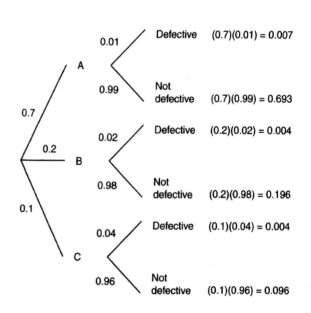

Chapter 16 – Random Variables

1. **Expected value.**

   a) $\mu = E(Y) = 10(0.3) + 20(0.5) + 30(0.2) = 19$

   b) $\mu = E(Y) = 2(0.3) + 4(0.4) + 6(0.2) + 8(0.1) = 4.2$

2. **Expected value.**

   a) $\mu = E(Y) = 0(0.2) + 1(0.4) + 2(0.4) = 1.2$

   b) $\mu = E(Y) = 100(0.1) + 200(0.2) + 300(0.5) + 400(0.2) = 280$

3. **Pick a card, any card.**

   a)

Win	$0	$5	$10	$30
P(amount won)	$\dfrac{26}{52}$	$\dfrac{13}{52}$	$\dfrac{12}{52}$	$\dfrac{1}{52}$

   b) $\mu = E(\text{amount won}) = \$0\left(\dfrac{26}{52}\right) + \$5\left(\dfrac{13}{52}\right) + \$10\left(\dfrac{12}{52}\right) + \$30\left(\dfrac{1}{52}\right) \approx \$4.13$

   c) Answers may vary. In the long run, the expected payoff of this game is $4.13 per play. Any amount less than $4.13 would be a reasonable amount to pay in order to play. Your decision should depend on how long you intend to play. If you are only going to play a few times, you should risk less.

4. **You bet!**

   a)

Win	$100	$50	$0
P(amount won)	$\dfrac{1}{6}$	$\left(\dfrac{5}{6}\right)\left(\dfrac{1}{6}\right) = \dfrac{5}{36}$	$\left(\dfrac{5}{6}\right)\left(\dfrac{5}{6}\right) = \dfrac{25}{36}$

   b) $\mu = E(\text{amount won}) = \$100\left(\dfrac{1}{6}\right) + \$50\left(\dfrac{5}{36}\right) + \$0\left(\dfrac{25}{36}\right) \approx \$23.61$

   c) Answers may vary. In the long run, the expected payoff of this game is $23.61 per play. Any amount less than $23.61 would be a reasonable amount to pay in order to play. Your decision should depend on how long you intend to play. If you are only going to play a few times, you should risk less.

5. **Kids.**

   a)

Kids	1	2	3
P(Kids)	0.5	0.25	0.25

b) $\mu = E(\text{Kids}) = 1(0.5) + 2(0.25) + 3(0.25) = 1.75$ kids

c)

Boys	0	1	2	3
P(boys)	0.5	0.25	0.125	0.125

$\mu = E(\text{Boys}) = 0(0.5) + 1(0.25) + 2(0.125) + 3(0.125) = 0.875$ boys

6. Carnival.

a)

Net winnings	$95	$90	$85	$80	-$20
number of darts	1 dart	2 darts	3 darts	4 darts (win)	4 darts (lose)
P(Amount won)	$\left(\dfrac{1}{10}\right)$ $= 0.1$	$\left(\dfrac{9}{10}\right)\left(\dfrac{1}{10}\right)$ $= 0.09$	$\left(\dfrac{9}{10}\right)^2\left(\dfrac{1}{10}\right)$ $= 0.081$	$\left(\dfrac{9}{10}\right)^3\left(\dfrac{1}{10}\right)$ $= 0.0729$	$\left(\dfrac{9}{10}\right)^4$ $= 0.6561$

b) $\mu = E(\text{number of darts}) = 1(0.1) + 2(0.09) + 3(0.081) + 4(0.0729) + 4(0.6561) \approx 3.44$ darts

c) $\mu = E(\text{winnings}) = \$95(0.1) + \$90(0.09) + \$85(0.081) + \$80(0.0729) - \$20(0.6561) \approx \$17.20$

7. Software.

Since the contracts are awarded independently, the probability that the company will get both contracts is $(0.3)(0.6) = 0.18$. Organize the disjoint events in a Venn diagram.

Profit	larger only $50,000	smaller only $20,000	both $70,000	neither $0
P(profit)	0.12	0.42	0.18	0.28

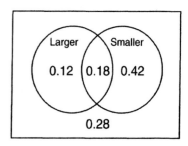

$\mu = E(\text{profit}) = \$50,000(0.12) + \$20,000(0.42) + \$70,000(0.18)$
$= \$27,000$

8. Racehorse.

Assuming that the two races are independent events, the probability that the horse wins both races is $(0.2)(0.3) = 0.06$. Organize the disjoint events in a Venn diagram.

Profit	1st only $30,000	2nd only $30,000	both $80,000	neither - $10,000
P(profit)	0.14	0.24	0.06	0.56

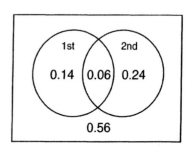

$\mu = E(\text{profit}) = \$30,000(0.14) + \$30,000(0.24)$
$\qquad\qquad + \$80,000(0.06) - \$10,000(0.56)$
$\qquad\qquad = \$10,600$

9. **Variation 1.**

   a)
   $$\sigma^2 = Var(Y) = (10-19)^2(0.3) + (20-19)^2(0.5) + (30-19)^2(0.2) = 49$$
   $$\sigma = SD(Y) = \sqrt{Var(Y)} = \sqrt{49} = 7$$

   b)
   $$\sigma^2 = Var(Y) = (2-4.2)^2(0.3) + (4-4.2)^2(0.4) + (6-4.2)^2(0.2) + (8-4.2)^2(0.1) = 3.56$$
   $$\sigma = SD(Y) = \sqrt{Var(Y)} = \sqrt{3.56} \approx 1.89$$

10. **Variation 2.**

    a)
    $$\sigma^2 = Var(Y) = (0-1.2)^2(0.2) + (1-1.2)^2(0.4) + (2-1.2)^2(0.4) = 0.56$$
    $$\sigma = SD(Y) = \sqrt{Var(Y)} = \sqrt{0.56} \approx 0.75$$

    b)
    $$\sigma^2 = Var(Y) = (100-280)^2(0.1) + (200-280)^2(0.2) + (300-280)^2(0.5) + (400-280)^2(0.2) = 7600$$
    $$\sigma = SD(Y) = \sqrt{Var(Y)} = \sqrt{7600} \approx 87.18$$

11. **Pick another card.**

    Answers may vary slightly (due to rounding of the mean)

    $$\sigma^2 = Var(\text{Won}) = (0-4.13)^2\left(\frac{26}{52}\right) + (5-4.13)^2\left(\frac{13}{52}\right)$$
    $$+ (10-4.13)^2\left(\frac{12}{52}\right) + (30-4.13)^2\left(\frac{1}{52}\right) \approx 29.5396$$
    $$\sigma = SD(\text{Won}) = \sqrt{Var(\text{Won})} \approx \sqrt{29.5396} \approx \$5.44$$

12. **The die.**

    Answers may vary slightly (due to rounding of the mean)

    $$\sigma^2 = Var(\text{Won}) = (100-23.61)^2\left(\frac{1}{6}\right) + (50-23.61)^2\left(\frac{5}{36}\right) + (0-23.61)^2\left(\frac{25}{36}\right) \approx 1456.4043$$
    $$\sigma = SD(\text{Won}) = \sqrt{Var(\text{Won})} \approx \sqrt{1456.4043} \approx \$38.16$$

13. **Kids.**

    $$\sigma^2 = Var(\text{Kids}) = (1-1.75)^2(0.5) + (2-1.75)^2(0.25) + (3-1.75)^2(0.25) = 0.6875$$
    $$\sigma = SD(\text{Kids}) = \sqrt{Var(\text{Kids})} = \sqrt{0.6875} \approx 0.83 \text{ kids}$$

14. **Darts.**

    $$\sigma^2 = Var(\text{Winnings}) = (95-17.20)^2(0.1) + (90-17.20)^2(0.09) + (85-17.20)^2(0.081)$$
    $$+ (80-17.20)^2(0.0729) + (-20-17.20)^2(0.6561) \approx 2650.057$$
    $$\sigma = SD(\text{Winnings}) = \sqrt{Var(\text{Winnings})} \approx \sqrt{2650.057} \approx \$51.48$$

15. Repairs.

a) $\mu = E(\text{Number of Repair Calls}) = 0(0.1) + 1(0.3) + 2(0.4) + 3(0.2) = 1.7$ calls

b)

$$\sigma^2 = Var(\text{Calls}) = (0 - 1.7)^2(0.1) + (1 - 1.7)^2(0.3) + (2 - 1.7)^2(0.4) + (3 - 1.7)^2(0.2) = 0.81$$

$$\sigma = SD(\text{Calls}) = \sqrt{Var(\text{Calls})} = \sqrt{0.81} = 0.9 \text{ calls}$$

16. Red lights.

a) $\mu = E(\text{Red lights}) = 0(0.05) + 1(0.25) + 2(0.35) + 3(0.15) + 4(0.15) + 5(0.05) = 2.25$ red lights

b)

$$\sigma^2 = Var(\text{Red lights}) = (0 - 2.25)^2(0.05) + (1 - 2.25)^2(0.25) + (2 - 2.25)^2(0.35)$$
$$+ (3 - 2.25)^2(0.15) + (4 - 2.25)^2(0.15) + (5 - 2.25)^2(0.05) = 1.5875$$

$$\sigma = SD(\text{Red lights}) = \sqrt{Var(\text{Red lights})} = \sqrt{1.5875} \approx 1.26 \text{ red lights}$$

17. Defects.

The percentage of cars with *no* defects is 61%.

$$\mu = E(\text{Defects}) = 0(0.61) + 1(0.21) + 2(0.11) + 3(0.07) = 0.64 \text{ defects}$$

$$\sigma^2 = Var(\text{Defects}) = (0 - 0.64)^2(0.61) + (1 - 0.64)^2(0.21)$$
$$+ (2 - 0.64)^2(0.11) + (3 - 0.64)^2(0.07) \approx 0.8704$$

$$\sigma = SD(\text{Defects}) = \sqrt{Var(\text{Defects})} \approx \sqrt{0.8704} \approx 0.93 \text{ defects}$$

18. Insurance.

a)

Profit	$100	- $9900	- $2900
P(Profit)	0.9975	0.0005	0.002

b) $\mu = E(\text{Profit}) = 100(0.9975) - 9900(0.0005) - 2900(0.002) = \$89$

c)

$$\sigma^2 = Var(\text{Profit}) = (100 - 89)^2(0.9975) + (-9900 - 89)^2(0.0005)$$
$$+ (-2900 - 89)^2(0.002) = 67,879$$

$$\sigma = SD(\text{Profit}) = \sqrt{Var(\text{Profit})} = \sqrt{67,879} \approx \$260.54$$

19. Contest.

a) The two games are not independent. The probability that you win the second depends on whether or not you win the first.

b)

$$P(\text{losing both games}) = P(\text{losing the first}) \, P(\text{losing the second} \mid \text{first was lost})$$
$$= (0.6)(0.7)$$
$$= 0.42$$

c)

$$P(\text{winning both games}) = P(\text{winning the first})\, P(\text{winning the second} \mid \text{first was won})$$
$$= (0.4)(0.2) = 0.08$$

d)

X	0	1	2
$P(X = x)$	0.42	0.50	0.08

e)

$$\mu = E(X) = 0(0.42) + 1(0.50) + 2(0.08) = 0.66 \text{ games}$$
$$\sigma^2 = Var(X) = (0 - 0.66)^2(0.42) + (1 - 0.66)^2(0.50) + (2 - 0.66)^2(0.08) = 0.3844$$
$$\sigma = SD(X) = \sqrt{Var(X)} = \sqrt{0.3844} = 0.62 \text{ games}$$

20. Contracts.

a) The contracts are not independent. The probability that your company wins the second contract depends on whether or not your company wins the first contract.

b)

$$P(\text{getting both contracts}) = P(\text{getting \#1})\, P(\text{getting \#2} \mid \text{got \#1})$$
$$= (0.8)(0.2)$$
$$= 0.16$$

c)

$$P(\text{getting no contract}) = P(\text{not getting \#1})\, P(\text{not getting \#2} \mid \text{didn't get \#1})$$
$$= (0.2)(0.7)$$
$$= 0.14$$

d)

X	0	1	2
$P(X = x)$	0.14	0.70	0.16

e) $\mu = E(X) = 0(0.14) + 1(0.70) + 2(0.16) = 1.02 \text{ contracts}$

$$\sigma^2 = Var(X) = (0 - 1.02)^2(0.14) + (1 - 1.02)^2(0.70) + (2 - 1.02)^2(0.16) = 0.2996$$
$$\sigma = SD(X) = \sqrt{Var(X)} = \sqrt{0.2996} \approx 0.55 \text{ contracts}$$

21. Batteries.

a)

Number good	0	1	2
P(number good)	$\left(\dfrac{3}{10}\right)\left(\dfrac{2}{9}\right) = \dfrac{6}{90}$	$\left(\dfrac{3}{10}\right)\left(\dfrac{7}{9}\right) + \left(\dfrac{7}{10}\right)\left(\dfrac{3}{9}\right) = \dfrac{42}{90}$	$\left(\dfrac{7}{10}\right)\left(\dfrac{6}{9}\right) = \dfrac{42}{90}$

b) $\mu = E(\text{number good}) = 0\left(\dfrac{6}{90}\right) + 1\left(\dfrac{42}{90}\right) + 2\left(\dfrac{42}{90}\right) = 1.4$ batteries

c)

$$\sigma^2 = Var(\text{number good}) = (0 - 1.4)^2\left(\dfrac{6}{90}\right) + (1 - 1.4)^2\left(\dfrac{42}{90}\right) + (2 - 1.4)^2\left(\dfrac{42}{90}\right) \approx 0.3733$$

$$\sigma = SD(\text{number good}) = \sqrt{Var(\text{number good})} \approx \sqrt{0.3733} \approx 0.61$$

22. Kittens.

a)

Number of males	0	1	2
P(number of males)	$\left(\dfrac{3}{7}\right)\left(\dfrac{2}{6}\right) = \dfrac{6}{42}$	$\left(\dfrac{4}{7}\right)\left(\dfrac{3}{6}\right) + \left(\dfrac{3}{7}\right)\left(\dfrac{4}{6}\right) = \dfrac{24}{42}$	$\left(\dfrac{4}{7}\right)\left(\dfrac{3}{6}\right) = \dfrac{12}{42}$

b) $\mu = E(\text{number of males}) = 0\left(\dfrac{6}{42}\right) + 1\left(\dfrac{24}{42}\right) + 2\left(\dfrac{12}{42}\right) \approx 1.14$ males

c) Answers may vary slightly (due to rounding of the mean)

$$\sigma^2 = Var(\text{number of males}) = (0 - 1.14)^2\left(\dfrac{6}{42}\right) + (1 - 1.14)^2\left(\dfrac{24}{42}\right) + (2 - 1.14)^2\left(\dfrac{12}{42}\right) \approx 0.4082$$

$$\sigma = SD(\text{number of males}) = \sqrt{Var(\text{number of males})} \approx \sqrt{0.4082} \approx 0.64 \text{ males}$$

23. Random variables.

a)

$$\mu = E(3X) = 3(E(X)) = 3(10) = 30$$
$$\sigma = SD(3X) = 3(SD(X)) = 3(2) = 6$$

b)

$$\mu = E(Y + 6) = E(Y) + 6 = 20 + 6 = 26$$
$$\sigma = SD(Y + 6) = SD(Y) = 5$$

c)

$$\mu = E(X + Y) = E(X) + E(Y) = 10 + 20 = 30$$
$$\sigma = SD(X + Y) = \sqrt{Var(X) + Var(Y)}$$
$$= \sqrt{2^2 + 5^2} \approx 5.39$$

d)

$$\mu = E(X - Y) = E(X) - E(Y) = 10 - 20 = -10$$
$$\sigma = SD(X - Y) = \sqrt{Var(X) + Var(Y)}$$
$$= \sqrt{2^2 + 5^2} \approx 5.39$$

24. Random variables.

a)

$$\mu = E(X - 20) = E(X) - 20 = 80 - 20 = 60$$
$$\sigma = SD(X - 20) = SD(X) = 12$$

b)

$$\mu = E(0.5Y) = 0.5(E(Y)) = 0.5(12) = 6$$
$$\sigma = SD(0.5Y) = 0.5(SD(Y)) = 0.5(3) = 1.5$$

c)

$$\mu = E(X + Y) = E(X) + E(Y) = 80 + 12 = 92$$
$$\sigma = SD(X + Y) = \sqrt{Var(X) + Var(Y)}$$
$$= \sqrt{12^2 + 3^2} \approx 12.37$$

d)

$$\mu = E(X - Y) = E(X) - E(Y) = 80 - 12 = 68$$
$$\sigma = SD(X - Y) = \sqrt{Var(X) + Var(Y)}$$
$$= \sqrt{12^2 + 3^2} \approx 12.37$$

25. Random variables.

a)
$$\mu = E(0.8Y) = 0.8(E(Y)) = 0.8(300) = 240$$
$$\sigma = SD(0.8Y) = 0.8(SD(Y)) = 0.8(16) = 12.8$$

b)
$$\mu = E(2X - 100) = 2(E(X)) - 100 = 140$$
$$\sigma = SD(2X - 100) = 2(SD(X)) = 2(12) = 24$$

c)
$$\mu = E(X + 2Y) = E(X) + 2(E(Y))$$
$$= 120 + 2(300) = 720$$
$$\sigma = SD(X + 2Y) = \sqrt{Var(X) + 2^2 Var(Y)}$$
$$= \sqrt{12^2 + 2^2(16^2)} \approx 34.18$$

d)
$$\mu = E(3X - Y) = 3(E(X)) - E(Y)$$
$$= 3(120) - 300 = 60$$
$$\sigma = SD(3X - Y) = \sqrt{3^2 Var(X) + Var(Y)}$$
$$= \sqrt{3^2(12^2) + 16^2} \approx 39.40$$

26. Random variables.

a)
$$\mu = E(2Y + 20) = 2(E(Y)) + 20$$
$$= 2(12) + 20 = 44$$
$$\sigma = SD(2Y + 20) = 2(SD(Y)) = 2(3) = 6$$

b)
$$\mu = E(3X) = 3(E(X)) = 3(80) = 240$$
$$\sigma = SD(3X) = 3(SD(X)) = 3(12) = 36$$

c)
$$\mu = E(0.25X + Y) = 0.25(E(X)) + E(Y)$$
$$= 0.25(80) + 12 = 32$$
$$\sigma = SD(0.25X + Y) = \sqrt{0.25^2 Var(X) + Var(Y)}$$
$$= \sqrt{0.25^2(12^2) + 3^2} \approx 4.24$$

d)
$$\mu = E(X - 5Y) = E(X) - 5(E(Y))$$
$$= 80 - 5(12) = 20$$
$$\sigma = SD(X - 5Y) = \sqrt{Var(X) + 5^2 Var(Y)}$$
$$= \sqrt{12^2 + 5^2(3^2)} \approx 19.21$$

27. Eggs.

a) $\mu = E(\text{Broken eggs in 3 dozen}) = 3(E(\text{Broken eggs in 1 dozen})) = 3(0.6) = 1.8$ eggs

b) $\sigma = SD(\text{Broken eggs in 3 dozen}) = \sqrt{0.5^2 + 0.5^2 + 0.5^2} \approx 0.87$ eggs

c) The cartons of eggs must be independent of each other.

28. Garden.

a) $\mu = E(\text{bad seeds in 5 packets}) = 5(E(\text{bad seeds in 1 packet})) = 5(2) = 10$ bad seeds

b) $\sigma = SD(\text{bad seeds in 5 packets}) = \sqrt{1.2^2 + 1.2^2 + 1.2^2 + 1.2^2 + 1.2^2} \approx 2.68$ bad seeds

c) The packets of seeds must be independent of each other. If you buy an assortment of seeds, this assumption is probably OK. If you buy all of one type of seed, the store probably has seed packets from the same batch or lot. If some are bad, the others might tend to be bad as well.

29. Repair calls.

$$\mu = E(\text{calls in 8 hours}) = 8(E(\text{calls in 1 hour}) = 8(1.7) = 13.6 \text{ calls}$$
$$\sigma = SD(\text{calls in 8 hours}) = \sqrt{8(Var(\text{calls in 1 hour}))} = \sqrt{8(0.9)^2} \approx 2.55 \text{ hours}$$

This is only valid if the hours are independent of one another.

30. Stop!

$\mu = E(\text{red lights in 5 days}) = 5(E(\text{red lights each day})) = 5(2.25) = 11.25 \text{ red lights}$

$\sigma = SD(\text{red lights in 5 days}) = \sqrt{5(Var(\text{red lights each day})} = \sqrt{5(1.26)^2} \approx 2.82 \text{ red lights}$

Standard deviation may vary slightly due to rounding of the standard deviation of the number of red lights each day, and may only be calculated if the days are independent of each other. This seems reasonable.

31. Fire!

a) The standard deviation is large because the profits on insurance are highly variable. Although there will be many small gains, there will occasionally be large losses, when the insurance company has to pay a claim.

b)

$\mu = E(\text{two policies}) = 2(E(\text{one policy})) = 2(150) = \$300$

$\sigma = SD(\text{two policies}) = \sqrt{2(Var(\text{one policy}))} = \sqrt{2(6000^2)} \approx \$8,485.28$

c)

$\mu = E(10,000 \text{ policies}) = 10,000(E(\text{one policy})) = 10,000(150) = \$1,500,000$

$\sigma = SD(10,000 \text{ policies}) = \sqrt{10,000(Var(\text{one policy}))} = \sqrt{10,000(6000^2)} = \$600,000$

d) If the company sells 10,000 policies, they are likely to be successful. A profit of \$0, is 2.5 standard deviations below the expected profit. This is unlikely to happen.
However, if the company sells fewer policies, then the likelihood of turning a profit decreases. In an extreme case, where only two policies are sold, a profit of \$0 is more likely, being only a small fraction of a standard deviation below the mean.

e) This analysis depends on each of the policies being independent from each other. This assumption of independence may be violated if there are many fire insurance claims as a result of a forest fire, or other natural disaster.

32. Casino.

a) The standard deviation of the slot machine payouts is large because most players will lose their dollar, but a few large payouts are expected. The payouts are highly variable.

b)

$\mu = E(\text{profit from 5 plays}) = 5(E(\text{profit from one play})) = 5(0.08) = \$0.40$

$\sigma = SD(\text{profit from 5 plays}) = \sqrt{5(Var(\text{profit from one play}))} = \sqrt{5(120^2)} \approx \$268.33$

c)

$\mu = E(\text{profit from 1000 plays}) = 1000(E(\text{profit from one play})) = 1000(0.08) = \$80$

$\sigma = SD(\text{profit from 1000 plays}) = \sqrt{1000(Var(\text{profit from one play}))} = \sqrt{1000(120^2)} \approx \$3,794.73$

d) If the machine is played only 1000 times a day, the chance of being profitable will be slightly more than 50%, since \$80 is approximately 0.02 standard deviations above 0. But if the casino has many slot machines, the chances of being profitable will go up.

33. Cereal.

a) $E(\text{large bowl} - \text{small bowl}) = E(\text{large bowl}) - E(\text{small bowl}) = 2.5 - 1.5 = 1$ ounce

b) $\sigma = SD(\text{large bowl} - \text{small bowl}) = \sqrt{Var(\text{large}) + Var(\text{small})} = \sqrt{0.4^2 + 0.3^2} = 0.5$ ounces

c)

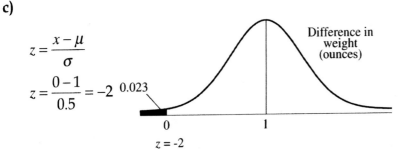

$$z = \frac{x - \mu}{\sigma}$$

$$z = \frac{0 - 1}{0.5} = -2$$

The small bowl will contain more cereal than the large bowl when the difference between the amounts is less than 0. According to the Normal model, the probability of this occurring is approximately 0.023.

d)

$\mu = E(\text{large bowl} + \text{small bowl}) = E(\text{large bowl}) + E(\text{small bowl}) = 2.5 + 1.5 = 4$ ounce

$\sigma = SD(\text{large bowl} + \text{small bowl}) = \sqrt{Var(\text{large}) + Var(\text{small})} = \sqrt{0.4^2 + 0.3^2} = 0.5$ ounces

e)

$$z = \frac{x - \mu}{\sigma}$$

$$z = \frac{4.5 - 4}{0.5} = 1$$

According to the Normal model, the probability that the total weight of cereal in the two bowls is more than 4.5 ounces is approximately 0.159.

f)

$\mu = E(\text{box} - \text{large} - \text{small}) = E(\text{box}) - E(\text{large}) - E(\text{small}) = 16.3 - 2.5 - 1.5 = 12.3$ ounces

$\sigma = SD(\text{box} - \text{large} - \text{small}) = \sqrt{Var(\text{box}) + Var(\text{large}) + Var(\text{small})}$

$$= \sqrt{0.2^2 + 0.3^2 + 0.4^2} \approx 0.54 \text{ ounces}$$

34. Pets.

a) $\mu = E(\text{dogs} - \text{cats}) = E(\text{dogs}) - E(\text{cats}) = 100 - 120 = -\$20$

b) $\sigma = SD(\text{dogs} - \text{cats}) = \sqrt{Var(\text{dogs}) + Var(\text{cats})} = \sqrt{30^2 + 35^2} \approx \$46.10$

c)

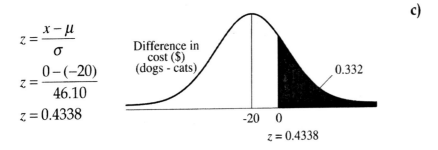

$$z = \frac{x - \mu}{\sigma}$$

$$z = \frac{0 - (-20)}{46.10}$$

$$z = 0.4338$$

35. More cereal.

a) $\mu = E(\text{box} - \text{large} - \text{small}) = E(\text{box}) - E(\text{large}) - E(\text{small}) = 16.2 - 2.5 - 1.5 = 12.2$ ounces

b)
$$\sigma = SD(\text{box} - \text{large} - \text{small}) = \sqrt{Var(\text{box}) + Var(\text{large}) + Var(\text{small})}$$
$$= \sqrt{0.1^2 + 0.3^2 + 0.4^2} \approx 0.51 \text{ ounces}$$

c)

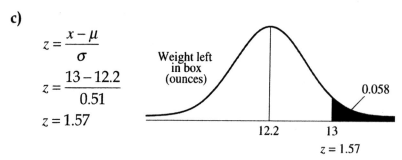

$z = \dfrac{x - \mu}{\sigma}$

$z = \dfrac{13 - 12.2}{0.51}$

$z = 1.57$

Weight left in box (ounces)

0.058

12.2   13

$z = 1.57$

According to the Normal model, the probability that the box contains more than 13 ounces is about 0.058.

36. More pets.

a) Let X = cost for a dog, and let Y = cost for a cat.

Total cost = $X + X + Y$

b)
$$\mu = E(X + X + Y) = E(X) + E(X) + E(Y) = 100 + 100 + 120 = \$320$$
$$\sigma = SD(X + X + Y) = \sqrt{Var(X) + Var(X) + Var(Y)}$$
$$= \sqrt{30^2 + 30^2 + 35^2} = \$55$$

Since the models for individual pets are Normal, the model for total costs is Normal with mean \$320 and standard deviation \$55.

c)

$z = \dfrac{x - \mu}{\sigma}$

$z = \dfrac{400 - 320}{55}$

$z = 1.45$

Total cost ($)

0.073

320   400

$z = 1.45$

According to the Normal model, the probability that the total cost of two dogs and a cat is more than \$400 is approximately 0.073.

37. Medley.

a)
$$\mu = E(\#1 + \#2 + \#3 + \#4) = E(\#1) + E(\#2) + E(\#3) + E(\#4)$$
$$= 50.72 + 55.51 + 49.43 + 44.91 = 200.57 \text{ seconds}$$
$$\sigma = SD(\#1 + \#2 + \#3 + \#4) = \sqrt{Var(\#1) + Var(\#2) + Var(\#3) + Var(\#4)}$$
$$= \sqrt{0.24^2 + 0.22^2 + 0.25^2 + 0.21^2} \approx 0.46 \text{ seconds}$$

b)

$$z = \frac{x-\mu}{\sigma}$$

$$z = \frac{199.48 - 200.57}{0.461}$$

$$z = -2.36$$

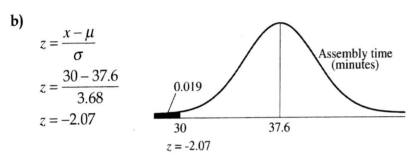

The team is not likely to swim faster than their best time. According to the Normal model, they are only expected to swim that fast or faster about 0.9% of the time.

38. Bikes.

a)

$$\mu = E(\text{unpack} + \text{assembly} + \text{tuning}) = E(\text{unpack}) + E(\text{assembly}) + E(\text{tuning})$$
$$= 3.5 + 21.8 + 12.3 = 37.6 \text{ minutes}$$

$$\sigma = SD(\text{unpack} + \text{assembly} + \text{tuning}) = \sqrt{Var(\text{unpack}) + Var(\text{assembly}) + Var(\text{tuning})}$$
$$= \sqrt{0.7^2 + 2.4^2 + 2.7^2} \approx 3.7 \text{ minutes}$$

b)

$$z = \frac{x-\mu}{\sigma}$$

$$z = \frac{30 - 37.6}{3.68}$$

$$z = -2.07$$

The bike is not likely to be ready on time. According to the Normal model, the probability that an assembly is completed in under 30 minutes is about 0.019.

39. Farmer's market.

a) Let A = price of a pound of apples, and let P = price of a pound of potatoes.

Profit $= 100A + 50P - 2$

b) $\mu = E(100A + 50P - 2) = 100(E(A)) + 50(E(P)) - 2 = 100(0.5) + 50(0.3) - 2 = \$63$

c) $\sigma = SD(100A + 50P - 2) = \sqrt{100^2(Var(A)) + 50^2(Var(P))} = \sqrt{100^2(0.2^2) + 50^2(0.1^2)} \approx \$20.62$

d) No assumptions are necessary to compute the mean. To compute the standard deviation, independent market prices must be assumed.

40. Bike sale.

a) Let B = number of basic bikes sold, and let D = number of deluxe bikes sold.

Net Profit $= 120B + 150D - 200$

b) $\mu = E(120B + 150D - 200) = 120(E(B)) + 150(E(D)) - 200 = 120(5.4) + 150(3.2) - 200 = \$928$

c)
$$\sigma = SD(120B + 150D - 200) = \sqrt{120^2(Var(B)) + 150^2(Var(D))}$$
$$= \sqrt{120^2(1.2^2) + 150^2(0.8^2)} \approx \$187.45$$

d) No assumptions are necessary to compute the mean. To compute the standard deviation, independent sales must be assumed.

Chapter 17 – Probability Models

1. **Bernoulli.**

   a) These are not Bernoulli trials. The possible outcomes are 1, 2, 3, 4, 5, and 6. There are more than two possible outcomes.

   b) These may be considered Bernoulli trials. There are only two possible outcomes, Type A and not Type A. Assuming the 120 donors are representative of the population, the probability of having Type A blood is 43%. The trials are not independent, because the population is finite, but the 120 donors represent less than 10% of the population of all possible donors.

   c) These are not Bernoulli trials. The probability of getting a heart changes as cards are dealt without replacement.

   d) These are not Bernoulli trials. We are sampling without replacement, so the trials are not independent. Samples without replacement may be considered Bernoulli trials if the sample size is less than 10% of the population, but our sample of 500 is more than 10% of 3000.

   e) These may be considered Bernoulli trails. There are only two possible outcomes, sealed properly and not sealed properly. The probability that a package is unsealed is constant, at about 10%, as long as the packages checked are a representative sample of all packages. Finally, the trials are not independent, since the total number of packages is finite, but the 24 packages checked probably represent less than 10% of the packages.

2. **Bernoulli 2.**

   a) These may be considered Bernoulli trials. There are only two possible outcomes, getting a 6 and not getting a 6. The probability of getting a 6 is constant at 1/6. The rolls are independent of one another, since the outcome of one die roll doesn't affect the other rolls.

   b) These are not Bernoulli trials. There are more than two possible outcomes for eye color.

   c) These can be considered Bernoulli trials. There are only two possible outcomes, properly attached buttons and improperly attached buttons. As long as the button problem occurs randomly, the probability of a doll having improperly attached buttons is constant at about 3%. The trails are not independent, since the total number of dolls is finite, but 37 dolls is probably less than 10% of all dolls.

   d) These are not Bernoulli trials. The trials are not independent, since the probability of picking a council member with a particular political affiliation changes depending on who has already been picked. The 10% condition is not met, since the sample of size 4 is more than 10% of the population of 19 people.

   e) These may be considered Bernoulli trials. There are only two possible outcomes, cheating and not cheating. Assuming that cheating patterns in this school are similar to the patterns in the nation, the probability that a student has cheated is constant, at 74%. The trials are not independent, since the population of all students is finite, but 481 is less than 10% of all students.

3. **Hoops.**

The player's shots may be considered Bernoulli trials. There are only two possible outcomes (make or miss), the probability of making a shot is constant (80%), and the shots are independent of one another (making, or missing, a shot does not affect the probability of making the next).

Let X = the number of shots until the first missed shot.
Let Y = the number of shots until the first made shot.

Since these problems deal with shooting until the first miss (or until the first made shot), a geometric model, either *Geom*(0.8) or *Geom*(0.2), is appropriate.

a) Use *Geom*(0.2). $P(X = 5) = (0.8)^4(0.2) = 0.08192$ (Four shots made, followed by a miss.)

b) Use *Geom*(0.8). $P(Y = 4) = (0.2)^3(0.8) = 0.0064$ (Three misses, then a made shot.)

c) Use *Geom*(0.8). $P(Y = 1) + P(Y = 2) + P(Y = 3) = (0.8) + (0.2)(0.8) + (0.2)^2(0.8) = 0.992$

4. **Chips.**

The selection of chips may be considered Bernoulli trials. There are only two possible outcomes (fail testing and pass testing). Provided that the chips selected are a representative sample of all chips, the probability that a chip fails testing is constant at 2%. The trails are not independent, since the population of chips is finite, but we won't need to sample more than 10% of all chips.

Let X = the number of chips required until the first bad chip.
The appropriate model is *Geom*(0.02).

a) $P(X = 5) = (0.98)^4(0.02) \approx 0.0184$ (Four good chips, then a bad one.)

b) $P(1 \le X \le 10) = (0.02) + (0.98)(0.02) + (0.98)^2(0.02) + \ldots + (0.98)^9(0.02) \approx 0.183$

(Use the geometric model on a calculator or computer for this one!)

5. **More hoops.**

As determined in Exercise 3, the shots can be considered Bernoulli trials, and since the player is shooting until the first miss, *Geom*(0.2) is the appropriate model.

$E(X) = \dfrac{1}{p} = \dfrac{1}{0.2} = 5$ shots   The player is expected to take 5 shots until the first miss.

6. **Chips ahoy.**

As determined in Exercise 4, the selection of chips can be considered Bernoulli trials, and since the company is selecting until the first bad chip, *Geom*(0.02) is the appropriate model.

$E(X) = \dfrac{1}{p} = \dfrac{1}{0.02} = 50$ chips   The first bad chip is expected to be the 50th chip selected.

7. **Blood.**

These may be considered Bernoulli trials. There are only two possible outcomes, Type AB and not Type AB. Provided that the donors are representative of the population, the probability of having Type AB blood is constant at 4%. The trials are not independent, since the population is finite, but we are selecting fewer than 10% of all potential donors. Since we are selecting people until the first success, the model *Geom*(0.04) may be used.

Let X = the number of donors until the first Type AB donor is found.

a) $E(X) = \dfrac{1}{p} = \dfrac{1}{0.04} = 25$ people   We expect the 25th person to be the first Type AB donor.

b)

P(a Type AB donor among the first 5 people checked)

$= P(X = 1) + P(X = 2) + P(X = 3) + P(X = 4) + P(X = 5)$

$= (0.04) + (0.96)(0.04) + (0.96)^2(0.04) + (0.96)^3(0.04) + (0.96)^4(0.04) \approx 0.185$

c)

P(a Type AB donor among the first 6 people checked)

$= P(X = 1) + P(X = 2) + P(X = 3) + P(X = 4) + P(X = 5) + P(X = 6)$

$= (0.04) + (0.96)(0.04) + (0.96)^2(0.04) + (0.96)^3(0.04) + (0.96)^4(0.04) + (0.96)^5(0.04) \approx 0.217$

d) P(no Type AB donor before the 10th person checked) $= P(X > 9) = (0.96)^9 \approx 0.693$
This one is a bit tricky. There is no implication that we actually find a donor on the 10th trial. We only care that nine trials passed with no Type AB donor.

8. **Colorblindness.**

These may be considered Bernoulli trials. There are only two possible outcomes, colorblind and not colorblind. As long as the men selected are representative of the population of all men, the probability of being colorblind is constant at about 8%. Trials are not independent, since the population is finite, but we won't be sampling more than 10% of the population.

Let X = the number of people checked until the first colorblind man is found.

Since we are selecting people until the first success, the geometric model, *Geom*(0.08), may be used.

a) $E(X) = \dfrac{1}{p} = \dfrac{1}{0.08} = 12.5$ people.   We expect to examine 12.5 people until finding the first colorblind person.

b) P(no colorblind men among the first 4) $= P(X > 4) = (0.92)^4 \approx 0.716$

c) P(first colorblind man is the sixth man checked) $= P(X = 6) = (0.92)^5(0.08) \approx 0.0527$

d)

P(she finds a colorblind man before the tenth man)

$= P(1 \le X \le 10)$

$= (0.08) + (0.92)(0.08) + (0.92)^2(0.08) + \ldots + (0.92)^8(0.08) \approx 0.528$

(Use the geometric model on a calculator or computer for this one!)

9. Lefties.

These may be considered Bernoulli trials. There are only two possible outcomes, left-handed and not left-handed. Since people are selected at random, the probability of being left-handed is constant at about 13%. The trails are not independent, since the population is finite, but a sample of 5 people is certainly fewer than 10% of all people.

Let X = the number of people checked until the first lefty is discovered.
Let Y = the number of lefties among $n = 5$.

a) Use *Geom*(0.13).

P(first lefty is the fifth person) $= P(X = 5) = (0.87)^4(0.13) \approx 0.0745$

b) Use *Binom*(5,0.13).

P(some lefties among the 5 people) $= 1 - P$(no lefties among the first 5 people)

$$= 1 - P(Y = 0)$$
$$= 1 - {}_5C_0(0.13)^0(0.87)^5$$
$$\approx 0.502$$

c) Use *Geom*(0.13).

P(first lefty is second or third person) $= P(X = 2) + P(X = 3) = (0.87)(0.13) + (0.87)^2(0.13) \approx 0.211$

d) Use *Binom*(5,0.13).

P(exactly 3 lefties in the group) $= P(Y = 3) = {}_5C_3(0.13)^3(0.87)^2 \approx 0.0166$

e) Use *Binom*(5,0.13).

P(at least 3 lefties in the group) $= P(Y = 3) + P(Y = 4) + P(Y = 5)$

$$= {}_5C_3(0.13)^3(0.87)^2 + {}_5C_4(0.13)^4(0.87)^1 + {}_5C_5(0.13)^5(0.87)^0$$
$$\approx 0.0179$$

f) Use *Binom*(5,0.13).

P(at most 3 lefties in the group) $= P(Y = 0) + P(Y = 1) + P(Y = 2) + P(Y = 3)$

$$= {}_5C_0(0.13)^0(0.87)^5 + {}_5C_1(0.13)^1(0.87)^4$$
$$+ {}_5C_2(0.13)^2(0.87)^3 + {}_5C_3(0.13)^3(0.87)^2$$
$$\approx 0.9987$$

10. Arrows.

These may be considered Bernoulli trials. There are only two possible outcomes, hitting the bull's-eye and not hitting the bull's-eye. The probability of hitting the bull's-eye is given, $p = 0.80$. The shots are assumed to be independent.

Let $X = $ the number of shots until the first bull's-eye.
Let $Y = $ the number of bull's-eyes in $n = 6$ shots.

a) Use *Geom*(0.80).
P(first bull's-eye is on the third shot) $= P(X = 3) = (0.20)^2(0.80) \approx 0.032$

b) Use *Binom*(6,0.80).
P(at least one miss out of 6 shots) $= 1 - P$(6 out of 6 hits)
$$= 1 - P(Y = 6)$$
$$= 1 - {}_6C_6(0.80)^6(0.20)^0$$
$$\approx 0.738$$

c) Use *Geom*(0.80).
P(first hit on fourth or fifth shot) $= P(X = 4) + P(X = 5) = (0.20)^3(0.80) + (0.20)^4(0.80) = 0.00768$

d) Use *Binom*(6,0.80).
P(at least four hits) $= P(Y = 4) + P(Y = 5) + P(Y = 6)$
$$= {}_6C_4(0.80)^4(0.20)^2 + {}_6C_5(0.80)^5(0.20)^1 + {}_6C_6(0.80)^6(0.20)^0$$
$$\approx 0.901$$

e) Use *Binom*(6,0.80).
P(exactly four hits) $= P(Y = 4)$
$$= {}_6C_4(0.80)^4(0.20)^2$$
$$\approx 0.246$$

f) Use *Binom*(6,0.80).
P(at most four hits) $= P(Y = 0) + P(Y = 1) + P(Y = 2) + P(Y = 3) + P(Y = 4)$
$$= {}_6C_0(0.80)^0(0.20)^6 + {}_6C_1(0.80)^1(0.20)^5 + {}_6C_2(0.80)^2(0.20)^4$$
$$+ {}_6C_3(0.80)^3(0.20)^3 + {}_6C_4(0.80)^4(0.20)^2$$
$$\approx 0.345$$

11. Lefties redux.

a) In Exercise 9, we determined that the selection of lefties could be considered Bernoulli trials. Since our group consists of 5 people, use *Binom*(5,0.13).

Let $Y = $ the number of lefties among $n = 5$.

$E(Y) = np = 5(0.13) = 0.65$ lefties

b) $SD(Y) = \sqrt{npq} = \sqrt{5(0.13)(0.87)} \approx 0.75$ lefties

c) Use *Geom*(0.13). Let X = the number of people checked until the first lefty is discovered.

$$E(X) = \frac{1}{p} = \frac{1}{0.13} \approx 7.69 \text{ people}$$

12. More arrows.

a) In Exercise 10, we determined that the shots could be considered Bernoulli trials. Since the archer is shooting 6 arrows, use *Binom*(6,0.80).

Let Y = the number of bull's-eyes in $n = 6$ shots.

$E(Y) = np = 6(0.80) = 4.8$ bull's-eyes.

b) $SD(Y) = \sqrt{npq} = \sqrt{6(0.80)(0.20)} \approx 0.98$ bull's-eyes.

c) Use *Geom*(0.80). Let X = the number of arrows shot until the first bull's-eye.

$$E(X) = \frac{1}{p} = \frac{1}{0.80} = 1.25 \text{ shots.}$$

13. Still more lefties.

a) In Exercise 9, we determined that the selection of lefties (and also righties) could be considered Bernoulli trials. Since our group consists of 12 people, and now we are considering the righties, use *Binom*(12,0.87).

Let Y = the number of righties among $n = 12$.

$E(Y) = np = 12(0.87) = 10.44$ righties

$SD(Y) = \sqrt{npq} = \sqrt{12(0.87)(0.13)} \approx 1.16$ righties

b)
$$
\begin{aligned}
P(\text{not all righties}) &= 1 - P(\text{all righties}) \\
&= 1 - P(Y = 12) \\
&= 1 - {}_{12}C_{12}(0.87)^{12}(0.13)^{0} \\
&\approx 0.812
\end{aligned}
$$

c)
$$
\begin{aligned}
P(\text{no more than 10 righties}) &= P(Y \leq 10) \\
&= P(Y = 0) + P(Y = 1) + P(Y = 2) + \ldots + P(Y = 10) \\
&= {}_{12}C_{0}(0.87)^{0}(0.13)^{12} + {}_{12}C_{1}(0.87)^{1}(0.13)^{11} + \ldots + {}_{12}C_{10}(0.87)^{10}(0.13)^{2} \\
&\approx 0.475
\end{aligned}
$$

d)
$$
\begin{aligned}
P(\text{exactly six of each}) &= P(Y = 6) \\
&= {}_{12}C_{6}(0.87)^{6}(0.13)^{6} \\
&\approx 0.00193
\end{aligned}
$$

e)

$$P(\text{majority righties}) = P(Y \geq 7)$$
$$= P(Y=7) + P(Y=8) + P(Y=9) + \ldots + P(Y=12)$$
$$= {}_{12}C_7(0.87)^7(0.13)^5 + {}_{12}C_8(0.87)^8(0.13)^4 + \ldots + {}_{12}C_{12}(0.87)^{12}(0.13)^0$$
$$\approx 0.998$$

14. Still more arrows.

a) In Exercise 10, we determined that the archer's shots could be considered Bernoulli trials. Since our archer is now shooting 10 arrows, use *Binom*(10,0.80).

Let Y = the number of bull's-eyes hit from $n = 10$ shots.

$E(Y) = np = 10(0.80) = 8$ bull's-eyes hit.

$SD(Y) = \sqrt{npq} = \sqrt{10(0.80)(0.20)} \approx 1.26$ bull's-eyes hit.

b)

$$P(\text{no misses out of 10 shots}) = P(\text{all hits out of 10 shots})$$
$$= P(Y=10)$$
$$= {}_{10}C_{10}(0.80)^{10}(0.20)^0$$
$$\approx 0.107$$

c)

$$P(\text{no more than 8 hits}) = P(Y \leq 8)$$
$$= P(Y=0) + P(Y=1) + P(Y=2) + \ldots + P(Y=8)$$
$$= {}_{10}C_0(0.80)^0(0.20)^{10} + {}_{10}C_1(0.80)^1(0.20)^9 + \ldots + {}_{10}C_8(0.80)^8(0.20)^2$$
$$\approx 0.624$$

d)

$$P(\text{exactly 8 out of 10 shots}) = P(Y=8)$$
$$= {}_{10}C_8(0.80)^8(0.20)^2$$
$$\approx 0.302$$

e)

$$P(\text{more hits than misses}) = P(Y \geq 6)$$
$$= P(Y=6) + P(Y=7) + \ldots + P(Y=10)$$
$$= {}_{10}C_6(0.80)^6(0.20)^4 + {}_{10}C_7(0.80)^7(0.20)^3 + \ldots + {}_{10}C_{10}(0.80)^{10}(0.20)^0$$
$$\approx 0.967$$

15. Tennis, anyone?

The first serves can be considered Bernoulli trials. There are only two possible outcomes, successful and unsuccessful. The probability of any first serve being good is given as $p = 0.70$. Finally, we are assuming that each serve is independent of the others. Since she is serving 6 times, use *Binom*(6,0.70).

Let X = the number of successful serves in $n = 6$ first serves.

a)

$P(\text{all six serves in }) = P(X = 6)$

$$= {}_6C_6(0.70)^6(0.30)^0$$
$$\approx 0.118$$

b)

$P(\text{exactly four serves in}) = P(X = 4)$

$$= {}_6C_4(0.70)^4(0.30)^2$$
$$\approx 0.324$$

c)

$P(\text{at least four serves in}) = P(X = 4) + P(X = 5) + P(X = 6)$

$$= {}_6C_4(0.70)^4(0.30)^2 + {}_6C_5(0.70)^5(0.30)^1 + {}_6C_6(0.70)^6(0.30)^0$$
$$\approx 0.744$$

d)

$P(\text{no more than four serves in}) = P(X = 0) + P(X = 1) + P(X = 2) + P(X = 3) + P(X = 4)$

$$= {}_6C_0(0.70)^0(0.30)^6 + {}_6C_1(0.70)^1(0.30)^5 + {}_6C_2(0.70)^2(0.30)^4$$
$$+ {}_6C_3(0.70)^3(0.30)^3 + {}_6C_4(0.70)^4(0.30)^2$$
$$\approx 0.580$$

16. Frogs.

The frog examinations can be considered Bernoulli trials. There are only two possible outcomes, having the trait and not having the trait. If the frequency of the trait has not changed, and the biologist collects a representative sample of frogs, then the probability of a frog having the trait is constant, at $p = 0.125$. The trials are not independent since the population of frogs is finite, but 12 frogs is fewer than 10% of all frogs. Since the biologist is collecting 12 frogs, use *Binom*(12,0.125).

Let $X =$ the number of frogs with the trait, from $n = 12$ frogs.

a)

$P(\text{no frogs have the trait}) = P(X = 0)$

$$= {}_{12}C_0(0.125)^0(0.875)^{12}$$
$$\approx 0.201$$

b)

$P(\text{at least two frogs}) = P(X \geq 2)$

$$= P(X = 2) + P(X = 3) + \ldots + P(X = 12)$$
$$= {}_{12}C_2(0.125)^2(0.875)^{10} + {}_{12}C_3(0.125)^3(0.875)^9 + \ldots + {}_{12}C_{12}(0.125)^{12}(0.875)^0$$
$$\approx 0.453$$

c)

$P(\text{three or four frogs have trait}) = P(X = 3) + P(X = 4)$

$$= {}_{12}C_3(0.125)^3(0.875)^9 + {}_{12}C_4(0.125)^4(0.875)^8$$
$$\approx 0.171$$

d)

$P(\text{no more than four}) = P(X \leq 4) = P(X = 0) + P(X = 1) + \ldots + P(X = 4)$

$$= {}_{12}C_0(0.125)^0(0.875)^{12} + {}_{12}C_1(0.125)^1(0.875)^{11} + \ldots + {}_{12}C_4(0.125)^4(0.875)^8$$
$$\approx 0.989$$

17. And more tennis.

The first serves can be considered Bernoulli trials. There are only two possible outcomes, successful and unsuccessful. The probability of any first serve being good is given as $p = 0.70$. Finally, we are assuming that each serve is independent of the others. Since she is serving 80 times, use *Binom*(80,0.70).

Let X = the number of successful serves in $n = 80$ first serves.

a) $E(X) = np = 80(0.70) = 56$ first serves in.

$SD(X) = \sqrt{npq} = \sqrt{80(0.70)(0.30)} \approx 4.10$ first serves in.

b) Since $np = 56$ and $nq = 24$ are both greater than 10, *Binom*(80,0.70) may be approximated by the Normal model, $N(56, 4.10)$.

c) According to the Normal model, in matches with 80 serves, she is expected to make between 51.9 and 60.1 first serves approximately 68% of the time, between 47.8 and 64.2 first serves approximately 95% of the time, and between 43.7 and 68.3 first serves approximately 99.7% of the time.

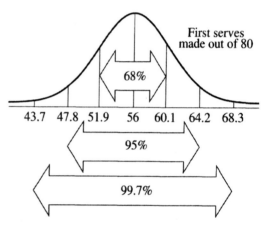

d) **Using *Binom*(80, 0.70):**
P(at least 65 first serves) = $P(X \geq 65)$

$$= P(X = 65) + P(X = 66) + \ldots + P(X = 80)$$

$$= {}_{80}C_{65}(0.70)^{65}(0.30)^{15} + {}_{80}C_{66}(0.70)^{66}(0.30)^{14} + \ldots + {}_{80}C_{80}(0.70)^{80}(0.30)^{0}$$

$$\approx 0.0161$$

According to the Binomial model, the probability that she makes at least 65 first serves out of 80 is approximately 0.0161.

Using $N(56, 4.10)$:

$z = \dfrac{x - \mu}{\sigma}$

$z = \dfrac{65 - 56}{4.10}$

$z \approx 2.195$

According to the Normal model, the probability that she makes at least 65 first serves out of 80 is approximately 0.0141.

$P(X \geq 65) \approx P(z > 2.195) \approx 0.0141$

18. More arrows.

These may be considered Bernoulli trials. There are only two possible outcomes, hitting the bull's-eye and not hitting the bull's-eye. The probability of hitting the bull's-eye is given, $p = 0.80$. The shots are assumed to be independent. Since she will be shooting 200 arrows, use *Binom*(200, 0.80).

Let $Y =$ the number of bull's-eyes in $n = 200$ shots.

a) $E(Y) = np = 200(0.80) = 160$ bull's-eyes.
   $SD(Y) = \sqrt{npq} = \sqrt{200(0.80)(0.20)} \approx 5.66$ bull's-eyes.

b) Since $np = 160$ and $nq = 40$ are both greater than 10, *Binom*(200, 0.80) may be approximated by the Normal model, $N(160, 5.66)$.

c) According to the Normal model, in matches with 200 arrows, she is expected to get between 154.34 and 165.66 bull's-eyes approximately 68% of the time, between 148.68 and 171.32 bull's-eyes approximately 95% of the time, and between 143.02 and 176.98 bull's-eyes approximately 99.7% of the time.

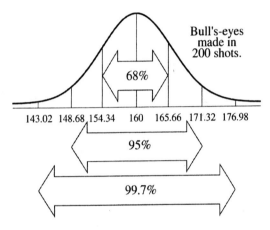

d) **Using *Binom*(200, 0.80):**
   $P(\text{at most } 140 \text{ hits}) = P(Y \le 140)$

   $$= P(Y = 0) + P(Y = 1) + \ldots + P(Y = 140)$$

   $$= {}_{200}C_0(0.80)^0(0.20)^{200} + {}_{200}C_1(0.80)^1(0.20)^{199} + \ldots + {}_{200}C_{140}(0.80)^{140}(0.70)^{60}$$

   $$\approx 0.0005$$

According to the Binomial model, the probability that she makes at most 140 bull's-eyes out of 200 is approximately 0.0005.

Using $N(160, 5.66)$:

$z = \dfrac{y - \mu}{\sigma}$

$z = \dfrac{140 - 160}{5.66}$

$z \approx -3.534$

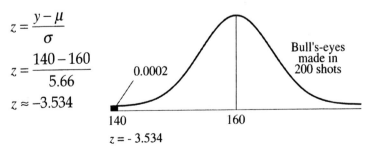

According to the Normal model, the probability that she hits at most 140 bull's-eyes out of 200 is approximately 0.0002.

$$P(Y \le 200) \approx P(z < -3.534) \approx 0.0002$$

Using either model, it is apparent that it is very unlikely that the archer would hit only 140 bull's-eyes out of 200.

19. Frogs, part II.

The frog examinations can be considered Bernoulli trials. There are only two possible outcomes, having the trait and not having the trait. If the frequency of the trait has not changed, and the biologist collects a representative sample of frogs, then the probability of a frog having the trait is constant, at $p = 0.125$. The trials are not independent since the population of frogs is finite, but 150 frogs is fewer than 10% of all frogs Since the biologist is collecting 150 frogs, use *Binom*(150,0.125).

Let X = the number of frogs with the trait, from $n = 150$ frogs.

a) $E(X) = np = 150(0.125) = 18.75$ frogs.

$SD(X) = \sqrt{npq} = \sqrt{150(0.125)(0.875)} \approx 4.05$ frogs.

b) Since $np = 18.75$ and $nq = 131.25$ are both greater than 10, *Binom*(200,0.125) may be approximated by the Normal model, $N(18.75, 4.05)$.

c) **Using *Binom*(150, 0.125):**
P(at least 22 frogs) $= P(X \geq 22)$

$$= P(X = 22) + \ldots + P(X = 150)$$

$$= {}_{150}C_{22}(0.125)^{22}(0.875)^{128} + \ldots + {}_{150}C_{150}(0.125)^{150}(0.875)^0$$

$$\approx 0.2433$$

According to the Binomial model, the probability that at least 22 frogs out of 150 have the trait is approximately 0.2433.

Using $N(18.75, 4.05)$:

$z = \dfrac{x - \mu}{\sigma}$

$z = \dfrac{22 - 18.75}{4.05}$

$z \approx 0.802$

$P(X \geq 22) \approx P(z > 0.802) \approx 0.2111$

According to the Normal model, the probability that at least 22 frogs out of 150 have the trait is approximately 0.2111.

Using either model, the probability that the biologist discovers 22 of 150 frogs with the trait simply as a result of natural variability is quite high. This doesn't prove that the trait has become more common.

20. Apples.

a) A binomial model and a normal model are both appropriate for modeling the number of cider apples that may come from the tree.

Let X = the number of cider apples found in the $n = 300$ apples from the tree.

The quality of the apples may be considered Bernoulli trials. There are only two possible outcomes, cider apple or not a cider apple. The probability that an apple must be used for a cider apple is constant, given as $p = 0.06$. The trials are not independent, since the population of apples is finite, but the apples on the tree are undoubtedly less than 10% of all the apples that the farmer has ever produced, so model with *Binom*(300, 0.06).

$E(X) = np = 300(0.06) = 18$ cider apples.

$SD(X) = \sqrt{npq} = \sqrt{300(0.06)(0.94)} \approx 4.11$ cider apples.

Since $np = 18$ and $nq = 282$ are both greater than 10, $Binom(300, 0.06)$ may be approximated by the Normal model, $N(18, 4.11)$.

b) Using $Binom(300, 0.06)$:

P(at most 12 cider apples) $= P(X \leq 12)$
$$= P(X = 0) + \ldots + P(X = 12)$$
$$= {}_{300}C_0(0.06)^0(0.94)^{300} + \ldots + {}_{300}C_{12}(0.06)^{12}(0.94)^{282}$$
$$\approx 0.085$$

According to the Binomial model, the probability that no more than 12 cider apples come from the tree is approximately 0.085.

Using $N(18, 4.11)$:

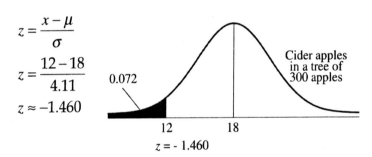

$z = \dfrac{x - \mu}{\sigma}$

$z = \dfrac{12 - 18}{4.11}$

$z \approx -1.460$

$z = -1.460$

$P(X \leq 12) \approx P(z < -1.460) \approx 0.072$

According to the Normal model, the probability that no more than 12 apples out of 300 are cider apples is approximately 0.072.

c) It is extremely unlikely that the tree will bear more than 50 cider apples. Using the Normal model, $N(18, 4.11)$, 50 cider apples is approximately 7.8 standard deviations above the mean.

21. Lefties again.

Let X = the number of righties among a class of $n = 188$ students.

Using $Binom(188, 0.87)$:
These may be considered Bernoulli trials. There are only two possible outcomes, right-handed and not right-handed. The probability of being right-handed is assumed to be constant at about 87%. The trials are not independent, since the population is finite, but a sample of 188 students is certainly fewer than 10% of all people. Therefore, the number of righties in a class of 188 students may be modeled by $Binom(188, 0.87)$.

If there are 171 or more righties in the class, some righties have to use a left-handed desk.

P(at least 171 righties) $= P(X \geq 171)$
$$= P(X = 171) + \ldots + P(X = 188)$$
$$= {}_{188}C_{171}(0.87)^{171}(0.13)^{17} + \ldots + {}_{188}C_{188}(0.87)^{188}(0.13)^0$$
$$\approx 0.061$$

According to the binomial model, the probability that a right-handed student has to use a left-handed desk is approximately 0.061.

Using N(163.56, 4.61):

$E(X) = np = 188(0.87) = 163.56$ righties.

$SD(X) = \sqrt{npq} = \sqrt{188(0.87)(0.13)} \approx 4.61$ righties.

Since $np = 163.56$ and $nq = 24.44$ are both greater than 10, *Binom*(188,0.87) may be approximated by the Normal model, *N*(163.56, 4.61).

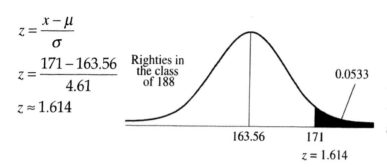

$z = \dfrac{x - \mu}{\sigma}$

$z = \dfrac{171 - 163.56}{4.61}$

$z \approx 1.614$

$P(X \geq 171) \approx P(z > 1.614) \approx 0.053$

According to the Normal model, the probability that there are at least 171 righties in the class of 188 is approximately 0.0533.

22. No-shows.

Let X = the number of passengers that show up for the flight of $n = 275$ passengers.

Using Binom(275, 0.95):
These may be considered Bernoulli trials. There are only two possible outcomes, showing up and not showing up. The airlines believe the probability of showing up is constant at about 95%. The trials are not independent, since the population is finite, but a sample of 275 passengers is certainly fewer than 10% of all passengers. Therefore, the number of passengers who show up for a flight of 275 may be modeled by *Binom*(275, 0.95).

If 266 or more passengers show up, someone has to get bumped off the flight.

P(at least 266 passengers) $= P(X \geq 266)$

$$= P(X = 266) + \ldots + P(X = 275)$$
$$= {}_{275}C_{266}(0.95)^{266}(0.05)^9 + \ldots + {}_{275}C_{275}(0.95)^{275}(0.05)^0$$
$$\approx 0.116$$

According to the binomial model, the probability someone on the flight must be bumped is approximately 0.116.

Using N(261.25, 3.61):

$E(X) = np = 275(0.95) = 261.25$ passengers.

$SD(X) = \sqrt{npq} = \sqrt{275(0.95)(0.05)} \approx 3.61$ passengers.

Since $np = 261.25$ and $nq = 13.75$ are both greater than 10, *Binom*(275,0.95) may be approximated by the Normal model, *N*(261.25, 3.61).

$$z = \frac{x - \mu}{\sigma}$$

$$z = \frac{266 - 261.25}{3.61}$$

$$z \approx 1.316$$

Passengers on a flight (275 booked)

$$P(X \geq 266) \approx P(z > 1.316) \approx 0.0941$$

According to the Normal model, the probability that at least 266 passengers show up is approximately 0.0941.

0.0941

261.25   266

$z = 1.316$

23. Annoying phone calls.

Let X = the number of sales made after making n = 200 calls.

Using *Binom*(200, 0.12):

These may be considered Bernoulli trials. There are only two possible outcomes, making a sale and not making a sale. The telemarketer was told that the probability of making a sale is constant at about p = 0.12. The trials are not independent, since the population is finite, but 200 calls is fewer than 10% of all calls. Therefore, the number of sales made after making 200 calls may be modeled by *Binom*(200, 0.12).

$$P(\text{at most } 10) = P(X \leq 10)$$

$$= P(X = 0) + \ldots + P(X = 10)$$

$$= {}_{200}C_0(0.12)^0(0.88)^{200} + \ldots + {}_{200}C_{10}(0.12)^{10}(0.88)^{190}$$

$$\approx 0.0006$$

According to the Binomial model, the probability that the telemarketer would make at most 10 sales is approximately 0.0006.

Using *N*(24, 4.60):

$E(X) = np = 200(0.12) = 24$ sales.

$SD(X) = \sqrt{npq} = \sqrt{200(0.12)(0.88)} \approx 4.60$ sales.

Since np = 24 and nq = 176 are both greater than 10, *Binom*(200,0.12) may be approximated by the Normal model, *N*(24, 4.60).

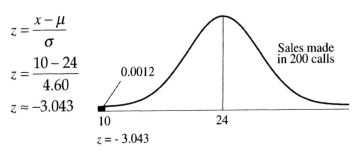

$$z = \frac{x - \mu}{\sigma}$$

$$z = \frac{10 - 24}{4.60}$$

$$z \approx -3.043$$

0.0012

Sales made in 200 calls

10   24

$z = -3.043$

$$P(X \leq 10) \approx P(z < -3.043) \approx 0.0012$$

According to the Normal model, the probability that the telemarketer would make at most 10 sales is approximately 0.0012.

Since the probability that the telemarketer made 10 sales, given that the 12% of calls result in sales is so low, it is likely that he was misled about the true success rate.

24. The euro.

Let X = the number of heads after spinning a Belgian euro n = 250 times.

Using *Binom*(250, 0.5):

These may be considered Bernoulli trials. There are only two possible outcomes, heads and tails. The probability that a fair Belgian euro lands heads is p = 0.5. The trials are independent, since the outcome of a spin does not affect other spins.

Therefore, *Binom*(250, 0.5) may be used to model the number of heads after spinning a Belgian euro 250 times.

$$P(\text{at least } 140) = P(X \geq 140)$$
$$= P(X = 140) + \ldots + P(X = 250)$$
$$= {}_{250}C_{140}(0.5)^{140}(0.5)^{110} + \ldots + {}_{250}C_{250}(0.5)^{250}(0.5)^{0}$$
$$\approx 0.0332$$

According to the Binomial model, the probability that a fair Belgian euro comes up heads at least 140 times is 0.0332.

Using *N*(125, 7.91):

$E(X) = np = 250(0.05) = 125$ heads.

$SD(X) = \sqrt{npq} = \sqrt{250(0.5)(0.5)} \approx 7.91$ heads.

Since np = 125 and nq = 125 are both greater than 10, *Binom*(250, 0.5) may be approximated by the Normal model, N(125, 7.91).

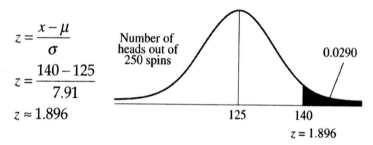

$z = \dfrac{x - \mu}{\sigma}$

$z = \dfrac{140 - 125}{7.91}$

$z \approx 1.896$

$P(X \geq 140) \approx P(z > 1.896) \approx 0.0290$

According to the Normal model, the probability that a fair Belgian euro lands heads at least 140 out of 250 spins is approximately 0.0290.

Since the probability that a fair Belgian euro lands heads at least 140 out of 250 spins is low, it is unlikely that the euro spins fairly. However, the probability is not extremely low, and we aren't sure of the source of the data, so it might be a good idea to spin it some more.

25. Seatbelts.

These stops may be considered Bernoulli trials. There are only two possible outcomes, belted or not belted. Police estimate that the probability that a driver is buckled is 80%. (The probability of not being buckled is therefore 20%.) Provided the drivers stopped are representative of all drivers, we can consider the probability constant. The trials are not independent, since the population of drivers is finite, but the police will not stop more than 10% of all drivers.

a) Let X = the number of cars stopped before finding a driver whose seat belt is not buckled. Use *Geom*(0.2) to model the situation.

$$E(X) = \frac{1}{p} = \frac{1}{0.2} = 5 \text{ cars.}$$

b) $P(\text{First unbelted driver is in the sixth car}) = P(X = 6) = (0.8)^5(0.2) \approx 0.066$

c) $P(\text{The first ten drivers are wearing seatbelts}) = (0.8)^{10} \approx .107$

d) Let Y = the number of drivers wearing their seatbelts in 30 cars. Use *Binom*(30, 0.8).

$$E(Y) = np = 30(0.8) = 24 \text{ drivers.}$$

$$SD(Y) = \sqrt{npq} = \sqrt{30(0.8)(0.2)} \approx 2.19 \text{ drivers.}$$

e) Let W = the number of drivers not wearing their seatbelts in 120 cars.

Using *Binom*(120, 0.2):

$P(\text{at least } 20) = P(W \geq 20)$

According to the Binomial model, the probability that at least 20 out of 120 drivers are not wearing their seatbelts is approximately 0.848.

$\qquad = P(W = 20) + \ldots + P(W = 120)$

$\qquad = {}_{120}C_{20}(0.2)^{20}(0.8)^{100} + \ldots + {}_{120}C_{120}(0.2)^{120}(0.8)^0$

$\qquad \approx 0.848$

Using *N*(24, 4.38):

$$E(W) = np = 120(0.2) = 24 \text{ drivers.}$$

$$SD(W) = \sqrt{npq} = \sqrt{120(0.2)(0.8)} \approx 4.38 \text{ drivers.}$$

Since $np = 24$ and $nq = 96$ are both greater than 10, *Binom*(120,0.2) may be approximated by the Normal model, *N*(24, 4.38).

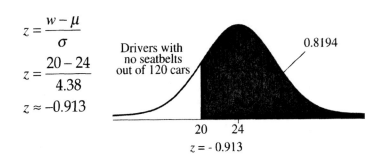

$$z = \frac{w - \mu}{\sigma}$$

$$z = \frac{20 - 24}{4.38}$$

$$z \approx -0.913$$

$P(W \geq 120) \approx P(z > -0.913) \approx 0.8194$

According to the Normal model, the probability that at least 20 out of 120 drivers stopped are not wearing their seatbelts is approximately 0.8194.

26. Rickets.

The selection of these children may be considered Bernoulli trials. There are only two possible outcomes, vitamin D deficient or not vitamin D deficient. Recent research indicates that 20% of British children are vitamin D-deficient. (The probability of not being vitamin D-deficient is therefore 80%.) Provided the students at this school are representative of all British children, we can consider the probability constant. The trials are not independent, since the population of British children is finite, but the children at this school represent fewer than 10% of all British children.

a) Let X = the number of students tested before finding a student who is vitamin D-deficient. Use *Geom*(0.2) to model the situation.

$P(\text{First vitamin D - deficient child is the eighth one tested}) = P(X = 8) = (0.8)^7(0.2) \approx 0.042$

b) $P(\text{The first ten children tested are okay}) = (0.8)^{10} \approx .107$

c) $E(X) = \dfrac{1}{p} = \dfrac{1}{0.2} = 5$ kids.

d) Let Y = the number of children who are vitamin D-deficient out of 50 children. Use *Binom*(50, 0.2).

$E(Y) = np = 50(0.2) = 10$ children.

$SD(Y) = \sqrt{npq} = \sqrt{50(0.2)(0.8)} \approx 2.83$ children.

e) **Using *Binom*(320, 0.2):**

$P(\text{no more than 50 children have the deficiency}) = P(X \le 50)$

$$= P(X = 0) + \ldots + P(X = 50)$$

$$= {}_{320}C_0(0.2)^0(0.8)^{320} + \ldots + {}_{320}C_{50}(0.2)^{50}(0.8)^{270}$$

$$\approx 0.027$$

According to the Binomial model, the probability that no more than 50 of the 320 children have the vitamin D deficiency is approximately 0.027.

Using *N*(64, 7.16):

$E(Y) = np = 320(0.2) = 64$ children.

$SD(Y) = \sqrt{npq} = \sqrt{320(0.2)(0.8)} \approx 7.16$ children.

Since $np = 64$ and $nq = 256$ are both greater than 10, *Binom*(320,0.2) may be approximated by the Normal model, *N*(64, 7.16).

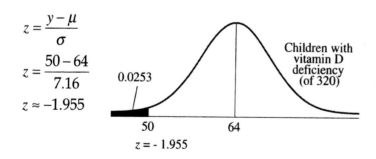

$z = \dfrac{y - \mu}{\sigma}$

$z = \dfrac{50 - 64}{7.16}$

$z \approx -1.955$

$P(Y \le 50) \approx P(z < -1.955) \approx 0.0253$

According to the Normal model, the probability that no more than 50 out of 320 children have the vitamin D deficiency is approximately 0.0253.

27. ESP.

Choosing symbols may be considered Bernoulli trials. There are only two possible outcomes, correct or incorrect. Assuming that ESP does not exist, the probability of a correct identification from a randomized deck is constant, at $p = 0.20$. The trials are independent, as long as the deck is shuffled after each attempt. Since 100 trials will be performed, use *Binom*(100, 0.2).

Let X = the number of symbols identified correctly out of 100 cards.

$E(X) = np = 100(0.2) = 20$ correct identifications.

$SD(X) = \sqrt{npq} = \sqrt{100(0.2)(0.8)} = 4$ correct identifications.

Answers may vary. In order be convincing, the "mind reader" would have to identify at least 32 out of 100 cards correctly, since 32 is three standard deviations above the mean. Identifying fewer cards than 32 could happen too often, simply due to chance.

28. True-False.

Guessing at answers may be considered Bernoulli trials. There are only two possible outcomes, correct or incorrect. If the student was guessing, the probability of a correct response is constant, at $p = 0.50$. The trials are independent, since the answer to one question should not have any bearing on the answer to the next. Since 50 questions are on the test use *Binom*(500, 0.5).

Let X = the number of questions answered correctly out of 50 questions.

$E(X) = np = 50(0.5) = 25$ correct answers.

$SD(X) = \sqrt{npq} = \sqrt{50(0.5)(0.5)} \approx 3.54$ correct answers.

Answers may vary. In order be convincing, the student would have to answer at least 36 out of 50 questions correctly, since 36 is approximately three standard deviations above the mean. Answering fewer than 36 questions correctly could happen too often, simply due to chance.

Review of Part IV

1. Quality Control.

Construct a Venn diagram of the disjoint outcomes.

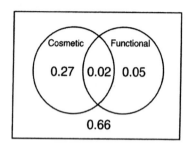

a) $P(\text{defect}) = P(\text{cosm.}) + P(\text{func.}) - P(\text{cosm. and func.})$
   $= 0.29 + 0.07 - 0.02 = 0.34$

Or, from the Venn: $0.27 + 0.02 + 0.05 = 0.34$

b) $P(\text{cosm. and no func.}) = P(\text{cosm.}) - P(\text{cosm. and func.})$
   $= 0.29 - 0.02 = 0.27$

Or, from the Venn: 0.27 (the region inside Cosmetic circle, yet outside Functional circle)

c) $P(\text{func.} \mid \text{cosm.}) = \dfrac{P(\text{func. and cosm.})}{P(\text{cosm.})} = \dfrac{0.02}{0.29} \approx 0.069$

From the Venn, consider only the region inside the Cosmetic circle. The probability that the car has a functional defect is 0.02 out of a total of 0.29 (the entire Cosmetic circle).

d) The two kinds of defects are not disjoint events, since 2% of cars have both kinds.

e) Approximately 6.9% of cars with cosmetic defects also have functional defects. Overall, the probability that a car has a cosmetic defect is 7%. The probabilities are estimates, so these are probably close enough to say that they two types of defects are independent.

2. Workers.

Organize the counts in a two-way table.

Job Type	Male	Female	Total
Management	7	6	13
Supervision	8	12	20
Production	45	72	117
Total	60	90	150

a) i) $P(\text{female}) = \dfrac{90}{150} = 0.6$

ii)
$P(\text{female or production}) = P(\text{female}) + P(\text{production}) - P(\text{female and production})$
$= \dfrac{90}{150} + \dfrac{117}{150} - \dfrac{72}{150}$
$= 0.9$

iii) Consider only the production row of the table. There are 72 women out of 117 production workers. $72/117 \approx 0.615$. Or, use the formula:

$P(\text{female} \mid \text{production}) = \dfrac{P(\text{female and production})}{P(\text{production})} = \dfrac{72/150}{117/150} \approx 0.615$

iv) Consider only the female column. There are 72 production workers out of a total of 90 women. 72/90 = 0.8. Or, use the formula:

$$P(\text{production} \mid \text{female}) = \frac{P(\text{production and female})}{P(\text{female})} = \frac{72/150}{90/150} = 0.8$$

b) These data suggest that holding a supervisory position is independent of gender.

$$P(\text{female} \mid \text{supervision}) = \frac{P(\text{female and supervison})}{P(\text{supervision})} = \frac{12/150}{20/150} = 0.6$$

60% of the plant employees are women, and 60% of the supervisors are women.

3. Airfares.

a) Let C = the price of a ticket to China
Let F = the price of a ticket to France.

Total price of airfare = $3C + 5F$

b) $\mu = E(3C + 5F) = 3E(C) + 5E(F) = 3(1000) + 5(500) = \$5500$

$\sigma = SD(3C + 5F) = \sqrt{3^2(Var(C)) + 5^2(Var(F))} = \sqrt{3^2(150^2) + 5^2(100^2)} \approx \$672.68$

c) $\mu = E(C - F) = E(C) - E(F) = 1000 - 500 = \$500$

$\sigma = SD(C - F) = \sqrt{Var(C) + Var(F)}$
$= \sqrt{150^2 + 100^2} \approx \$180.28$

d) No assumptions are necessary when calculating means. When calculating standard deviations, we must assume that ticket prices are independent of each other for different countries but all tickets to the same country are at the same price.

4. Bipolar.

Let X = the number of people with bipolar disorder in a city of n = 10,000 residents.

These may be considered Bernoulli trials. There are only two possible outcomes, having bipolar disorder or not having bipolar disorder. Psychiatrists estimate that the probability that a person has bipolar is about 1 in 100, so p = 0.01. We will assume that the cases of bipolar disorder are distributed randomly throughout the populations. The trials are not independent, since the population is finite, but 10,000 people represent fewer than 10% of all people. Therefore, the number of people with bipolar disorder in a city of 10,000 may be modeled by $Binom(10000, 0.01)$.

Since np = 100 and nq = 9900 are both greater than 10, $Binom(10000,0.01)$ may be approximated by the Normal model, $N(100, 9.95)$.

$E(X) = np = 10{,}000(0.01) = 100$ residents.

$SD(X) = \sqrt{npq} = \sqrt{10{,}000(0.01)(0.99)} \approx 9.95$ residents.

We expect 100 city residents to have bipolar disorder. According to the Normal model, 200 cases would be over 10 standard deviations above this mean. The probability of this occurring is essentially zero.

Technology can compute the probability according to the Binomial model. Again, the probability that 200 cases of bipolar disorder exist in the city is essentially zero. We use the Normal model in this case, since it gives us a more intuitive idea of just how unlikely this event is.

5. **A game.**

    a) Let X = net amount won

X	$0	$2	-$2
$P(X)$	0.10	0.40	0.50

$\mu = E(X) = 0(0.10) + 2(0.40) - 2(0.50) = -\$0.20$

$\sigma^2 = Var(X) = (0 - (-0.20))^2(0.10) + (2 - (-0.20))^2(0.40) + (-2 - (-0.20))^2(0.50) = 3.56$

$\sigma = SD(X) = \sqrt{Var(X)} = \sqrt{3.56} \approx \$1.89$

    b) $X + X$ = the total winnings for two plays.

$\mu = E(X + X) = E(X) + E(X) = (-0.20) + (-0.20) = -\$0.40$

$\sigma = SD(X + X) = \sqrt{Var(X) + Var(X)}$

$\qquad\qquad = \sqrt{1.89^2 + 1.89^2} \approx \$2.67$

6. **Emergency switch.**

    Construct a Venn diagram of the disjoint outcomes.

    a) From the Venn diagram, 3% of the workers were unable to operate the switch with either hand.

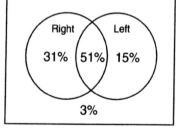

    b) $P(\text{left} \mid \text{right}) = \dfrac{P(\text{ left and right})}{P(\text{right})} = \dfrac{0.51}{0.82} \approx 0.622$

    About 62% of the workers who could operate the switch with their right hands could also operate it with left hands. Overall, the probability that a worker could operate the switch with their right hands was 66%. Workers who could operate the switch with their right hands were less likely to be able to operate the switch with their left hand, so success is not independent of hand.

    c) Success with right and left hands are not disjoint events. 51% of the workers had success with both hands.

7. **Twins.**

    The selection of these women can be considered Bernoulli trials. There are two possible outcomes, twins or no twins. As long as the women selected are representative of the population of all pregnant women, then $p = 1/90$. (If the women selected are representative of the population of women taking Clomid, then $p = 1/10$.) The trials are not independent since the population of all women is finite, but 10 women are fewer than 10% of the population of women.

Let X = the number of twin births from n = 10 pregnant women.

Let Y = the number of twin births from n = 10 pregnant women taking Clomid.

a) Use *Binom*(10, 1/90)

$$P(\text{at least one has twins}) = 1 - P(\text{none have twins})$$
$$= 1 - P(X = 0)$$
$$= 1 - {}_{10}C_0 \left(\frac{1}{90} \right)^0 \left(\frac{89}{90} \right)^{10}$$
$$\approx 0.106$$

b) Use *Binom*(10, 1/10)

$$P(\text{at least one has twins}) = 1 - P(\text{none have twins})$$
$$= 1 - P(Y = 0)$$
$$= 1 - {}_{10}C_0 \left(\frac{1}{10} \right)^0 \left(\frac{9}{10} \right)^{10}$$
$$\approx 0.651$$

c) Use *Binom*(5, 1/90) and *Binom*(5, 1/90).

$$P(\text{at least one has twins}) = 1 - P(\text{no twins without Clomid})P(\text{no twins with Clomid})$$
$$= 1 - \left({}_5C_0 \left(\frac{1}{90} \right)^0 \left(\frac{89}{90} \right)^5 \right) \left({}_5C_0 \left(\frac{1}{9} \right)^0 \left(\frac{9}{10} \right)^5 \right)$$
$$\approx 0.442$$

8. Deductible.

$$\mu = E(\text{cost}) = 500(0.005) = \$2.50$$

$$\sigma^2 = Var(\text{cost}) = (2.50 - 500)^2(0.005) + (2.50 - 0)^2(0.995) = 1243.75$$

$$\sigma = SD(\text{cost}) = \sqrt{Var(\text{cost})} = \sqrt{1243.75} \approx \$35.27$$

Expected (extra) cost of the cheaper policy with the deductible is $2.50, much less than the $12 for the no deductible surcharge, so on average she will save money by going with the deductible. The standard deviation, at $35.27, is quite high compared to the $12 surcharge, indicating a high amount of variability. The value of the car shouldn't influence the decision.

9. More twins.

In Exercise 7, it was determined that these were Bernoulli trials. Use *Binom*(5, 0.10).

Let X = the number of twin births from $n = 5$ pregnant women taking Clomid.

a)

$P(\text{none have twins}) = P(X = 0)$
$$= {}_5C_0(0.1)^0(0.9)^5$$
$$\approx 0.590$$

b)

$P(\text{exactly one has twins}) = P(X = 1)$
$$= {}_5C_1(0.1)^1(0.9)^4$$
$$\approx 0.328$$

c)

$P(\text{at least three will have twins}) = P(X = 3) + P(X = 4) + P(X = 5)$
$$= {}_5C_3(0.1)^3(0.9)^2 + {}_5C_4(0.1)^4(0.9)^1 + {}_5C_5(0.1)^5(0.9)^0$$
$$= 0.00856$$

10. At fault.

If we assume that these drivers are representative of all drivers insured by the company, then these insurance policies can be considered Bernoulli trials. There are only two possible outcomes, accident or no accident. The probability of having an accident is constant, $p = 0.005$. The trials are not independent, since the populations of all drivers is finite, but 1355 drivers represent fewer than 10% of all drivers. Use *Binom*(1355, 0.005).

a) Let X = the number of drivers who have an at-fault accident out of $n = 1355$.

$E(X) = np = 1,355(0.005) = 6.775$ drivers.

$SD(X) = \sqrt{npq} = \sqrt{1,355(0.005)(0.995)} \approx 2.60$ drivers.

b) Since $np = 6.775 < 10$, the Normal model cannot be used to model the number of drivers who are expected to have accidents. The Success/Failure condition is not satisfied.

11. Twins, part III.

In Exercise 7, it was determined that these were Bernoulli trials. Use *Binom*(152, 0.10).

Let X = the number of twin births from $n = 152$ pregnant women taking Clomid.

a) $E(X) = np = 152(0.10) = 15.2$ births.

$SD(X) = \sqrt{npq} = \sqrt{152(0.10)(0.90)} \approx 3.70$ births.

b) Since $np = 15.2$ and $nq = 136.8$ are both greater than 10, the Success/Failure condition is satisfied and *Binom*(152, 0.10) may be approximated by $N(15.2, 3.70)$.

c) Using *Binom*(152, 0.10):
$P(\text{no more than } 10) = P(X \le 10)$
$$= P(X = 0) + \ldots + P(X = 10)$$
$$= {}_{152}C_0(0.10)^0(0.90)^{152} + \ldots + {}_{152}C_{10}(0.10)^{10}(0.90)^{142}$$
$$\approx 0.097$$

According to the Binomial model, the probability that no more than 10 women would have twins is approximately 0.097.

Using N(15.2, 3.70):

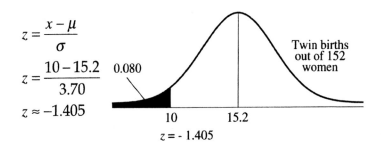

$$z = \frac{x - \mu}{\sigma}$$

$$z = \frac{10 - 15.2}{3.70}$$

$$z \approx -1.405$$

0.080

Twin births out of 152 women

10 15.2

$z = -1.405$

$$P(X \le 10) \approx P(z < -1.405) \approx 0.080$$

According to the Normal model, the probability that no more than 10 women would have twins is approximately 0.080.

12. Child's play.

a) Let X = the number indicated on the spinner

X	5	10	20
$P(X)$	0.5	0.25	0.25

b) $\mu = E(X) = 5(0.5) + 10(0.25) + 20(0.25) = 10$

$\sigma^2 = Var(X) = (5-10)^2(0.5) + (10-10)^2(0.25) + (20-10)^2(0.25) = 37.5$

$\sigma = SD(X) = \sqrt{Var(X)} = \sqrt{37.5} \approx 6.12$

c) Let Y = the number indicated on the die

Y	0	1	2	3	4
$P(Y)$	$\frac{1}{3}$	$\frac{1}{6}$	$\frac{1}{6}$	$\frac{1}{6}$	$\frac{1}{6}$

d) $\mu = E(Y) = 0\left(\frac{1}{3}\right) + 1\left(\frac{1}{6}\right) + 2\left(\frac{1}{6}\right) + 3\left(\frac{1}{6}\right) + 4\left(\frac{1}{6}\right) = \frac{10}{6} \approx 1.67$

$Var(Y) = \left(0 - \frac{10}{6}\right)^2\left(\frac{1}{3}\right) + \left(1 - \frac{10}{6}\right)^2\left(\frac{1}{6}\right) + \left(2 - \frac{10}{6}\right)^2\left(\frac{1}{6}\right) + \left(3 - \frac{10}{6}\right)^2\left(\frac{1}{6}\right) + \left(4 - \frac{10}{6}\right)^2\left(\frac{1}{6}\right) \approx 2.22$

$\sigma = SD(Y) = \sqrt{Var(Y)} = \sqrt{2.22} \approx 1.49$

e) $\mu = E(X+Y) = E(X) + E(Y) \approx 10 + 1.67 \approx 11.67$ spaces

$\sigma = SD(X+Y) = \sqrt{Var(X) + Var(Y)}$
$\approx \sqrt{37.5 + 2.22} \approx 6.30$ spaces

13. Language.

Assuming that the freshman composition class consists of 25 randomly selected people, these may be considered Bernoulli trials. There are only two possible outcomes, having a specified language center or not having the specified language center. The probabilities of the specified language centers are constant at 80%, 10%, or 10%, for right, left, and two-sided language center, respectively. The trials are not independent, since the population of people is finite, but we will select fewer than 10% of all people.

a) Let L = the number of people with left-brain language control from n = 25 people.

Use *Binom*(25, 0.80).

$$P(\text{no more than 15}) = P(L \leq 15)$$
$$= P(L=0) + \ldots + P(L=15)$$
$$= {}_{25}C_0 (0.80)^0 (0.20)^{25} + \ldots + {}_{25}C_{15}(0.80)^{15}(0.20)^{10}$$
$$\approx 0.0173$$

According to the Binomial model, the probability that no more than 15 students in a class of 25 will have left-brain language centers is approximately 0.0173.

b) Let T = the number of people with two-sided language control from n = 5 people.

Use *Binom*(5, 0.10).

$$P(\text{none have two-sided language control}) = P(T = 0)$$
$$= {}_5C_0 (0.10)^0 (0.90)^5$$
$$\approx 0.590$$

c) Use Binomial models:

$$E(\text{left}) = np_L = 1200(0.80) = 960 \text{ people}$$
$$E(\text{right}) = np_R = 1200(0.10) = 120 \text{ people}$$
$$E(\text{two} - \text{sided}) = np_T = 1200(0.10) = 120 \text{ people}$$

d) Let R = the number of people with right-brain language control.

$$E(R) = np_R = 1200(0.10) = 120 \text{ people}$$

$$SD(R) = \sqrt{np_R q_R} = \sqrt{1200(0.10)(0.90)} \approx 10.39 \text{ people.}$$

e) Since $np_R = 120$ and $nq_R = 1080$ are both greater than 10, the Normal model, $N(120, 10.39)$, may be used to approximate *Binom*(1200, 0.10). According to the Normal model, about 68% of randomly selected groups of 1200 people could be expected to have between 109.61 and 130.39 people with right-brain language control. About 95% of randomly selected groups of 1200 people could be expected to have between 99.22 and 140.78 people with right-brain language control. About 99.7% of randomly selected groups of 1200 people could be expected to have between 88.83 and 151.17 people with right-brain language control.

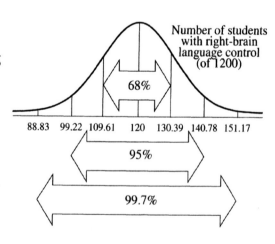

14. Play again.

$$\mu = E(X - Y) = E(X) - E(Y) \approx 10 - 1.67 \approx 8.33 \text{ spaces}$$

$$\sigma = SD(X - Y) = \sqrt{Var(X) + Var(Y)}$$
$$\approx \sqrt{37.5 + 2.22} \approx 6.30 \text{ spaces}$$

15. Beanstalks.

a) The greater standard deviation for men's heights indicates that men's heights are more variable than women's heights.

b) Admission to a Beanstalk Club is based upon extraordinary height for both men and women, but men are slightly more likely to qualify. The qualifying height for women is about 2.4 standard deviations above the mean height of women, while the qualifying height for men is about 1.75 standard deviations above the mean height for men.

c) Let M = the height of a randomly selected man from $N(69.1, 2.8)$.

Let W = the height of a randomly selected woman from $N(64.0, 2.5)$.

$M - W$ = the difference in height of a randomly selected man and woman.

d) $E(M-W) = E(M) - E(W) = 69.1 - 64.0 = 5.1$ inches

e) $SD(M-W) = \sqrt{Var(M) + Var(W)} = \sqrt{2.8^2 + 2.5^2} \approx 3.75$ inches

f) Since each distribution is described by a Normal model, the distribution of the difference in height between a randomly selected man and woman is $N(5.1, 3.75)$.

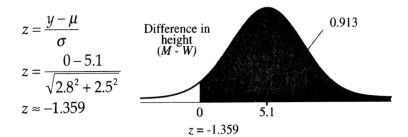

$$z = \frac{y - \mu}{\sigma}$$

$$z = \frac{0 - 5.1}{\sqrt{2.8^2 + 2.5^2}}$$

$$z \approx -1.359$$

According to the Normal model, the probability that a randomly selected man is taller than a randomly selected woman is approximately 0.913.

g) If people chose spouses independent of height, we would expect 91.3% of married couples to consist of a taller husband and shorter wife. The 92% that was seen in the survey is close to 91.3%, and the difference may be due to natural sampling variability. Unless this survey is very large, there is not sufficient evidence of association between height and choice of partner.

16. Stocks.

a) $P(\text{market will rise for 3 consecutive years}) = (0.73)^3 \approx 0.389$

b) Use *Binom*(5, 0.73).

$P(\text{market will rise in 3 out of 5 years}) = {_5}C_3 (0.73)^3 (0.27)^2 \approx 0.284$

c) $P(\text{fall in at least 1 of next 5 years}) = 1 - P(\text{no fall in 5 years}) = 1 - (0.73)^5 \approx 0.793$

d) Let X = the number of years in which the market rises. Use *Binom*(10, 0.73).
$P(\text{rises in the majority of years in a decade}) = P(X \geq 6)$

$$= P(X = 6) + \ldots + P(X = 10)$$
$$= {_{10}}C_6 (0.73)^6 (0.27)^4 + \ldots + {_{10}}C_{10} (0.73)^{10} (0.27)^0$$
$$\approx 0.896$$

17. Multiple choice.

Guessing at questions can be considered Bernoulli trials. There are only two possible outcomes, correct or incorrect. If you are guessing, the probability of success is $p = 0.25$, and the questions are independent. Use *Binom*(50, 0.25) to model the number of correct guesses on the test.

a) Let $X = $ the number of correct guesses.
$P(\text{at least 30 of 50 correct}) = P(X \geq 30)$
$$= P(X = 30) + \ldots + P(X = 50)$$
$$= {}_{50}C_{30}(0.25)^{30}(0.75)^{20} + \ldots + {}_{50}C_{50}(0.25)^{50}(0.75)^{0}$$
$$\approx 0.00000016$$

You are **very** unlikely to pass by guessing on every question.

b) Use *Binom*(50, 0.70).
$P(\text{at least 30 of 50 correct}) = P(X \geq 30)$
$$= P(X = 30) + \ldots + P(X = 50)$$
$$= {}_{50}C_{30}(0.70)^{30}(0.30)^{20} + \ldots + {}_{50}C_{50}(0.70)^{50}(0.30)^{0}$$
$$\approx 0.952$$

According to the Binomial model, your chances of passing are about 95.2%.

c) Use *Geom*(0.70).
$P(\text{first correct on third question}) = (0.30)^2(0.70) = 0.063$

18. Stock strategy.

a) This does not confirm the advice. Stocks have risen 75% of the time after a two-year fall, but there have only been eight occurrences of the two-year fall. The sample size is very small, and therefore highly variable.

b) Stocks have actually risen in 73% of years. This is not much different from the strategy of the advisors, which yielded a rise in 75% of years (from a very small sample of years.)

19. Insurance.

The company is expected to pay $100,000 only 2.6% of the time, while always gaining $520 from every policy sold. When they pay, they actually only pay $99,480.

$E(\text{profit}) = \$520(0.974) - \$99,480(0.026) = -\$2080$.

The expected profit is actually a **loss** of $2080 per policy. The company had better raise its premiums if it hopes to stay in business.

20. Teen smoking.

Randomly selecting high school students can be considered Bernoulli trials. There are only two possible outcomes, smoker or nonsmoker. The probability that a student is a smoker is $p = 0.30$. The trials are not independent, since the population is finite, but we are not sampling more than 10% of all high school students.

a) $P(\text{none of the first 4 are smokers}) = (0.7)^4 = 0.2401$

b) Use $Geom(0.3)$.

$P(\text{first smoker is the sixth person}) = (0.7)^5(0.3) \approx 0.050$

c) Use $Binom(10, 0.3)$. Let X = the number of smokers among $n = 10$ students.

$P(\text{no more than 2 smokers of 10}) = P(X \le 2)$

$$= P(X = 0) + P(X = 1) + P(X = 2)$$
$$= {}_{10}C_0(0.30)^0(0.70)^{10} + {}_{10}C_1(0.30)^1(0.70)^9 + {}_{10}C_2(0.30)^2(0.70)^8$$
$$\approx 0.383$$

21. Passing stats.

Organize the information in a tree diagram.

a)

$P(\text{Passing Statistics})$

$= P(\text{Scedastic and Pass})$

$+ P(\text{Kurtosis and Pass})$

$\approx 0.4667 + 0.25$

≈ 0.717

b)

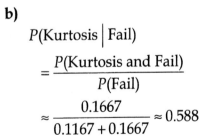

$P(\text{Kurtosis} \mid \text{Fail})$

$= \dfrac{P(\text{Kurtosis and Fail})}{P(\text{Fail})}$

$\approx \dfrac{0.1667}{0.1167 + 0.1667} \approx 0.588$

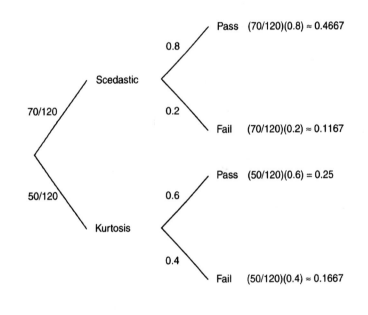

22. Teen smoking II.

In Exercise 22, it was determined that the selection of students could be considered to be Bernoulli trials.

a) Use $Binom(120, 0.30)$ to model the number of smokers out $n = 120$ students.

$E(\text{number of smokers}) = np = 120(0.30) = 36$ smokers.

b) $SD(\text{number of smokers}) = \sqrt{npq} = \sqrt{120(0.30)(0.70)} \approx 5.02$ smokers.

c) Since $np = 36$ and $nq = 84$ are both greater than 10, the Success/Failure condition is satisfied and $Binom(120, 0.30)$ may be approximated by $N(36, 5.02)$.

d) According to the Normal model, approximately 68% of samples of size $n = 120$ are expected to have between 30.98 and 41.02 smokers, approximately 95% of the samples are expected to have between 25.96 and 46.04 smokers, and approximately 99.7% of the samples are expected to have between 20.94 and 51.06 smokers.

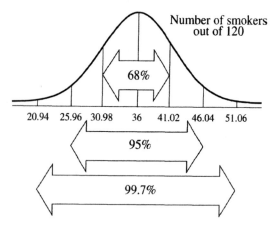

23. Random variables.

a)
$$\mu = E(X + 50) = E(X) + 50 = 50 + 50 = 100$$
$$\sigma = SD(X + 50) = SD(X) = 8$$

b)
$$\mu = E(10Y) = 10E(Y) = 10(100) = 1000$$
$$\sigma = SD(10Y) = 10SD(Y) = 60$$

c)
$$\mu = E(X + 0.5Y) = E(X) + 0.5E(Y)$$
$$= 50 + 0.5(100) = 100$$
$$\sigma = SD(X + 0.5Y) = \sqrt{Var(X) + 0.5^2 Var(Y)}$$
$$= \sqrt{8^2 + 0.5^2(6^2)} \approx 8.54$$

d)
$$\mu = E(X - Y) = E(X) - E(Y) = 50 - 100 = -50$$
$$\sigma = SD(X - Y) = \sqrt{Var(X) + Var(Y)}$$
$$= \sqrt{8^2 + 6^2} = 10$$

24. Merger.

Small companies may run in to trouble in the insurance business. Even if the expected profit from each policy is large, the profit is highly variable. There is a small chance that a company would have to make several huge payouts, resulting in an overall loss, not a profit. By combining two small companies together, the company takes in profit from more policies, making the larger company more resistant to the possibility of a large payout. This is because the total profit is increasing by the expected profit from each additional policy, but the standard deviation is increasing by the square root of the sum of the variances. The larger a company gets, the more the expected profit outpaces the variability associated with that profit.

25. Youth survey.

a) Many boys play computer games and use email, so the probabilities can total more than 100%. There is no evidence that there is a mistake in the report.

b) Playing computer games and using email are not disjoint. If they were, the probabilities would total 100% or less.

c) Emailing friends and gender are not independent. 76% of girls emailed friends in the last week, but only 65% of boys emailed. If emailing were independent of gender, the probabilities would be the same.

d) Let X = the number of students chosen until the first student is found who does not use the Internet. Use *Geom*(0.07). $P(X = 5) = (0.93)^4 (0.07) \approx 0.0524$.

26. Meals.

Let X = the amount the student spends daily.

a) $\mu = E(X + X) = E(X) + E(X) = 13.50 + 13.50 = \$27.00$

$$\sigma = SD(X + X) = \sqrt{Var(X) + Var(X)}$$
$$= \sqrt{7^2 + 7^2} \approx \$9.90$$

b) In order to calculate the standard deviation, we must assume that spending on different days is independent. This is probably not valid, since the student might tend to spend less on a day after he has spent a lot. He might not even have money left to spend!

c) $\mu = E(X + X + X + X + X + X + X) = 7E(X) = 7(\$13.50) = \$94.50$

$$\sigma = SD(X + X + X + X + X + X + X)$$
$$= \sqrt{Var(X) + Var(X) + Var(X) + Var(X) + Var(X) + Var(X) + Var(X)}$$
$$= \sqrt{7(7^2)} \approx \$18.52$$

d) Assuming once again that spending on different days is independent, it is unlikely that the student will spend less than \$50. This level of spending is about 2.4 standard deviations below the weekly mean. Don't try to approximate the probability! We don't know the shape of this distribution.

27. Travel to Kyrgyzstan.

a) If you spend an average of 4237 soms per day, you can stay about $\dfrac{90{,}000}{4237} \approx 21$ days.

b) Assuming that your daily spending is independent, the standard deviation is the square root of the sum of the variances for 21 days.

$$\sigma = \sqrt{21(360)^2} \approx 1649.73 \text{ soms}$$

c) The standard deviation in your total expenditures is about 1650 soms, so if you don't think you will exceed your expectation by more than 2 standard deviations, bring an extra 3300 soms. This gives you a cushion of about 157 soms for each of the 21 days.

28. Picking melons.

a) $\mu = E(\text{First} - \text{Second}) = E(\text{First}) - E(\text{Second}) = 22 - 18 = 4$ lbs.

b) $\sigma = SD(\text{First} - \text{Second}) = \sqrt{Var(\text{First}) + Var(\text{Second})} = \sqrt{2.5^2 + 2^2} \approx 3.20$ lbs.

c)

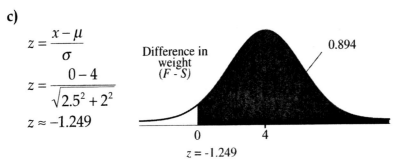

$z = \dfrac{x - \mu}{\sigma}$

$z = \dfrac{0 - 4}{\sqrt{2.5^2 + 2^2}}$

$z \approx -1.249$

According to the Normal model, the probability that a melon from the first store weighs more than a melon from the second store is approximately 0.894.

29. Home sweet home.

Since the homes are randomly selected, these can be considered Bernoulli trials. There are only two possible outcomes, owning the home or not owning the home. The probability of any randomly selected resident home being owned by the current resident is 0.66. The trials are not independent, since the population is finite, but as long as the city has more than 8200 homes, we are not sampling more than 10% of the population. The Binomial model, *Binom*(820, 0.66), can be used to model the number of homeowners among the 820 homes surveyed. Let H = the number of homeowners found in n = 820 homes.

$E(H) = np = 820(0.66) = 541.2$ homes.

$SD(H) = \sqrt{npq} = \sqrt{820(0.66)(0.34)} \approx 13.56$ homes.

The 523 homeowners found in the candidate's survey represent a number of homeowners that is only about 1.34 standard deviations below the expected number of homeowners. It is not particularly unusual to be 1.34 standard deviations below the mean. There is little support for the candidate's claim of a low level of home ownership.

30. Buying watermelons.

The mean price of a watermelon at the first store is 22(0.32) = \$7.04.
At the second store the mean price is 18(0.25) = \$4.50.
The difference in the price of the watermelons is expected to be \$7.04 – \$4.50 = \$2.54.

The standard deviation in price at the first store is 2.5(0.32) = \$0.80.
At the second store, the standard deviation in price is 2(0.25) = \$0.50.
The standard deviation of the difference is $\sqrt{0.80^2 + 0.50^2} \approx \$0.94$.

31. Who's the boss?

a) $P(\text{first three owned by women}) = (0.26)^3 \approx 0.018$

b) $P(\text{none of the first four are owned by women}) = (0.74)^4 \approx 0.300$

c) $P(\text{sixth firm called is owned by women} \mid \text{none of the first five were}) = 0.26$

Since the firms are chosen randomly, the fact that the first five firms were owned by men has no bearing on the ownership of the sixth firm.

32. Jerseys.

a) $P(\text{all four kids get the same color}) = 4\left(\dfrac{1}{4}\right)^4 \approx 0.0156$

(There are four different ways for this to happen, one for each color.)

b) $P(\text{all four kids get white}) = \left(\dfrac{1}{4}\right)^4 \approx 0.0039$

c) $P(\text{all four kids get white}) = \left(\dfrac{1}{6}\right)\left(\dfrac{1}{4}\right)^3 \approx 0.0026$

33. When to stop?

a) Since there are only two outcomes, 6 or not 6, the probability of getting a 6 is 1/6, and the trials are independent, these are Bernoulli trials. Use *Geom*(1/6).

$$\mu = \frac{1}{p} = \frac{1}{1/6} = 6 \text{ rolls}$$

b) If 6's are not allowed, the mean of each die roll is $\frac{1+2+3+4+5}{5} = 3$. You would expect to get 15 if you rolled 5 times.

c) $P(5 \text{ rolls without a } 6) = \left(\frac{5}{6}\right)^5 \approx 0.402$

34. Plan B.

a) If 6's are not allowed, the mean of each die roll is $\frac{1+2+3+4+5}{5} = 3$.

b) Let X = your current score. You expect to lose it all $\frac{1}{6}$ of the time, so your expected loss per roll is $\frac{1}{6}X$.

c) Expected gain equals expected loss when $\frac{1}{6}X = 3$. So, $X = 18$.

d) Roll until you get 18 points, then stop.

35. Technology on campus.

Construct a Venn diagram of the disjoint outcomes.

a)

$P(\text{neither tech.}) = 1 - P(\text{either tech.})$
$= 1 - \left[P(\text{calculator}) + P(\text{computer}) - P(\text{both})\right]$
$= 1 - \left[0.51 + 0.31 - 0.16\right]$
$= 0.34$

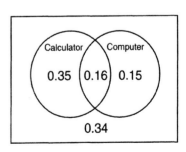

Or, from the Venn: 0.34 (the region outside both circles) This is MUCH easier.

34% of students use neither type of technology.

b) $P(\text{calc. and no comp.}) = P(\text{calc.}) - P(\text{calc. and comp.}) = 0.51 - 0.16 = 0.35$

Or, from the Venn: 0.35 (region inside the Calculator circle, outside the Computer circle)

35% of students use calculators, but not computers.

c) $P(\text{computer} \mid \text{calculator}) = \dfrac{P(\text{comp. and calc.})}{P(\text{calc.})} = \dfrac{0.16}{0.51} \approx 0.314$

About 31.4% of calculator users have computer assignments.

d) The percentage of computer users overall is 31%, while 31.4% of calculator users were computer users. These are very close. There is no indication of an association between computer use and calculator use.

36. Dogs.

Since the outcomes are disjoint, probabilities may be added and subtracted as needed.

a) $P(\text{no dogs}) = (0.77)(0.77) = 0.5929$

b) $P(\text{some dogs}) = 1 - P(\text{no dogs}) = 1 - (0.77)(0.77) = 0.4071$

c) $P(\text{both dogs}) = (0.23)(0.23) = 0.0529$

d) $P(\text{more than one dog in each}) = (0.05)(0.05) \approx 0.0025$

37. Socks.

Since we are sampling without replacement, use conditional probabilities throughout.

a) $P(2 \text{ blue}) = \left(\dfrac{4}{12}\right)\left(\dfrac{3}{11}\right) = \dfrac{12}{132} = \dfrac{1}{11}$

b) $P(\text{no grey}) = \left(\dfrac{7}{12}\right)\left(\dfrac{6}{11}\right) = \dfrac{42}{132} = \dfrac{7}{22}$

c) $P(\text{at least one black}) = 1 - P(\text{no black}) = 1 - \left(\dfrac{9}{12}\right)\left(\dfrac{8}{11}\right) = \dfrac{60}{132} = \dfrac{5}{11}$

d) $P(\text{green}) = 0$ (There aren't any green socks in the drawer.)

e) $P(\text{match}) = P(2 \text{ blue}) + P(2 \text{ grey}) + P(2 \text{ black}) = \left(\dfrac{4}{12}\right)\left(\dfrac{3}{11}\right) + \left(\dfrac{5}{12}\right)\left(\dfrac{4}{11}\right) + \left(\dfrac{3}{12}\right)\left(\dfrac{2}{11}\right) = \dfrac{38}{132} = \dfrac{19}{66}$

38. Coins.

Coin flips are Bernoulli trials. There are only two possible outcomes, the probability of each outcome is constant, and the trials are independent.

a) Use *Binom*(36, 0.5). Let H = the number of heads in $n = 36$ flips.

$\mu = E(H) = np = 36(0.5) = 18$ heads.

$\sigma = SD(H) = \sqrt{npq} = \sqrt{36(0.5)(0.5)} = 3$ heads.

b) Two standard deviations above the mean corresponds to 6 "extra" heads observed.

c) The standard deviation of the number of heads when 100 coins are flipped is

$\sigma = \sqrt{npq} = \sqrt{100(0.5)(0.5)} = 5$ heads. Getting 6 "extra" heads is not unusual.

d) Following the "two standard deviations" measurement, 10 or more "extra" heads would be unusual.

e) What appears surprising in the short run becomes expected in a larger number of flips. The "Law of Averages" is refuted, because the coin does not compensate in the long run. A coin that is flipped many times is actually less likely to show exactly half heads than a coin flipped only a few times. The Law of Large Numbers is confirmed, because the percentage of heads observed gets closer to the percentage expected due to probability.

39. The Drake equation.

a) $N \cdot f_p$ represents the number of stars in the Milky Way Galaxy expected to have planets.

b) $N \cdot f_p \cdot n_e \cdot f_l$ represents the number of planets in the Milky Way Galaxy expected to have intelligent life.

c) $f_l \cdot f_i$ is the probability that a planet has a suitable environment and has intelligent life.

d) $f_l = P(\text{life} \mid \text{suitable environment})$. This is the probability that life develops, if a planet has a suitable environment.

$f_i = P(\text{intelligence} \mid \text{life})$. This is the probability that the life develops intelligence, if a planet already has life.

$f_c = P(\text{communication} \mid \text{intelligence})$. This is the probability that radio communication develops, if a planet already has intelligent life.

40. Recalls.

Organize the information in a tree diagram.

a)

$$P(\text{recall}) = P(\text{American recall})$$
$$+ P(\text{Japanese recall})$$
$$+ P(\text{German recall})$$
$$= 0.014 + 0.002 + 0.001$$
$$= 0.017$$

b)

$$P(\text{American} \mid \text{recall}) = \frac{P(\text{American and recall})}{P(\text{recall})}$$
$$= \frac{0.014}{0.014 + 0.002 + 0.001}$$
$$\approx 0.824$$

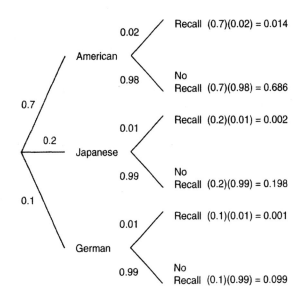

41. Pregnant?

Organize the information in a tree diagram.

$P(\text{pregnant} \mid \text{positive test})$

$= \dfrac{P(\text{pregnant and positive test})}{P(\text{positive test})}$

$= \dfrac{0.686}{0.686 + 0.006}$

≈ 0.991

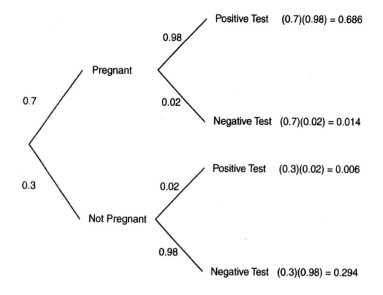

Pregnant
0.98 → Positive Test (0.7)(0.98) = 0.686
0.02 → Negative Test (0.7)(0.02) = 0.014
0.7

Not Pregnant
0.02 → Positive Test (0.3)(0.02) = 0.006
0.98 → Negative Test (0.3)(0.98) = 0.294
0.3

42. Door prize.

a) The probability that the first person in line wins is 1 out of 100, or 0.01.

b) If you are third in line, the two people ahead of you must not win in order for you to win. The probability is $(0.99)(0.99)(0.01) = 0.009801$.

c) There must be 100 losers in a row. The probability is $(0.99)^{100} \approx 0.366$.

d) The first person in line has the greatest chance of winning at $p = 0.01$. The probability of winning decreases from there, since winning is dependent upon everyone else in front of you in line losing.

e) Position is irrelevant now. Everyone has the same chance of winning, $p = 0.01$. One way to visualize this is to imagine that one ball is handed out to each person. Only one person out of the 100 people has the red ball. It might be you!

If you insist that the probabilities are still conditional, since you are sampling without replacement, look at it this way:

Consider $P(\text{sixth person wins}) = \left(\dfrac{99}{100}\right)\left(\dfrac{98}{99}\right)\left(\dfrac{97}{98}\right)\left(\dfrac{96}{97}\right)\left(\dfrac{95}{96}\right)\left(\dfrac{1}{95}\right) = \dfrac{1}{100}$

Chapter 18 – Sampling Distribution Models

1. **Coin tosses.**

   a) The histogram of these proportions is expected to be symmetric, but **not** because of the Central Limit Theorem. The sample of 16 coin flips is not large. The distribution of these proportions is expected to be symmetric because the probability that the coin lands heads is the same as the probability that the coin lands tails.

   b) The histogram is expected to have its center at 0.5, the probability that the coin lands heads.

   c) The standard deviation of data displayed in this histogram should be approximately equal to the standard deviation of the sampling distribution model, $\sqrt{\dfrac{pq}{n}} = \sqrt{\dfrac{(0.5)(0.5)}{16}} = 0.125$.

   d) The expected number of heads, $np = 16(0.5) = 8$, which is less than 10. The Success/Failure condition is not met. The Normal model is not appropriate in this case.

2. **M&M's.**

   a) The histogram of the proportions of green candies in the bags would probably be skewed slightly to the right, for the simple reason that the proportion of green M&M's could never fall below 0 on the left, but has the potential to be higher on the right.

   b) The Normal model cannot be used to approximate the histogram, since the expected number of green M&M's is $np = 50(0.10) = 5$, which is less than 10. The Success/Failure condition is not met.

   c) The histogram should be centered around the expected proportion of green M&M's, at about 0.10.

   d) The histogram should have standard deviation approximately equal to the standard deviation of the sampling distribution model, $\sqrt{\dfrac{pq}{n}} = \sqrt{\dfrac{(0.1)(0.9)}{50}} \approx 0.042$.

3. **More coins.**

   a) $\mu_{\hat{p}} = p = 0.5$ and $\sigma(\hat{p}) = \sqrt{\dfrac{pq}{n}} = \sqrt{\dfrac{(0.5)(0.5)}{25}} = 0.1$

   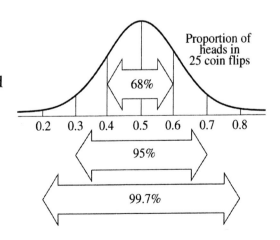

   About 68% of the sample proportions are expected to be between 0.4 and 0.6, about 95% are expected to be between 0.3 and 0.7, and about 99.7% are expected to be between 0.2 and 0.8.

b) First of all, coin flips are independent of one another. There is no need to check the 10% Condition. Second, $np = nq = 12.5$, so both are greater than 10. The Success/Failure condition is met, so the sampling distribution model is $N(0.5, 0.1)$.

c) $\mu_{\hat{p}} = p = 0.5$ and $\sigma(\hat{p}) = \sqrt{\dfrac{pq}{n}} = \sqrt{\dfrac{(0.5)(0.5)}{64}} = 0.0625$

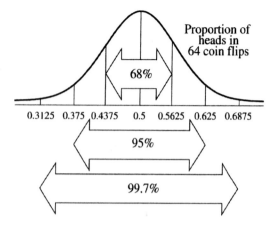

Proportion of heads in 64 coin flips

About 68% of the sample proportions are expected to be between 0.4375 and 0.5625, about 95% are expected to be between 0.375 and 0.625, and about 99.7% are expected to be between 0.3125 and 0.6875.

Coin flips are independent of one another, and $np = nq = 32$, so both are greater than 10. The Success/Failure condition is met, so the sampling distribution model is $N(0.5, 0.0625)$.

d) As the number of tosses increases, the sampling distribution model will still be Normal and centered at 0.5, but the standard deviation will decrease. The sampling distribution model will be less spread out.

4. Bigger bag.

a) **10% condition:** The 200 M&M's in the bag can be considered representative of all M&M's and 200 is certainly less than 10% of all M&M's.

Success/Failure condition: $np = 20$ and $nq = 180$ are both greater than 10.

b) The sampling distribution model is Normal, with:

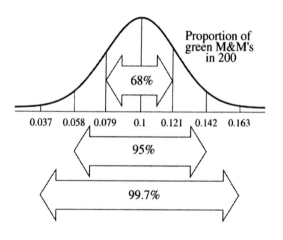

Proportion of green M&M's in 200

$\mu_{\hat{p}} = p = 0.1$ and $\sigma(\hat{p}) = \sqrt{\dfrac{pq}{n}} = \sqrt{\dfrac{(0.1)(0.9)}{200}} \approx 0.021$

About 68% of the sample proportions are expected to be between 0.079 and 0.121, about 95% are expected to be between 0.058 and 0.142, and about 99.7% are expected to be between 0.037 and 0.163.

c) If the bags contained more candies, the sampling distribution model would still be Normal and centered at 0.1, but the standard deviation would decrease. The sampling distribution model would be less spread out.

5. Just (un)lucky.

For 200 flips, the sampling distribution model is Normal with $\mu_{\hat{p}} = p = 0.5$ and $\sigma(\hat{p}) = \sqrt{\dfrac{pq}{n}} = \sqrt{\dfrac{(0.5)(0.5)}{200}} \approx 0.0354$. Her sample proportion of $\hat{p} = 0.42$ is about 2.26 standard deviations below the expected proportion, which is unusual, but not extraordinary. According to the Normal model, we expect sample proportions this low or lower about 1.2% of the time.

6. **Too many green ones?**

   For 500 candies, the sampling distribution model is Normal with $\mu_{\hat{p}} = p = 0.1$ and

   $\sigma(\hat{p}) = \sqrt{\dfrac{pq}{n}} = \sqrt{\dfrac{(0.1)(0.9)}{500}} \approx 0.01342$. The sample proportion of $\hat{p} = 0.12$ is about 1.49

   standard deviations above the expected proportion, which is not at all unusual. According to the Normal model, we expect sample proportions this high or higher about 6.8% of the time.

7. **Speeding.**

   a) $\mu_{\hat{p}} = p = 0.70$

   $\sigma(\hat{p}) = \sqrt{\dfrac{pq}{n}} = \sqrt{\dfrac{(0.7)(0.3)}{80}} \approx 0.051$.

   About 68% of the sample proportions are expected to be between 0.649 and 0.751, about 95% are expected to be between 0.598 and 0.802, and about 99.7% are expected to be between 0.547 and 0.853.

   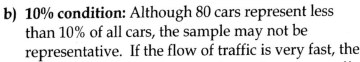

   b) **10% condition:** Although 80 cars represent less than 10% of all cars, the sample may not be representative. If the flow of traffic is very fast, the speed of the other cars around may have some effect on the speed of each driver. Likewise, if traffic is slow, the police may find a smaller proportion of speeders than they expect.

   Success/Failure condition: $np = 56$ and $nq = 24$ are both greater than 10.

   The Normal model may not be appropriate. Use caution. (And don't speed!)

8. **Smoking.**

   10% condition: 50 people are selected at random, and 50 is less than 10% of all people.

   Success/Failure condition: $np = 13.2$ and $nq = 36.8$ are both greater than 10.

   The sampling distribution model is Normal, with:
   $\mu_{\hat{p}} = p = 0.264$

   $\sigma(\hat{p}) = \sqrt{\dfrac{pq}{n}} = \sqrt{\dfrac{(0.264)(0.736)}{50}} \approx 0.062$

   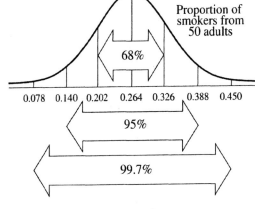

   There is an approximate chance of 68% that between 20.2% and 32.6% are smokers, an approximate chance of 95% that between 14.0% and 38.8% are smokers, and an approximate chance of 99.7% that between 7.8% and 45.0% are smokers.

9. **Loans.**

   a) $\mu_{\hat{p}} = p = 7\%$

   $\sigma(\hat{p}) = \sqrt{\dfrac{pq}{n}} = \sqrt{\dfrac{(0.07)(0.93)}{200}} \approx 1.8\%$

b) **10% condition:** Assume that the 200 people are a representative sample of all loan recipients. A sample of this size is less than 10% of all loan recipients.

Success/Failure condition: $np = 14$ and $nq = 186$ are both greater than 10.

Therefore, the sampling distribution model for the proportion of 200 loan recipients who will not make payments on time is $N(0.07, 0.018)$.

c) According to the Normal model, the probability that over 10% of these clients will not make timely payments is approximately 0.048.

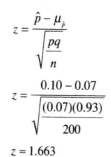

$$z = \frac{\hat{p} - \mu_{\hat{p}}}{\sqrt{\dfrac{pq}{n}}}$$

$$z = \frac{0.10 - 0.07}{\sqrt{\dfrac{(0.07)(0.93)}{200}}}$$

$$z \approx 1.663$$

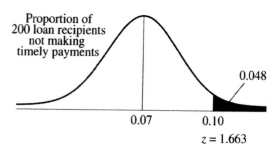

10. Contacts.

a) **10% condition:** 100 students are selected at random, and 100 is less than 10% of all of the students at the university, provided the university has more than 1000 students.

Success/Failure condition: $np = 30$ and $nq = 70$ are both greater than 10.

Therefore, the sampling distribution model for $\hat{p}$ is Normal, with:

$$\mu_{\hat{p}} = p = 0.30$$

$$\sigma(\hat{p}) = \sqrt{\frac{pq}{n}} = \sqrt{\frac{(0.30)(0.70)}{100}} \approx 0.046$$

b) According to the Normal model, the probability that more than one-third of the students in this sample wear contacts is approximately 0.234.

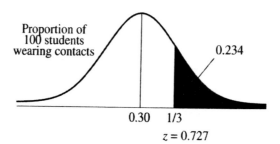

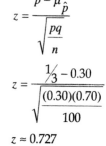

$$z = \frac{\hat{p} - \mu_{\hat{p}}}{\sqrt{\dfrac{pq}{n}}}$$

$$z = \frac{\frac{1}{3} - 0.30}{\sqrt{\dfrac{(0.30)(0.70)}{100}}}$$

$$z \approx 0.727$$

11. Polling.

10% condition: We will assume that the 400 voters polled were chosen randomly, and represent less than 10% of potential voters.

Success/Failure condition: $np = 208$ and $nq = 192$ are both greater than 10.

Therefore, the sampling distribution model for $\hat{p}$ is Normal, with:

$$\mu_{\hat{p}} = p = 0.52$$

$$\sigma(\hat{p}) = \sqrt{\frac{pq}{n}} = \sqrt{\frac{(0.52)(0.48)}{400}} \approx 0.025$$

According to the Normal model, the probability that the newspaper's sample will lead them to predict defeat (that is, predict budget support below 50%) is approximately 0.212.

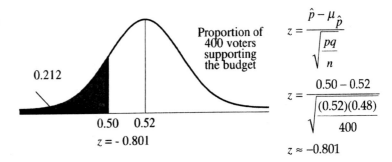

$$z = \frac{\hat{p} - \mu_{\hat{p}}}{\sqrt{\dfrac{pq}{n}}}$$

$$z = \frac{0.50 - 0.52}{\sqrt{\dfrac{(0.52)(0.48)}{400}}}$$

$$z \approx -0.801$$

12. Seeds.

10% condition: We will assume that the 160 seeds in a pack are a random sample. The 160 seeds represent less than 10% of all seeds. The seeds in each pack may not be a random sample, so proceed with caution.

Success/Failure condition: $np = 147.2$ and $nq = 12.8$ are both greater than 10.

Therefore, the sampling distribution model for $\hat{p}$ is Normal, with:

$$\mu_{\hat{p}} = p = 0.92$$

$$\sigma(\hat{p}) = \sqrt{\frac{pq}{n}} = \sqrt{\frac{(0.92)(0.08)}{160}} \approx 0.0215$$

According to the Normal model, the probability that more than 95% of the seeds will germinate is approximately 0.081.

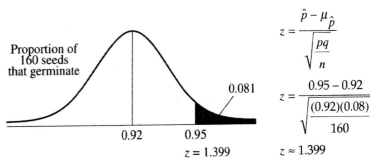

$$z = \frac{\hat{p} - \mu_{\hat{p}}}{\sqrt{\dfrac{pq}{n}}}$$

$$z = \frac{0.95 - 0.92}{\sqrt{\dfrac{(0.92)(0.08)}{160}}}$$

$$z \approx 1.399$$

13. Apples.

10% condition: A random sample of 150 apples is taken from each truck, and 150 is less than 10% of all apples.

Success/Failure Condition: $np = 12$ and $nq = 138$ are both greater than 10.

Therefore, the sampling distribution model for $\hat{p}$ is Normal, with:

$$\mu_{\hat{p}} = p = 0.08$$

$$\sigma(\hat{p}) = \sqrt{\frac{pq}{n}} = \sqrt{\frac{(0.08)(0.92)}{150}} \approx 0.0222$$

According to the Normal model, the probability that less than 5% of the apples in the sample are unsatisfactory is approximately 0.088.

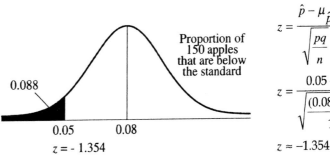

$$z = \frac{\hat{p} - \mu_{\hat{p}}}{\sqrt{\dfrac{pq}{n}}}$$

$$z = \frac{0.05 - 0.08}{\sqrt{\dfrac{(0.08)(0.92)}{150}}}$$

$$z \approx -1.354$$

14. Genetic Defect.

10% condition: We will assume that the 732 newborns are representative of all newborns. These 732 newborns certainly represent less than 10% of all newborns.

Success/Failure condition: $np = 29.28$ and $nq = 702.72$ are both greater than 10.

Therefore, the sampling distribution model for $\hat{p}$ is Normal, with:

$$\mu_{\hat{p}} = p = 0.04$$

$$\sigma(\hat{p}) = \sqrt{\frac{pq}{n}} = \sqrt{\frac{(0.04)(0.96)}{732}} \approx 0.0072$$

In order to get the 20 newborns for the study, the researchers hope to find at least $\hat{p} = \dfrac{20}{732} \approx 0.0273$ as the proportion of newborns in the sample with juvenile diabetes.

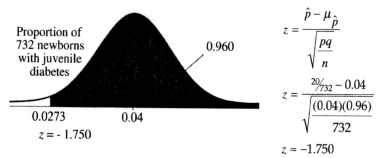

According to the Normal model, the probability that the researchers find at least 20 newborns with juvenile diabetes is approximately 0.960.

Proportion of 732 newborns with juvenile diabetes

0.960

0.0273 0.04

$z = -1.750$

$$z = \frac{\hat{p} - \mu_{\hat{p}}}{\sqrt{\dfrac{pq}{n}}}$$

$$z = \frac{\frac{20}{732} - 0.04}{\sqrt{\dfrac{(0.04)(0.96)}{732}}}$$

$$z \approx -1.750$$

15. Nonsmokers.

10% condition: We will assume that the 120 customers (to fill the restaurant to capacity) are representative of all customers, and represent less than 10% of all potential customers.

Success/Failure condition: $np = 72$ and $nq = 48$ are both greater than 10.

Therefore, the sampling distribution model for $\hat{p}$ is Normal, with:

$$\mu_{\hat{p}} = p = 0.60$$

$$\sigma(\hat{p}) = \sqrt{\frac{pq}{n}} = \sqrt{\frac{(0.60)(0.40)}{120}} \approx 0.0447$$

Answers may vary. We will use 3 standard deviations above the expected proportion of customers who demand nonsmoking seats to be "very sure".

$$\mu_{\hat{p}} + 3\left(\sqrt{\frac{pq}{n}}\right) \approx 0.60 + 3(0.0447) \approx 0.734$$

Since $120(0.734) = 88.08$, the restaurant needs at least 89 seats in the nonsmoking section.

16. Meals.

10% condition: We will assume that the 180 customers are representative of all customers, and represent less than 10% of all potential customers.

Success/Failure condition: $np = 36$ and $nq = 144$ are both greater than 10.

Therefore, the sampling distribution model for $\hat{p}$ is Normal, with:

$\mu_{\hat{p}} = p = 0.20$

$\sigma(\hat{p}) = \sqrt{\dfrac{pq}{n}} = \sqrt{\dfrac{(0.20)(0.80)}{180}} \approx 0.0298$

Answers may vary. We will use 2 standard deviations above the expected proportion of customers who order the steak special to be "pretty sure".

$\mu_{\hat{p}} + 2\left(\sqrt{\dfrac{pq}{n}}\right) \approx 0.20 + 2(0.0298) \approx 0.2596$

Since $180(0.2596) = 46.728$, the chef needs at least 47 steaks on hand.

17. Sampling.

a) The sampling distribution model for the sample mean is $N\left(\mu, \dfrac{\sigma}{\sqrt{n}}\right)$.

b) If we choose a larger sample, the mean of the sampling distribution model will remain the same, but the standard deviation will be smaller.

18. Sampling, part II.

a) The sampling distribution model for the sample mean will be skewed to the left as well, centered at μ, with standard deviation $\dfrac{\sigma}{\sqrt{n}}$.

b) When the sample size is increased, the sampling distribution model becomes more Normal in shape, centered at μ, with standard deviation $\dfrac{\sigma}{\sqrt{n}}$.

c) As we make the sample size larger, the sampling distribution model becomes increasingly Normal in shape, centered at μ, with standard deviation $\dfrac{\sigma}{\sqrt{n}}$. The distribution becomes less variable as the sample size increases.

19. GPAs.

Random sampling condition: Assume that the students are randomly assigned to seminars.

Independence assumption: It is reasonable to think that GPAs for randomly selected students are mutually independent.

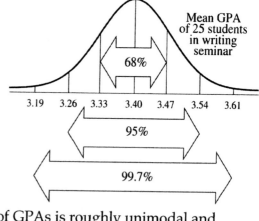

10% condition: The 25 students in the seminar certainly represent less than 10% of the population of students.

The mean GPA for the freshmen was $\mu = 3.4$, with standard deviation $\sigma = 0.35$. Since the distribution of GPAs is roughly unimodal and symmetric, and the sample size is not too small, the Central Limit Theorem tells us that we can model the sampling distribution of the mean GPA with a Normal model, with

$\mu_{\bar{y}} = 3.4$ and standard deviation $\sigma(\bar{y}) = \dfrac{0.35}{\sqrt{25}} \approx 0.07$.

The sampling distribution model for the sample mean GPA is approximately $N(3.4, 0.07)$.

20. Home values.

Random sampling condition: Homes were selected at random.

Independence assumption: It is reasonable to think that assessments for randomly selected homes are mutually independent.

10% condition: The 100 homes in the sample certainly represent less than 10% of the population of all homes in the city. A small city will likely have more than 1,000 homes.

The mean home value was $\mu = \$140,000$, with standard deviation $\sigma = \$60,000$. Since the sample is large, the Central Limit Theorem tells us that we can model the sampling distribution of the mean home value with a Normal model, with $\mu_{\bar{y}} = \$140,000$ and standard deviation

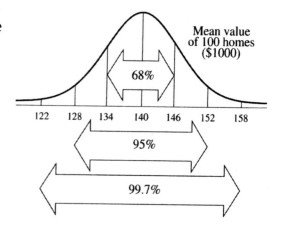

$\sigma(\bar{y}) = \dfrac{60,000}{\sqrt{100}} = \$6000$.

The sampling distribution model for the sample mean home values is approximately $N(140000, 6000)$.

21. Pregnancy.

a)

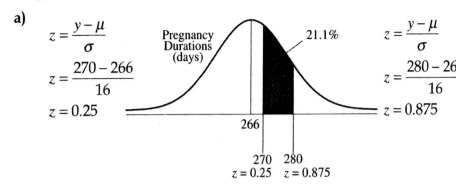

$z = \dfrac{y - \mu}{\sigma}$

$z = \dfrac{270 - 266}{16}$

$z = 0.25$

$z = \dfrac{y - \mu}{\sigma}$

$z = \dfrac{280 - 266}{16}$

$z = 0.875$

According to the Normal model, approximately 21.1% of all pregnancies are expected to last between 270 and 280 days.

b)

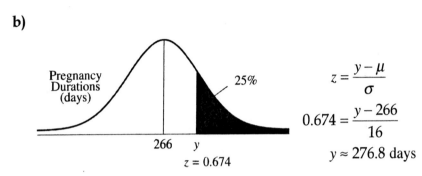

$z = \dfrac{y - \mu}{\sigma}$

$0.674 = \dfrac{y - 266}{16}$

$y \approx 276.8 \text{ days}$

According to the Normal model, the longest 25% of pregnancies are expected to last approximately 276.8 days or more.

c) **Random sampling condition:** Assume that the 60 women the doctor is treating can be considered a representative sample of all pregnant women.

Independence assumption: It is reasonable to think that the durations of the patients' pregnancies are mutually independent.

10% condition: The 60 women that the doctor is treating certainly represent less than 10% of the population of all women.

The mean duration of the pregnancies was $\mu = 266$ days, with standard deviation $\sigma = 16$ days. Since the distribution of pregnancy durations is Normal, we can model the sampling distribution of the mean pregnancy duration with a Normal model, with $\mu_{\bar{y}} = 266$ days and standard deviation $\sigma(\bar{y}) = \dfrac{16}{\sqrt{60}} \approx 2.07$ days.

d) According to the Normal model, the probability that the mean pregnancy duration is less than 260 days is 0.002.

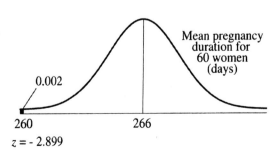

22. Rainfall.

a) According to the Normal model, Ithaca is expected
   to get more than 40 inches of rain in approximately
   13.7% of years.

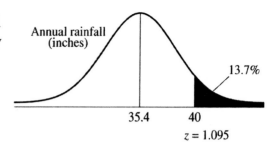

b) According to the Normal
   model, Ithaca is expected to get
   less than 31.9 inches of rain in
   driest 20% of years.

$$z = \frac{y - \mu}{\sigma}$$

$$-0.842 = \frac{y - 35.4}{4.2}$$

$$y \approx 31.9$$

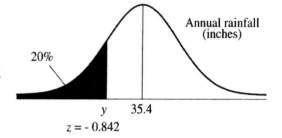

c) **Random sampling condition:** Assume that the 4 years in which the student was in Ithaca
   can be considered a representative sample of all years.

   Independence assumption: It is reasonable to think that the rainfall totals for the years are
   mutually independent.

   10% condition: The 4 years in which the student was in Ithaca certainly represent less than
   10% of all years.

   The mean rainfall was $\mu = 35.4$ inches, with standard deviation $\sigma = 4.2$ inches. Since the
   distribution of yearly rainfall is Normal, we can model the sampling distribution of the
   mean annual rainfall with a Normal model, with $\mu_{\bar{y}} = 35.4$ inches and standard deviation

   $$\sigma(\bar{y}) = \frac{4.2}{\sqrt{4}} = 2.1 \text{ inches}.$$

d) According to the Normal model, the probability
   that those four years averaged less than 30 inches
   of rain is 0.005.

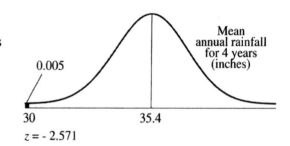

23. Pregnant again.

a) The distribution of pregnancy durations may be skewed to the left since there are more
   premature births than very long pregnancies. Modern practice of medicine stops
   pregnancies at about 2 weeks past normal due date by inducing labor or performing a
   Caesarean section.

b) We can no longer answer the questions posed in parts a and b. The Normal model is not
   appropriate for skewed distributions. The answer to part c is still valid. The Central Limit
   Theorem guarantees that the sampling distribution model is Normal when the sample size
   is large.

24. At work.

a) The distribution of length of time people work at a job is likely to be skewed to the right, because some people stay at the same job for much longer than the mean plus two or three standard deviations. Additionally, the left tail cannot be long, because a person cannot work at a job for less than 0 years.

b) The Central Limit Theorem guarantees that the distribution of the mean time is Normally distributed for large sample sizes, as long as the assumptions and conditions are satisfied. The CLT doesn't help us with the distribution of individual times.

25. Dice and dollars.

a) Let X = the number of dollars won in one play.

$$\mu = E(X) = 0\left(\frac{3}{6}\right) + 1\left(\frac{2}{6}\right) + 10\left(\frac{1}{6}\right) = \$2$$

$$\sigma^2 = Var(X) = (0-2)^2\left(\frac{3}{6}\right) + (1-2)^2\left(\frac{2}{6}\right) + (10-2)^2\left(\frac{1}{6}\right) = 13$$

$$\sigma = SD(X) = \sqrt{Var(X)} = \sqrt{13} \approx \$3.61$$

b) $X + X$ = the total winnings for two plays.

$$\mu = E(X + X) = E(X) + E(X) = 2 + 2 = \$4$$

$$\sigma = SD(X + X) = \sqrt{Var(X) + Var(X)}$$
$$= \sqrt{13 + 13} \approx \$5.10$$

c) In order to win at least $100 in 40 plays, you must average at least $\dfrac{100}{40} = \$2.50$ per play.

The expected value of the winnings is $\mu = \$2$, with standard deviation $\sigma = \$3.61$. Rolling a die is random and the outcomes are mutually independent, so the Central Limit Theorem guarantees that the sampling distribution model is Normal with with $\mu_{\bar{x}} = \$2$ and standard deviation $\sigma(\bar{x}) = \dfrac{\$3.61}{\sqrt{40}} \approx \$0.571$.

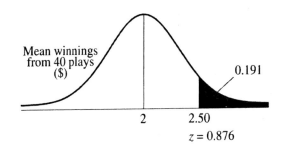

According to the Normal model, the probability that you win at least $100 in 40 plays is approximately 0.191.

(This is equivalent to using $N(80, 22.83)$ to model your total winnings.)

26. New game.

a) Let X = the amount of money won.

X	$40	$0	– $10
$P(X)$	$\dfrac{1}{6}$	$\left(\dfrac{5}{6}\right)\left(\dfrac{1}{6}\right) = \dfrac{5}{36}$	$\left(\dfrac{5}{6}\right)\left(\dfrac{5}{6}\right) = \dfrac{25}{36}$

b) $\mu = E(X) = 40\left(\dfrac{1}{6}\right) + 0\left(\dfrac{5}{36}\right) - 10\left(\dfrac{25}{36}\right) \approx -\$0.28$

$\sigma^2 = Var(X) = (40 - (-0.28))^2\left(\dfrac{1}{6}\right) + (0 - (-0.28))^2\left(\dfrac{5}{36}\right) + (-10 - (-0.28))^2\left(\dfrac{25}{36}\right) \approx 336.034$

$\sigma = SD(X) = \sqrt{Var(X)} = \sqrt{336.034} \approx \$18.33$

c) $\mu = E(X + X + X + X + X) = 5E(X) = 5(-0.28) = -\$1.40$

$\sigma = SD(X + X + X + X + X) = \sqrt{5(Var(X))} = \sqrt{5(336.034)} \approx \$40.99$

d) In order for the person running the game to make a profit, the average winnings of the 100 people must be less than $0.

The expected value of the winnings is $\mu = -\$0.28$, with standard deviation $\sigma = \$18.33$. Rolling a die is random and the outcomes are mutually independent, so the Central Limit Theorem guarantees that the sampling distribution model is Normal with with $\mu_{\bar{x}} = -\$0.28$

and standard deviation $\sigma(\bar{x}) = \dfrac{18.33}{\sqrt{100}} \approx \$1.833$.

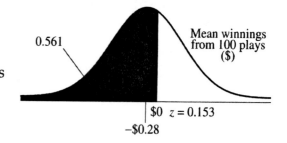

According to the Normal model, the probability that the person running the game makes a profit is approximately 0.561.

27. Pollution.

a) Random sampling condition: Assume that the 80 cars can be considered a representative sample of all cars of this type.

Independence assumption: It is reasonable to think that the CO emissions for these cars are mutually independent.

10% condition: The 80 cars in the fleet certainly represent less than 10% of all cars of this type.

The mean CO level was $\mu = 2.9$ gm/mi, with standard deviation $\sigma = 0.4$ gm/mi. Since the sample size is large, the CLT allows us to model the sampling distribution of the $\bar{y}$ with a Normal model, with $\mu_{\bar{y}} = 2.9$ gm/mi and standard deviation $\sigma(\bar{y}) = \dfrac{0.4}{\sqrt{80}} = 0.045$ gm/mi.

b) According to the Normal model, the probability that $\bar{y}$ is between 3.0 and 3.1 gm/mi is approximately 0.0131.

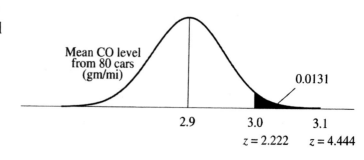

c) According to the Normal model, there is only a 5% chance that the fleet's mean CO level is greater than approximately 2.97 gm/mi.

$$z = \frac{\bar{y} - \mu_{\bar{y}}}{\sigma(\bar{y})}$$

$$1.645 = \frac{\bar{y} - 2.9}{0.045}$$

$$\bar{y} \approx 2.97$$

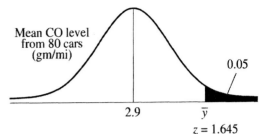

28. Potato chips.

a) According to the Normal model, only about 4.78% of the bags sold are underweight.

b) P(none of the 3 bags are underweight)
$= (1 - 0.0478)^3 \approx 0.863.$

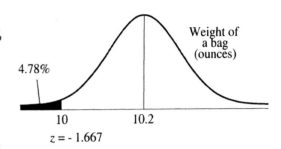

c) **Random sampling condition:** Assume that the 3 bags can be considered a representative sample of all bags.

Independence assumption: It is reasonable to think that the weights of these bags are mutually independent.

10% condition: The 3 bags certainly represent less than 10% of all bags.

The mean weight is $\mu = 10.2$ ounces, with standard deviation $\sigma = 0.12$ ounces. Since the distribution of weights is believed to be Normal, we can model the sampling distribution of $\bar{y}$ with a Normal model, with $\mu_{\bar{y}} = 10.2$ ounces and standard deviation

$$\sigma(\bar{y}) = \frac{0.12}{\sqrt{3}} \approx 0.069 \text{ ounces.}$$

According to the Normal model, the probability that the mean weight of the 3 bags is less than 10 ounces is approximately 0.0019.

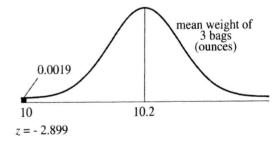

d) For 24 bags, the standard deviation of the sampling distribution model is $\sigma(\bar{y}) = \frac{0.12}{\sqrt{24}} \approx 0.024$ ounces. Now, an average of 10 ounces is over 8 standard deviations below the mean of the sampling distribution model. The probability of this happening is essentially 0.

29. Tips.

a) Since the distribution of tips is skewed to the right, we can't use the Normal model to determine the probability that a given party will tip at least $20.

b) No. A sample of 4 parties is probably not a large enough sample for the CLT to allow us to use the Normal model to estimate the distribution of averages.

c) A sample of 10 parties may not be large enough to allow the use of a Normal model to describe the distribution of averages. It would be risky to attempt to estimate the probability that his next 10 parties tip an average of $15. However, since the distribution of tips has $\mu = \$9.60$, with standard deviation $\sigma = \$5.40$, we still know that the mean of the sampling distribution model is $\mu_{\bar{y}} = \$9.60$ with standard deviation $\sigma(\bar{y}) = \dfrac{5.40}{\sqrt{10}} \approx \$1.71$.

We don't know the exact shape of the distribution, but we can still assess the likelihood of specific means. A mean tip of $15 is over 3 standard deviations above the expected mean tip for 10 parties. That's not very likely to happen.

30. Groceries.

a) Since the distribution of purchases is skewed, we can't use the Normal model to determine the probability that a given purchase is at least $40.

b) A sample of 10 customers may not be large enough for the CLT to allow the use of a Normal model for the sampling distribution model. If the distribution of purchases is only slightly skewed, the sample may be large enough, but if the distribution is heavily skewed, it would be very risky to attempt to determine the probability.

c) **Random sampling condition:** Assume that the 50 purchases can be considered a representative sample of all purchases.

Independence assumption: It is reasonable to think that the purchases are mutually independent, unless there is a sale or other incentive to purchase more.

10% condition: The 50 purchases certainly represent less than 10% of all purchases.

The mean purchase is $\mu = \$32$, with standard deviation $\sigma = \$20$. Since the sample is large, the CLT allows us to model the sampling distribution of $\bar{y}$ with a Normal model, with $\mu_{\bar{y}} = \$32$ and standard deviation $\sigma(\bar{y}) = \dfrac{20}{\sqrt{50}} \approx \$2.83$.

According to the Normal model, the probability that the mean purchase of the 50 customers is at least $40 is approximately 0.0023.

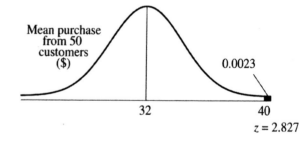

31. More tips.

a) **Random sampling condition:** Assume that the tips from 40 parties can be considered a representative sample of all tips.

Independence assumption: It is reasonable to think that the tips are mutually independent, unless the service is particularly good or bad during this weekend.

10% condition: The tips of 40 parties certainly represent less than 10% of all tips.

The mean tip is $\mu = \$9.60$, with standard deviation $\sigma = \$5.40$. Since the sample size is large, the CLT allows us to model the sampling distribution of $\bar{y}$ with a Normal model, with $\mu_{\bar{y}} = \$9.60$ and standard deviation $\sigma(\bar{y}) = \dfrac{5.40}{\sqrt{40}} \approx \$0.8538$.

In order to earn at least $500, the waiter would have to average $\dfrac{500}{40} = \$12.50$ per party.

According to the Normal model, the probability that the waiter earns at least $500 in tips in a weekend is approximately 0.0003.

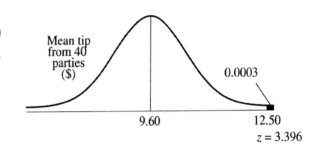

b) According to the Normal model, the waiter can expect to have a mean tip of about $10.6942, which corresponds to about $427.77 for 40 parties, in the best 10% of such weekends.

$$z = \frac{\bar{y} - \mu_{\bar{y}}}{\sigma(\bar{y})}$$

$$1.2816 = \frac{\bar{y} - 9.60}{0.8538}$$

$$\bar{y} \approx 10.6942$$

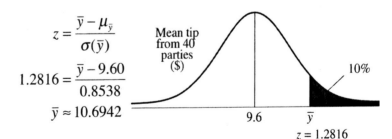

32. More groceries.

a) Assumptions and conditions for the use of the CLT were verified in Exercise 30.

The mean purchase is $\mu = \$32$, with standard deviation $\sigma = \$20$. Since the sample is large, the CLT allows us to model the sampling distribution of $\bar{y}$ with a Normal model, with $\mu_{\bar{y}} = \$32$ and standard deviation $\sigma(\bar{y}) = \dfrac{20}{\sqrt{312}} \approx \$1.1323$.

In order to have revenues of at least $10,000, the mean purchase must be at least $\dfrac{10,000}{312} = \$32.0513$.

According to the Normal model, the probability of having a mean purchase at least that high (and therefore at total revenue of at least $10,000) is 0.482.

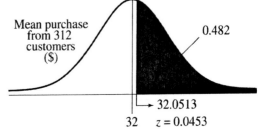

b) According to the Normal model, the mean purchase on the worst 10% of such days is approximately $30.548928, so 312 customers are expected to spend about $9531.27.

$$z = \frac{\bar{y} - \mu_{\bar{y}}}{\sigma(\bar{y})}$$

$$-1.281552 = \frac{\bar{y} - 32}{\frac{20}{\sqrt{312}}}$$

$$\bar{y} \approx 30.548928 \quad z = -1.281552$$

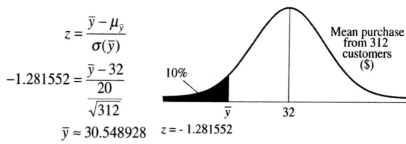

Mean purchase from 312 customers ($)

10%

$\bar{y}$ 32

(Be sure to carry as much accuracy as you can, since all of the multiplying can magnify error. Use exact values and technology whenever possible. That's just hard to show here.)

33. IQs.

a) According to the Normal model, the probability that the IQ of a student from East State is at least 125 is approximately 0.734.

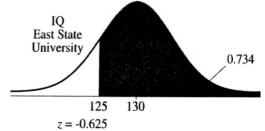

IQ
East State
University

0.734

125   130

$z = -0.625$

b) First, we will need to generate a model for the difference in IQ between the two schools. Since we are choosing at random, it is reasonable to believe that the students' IQs are independent, which allows us to calculate the standard deviation of the difference.

$$\mu = E(E - W) = E(E) - E(W) = 130 - 120 = 10$$

$$\sigma = SD(E - W) = \sqrt{Var(E) + Var(W)}$$

$$= \sqrt{8^2 + 10^2} \approx 12.806$$

Since both distributions are Normal, the distribution of the difference is $N(10, 12.806)$.

According to the Normal model, the probability that the IQ of a student at ESU is at least 5 points higher than a student at WSU is approximately 0.652.

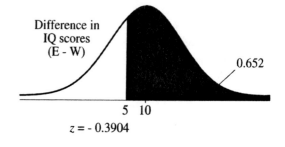

Difference in
IQ scores
(E - W)

0.652

5   10

$z = -0.3904$

c) **Random sampling condition:** Students are randomly sampled from WSU.

Independence assumption: It is reasonable to think that the IQs are mutually independent.

10% condition: The 3 students certainly represent less than 10% of students.

The mean IQ is $\mu_w = 120$, with standard deviation $\sigma_w = 10$. Since the distribution IQs is Normal, we can model the sampling distribution of $\bar{w}$ with a Normal model, with $\mu_{\bar{w}} = 120$ with standard deviation

$$\sigma(\bar{w}) = \frac{10}{\sqrt{3}} \approx 5.7735.$$

According to the Normal model, the probability that the mean IQ of the 3 WSU students is above 125 is approximately 0.193.

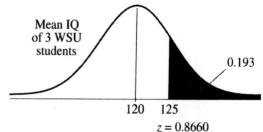

Mean IQ
of 3 WSU
students

0.193

120   125

$z = 0.8660$

d) As in part c, the sampling distribution of $\bar{e}$, the mean IQ of 3 ESU students, can be modeled with a Normal model, with $\mu_{\bar{e}} = 130$ with standard deviation $\sigma(\bar{e}) = \dfrac{8}{\sqrt{3}} \approx 4.6188$.

The distribution of the difference in mean IQ is Normal, with the following parameters:

$$\mu_{\bar{e}-\bar{w}} = E(\bar{e} - \bar{w}) = E(\bar{e}) - E(\bar{w}) = 130 - 120 = 10$$

$$\sigma_{\bar{e}-\bar{w}} = SD(\bar{e} - \bar{w}) = \sqrt{Var(\bar{e}) + Var(\bar{w})}$$

$$= \sqrt{\left(\frac{10}{\sqrt{3}}\right)^2 + \left(\frac{8}{\sqrt{3}}\right)^2} \approx 7.3937$$

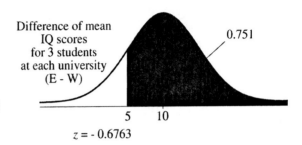

According to the Normal model, the probability that the mean IQ of 3 ESU students is at least 5 points higher than the mean IQ of 3 WSU students is approximately 0.751.

34. Milk.

a) According to the Normal model, the probability that an Ayrshire averages more than 50 pounds of milk per day is approximately 0.309.

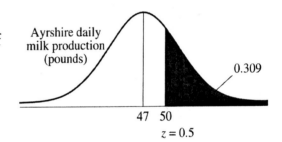

b) First, we will need to generate a model for the difference in milk production between the two cows. Since we are choosing at random, it is reasonable to believe that the cows' milk productions are independent, which allows us to calculate the standard deviation of the difference.

$$\mu = E(A - J) = E(A) - E(J) = 47 - 43 = 4 \text{ pounds}$$

$$\sigma = SD(A - J) = \sqrt{Var(A) + Var(J)}$$

$$= \sqrt{6^2 + 5^2} \approx 7.8102 \text{ pounds}$$

Since both distributions are Normal, the distribution of the difference is $N(4, 7.8102)$.

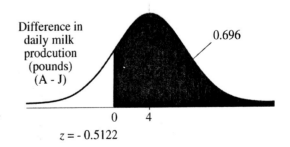

According to the Normal model, the probability that the Ayrshire gives more milk than the Jersey is approximately 0.696.

c) **Random sampling condition:** Assume that the farmer's 20 Jerseys are a representative sample of all Jerseys.

Independence assumption: It is reasonable to think that the cows have mutually independent milk production.

10% condition: The 20 cows certainly represent less than 10% of cows.

The mean milk production is $\mu_j = 43$ pounds, with standard deviation $\sigma_j = 5$. Since the distribution of milk production is Normal, we can model the sampling distribution of $\bar{j}$ with a Normal model, with $\mu_{\bar{j}} = 43$ pounds with standard deviation $\sigma(\bar{j}) = \dfrac{5}{\sqrt{20}} \approx 1.1180$.

According to the Normal model, the probability that the mean milk production of the 20 Jerseys is above 45 pounds of milk per day is approximately 0.037.

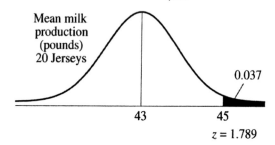

d) As in part c, the sampling distribution of $\bar{a}$, the mean milk production of 10 Ayrshires, can be modeled with a Normal model, with $\mu_{\bar{a}} = 47$ pounds with standard deviation

$$\sigma(\bar{a}) = \frac{6}{\sqrt{10}} \approx 1.8974 \text{ pounds.}$$

The distribution of the difference in mean milk production is Normal, with the following parameters:

$$\mu_{\bar{a}-\bar{j}} = E(\bar{a} - \bar{j}) = E(\bar{a}) - E(\bar{j}) = 47 - 43 = 4 \text{ pounds}$$

$$\sigma_{\bar{a}-\bar{j}} = SD(\bar{a} - \bar{j}) = \sqrt{Var(\bar{a}) + Var(\bar{j})}$$

$$= \sqrt{\left(\frac{6}{\sqrt{10}}\right)^2 + \left(\frac{5}{\sqrt{20}}\right)^2} \approx 2.2023 \text{ pounds}$$

According to the Normal model, the probability that the mean milk production of 10 Ayrshires is at least 5 pounds higher than the mean milk production of 20 Jerseys is approximately 0.325.

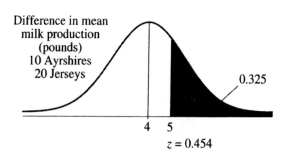

Chapter 19 – Confidence Intervals for Proportions

1. **Margin of error.**

   He believes the true proportion of voters with a certain opinion is within 4% of his estimate, with some degree of confidence, perhaps 95% confidence.

2. **Margin of error.**

   He believes the true percentage of children who are exposed to lead-base paint is within 3% of his estimate, with some degree of confidence, perhaps 95% confidence.

3. **Conditions.**

   a) *Population* – all cars; *sample* – 134 cars actually stopped at the checkpoint; p – proportion of all cars with safety problems; $\hat{p}$ – proportion of cars in the sample that actually have safety problems (10.4%).

   Plausible independence condition: There is no reason to believe that the safety problems of cars are related to each other.

   Randomization condition: This sample is not random, so hopefully the cars stopped are representative of cars in the area.

   10% condition: The 134 cars stopped represent a small fraction of all cars, certainly less than 10%.

   Success/Failure condition: $n\hat{p} = 14$ and $n\hat{q} = 120$ are both greater than 10, so the sample is large enough.

   A one-proportion z-interval can be created for the proportion of all cars in the area with safety problems.

   b) *Population* – the general public; *sample* – 602 viewers that logged on to the Web site; p –proportion of the general public that support prayer in school; $\hat{p}$- proportion of viewers that logged on to the Web site and voted that support prayer in schools (81.1%).

   Randomization condition: This sample is not random, but biased by voluntary response. It would be very unwise to attempt to use this sample to infer anything about the opinion of the general public related to school prayer.

   c) *Population* – parents at the school; *sample* – 380 parents who returned surveys; p – proportion of all parents in favor of uniforms; $\hat{p}$ – proportion of those who responded that are in favor of uniforms (60%).

   Randomization condition: This sample is not random, but rather biased by nonresponse. There may be lurking variables that affect the opinions of parents who return surveys (and the children who deliver them!).

   It would be very unwise to attempt to use this sample to infer anything about the opinion of the parents about uniforms.

   d) *Population* – all freshmen enrollees at the college (not just one year); *sample* – 1632 freshmen during the specified year; p – proportion of all students who will graduate on time; $\hat{p}$ – proportion of students from that year who graduate on time (85.05%).

   Plausible independence condition: It is reasonable to think that the abilities of students to graduate on time are mutually independent.

Randomization condition: This sample is not random, but this year's freshmen class is probably representative of freshman classes in other years.

10% condition: The 1632 students in that years freshmen class represent less than 10% of all possible students.

Success/Failure condition: $n\hat{p} = 1388$ and $n\hat{q} = 244$ are both greater than 10, so the sample is large enough.

A one-proportion z-interval can be created for the proportion of freshmen that graduate on time from this college.

4. **More conditions.**

a) *Population* – all customers who recently bought new cars; *sample* – 167 people surveyed about their experience; p – proportion of all new car buyers who are dissatisfied with the salesperson; $\hat{p}$ – proportion of new car buyers surveyed who are dissatisfied with the salesperson (3%).

Success/Failure condition: $n\hat{p} = 167(0.03) = 5$ and $n\hat{q} = 162$. Since only 5 people were dissatisfied, the sample is **not** large enough to use a confidence interval to estimate the proportion of dissatisfied car buyers.

b) *Population* – all college students; *sample* – 2883 who were asked about their cell phones at the football stadium; p – proportion of all college students with cell phones; $\hat{p}$ – proportion of college students at the football stadium with cell phones (8.4%).

Plausible independence condition: Whether or not a student has a cell phone shouldn't affect the probability that another does.

Randomization condition: This sample is not random. The best we can hope for is that the students at the football stadium are representative of all college students.

10% condition: The 2883 students at the football stadium represent less than 10% of all college students.

Success/Failure condition: $n\hat{p} = 243$ and $n\hat{q} = 2640$ are both greater than 10, so the sample is large enough.

Extreme caution should be used when using a one-proportion z-interval to estimate the proportion of college students with cell phones. The students at the football stadium may not be representative of all students.

c) *Population* – potato plants in the U.S.; *sample* – 240 potato plants in a field in Maine; p – proportion of all potato plants in the U.S. that show signs of blight; $\hat{p}$ – proportion of potato plants in the sample that show signs of blight (2.9%).

Plausible independence condition: It is **not** reasonable to think that signs of blight are independent. Blight is a contagious disease!

Randomization condition: Although potato plants are randomly selected from the field in Maine, it doesn't seem reasonable to assume that these potato plants are representative of all potato plants in the U.S.

Success/Failure condition: $n\hat{p} = 7$ and $n\hat{q} = 233$. There are only 7 (less than 10!) plants with signs of blight. The sample is not large enough.

Three conditions are not met! Don't use a confidence interval to attempt to estimate the percentage of potato plants in the U.S. that show signs of blight.

d) *Population* – all employees at the company; *sample* – all employees during the specified year; *p* – proportion of all employees who will have an injury on the job in a year; $\hat{p}$ – proportion of employees who had an injury on the job during the specified year.
Plausible independence condition: It is reasonable to think that the injuries are mutually independent.
Randomization condition: This sample is not random, but this year's employees are probably representative of employees in other years, with regards to injury on the job.
10% condition: The 309 employees represent less than 10% of all possible employees over many years.
Success/Failure condition: $n\hat{p} = 12$ and $n\hat{q} = 297$ are both greater than 10, so the sample is large enough.
A one-proportion z-interval can be created for the proportion of employees who are expected to suffer an injury on the job in future years, provided that this year is representative of future years.

5. **Conclusions.**

   a) Not correct. This statement implies certainty. There is no level of confidence in the statement.

   b) Not correct. Different samples will give different results. Many fewer than 95% of samples are expected to have *exactly* 88% on-time orders.

   c) Not correct. A confidence interval should say something about the unknown population proportion, not the sample proportion in different samples.

   d) Not correct. We *know* that 88% of the orders arrived on time. There is no need to make an interval for the sample proportion.

   e) Not correct. The interval should be about the proportion of on-time orders, not the days.

6. **More conclusions.**

   a) Not correct. This statement implies certainty. There is no level of confidence in the statement.

   b) Not correct. We *know* that 56% of the spins in this experiment landed heads. There is no need to make an interval for the sample proportion.

   c) Not correct. The interval should be about the proportion of heads, not the spins.

   d) Not correct. The interval should be about the proportion of heads, not the spins.

   e) Not correct. The interval should be about the proportion of heads, not the percentage of euros.

7. **Confidence intervals.**

   a) False. For a given sample size, higher confidence means a *larger* margin of error.

   b) True. Larger samples lead to smaller standard errors, which lead to smaller margins of error.

   c) True. Larger samples are less variable, which makes us more confident that a given confidence interval succeeds in catching the population proportion.

    d) False. The margin of error decreases as the square root of the sample size increases. Halving the margin of error requires a sample four times as large as the original.

8. Confidence intervals, again.

    a) True. The smaller the margin of error is, the less confidence we have in the ability of our interval to catch the population proportion.

    b) True. Larger samples are less variable, which translates to a smaller margin of error. We can be more precise at the same level of confidence.

    c) True. Smaller samples are more variable, leading us to be less confident in the ability of our interval to catch the true population proportion.

    d) True. The margin of error decreases as the square root of the sample size increases.

9. Cars.

    We are 90% confident that between 29.9% and 47.0% of cars are made in Japan.

10. Parole.

    We are 95% confident that between 56.1% and 62.5% of paroles are granted by the Nebraska Board of Parole.

11. Ghosts.

    a) $ME = z^* \times SE(\hat{p}) = z^* \times \sqrt{\dfrac{\hat{p}\hat{q}}{n}} = 1.645 \times \sqrt{\dfrac{(0.38)(0.62)}{1012}} \approx 2.5\%$

    b) The pollsters are 90% confident that the true proportion of adults who believe in ghosts is within 2.5% of the estimated 38%.

    c) A 99% confidence interval requires a larger margin of error. In order to increase confidence, the interval must be wider.

    d) $ME = z^* \times SE(\hat{p}) = z^* \times \sqrt{\dfrac{\hat{p}\hat{q}}{n}} = 2.576 \times \sqrt{\dfrac{(0.38)(0.62)}{1012}} \approx 3.9\%$

    e) Smaller margins of error will give us less confidence in the interval.

12. Cloning.

    a) $ME = z^* \times SE(\hat{p}) = z^* \times \sqrt{\dfrac{\hat{p}\hat{q}}{n}} = 1.960 \times \sqrt{\dfrac{(0.08)(0.92)}{1012}} \approx 1.7\%$

    b) The pollsters are 95% confident that the true proportion of adults who approve of attempts to clone a human is within 1.7% of the estimated 8%.

    c) A 90% confidence interval requires a smaller margin of error. If confidence is decreased, a smaller interval is allowed.

    d) $ME = z^* \times SE(\hat{p}) = z^* \times \sqrt{\dfrac{\hat{p}\hat{q}}{n}} = 1.645 \times \sqrt{\dfrac{(0.08)(0.92)}{1012}} \approx 1.4\%$

    e) Smaller samples generally produce larger intervals. Smaller samples are more variable, which increases the margin of error.

13. Teenage drivers.

a) **Plausible independence condition:** There is no reason to believe that accidents selected at random would be related to one another.
Randomization condition: The insurance company randomly selected 582 accidents.
10% condition: 582 accidents represent less than 10% of all accidents.
Success/Failure condition: $n\hat{p} = 91$ and $n\hat{q} = 491$ are both greater than 10, so the sample is large enough.

Since the conditions are met, we can use a one-proportion z-interval to estimate the percentage of accidents involving teenagers.

$$\hat{p} \pm z^* \sqrt{\frac{\hat{p}\hat{q}}{n}} = \left(\frac{91}{582}\right) \pm 1.960 \sqrt{\frac{\left(\frac{91}{582}\right)\left(\frac{491}{582}\right)}{582}} = (12.7\%, 18.6\%)$$

b) We are 95% confident that between 12.7% and 18.6% of all accidents involve teenagers.

c) About 95% of random samples of size 582 will produce intervals that contain the true proportion of accidents involving teenagers.

d) Our confidence interval contradicts the assertion of the politician. The figure quoted by the politician, 1 out of every 5, or 20%, is outside the interval.

14. Junk mail.

a) **Plausible independence condition:** There is no reason to believe that one randomly selected person's response will affect another's.
Randomization condition: The company randomly selected 1000 recipients.
10% condition: 1000 recipients is less than 10% of the population of 200,000 people.
Success/Failure condition: $n\hat{p} = 123$ and $n\hat{q} = 877$ are both greater than 10, so the sample is large enough.

Since the conditions are met, we can use a one-proportion z-interval to estimate the percentage of people who will respond to the new flyer.

$$\hat{p} \pm z^* \sqrt{\frac{\hat{p}\hat{q}}{n}} = \left(\frac{123}{1000}\right) \pm 1.645 \sqrt{\frac{\left(\frac{123}{1000}\right)\left(\frac{877}{1000}\right)}{1000}} = (10.6\%, 14.0\%)$$

b) We are 90% confident that between 10.6% and 14.0% of people will respond to the new flyer.

c) About 90% of random samples of size 1000 will produce intervals that contain the true proportion of people who will respond to the new flyer.

d) Our confidence interval suggests that the company should do the mass mailing. The entire interval is well above the cutoff of 5%.

15. Safe food.

The grocer can conclude nothing about the opinions of all his customers from this survey. Those customers who bothered to fill out the survey represent a voluntary response sample, consisting of people who felt strongly one way or another about irradiated food. The random condition was not met.

16. Local news.

The city council can conclude nothing about general public support for the mayor's initiative. Those who showed up for the meeting are probably a biased group. In addition, a show of hands vote may influence people, affecting the independence of the votes.

17. Marijuana, again.

a) **Plausible independence condition:** There is no reason to believe that one randomly selected person's response will affect another's.
Randomization condition: The pollsters randomly selected 505 respondents.
10% condition: 505 respondents is less than 10% of all adults.
Success/Failure condition: $n\hat{p} = 505(0.31) = 157$ and $n\hat{q} = 505(0.69) = 348$ are both greater than 10, so the sample is large enough.

Since the conditions are met, we can use a one-proportion z-interval to estimate the percentage of people who think "the use of marijuana should be made legal."

$$\hat{p} \pm z^* \sqrt{\frac{\hat{p}\hat{q}}{n}} = (0.31) \pm 1.960 \sqrt{\frac{(0.31)(0.69)}{505}} = (27.0\%, 35.0\%)$$

We are 95% confident that between 27% and 35% of people think "the use of marijuana should be made legal."

b) The percentage of 47% does not fall within the interval created in part a.

18. Drinking.

a) $\hat{p} = \dfrac{21}{110} \approx 0.191$. About 19.1% of this city's youth reported having been drunk.

b) This estimate is from one sample. Other samples will give different proportions. We need to create a confidence interval.

c) **Plausible independence condition:** There is no reason to believe that one randomly selected student's response will affect another's. The survey was anonymous.
Randomization condition: The health agency randomly selected 110 respondents.
10% condition: 110 students is less than 10% of the 1212 students.
Success/Failure condition: $n\hat{p} = 21$ and $n\hat{q} = 89$ are both greater than 10, so the sample is large enough.

Since the conditions are met, we can use a one-proportion z-interval to estimate the proportion of the city's youth who have been drunk.

$$\hat{p} \pm z^* \sqrt{\frac{\hat{p}\hat{q}}{n}} = \left(\frac{21}{110}\right) \pm 1.960 \sqrt{\frac{\left(\frac{21}{110}\right)\left(\frac{89}{110}\right)}{110}} = (11.7\%, 26.4\%)$$

We are 95% confident that between 11.7% and 26.4% of the city's youth have been drunk.

d) There is reason to believe that the national level of 30% is not true of the middle school students in this city. The national level of 30% is above the interval.

19. Marijuana poll.

a) There may be response bias based on the wording of the question.

b) $\hat{p} \pm z^* \sqrt{\dfrac{\hat{p}\hat{q}}{n}} = (0.39) \pm 1.960 \sqrt{\dfrac{(0.39)(0.61)}{1010}} = (36.0\%, 42.0\%)$

c) The margin of error based on the pooled sample is smaller, since the sample size is larger.

20. Gambling.

a) The interval based on the survey conducted by the college Statistics class will have the larger margin of error, since the sample size is smaller.

b) **Plausible independence condition:** There is no reason to believe that one randomly selected voter's response will influence another.
Randomization condition: Both samples were random.
10% condition: Both samples are probably less than 10% of the city's voters, provided the city has more than 12,000 voters.
Success/Failure condition:
For the newspaper, $n_1\hat{p}_1 = (1200)(0.53) = 636$ and $n_1\hat{q}_1 = (1200)(0.47) = 564$
For the Statistics class, $n_2\hat{p}_2 = (450)(0.54) = 243$ and $n_2\hat{q}_2 = (450)(0.46) = 207$
All the expected successes and failures are greater than 10, so the samples are large enough.

Since the conditions are met, we can use one-proportion z-intervals to estimate the proportion of the city's voters that support the gambling initiative.

Newspaper poll: $\quad \hat{p}_1 \pm z^* \sqrt{\dfrac{\hat{p}_1\hat{q}_1}{n_1}} = (0.53) \pm 1.960 \sqrt{\dfrac{(0.53)(0.47)}{1200}} = (50.2\%, 55.8\%)$

Statistics class poll: $\hat{p}_2 \pm z^* \sqrt{\dfrac{\hat{p}_2\hat{q}_2}{n_2}} = (0.54) \pm 1.960 \sqrt{\dfrac{(0.54)(0.46)}{450}} = (49.4\%, 58.6\%)$

c) The Statistics class should conclude that the outcome is too close to call, because 50% is in their interval.

21. Rickets.

a) **Plausible independence condition:** It is reasonable to think that the randomly selected children are mutually independent in regards to vitamin D deficiency.
Randomization condition: The 2,700 children were chosen at random.
10% condition: 2,700 children are less than 10% of all English children.
Success/Failure condition: $n\hat{p} = (2,700)(0.20) = 540$ and $n\hat{q} = (2,700)(0.80) = 2160$ are both greater than 10, so the sample is large enough.

Since the conditions are met, we can use a one-proportion z-interval to estimate the proportion of the English children with vitamin D deficiency.

$\hat{p} \pm z^* \sqrt{\dfrac{\hat{p}\hat{q}}{n}} = (0.20) \pm 2.326 \sqrt{\dfrac{(0.20)(0.80)}{2700}} = (18.2\%, 21.8\%)$

b) We are 98% confident that between 18.2% and 21.8% of English children are deficient in vitamin D.

c) About 98% of random samples of size 2,700 will produce confidence intervals that contain the true proportion of English children that are deficient in vitamin D.

22. Pregnancy.

a) **Plausible independence condition:** There is no reason to believe that one woman's ability to conceive would affect others.
Randomization condition: These women are not chosen at random. Assume that they are representative of all women under 40 that had previously been unable to conceive.
10% condition: 207 women is less than 10% of all such women.
Success/Failure condition: $n\hat{p} = 49$ and $n\hat{q} = 158$ are both greater than 10, so the sample is large enough.

Since the conditions are met, we can use a one-proportion z-interval to estimate the proportion of the births to women at the clinic.

$$\hat{p} \pm z^* \sqrt{\frac{\hat{p}\hat{q}}{n}} = \left(\frac{49}{207}\right) \pm 1.645 \sqrt{\frac{\left(\frac{49}{207}\right)\left(\frac{158}{207}\right)}{207}} = (18.8\%, 28.5\%)$$

b) We are 90% confident that between 18.8% and 28.5% of women under 40 who are treated at this clinic will give birth.

c) About 90% of random samples of size 207 will produce confidence intervals that contain the true proportion of women under 40 who are treated at this clinic that will give birth.

d) It would not be misleading for the clinic to advertise a 25% success rate, since 25% is in the interval.

23. Deer ticks.

a) **Plausible independence condition:** Deer ticks are parasites. A deer carrying the parasite may spread it to others. Ticks may not be distributed evenly throughout the population.
Randomization condition: The sample is not random and may not represent all deer.
10% condition: 153 deer are less than 10% of all deer.
Success/Failure condition: $n\hat{p} = 32$ and $n\hat{q} = 121$ are both greater than 10, so the sample is large enough.

The conditions are not satisfied, so we should use caution when a one-proportion z-interval is used to estimate the proportion of deer carrying ticks.

$$\hat{p} \pm z^* \sqrt{\frac{\hat{p}\hat{q}}{n}} = \left(\frac{32}{153}\right) \pm 1.645 \sqrt{\frac{\left(\frac{32}{153}\right)\left(\frac{121}{153}\right)}{153}} = (15.5\%, 26.3\%)$$

We are 90% confident that between 15.5% and 26.3% of deer have ticks.

b) In order to cut the margin of error in half, they must sample 4 times as many deer. 4(153) = 612 deer.

c) The incidence of deer ticks is not plausibly independent, and the sample may not be representative of all deer, since females and young deer are usually not hunted.

24. Pregnancy.

a) In order to cut the margin of error in half, they must use 4 times as many patient results. 4(207) = 828.

b) A sample this large may be more than 10% of the population of all potential patients.

25. Graduation.

a)

$$ME = z^* \sqrt{\frac{\hat{p}\hat{q}}{n}}$$

$$0.06 = 1.645 \sqrt{\frac{(0.25)(0.75)}{n}}$$

$$n = \frac{(1.645)^2(0.25)(0.75)}{(0.06)^2}$$

$$n \approx 141 \text{ people}$$

In order to estimate the proportion of non-graduates in the 25-to 30-year-old age group to within 6% with 90% confidence, we would need a sample of at least 141 people. All decimals in the final answer must be rounded up, to the next person.
(For a more cautious answer, let $\hat{p} = \hat{q} = 0.5$. This method results in a required sample of 188 people.)

b)

$$ME = z^* \sqrt{\frac{\hat{p}\hat{q}}{n}}$$

$$0.04 = 1.645 \sqrt{\frac{(0.25)(0.75)}{n}}$$

$$n = \frac{(1.645)^2(0.25)(0.75)}{(0.04)^2}$$

$$n \approx 318 \text{ people}$$

In order to estimate the proportion of non-graduates in the 25-to 30-year-old age group to within 4% with 90% confidence, we would need a sample of at least 318 people. All decimals in the final answer must be rounded up, to the next person.
(For a more cautious answer, let $\hat{p} = \hat{q} = 0.5$. This method results in a required sample of 423 people.)
Alternatively, the margin of error is now 2/3 of the original, so the sample size must be increased by a factor of 9/4. 141(9/4) ≈ 318 people.

c)

$$ME = z^* \sqrt{\frac{\hat{p}\hat{q}}{n}}$$

$$0.03 = 1.645 \sqrt{\frac{(0.25)(0.75)}{n}}$$

$$n = \frac{(1.645)^2(0.25)(0.75)}{(0.03)^2}$$

$$n \approx 564 \text{ people}$$

In order estimate the proportion of non-graduates in the 25-to 30-year-old age group to within 3% with 90% confidence, we would need a sample of at least 564 people. All decimals in the final answer must be rounded up, to the next person.
(For a more cautious answer, let $\hat{p} = \hat{q} = 0.5$. This method results in a required sample of 752 people.)

Alternatively, the margin of error is now half that of the original, so the sample size must be increased by a factor of 4. 141(4) ≈ 564 people.

26. Hiring.

a)

$$ME = z^* \sqrt{\frac{\hat{p}\hat{q}}{n}}$$

$$0.05 = 2.326 \sqrt{\frac{(0.5)(0.5)}{n}}$$

$$n = \frac{(2.326)^2 (0.5)(0.5)}{(0.05)^2}$$

$$n \approx 542 \text{ businesses}$$

In order to estimate the percentage of businesses planning to hire additional employees within the next 60 days to within 5% with 98% confidence, we would need a sample of at least 542 businesses. All decimals in the final answer must be rounded up, to the next business.

b)

$$ME = z^* \sqrt{\frac{\hat{p}\hat{q}}{n}}$$

$$0.03 = 2.326 \sqrt{\frac{(0.5)(0.5)}{n}}$$

$$n = \frac{(2.326)^2 (0.5)(0.5)}{(0.03)^2}$$

$$n \approx 1503 \text{ businesses}$$

In order to estimate the percentage of businesses planning to hire additional employees within the next 60 days to within 3% with 98% confidence, we would need a sample of at least 1503 businesses. All decimals in the final answer must be rounded up, to the next business.

(Alternatively, the margin of error is being decreased to 3/5 of its original size, so the sample size must increase by a factor of 25/9. 542(25/9) ≈ 1506 businesses. A bit off, because 542 was rounded, but close enough!

c)

$$ME = z^* \sqrt{\frac{\hat{p}\hat{q}}{n}}$$

$$0.01 = 2.326 \sqrt{\frac{(0.5)(0.5)}{n}}$$

$$n = \frac{(2.326)^2 (0.5)(0.5)}{(0.01)^2}$$

$$n \approx 13,526 \text{ businesses}$$

In order to estimate the percentage of businesses planning to hire additional employees within the next 60 days to within 1% with 98% confidence, we would need a sample of at least 13,526 businesses.

(Alternatively, the margin of error has been decreased to 1/5 of its original size, so a sample 25 times as large would be needed. 25(542) = 13,550. Close enough!

It would probably be very expensive and time consuming to sample that many businesses.

27. Graduation, again.

$$ME = z^* \sqrt{\frac{\hat{p}\hat{q}}{n}}$$

$$0.02 = 1.960 \sqrt{\frac{(0.25)(0.75)}{n}}$$

$$n = \frac{(1.960)^2 (0.25)(0.75)}{(0.02)^2}$$

$$n \approx 1{,}801 \text{ people}$$

In order to estimate the proportion of non-graduates in the 25-to 30-year-old age group to within 2% with 95% confidence, we would need a sample of at least 1,801 people. All decimals in the final answer must be rounded up, to the next person.
(For a more cautious answer, let $\hat{p} = \hat{q} = 0.5$. This method results in a required sample of 2,401 people.)

28. Better hiring info.

$$ME = z^* \sqrt{\frac{\hat{p}\hat{q}}{n}}$$

$$0.04 = 2.576 \sqrt{\frac{(0.5)(0.5)}{n}}$$

$$n = \frac{(2.576)^2 (0.5)(0.5)}{(0.04)^2}$$

$$n \approx 1{,}037 \text{ businesses}$$

In order to estimate the percentage of businesses planning to hire additional employees within the next 60 days to within 4% with 99% confidence, we would need a sample of at least 1,037 businesses. All decimals in the final answer must be rounded up, to the next business.

29. Pilot study.

$$ME = z^* \sqrt{\frac{\hat{p}\hat{q}}{n}}$$

$$0.03 = 1.645 \sqrt{\frac{(0.15)(0.85)}{n}}$$

$$n = \frac{(1.645)^2 (0.15)(0.85)}{(0.03)^2}$$

$$n \approx 384 \text{ cars}$$

Use $\hat{p} = \frac{9}{60} = 0.15$ from the pilot study as an estimate.

In order to estimate the percentage of cars with faulty emissions systems to within 3% with 90% confidence, the state's environmental agency will need a sample of at least 384 cars. All decimals in the final answer must be rounded up, to the next car.

30. Another pilot study.

$$ME = z^* \sqrt{\frac{\hat{p}\hat{q}}{n}}$$

$$0.04 = 2.326 \sqrt{\frac{(0.22)(0.78)}{n}}$$

$$n = \frac{(2.326)^2 (0.22)(0.78)}{(0.04)^2}$$

$$n \approx 581 \text{ adults}$$

Use $\hat{p} = 0.22$ from the pilot study as an estimate.

In order to estimate the percentage of adults with higher than normal levels of glucose in their blood to within 4% with 98% confidence, the researchers will need a sample of at least 581 adults. All decimals in the final answer must be rounded up, to the next adult.

31. Approval rating.

$$ME = z^* \sqrt{\frac{\hat{p}\hat{q}}{n}}$$

$$0.025 = z^* \sqrt{\frac{(0.65)(0.35)}{972}}$$

$$z^* = \frac{0.025}{\sqrt{\frac{(0.65)(0.35)}{972}}}$$

$$z^* \approx 1.634$$

Since $z^* \approx 1.634$, which is close to 1.645, the pollsters were probably using 90% confidence. The slight difference in the z^* values is due to rounding of the governor's approval rating.

32. Amendment.

a) This poll is inconclusive because the confidence interval, 52% ± 3% contains 50%. The true proportion of voters in favor of the constitutional amendment is estimated to be between 49% (minority) to 55% (majority). We can't be sure whether or not the majority of voters support the amendment or not.

b)

$$ME = z^* \sqrt{\frac{\hat{p}\hat{q}}{n}}$$

$$0.03 = z^* \sqrt{\frac{(0.52)(0.48)}{1505}}$$

$$z^* = \frac{0.03}{\sqrt{\frac{(0.52)(0.48)}{1505}}}$$

$$z^* \approx 2.3295$$

Since $z^* \approx 2.3295$, which is close to 2.326, the pollsters were probably using 98% confidence. The slight difference in the z^* values is due to rounding of the amendment's approval rating.

Chapter 20 – Testing Hypotheses about Proportions

1. **Hypotheses.**

   a) H_0 : The governor's "negatives" are 30%. ($p = 0.30$)
   H_A : The governor's "negatives" are less than 30%. ($p < 0.30$)

   b) H_0 : The proportion of heads is 50%. ($p = 0.50$)
   H_A : The proportion of heads is not 50%. ($p \neq 0.50$)

   c) H_0 : The proportion of people who quit smoking is 20%. ($p = 0.20$)
   H_A : The proportion of people who quit smoking is greater than 20%. ($p > 0.20$)

2. **More hypotheses.**

   a) H_0 : The proportion of high school graduates is 40%. ($p = 0.40$)
   H_A : The proportion of high school graduates is not 40%. ($p \neq 0.40$)

   b) H_0 : The proportion of cars needing transmission repair is 20%. ($p = 0.20$)
   H_A : The proportion of cars needing transmission repair is less than 20%. ($p < 0.20$)

   c) H_0 : The proportion of people who like the flavor is 60%. ($p = 0.60$)
   H_A : The proportion of people who like the flavor is greater than 60%. ($p > 0.60$)

3. **Negatives.**

   Statement d is the correct interpretation of a *P*-value.

4. **Dice.**

   Statement d is the correct interpretation of a *P*-value.

5. **Relief.**

   It is *not* reasonable to conclude that the new formula and the old one are equally effective. Furthermore, our inability to make that conclusion has nothing to do with the *P*-value. We can not prove the null hypothesis (that the new formula and the old formula are equally effective), but can only fail to find evidence that would cause us to reject it. All we can say about this *P*-value is that there is a 27% chance of seeing the observed effectiveness from natural sampling variation if the new formula and the old one are equally effective.

6. **Cars.**

   It is reasonable to conclude that a greater proportion of high schoolers have cars. If the proportion were no higher than it was a decade ago, there is only a 1.7% chance of seeing such a high sample proportion just from natural sampling variability.

7. **Blinking timers.**

   1) Null and alternative hypotheses should involve p, not $\hat{p}$.

   2) The company wants *at least* 90% of its customers to succeed. H_A should be $p > 0.90$.

   3) The student failed to check $nq = (200)(0.10) = 20 > 10$.

4) $SD(\hat{p}) = \sqrt{\dfrac{pq}{n}} = \sqrt{\dfrac{(0.90)(0.10)}{200}} \approx 0.021$. The student used $\hat{p}$ and $\hat{q}$.

5) Value of z is incorrect. The correct value is $z = \dfrac{0.94 - 0.90}{0.021} \approx 1.90$.

6) *P*-value is incorrect. $P = P(z > 1.90) = 0.029$

7) For the *P*-value given, an incorrect conclusion is drawn. A *P*-value of 0.009 provides evidence that the new system works. We can never make a conclusion like the one given, regardless of the *P*-value. The most we can ever say (for high *P*-values) is that there is no evidence that the new system works. The correct conclusion for the corrected *P*-value is: Since the *P*-value is low, there is strong evidence that the new system works.

8. Got milk?

1) Null and alternative hypotheses should involve p, not $\hat{p}$.

2) The question asks if there is evidence that the 90% figure is *not accurate*, so a two-sided alternative hypothesis should be used. H_A should be $p \neq 0.90$.

3) One of the conditions checked appears to be $n > 10$, which is not a condition for hypothesis tests. The Success/Failure Condition checks $np = (750)(0.90) = 675 > 10$ and $nq = (750)(0.10) = 75 > 10$.

4) $SD(\hat{p}) = \sqrt{\dfrac{pq}{n}} = \sqrt{\dfrac{(0.90)(0.10)}{750}} \approx 0.011$. The student used rounded values of $\hat{p}$ and $\hat{q}$.

5) Value of z is incorrect. The correct value is $z = \dfrac{0.876 - 0.90}{0.011} \approx -2.18$.

6) The *P*-value calculated is in the wrong direction. To test the given hypothesis, the lower-tail probability should have been calculated. The correct, two-tailed *P*-values is $P = 2P(z < -2.18) = 0.029$.

7) The *P*-value is misinterpreted. Since the *P*-value is so low, there is moderately strong evidence that the proportion of adults who drink milk is different than the claimed 90%. In fact, our sample suggests that the proportion may be lower. There is only a 2.9% chance of observing a $\hat{p}$ as far from 0.90 as this simply from natural sampling variation.

9. Dowsing.

a) H_0 : The percentage of successful wells drilled by the dowser is 30%. ($p = 0.30$)
   H_A : The percentage of successful wells drilled by the dowser is greater than 30%. ($p > 0.30$)

b) **Plausible independence condition:** There is no reason to think that finding water in one well will affect the probability that water is found in another, unless the wells are close enough to be fed by the same underground water source.
   Randomization condition: This sample is not random, so hopefully the customers you check with are representative of all of the dowser's customers.
   10% condition: The 80 customers sampled may be considered less than 10% of all possible customers.
   Success/Failure condition: $np = (80)(0.30) = 24$ and $nq = (80)(0.70) = 56$ are both greater than 10, so the sample is large enough.

c) The sample of customers may not be representative of all customers, so we will proceed cautiously. A Normal model can be used to model the sampling distribution of the proportion, with $\mu_{\hat{p}} = p = 0.30$ and $\sigma(\hat{p}) = \sqrt{\dfrac{pq}{n}} = \sqrt{\dfrac{(0.30)(0.70)}{80}} \approx 0.0512.$

We can perform a one-proportion z-test. The observed proportion of successful wells is $\hat{p} = \dfrac{27}{80} = 0.3375.$

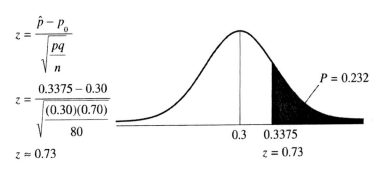

$$z = \dfrac{\hat{p} - p_0}{\sqrt{\dfrac{pq}{n}}}$$

$$z = \dfrac{0.3375 - 0.30}{\sqrt{\dfrac{(0.30)(0.70)}{80}}}$$

$$z \approx 0.73$$

d) If his dowsing has the same success rate as standard drilling methods, there is more than a 23% chance of seeing results as good as those of the dowser, or better, by natural sampling variation.

e) With a P-value of 0.232, we fail to reject the null hypothesis. There is no evidence to suggest that the dowser has a success rate any higher than 30%.

10. Autism.

a) H_0: The percentage of children with autism is 5%. $(p = 0.05)$
H_A: The percentage of with autism is greater than 5%. $(p > 0.05)$

b) **Plausible independence condition:** There is no reason to think that one child being autistic would affect the probability that other children are autistic.
Randomization condition: This sample may not be random, but autism is plausibly independent. The sample is probably representative of all children, with regards to autism.
10% condition: The sample of 384 children is less than 10% of all children.
Success/Failure condition: $np = (384)(0.05) = 19.2$ and $nq = (384)(0.95) = 364.8$ are both greater than 10, so the sample is large enough.

c) The conditions have been satisfied, so a Normal model can be used to model the sampling distribution of the proportion, with $\mu_{\hat{p}} = p = 0.05$ and $\sigma(\hat{p}) = \sqrt{\dfrac{pq}{n}} = \sqrt{\dfrac{(0.05)(0.95)}{384}} \approx 0.0111.$

We can perform a one-proportion z-test.

$$z = \dfrac{\hat{p} - p_0}{\sqrt{\dfrac{pq}{n}}}$$

$$z = \dfrac{0.1198 - 0.05}{\sqrt{\dfrac{(0.05)(0.95)}{384}}}$$

$$z \approx 6.28$$

The observed proportion of autistic children is $\hat{p} = \dfrac{46}{384} \approx 0.1198.$

The value of z is approximately 6.28, meaning that the observed proportion of autistic children is over 6 standard deviations above the hypothesized proportion. The P-value associated with this z score is 2×10^{-10}, essentially 0.

d) If 5% of children are autistic, the chance of observing 46 autistic children in a random sample of 384 children is essentially 0.

e) With a *P*-value of this low, we reject the null hypothesis. There is strong evidence that more than 5% of children are autistic.

f) We don't know that environmental chemicals cause autism. We merely have evidence that suggests that a greater percentage of children are autistic now, compared to the 1980s.

11. Smoking.

a) **Plausible independence condition:** There is no reason to believe that one randomly selected adult's response will affect another's, with regards to smoking.
Randomization condition: The health survey used 881 randomly selected adults.
10% condition: 881 adults is less than 10% of all adults.
Success/Failure condition: $n\hat{p} = 458$ and $n\hat{q} = 423$ are both greater than 10, so the sample is large enough.

Since the conditions are met, we can use a one-proportion *z*-interval to estimate the proportion of the adults who have never smoked.

$$\hat{p} \pm z^* \sqrt{\frac{\hat{p}\hat{q}}{n}} = (0.52) \pm 1.960 \sqrt{\frac{(0.52)(0.48)}{881}} = (48.7\%, 55.3\%)$$

We are 95% confident that between 48.7% and 55.3% of adults have never smoked.

b) H_0 : The percentage of adults who smoke is 44%. ($p = 0.44$)
H_A : The percentage of adults who smoke is not 44%. ($p \neq 0.44$)

Since 44% is not in the 95% confidence interval, we will reject the null hypothesis. There is strong evidence that, in 1995, the percentage of adults who have never smoked was not 44%. In fact, our sample indicates an increase (since the 1960s) in the percentage of adults who have never smoked.

12. Satisfaction.

a) **Plausible independence condition:** There is no reason to believe that one randomly selected customer's response will affect another's, with regards to complaints.
Randomization condition: The survey used 350 randomly selected customers.
10% condition: 350 customers are less than 10% of all possible customers.
Success/Failure condition: $n\hat{p} = 10$ and $n\hat{q} = 340$ are both greater than (or equal to!) 10, so the sample is large enough.

Since the conditions are met, we can use a one-proportion *z*-interval to estimate the proportion of the customers who have complaints.

$$\hat{p} \pm z^* \sqrt{\frac{\hat{p}\hat{q}}{n}} = \left(\frac{10}{350}\right) \pm 1.960 \sqrt{\frac{\left(\frac{10}{350}\right)\left(\frac{340}{350}\right)}{350}} = (1.1\%, 4.6\%)$$

We are 95% confident that between 1.1% and 4.6% of customers have complaints.

b) H_0 : The percentage of customers with complaints is 5%. $(p = 0.05)$
H_A : The percentage of customers with complaints is less than 5%. $(p < 0.05)$

Since 5% is not in the 95% confidence interval, we will reject the null hypothesis (at $\alpha = 0.025$. You'll learn more about α in the next chapter.). There is strong evidence that less than 5% of customers have complaints. This is evidence that the company has met its goal.

13. Pollution.

H_0 : The percentage of cars with faulty emissions is 20%. $(p = 0.20)$
H_A : The percentage of cars with faulty emissions is greater than 20%. $(p > 0.20)$

Two conditions are not satisfied. 22 is greater than 10% of the population of 150 cars, and $np = (22)(0.20) = 4.4$, which is not greater than 10. It's probably not a good idea to proceed with a hypothesis test.

14. Scratch and dent.

H_0 : The percentage of damaged machines is 2%, and the warehouse is meeting the company goal. $(p = 0.02)$
H_A : The percentage of damaged machines is greater than 2%, and the warehouse is failing to meet the company goal. $(p > 0.02)$

An important condition is not satisfied. $np = (60)(0.02) = 1.2$, which is not greater than 10. The Normal model is not appropriate for modeling the sampling distribution. It's probably not a good idea to proceed with a hypothesis test.

15. Twins.

H_0 : The percentage of twin births to teenage girls is 3%. $(p = 0.03)$
H_A : The percentage of twin births to teenage girls is less than 3%. $(p < 0.03)$

Plausible independence condition: One mother having twins will not affect another. Observations are plausibly independent.
Randomization condition: This sample may not be random, but it is reasonable to think that this hospital has a representative sample of teen mothers, with regards to twin births.
10% condition: The sample of 469 teenage mothers is less than 10% of all such mothers.
Success/Failure condition: $np = (469)(0.03) = 14.07$ and $nq = (469)(0.97) = 454.93$ are both greater than 10, so the sample is large enough.

The conditions have been satisfied, so a Normal model can be used to model the sampling distribution of the proportion, with $\mu_{\hat{p}} = p = 0.03$ and $\sigma(\hat{p}) = \sqrt{\dfrac{pq}{n}} = \sqrt{\dfrac{(0.03)(0.97)}{469}} \approx 0.0079$.

We can perform a one-proportion z-test. The observed proportion of twin births to teenage mothers is $\hat{p} = \dfrac{7}{469} \approx 0.015$.

Since the *P*-value = 0.0278 is low, we reject the null hypothesis. There is evidence that the proportion of twin births for teenage mothers at this large city hospital is lower than the proportion of twin births for all mothers.

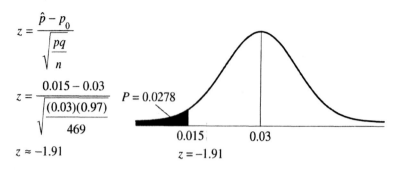

$$z = \frac{\hat{p} - p_0}{\sqrt{\dfrac{pq}{n}}}$$

$$z = \frac{0.015 - 0.03}{\sqrt{\dfrac{(0.03)(0.97)}{469}}}$$

$$z \approx -1.91$$

$P = 0.0278$

0.015   0.03

$z = -1.91$

16. Football.

H_0 : The percentage of home team wins is 50%. ($p = 0.50$)
H_A : The percentage of home team wins is greater than 50%. ($p > 0.50$)

Plausible independence condition: Results of one game should not affect others.
Randomization condition: This season should be representative of other seasons, with regards to home team wins.
10% condition: 240 games represent less than 10% of all games, in all seasons.
Success/Failure condition: $np = (240)(0.50) = 120$ and $nq = (240)(0.50) = 120$ are both greater than 10, so the sample is large enough.

The conditions have been satisfied, so a Normal model can be used to model the sampling distribution of the proportion, with $\mu_{\hat{p}} = p = 0.50$ and $\sigma(\hat{p}) = \sqrt{\dfrac{pq}{n}} = \sqrt{\dfrac{(0.5)(0.5)}{240}} \approx 0.0323$.

We can perform a one-proportion *z*-test. The observed proportion of home team wins is $\hat{p} = \dfrac{138}{240} = 0.575$.

Since the *P*-value = 0.0101 is low, we reject the null hypothesis. There is strong evidence that the proportion of home teams wins is greater than 50%. This provides evidence of a home team advantage.

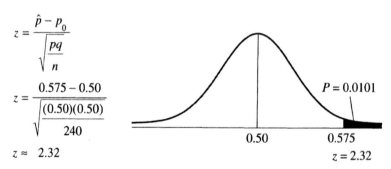

$$z = \frac{\hat{p} - p_0}{\sqrt{\dfrac{pq}{n}}}$$

$$z = \frac{0.575 - 0.50}{\sqrt{\dfrac{(0.50)(0.50)}{240}}}$$

$$z \approx 2.32$$

$P = 0.0101$

0.50   0.575

$z = 2.32$

17. Webzine.

H_0 : The percentage of readers interested in an online edition is 25%. ($p = 0.25$)
H_A : The percentage of readers interested in an online edition is greater than 25%. ($p > 0.25$)

Plausible independence condition: Interest of one reader should not affect interest of other readers.
Randomization condition: The magazine conducted an SRS of 500 current readers.
10% condition: 500 readers are less than 10% of all potential subscribers.
Success/Failure condition: $np = (500)(0.25) = 125$ and $nq = (500)(0.75) = 375$ are both greater than 10, so the sample is large enough.

The conditions have been satisfied, so a Normal model can be used to model the sampling distribution of the proportion, with $\mu_{\hat{p}} = p = 0.25$ and $\sigma(\hat{p}) = \sqrt{\dfrac{pq}{n}} = \sqrt{\dfrac{(0.25)(0.75)}{500}} \approx 0.0194$.

We can perform a one-proportion z-test. The observed proportion of interested readers is $\hat{p} = \dfrac{137}{500} = 0.274$.

Since the *P*-value = 0.1076 is high, we fail to reject the null hypothesis. There is little evidence to suggest that the proportion of interested readers is greater than 25%. The magazine should not publish the online edition.

$$z = \dfrac{\hat{p} - p_0}{\sqrt{\dfrac{pq}{n}}}$$

$$z = \dfrac{0.274 - 0.25}{\sqrt{\dfrac{(0.25)(0.75)}{500}}}$$

$$z \approx 1.24$$

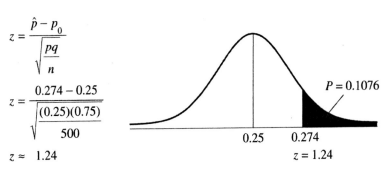

$P = 0.1076$

0.25   0.274

$z = 1.24$

18. Seeds.

H_0 : The germination rate of the green bean seeds is 92%. ($p = 0.92$)
H_A : The germination rate of the green bean seeds is less than 92%. ($p < 0.92$)

Plausible independence condition: Seeds in a single packet may not germinate independently. They have been treated identically with regards to moisture exposure, temperature, etc. They may have higher or lower germination rates than seeds in general.
Randomization condition: The cluster sample of one bag of seeds was not random.
10% condition: 200 seeds is less than 10% of all seeds.
Success/Failure condition: $np = (200)(0.92) = 184$ and $nq = (200)(0.08) = 16$ are both greater than 10, so the sample is large enough.

The conditions have *not* been satisfied. We will assume that the seeds in the bag are representative of all seeds, and cautiously use a Normal model to model the sampling distribution of the proportion, with $\mu_{\hat{p}} = p = 0.92$ and $\sigma(\hat{p}) = \sqrt{\dfrac{pq}{n}} = \sqrt{\dfrac{(0.92)(0.08)}{200}} \approx 0.0192$.

We can perform a one-proportion z-test. The observed proportion of germinated seeds is $\hat{p} = \dfrac{171}{200} = 0.85$.

Since the *P*-value = 0.0004 is very low, we reject the null hypothesis. There is strong evidence that the germination rate of the seeds in less than 92%. We should use extreme caution in generalizing these results to all seeds, but the manager should be safe, and avoid selling faulty seeds.
The seeds should be thrown out.

$$z = \dfrac{\hat{p} - p_0}{\sqrt{\dfrac{pq}{n}}}$$

$$z = \dfrac{0.855 - 0.92}{\sqrt{\dfrac{(0.92)(0.08)}{200}}}$$

$$z \approx -3.39$$

$P = 0.0004$

0.855   0.92

$z = -3.39$

19. Women executives.

H_0 : The proportion of female executives is similar to the overall proportion of female employees at the company. ($p = 0.40$)
H_A : The proportion of female executives is lower than the overall proportion of female employees at the company. ($p < 0.40$)

Plausible independence condition: It is reasonable to think that executives at this company were chosen independently.
Randomization condition: The executives were not chosen randomly, but it is reasonable to think of these executives as representative of all potential executives over many years.
10% condition: 43 executives are less than 10% of all possible executives at the company.
Success/Failure condition: $np = (43)(0.40) = 17.2$ and $nq = (43)(0.60) = 25.8$ are both greater than 10, so the sample is large enough.

The conditions have been satisfied, so a Normal model can be used to model the sampling distribution of the proportion, with $\mu_{\hat{p}} = p = 0.40$ and $\sigma(\hat{p}) = \sqrt{\dfrac{pq}{n}} = \sqrt{\dfrac{(0.40)(0.60)}{43}} \approx 0.0747$.

We can perform a one-proportion z-test. The observed proportion of female executives is $\hat{p} = \dfrac{13}{43} \approx 0.302$.

Since the *P*-value = 0.0955 is high, we fail to reject the null hypothesis. There is little evidence to suggest proportion of female executives is any different from the overall proportion of 40% female employees at the company.

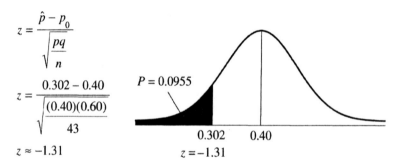

$$z = \frac{\hat{p} - p_0}{\sqrt{\dfrac{pq}{n}}}$$

$$z = \frac{0.302 - 0.40}{\sqrt{\dfrac{(0.40)(0.60)}{43}}}$$

$$z \approx -1.31$$

$P = 0.0955$

0.302   0.40

$z = -1.31$

20. Jury.

H_0 : The proportion of Hispanics called for jury duty is similar to the proportion of Hispanics in the county, 19%. ($p = 0.19$)
H_A : The proportion of Hispanics called for jury duty is less than the proportion of Hispanics in the county, 19%. ($p < 0.19$)

Plausible independence condition/Randomization condition: Assume that potential jurors were called randomly from all of the residents in the county. This is really what we are testing. If we reject the null hypothesis, we will have evidence that jurors are not called randomly.
10% condition: 72 people are less than 10% of all potential jurors in the county.
Success/Failure condition: $np = (72)(0.19) = 13.68$ and $nq = (72)(0.81) = 58.32$ are both greater than 10, so the sample is large enough.

The conditions have been satisfied, so a Normal model can be used to model the sampling distribution of the proportion, with $\mu_{\hat{p}} = p = 0.19$ and $\sigma(\hat{p}) = \sqrt{\dfrac{pq}{n}} = \sqrt{\dfrac{(0.19)(0.81)}{72}} \approx 0.0462$.

We can perform a one-proportion z-test. The observed proportion of Hispanics called for jury duty is $\hat{p} = \dfrac{9}{72} \approx 0.125$.

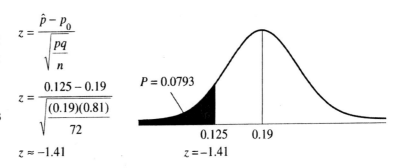

Since the P-value = 0.0793 is somewhat high, we fail to reject the null hypothesis. We are not convinced that Hispanics are underrepresented in the jury selection system. However, this P-value isn't extremely high. There is some evidence that the selection process may be biased. We should examine some other groups called for jury duty and take a closer look.

$$z = \frac{\hat{p} - p_0}{\sqrt{\dfrac{pq}{n}}}$$

$$z = \frac{0.125 - 0.19}{\sqrt{\dfrac{(0.19)(0.81)}{72}}}$$

$$z \approx -1.41$$

21. Dropouts.

H_0 : The proportion of dropouts at this high school is similar to 10.9%, the proportion of dropouts nationally. ($p = 0.109$)
H_A : The proportion of dropouts at this high school is greater than 10.9%, the proportion of dropouts nationally. ($p > 0.109$)

Plausible independence condition/Randomization condition: Assume that the students at this high school are representative of all students nationally. This is really what we are testing. The dropout rate at this high school has traditionally been close to the national rate. If we reject the null hypothesis, we will have evidence that the dropout rate at this high school is no longer close to the national rate.
10% condition: 1792 students are less than 10% of all students nationally.
Success/Failure condition: $np = (1782)(0.109) = 194.238$ and $nq = (1782)(0.891) = 1587.762$ are both greater than 10, so the sample is large enough.

The conditions have been satisfied, so a Normal model can be used to model the sampling distribution of the proportion, $\mu_{\hat{p}} = p = 0.109$ and $\sigma(\hat{p}) = \sqrt{\dfrac{pq}{n}} = \sqrt{\dfrac{(0.109)(0.891)}{1782}} \approx 0.0074$.

We can perform a one-proportion z-test. The observed proportion of dropouts is $\hat{p} = \dfrac{210}{1782} \approx 0.117845$.

Since the P-value = 0.115 is high, we fail to reject the null hypothesis. There is little evidence of an increase in dropout rate from 10.9%.

$$z = \frac{\hat{p} - p_0}{\sqrt{\dfrac{pq}{n}}}$$

$$z = \frac{0.117845 - 0.109}{\sqrt{\dfrac{(0.109)(0.891)}{1782}}}$$

$$z \approx 1.198$$

22. Acid rain.

H_0 : The proportion of trees with acid rain damage in Hopkins Forest is 15%, the proportion of trees with acid rain damage in the Northeast. ($p = 0.15$)

H_A : The proportion of trees with acid rain damage in Hopkins Forest is greater than 15%, the proportion of trees with acid rain damage in the Northeast. ($p > 0.15$)

Plausible independence condition/Randomization condition: Assume that the trees in Hopkins Forest are representative of all trees in the Northeast. This is really what we are testing. If we reject the null hypothesis, we will have evidence that the proportion of trees with acid rain damage is greater in Hopkins Forest than the proportion in the Northeast.

10% condition: 100 trees are less than 10% of all trees.

Success/Failure condition: $np = (100)(0.15) = 15$ and $nq = (100)(0.85) = 85$ are both greater than 10, so the sample is large enough.

The conditions have been satisfied, so a Normal model can be used to model the sampling distribution of the proportion, with $\mu_{\hat{p}} = p = 0.109$ and $\sigma(\hat{p}) = \sqrt{\dfrac{pq}{n}} = \sqrt{\dfrac{(0.15)(0.85)}{100}} \approx 0.0357$.

We can perform a one-proportion z-test. The observed proportion of damaged trees is

$\hat{p} = \dfrac{25}{100} = 0.25$.

Since the *P*-value = 0.0026 is low, we reject the null hypothesis. There is strong evidence that the trees in Hopkins forest have a greater proportion of acid rain damage than the 15% reported for the Northeast.

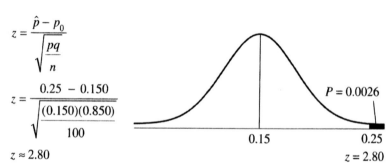

$$z = \dfrac{\hat{p} - p_0}{\sqrt{\dfrac{pq}{n}}}$$

$$z = \dfrac{0.25 - 0.150}{\sqrt{\dfrac{(0.150)(0.850)}{100}}}$$

$z \approx 2.80$

23. Lost luggage.

H_0 : The proportion of lost luggage returned the next day is 90%. ($p = 0.90$)

H_A : The proportion of lost luggage returned the next day is lower than 90%. ($p < 0.90$)

Plausible independence condition: It is reasonable to think that the people surveyed were independent with regards to their luggage woes.

Randomization condition: Although not stated, we will hope that the survey was conducted randomly, or at least that these air travelers are representative of all air travelers for that airline.

10% condition: 122 air travelers are less than 10% of all air travelers on the airline.

Success/Failure condition: $np = (122)(0.90) = 109.8$ and $nq = (122)(0.10) = 12.2$ are both greater than 10, so the sample is large enough.

The conditions have been satisfied, so a Normal model can be used to model the sampling distribution of the proportion, with $\mu_{\hat{p}} = p = 0.90$ and $\sigma(\hat{p}) = \sqrt{\dfrac{pq}{n}} = \sqrt{\dfrac{(0.90)(0.10)}{122}} \approx 0.0272$.

We can perform a one-proportion z-test. The observed proportion of dropouts is

$\hat{p} = \dfrac{103}{122} \approx 0.844$.

Since the *P*-value = 0.0201 is low, we reject the null hypothesis. There is evidence that the proportion of lost luggage returned the next day is lower than the 90% claimed by the airline.

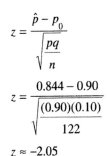

$z = \dfrac{\hat{p} - p_0}{\sqrt{\dfrac{pq}{n}}}$

$z = \dfrac{0.844 - 0.90}{\sqrt{\dfrac{(0.90)(0.10)}{122}}}$

$z \approx -2.05$

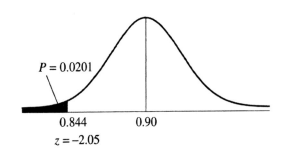

$P = 0.0201$

0.844   0.90

$z = -2.05$

24. TV ads.

H_0 : The proportion of respondents who recognize the name is 40%. ($p = 0.40$)
H_A : The proportion of respondents who recognize the name is more than 40%. ($p > 0.40$)

Plausible independence condition: There is no reason to believe that the responses of randomly selected people would influence others.
Randomization condition: The pollster contacted the 420 adults randomly.
10% condition: A sample of 420 adults is less than 10% of all adults.
Success/Failure condition: $np = (420)(0.40) = 168$ and $nq = (420)(0.60) = 252$ are both greater than 10, so the sample is large enough.

The conditions have been satisfied, so a Normal model can be used to model the sampling distribution of the proportion, with $\mu_{\hat{p}} = p = 0.40$ and $\sigma(\hat{p}) = \sqrt{\dfrac{pq}{n}} = \sqrt{\dfrac{(0.40)(0.60)}{420}} \approx 0.0239$.

We can perform a one-proportion z-test. The observed proportion of dropouts is

$\hat{p} = \dfrac{181}{420} \approx 0.431$.

Since the *P*-value = 0.0977 is fairly high, we fail to reject the null hypothesis. There is little evidence that more than 40% of the public recognizes the product.
Don't run commercials during the Super Bowl!

$z = \dfrac{\hat{p} - p_0}{\sqrt{\dfrac{pq}{n}}}$

$z = \dfrac{0.431 - 0.40}{\sqrt{\dfrac{(0.40)(0.60)}{420}}}$

$z \approx 1.29$

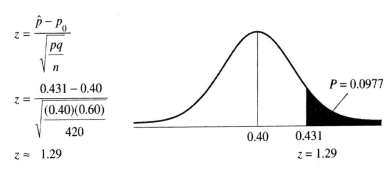

$P = 0.0977$

0.40   0.431

$z = 1.29$

Chapter 21 – More About Tests

1. **P-value.**

   If the effectiveness of the new poison ivy treatment is the same as the effectiveness of the old treatment, the chance of observing an effectiveness this large or larger in a sample of the same size is 4.7% by natural sampling variation alone.

2. **Another P-value.**

   If the rate of seat belt usage after the campaign is the same as the rate of seat belt usage before the campaign, there is a 17% chance of observing a rate of seat belt usage after the campaign this large or larger in a sample of the same size by natural sampling variation alone.

3. **Alpha.**

   Since the null hypothesis was rejected at $\alpha = 0.05$, the P-value for the researcher's test must have been less than 0.05. He would have made the same decision at $\alpha = 0.10$, since the P-value must also be less than 0.10. We can't be certain whether or not he would have made the same decision at $\alpha = 0.01$, since we only know that the P-value was less than 0.05. It may have been less than 0.01, but we can't be sure.

4. **Alpha again.**

   Since the environmentalists failed to reject the null hypothesis at $\alpha = 0.05$, the P-value for the environmentalists' test must have been greater than 0.05. We can't be certain whether or not they would have made the same decision at $\alpha = 0.10$, since we only know that the P-value was greater than 0.05. It may have been greater than 0.10 as well, but we can't be sure. They would have made them same decision at $\alpha = 0.01$, since the P-value must also be greater than 0.01.

5. **Significant?**

   a) If 90% of children have really been vaccinated, there is only a 1.1% chance of observing 89.4% of children (in a sample of 13,000) vaccinated by natural sampling variation alone.

   b) We conclude that the proportion of children who have been vaccinated is below 90%, but a 95% confidence interval would show that the true proportion is between 88.9% and 89.9%. Most likely a decrease from 90% to 89.9% would not be considered important. The 90% figure was probably an approximate figure anyway.

6. **Significant?**

   a) If 15.9% is the true percentage of children who did not attain the grade level standard, there is only a 2.3% chance of observing 15.1% of children (in a sample of 8500) not attaining grade level by natural sampling variation alone.

   b) Under old methods, 1352 students would not be expected to read at grade level. With the new program, 1284 would not be expected to read at grade level. This is only a decrease of 68 students. The costs of switching to the new program might outweigh the potential benefit. It is also important to realize that this is only a *potential* benefit.

7. **Testing cars.**

   H_0 : The shop is meeting the emissions standards.
   H_A : The shop is not meeting the emissions standards.

   a) Type I error is when the regulators decide that the shop is not meeting standards when they actually are meeting the standards.

   b) Type II error is when the regulators certify the shop when they are not meeting the standards.

   c) Type I would be more serious to the shop owners. They would lose their certification, even though they are meeting the standards.

   d) Type II would be more serious to environmentalists. Shops are allowed to operate, even though they are allowing polluting cars to operate.

8. **Quality control.**

   H_0 : The assembly process is working fine.
   H_A : The assembly process is producing defective items.

   a) Type I error is when the production managers decide that there has been an increase in the number of defective items and stop the assembly line, when the assembly process is working fine.

   b) Type II error is when the production managers decide that the assembly process is working fine, but defective items are being produced.

   c) The factory owner would probably consider Type II error to be more serious, depending of the costs of shutting the line down. Generally, because of warranty costs and lost customer loyalty, defects that are caught in the factory are much cheaper to fix than defects found after items are sold.

   d) Customers would consider Type II error to be more serious, since customers don't want to buy defective items.

9. **Cars again.**

   a) The power of the test is the probability of detecting that the shop is not meeting standards when they are not.

   b) The power of the test will be greater when 40 cars are tested. A larger sample size increases the power of the test.

   c) The power of the test will be greater when the level of significance is 10%. There is a greater chance that the null hypothesis will be rejected.

   d) The power of the test will be greater when the shop is out of compliance "a lot". Larger problems are easier to detect.

10. **Production.**

    a) The power of the test is the probability that the assembly process is stopped when defective items are being produced.

b) An advantage of testing more items is an increase in the power of the test to detect a problem. The disadvantages of testing more items are the additional cost and time spent testing.

c) An advantage of lowering the alpha level is that the probability of stopping the assembly process when everything is working fine (committing a Type I error) is decreased. A disadvantage is that the power of the test to detect defective items is also decreased.

d) The power of the test will increase as a day passes. Bigger problems are easier to detect.

11. Equal opportunity?

H_0 : The company is not discriminating against minorities.
H_A : The company is discriminating against minorities.

a) This is a one-tailed test. They wouldn't sue if "too many" minorities were hired.

b) Type I error would be deciding that the company is discriminating against minorities when they are not discriminating.

c) Type II error would be deciding that the company is not discriminating against minorities when they actually are discriminating.

d) The power of the test is the probability that discrimination is detected when it is actually occurring.

e) The power of the test will increase when the level of significance is increased from 0.01 to 0.05.

f) The power of the test is lower when the lawsuit is based on 37 employees instead of 87. Lower sample size leads to less power.

12. Stop signs.

H_0 : The new signs provide the same visibility than the old signs.
H_A : The new signs provide greater visibility than the old signs.

a) The test is one-tailed, because we are only interested in whether or not the signs are more visible. If the new design is less visible, we don't care how much less visible it is.

b) Type I error happens when the engineers decide that the new signs are more visible when they are not more visible.

c) Type II error happens when the engineers decide that the new signs are not more visible when they actually are more visible.

d) The power of the test is the probability that the engineers detect a sign that is truly more visible.

e) When the level of significance is dropped from 5% to 1%, power decreases. The null hypothesis is harder to reject, since more evidence is required.

f) If a sample of size 20 is used instead of 50, power will decrease. A smaller sample size has more variability, lowering the ability of the test to detect falsehoods.

13. Dropouts.

a) The test is one-tailed. We are testing to see if a decrease in the dropout rate is associated with the software.

b) H_0 : The dropout rate does not change following the use of the software. ($p = 0.13$)
H_A : The dropout rate decreases following the use of the software. ($p < 0.13$)

c) The professor makes a Type I error if he buys the software when the dropout rate has not actually decreased.

d) The professor makes a Type II error if he doesn't buy the software when the dropout rate has actually decreased.

e) The power of the test is the probability of buying the software when the dropout rate has actually decreased.

14. Ads.

a) H_0 : The percentage of residents that have heard the ad and recognize the product is 20%. ($p = 0.20$)
H_A : The percentage of residents that have heard the ad and recognize the product is greater than 20%. ($p > 0.20$)

b) The company wants more evidence that the ad is effective before deciding it really is. By lowering the level of significance from 10% to 5%, the probability of Type I error is decreased. The company is less likely to think that the ad is effective when it actually is not effective.

c) The power of the test is the probability of correctly deciding more than 20% have heard the ad and recognize the product when it's true.

d) The power of the test will be higher for a level of significance of 10%. There is a greater chance of rejecting the null hypothesis.

e) Increasing the sample size to 600 will lower the risk of Type II error. A larger sample size decreases variability, which helps us notice what is really going on. The company will be more likely to notice when the ad really works.

15. Dropouts, part II.

a) H_0 : The dropout rate does not change following the use of the software. ($p = 0.13$)
H_A : The dropout rate decreases following the use of the software. ($p < 0.13$)

Plausible independence assumption: One student's decision about dropping out should not influence another's decision.
Randomization condition: This year's class of 203 students is probably representative of all stats students.
10% condition: A sample of 203 students is less than 10% of all students.
Success/Failure condition: $np = (203)(0.13) = 26.39$ and $nq = (203)(0.87) = 176.61$ are both greater than 10, so the sample is large enough.

The conditions have been satisfied, so a Normal model can be used to model the sampling distribution of the proportion, with $\mu_{\hat{p}} = p = 0.13$ and $\sigma(\hat{p}) = \sqrt{\dfrac{pq}{n}} = \sqrt{\dfrac{(0.13)(0.87)}{203}} \approx 0.0236$.

We can perform a one-proportion z-test. The observed proportion of dropouts is $\hat{p} = \dfrac{11}{203} \approx 0.054$.

Since the P-value = 0.0007 is very low, we reject the null hypothesis. There is strong evidence that the dropout rate has dropped since use of the software program was implemented. As long as the professor feels confident that this class of stats students is representative of all potential students, then he should buy the program.

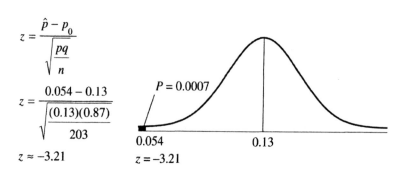

$$z = \dfrac{\hat{p} - p_0}{\sqrt{\dfrac{pq}{n}}}$$

$$z = \dfrac{0.054 - 0.13}{\sqrt{\dfrac{(0.13)(0.87)}{203}}}$$

$$z \approx -3.21$$

If you used a 95% confidence interval to assess the effectiveness of the program:

$$\hat{p} \pm z^* \sqrt{\dfrac{\hat{p}\hat{q}}{n}} = \left(\dfrac{11}{203}\right) \pm 1.960 \sqrt{\dfrac{\left(\frac{11}{203}\right)\left(\frac{192}{203}\right)}{203}} = (2.3\%, 8.5\%)$$

We are 95% confident that the dropout rate is between 2.3% and 8.5%. Since 15% is not contained in the interval, this provides evidence that the dropout rate has changed following the implementation of the software program.

b) The chance of observing 11 or fewer dropouts in a class of 203 is only 0.07% if the dropout rate is really 13%.

16. Testing the ads.

a) H_0 : The percentage of residents that remember the ad is 20%. ($p = 0.20$)
H_A : The percentage of residents that remember the ad is greater than 20%. ($p > 0.20$)

Plausible independence assumption: It is reasonable to think that randomly selected residents would remember the ad independently of one another.
Randomization condition: The sample consisted of 600 randomly selected residents.
10% condition: The sample of 600 is less than 10% of the population of the city.
Success/Failure condition: $np = (600)(0.20) = 120$ and $nq = (600)(0.80) = 480$ are both greater than 10, so the sample is large enough.

The conditions have been satisfied, so a Normal model can be used to model the sampling distribution of the proportion, with $\mu_{\hat{p}} = p = 0.20$ and $\sigma(\hat{p}) = \sqrt{\dfrac{pq}{n}} = \sqrt{\dfrac{(0.20)(0.80)}{600}} \approx 0.0163$.

We can perform a one-proportion z-test. The observed proportion of residents who remembered the ad is $\hat{p} = \dfrac{133}{600} \approx 0.222$.

Since the *P*-value = 0.0923 is somewhat high, we fail to reject the null hypothesis. There is little evidence that more than 20% of people remember the ad. The company should not renew the contract.

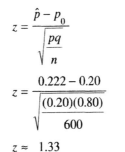

$$z = \frac{\hat{p} - p_0}{\sqrt{\dfrac{pq}{n}}}$$

$$z = \frac{0.222 - 0.20}{\sqrt{\dfrac{(0.20)(0.80)}{600}}}$$

$$z \approx 1.33$$

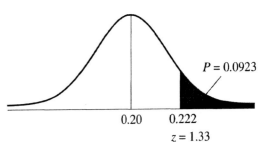

b) There is a 9.23% chance of having 133 or fewer of 600 people in a random sample remember the ad, if in fact only 20% of people do.

17. Hoops.

H_0 : The player's foul-shot percentage is only 60%. ($p = 0.60$)
H_A : The player's foul-shot percentage is better than 60%. ($p > 0.60$)

a) The player's shots can be considered Bernoulli trials. There are only two possible outcomes, make the shot and miss the shot. The probability of making any shot is constant at $p = 0.60$. Assume that the shots are independent of each other. Use *Binom*(10, 0.60).

Let X = the number of shots made out of $n = 10$.

$$P(\text{makes at least 9 out of 10}) = P(X \geq 9)$$
$$= P(X = 9) + P(X = 10)$$
$$= {}_{10}C_9 (0.60)^9 (0.40)^1 + {}_{10}C_{10} (0.60)^{10} (0.40)^0$$
$$\approx 0.0464$$

b) The coach made a Type I error.

c) The power of the test can be calculated for specific values of the new probability of success. Each true value of p has a power calculation associated with it. In this case, we are finding the power of the test to detect an 80% foul-shooter. Use *Binom*(10, 0.80).

Let X = the number of shots made out of $n = 10$.

$$P(\text{makes at least 9 out of 10}) = P(X \geq 9)$$
$$= P(X = 9) + P(X = 10)$$
$$= {}_{10}C_9 (0.80)^9 (0.20)^1 + {}_{10}C_{10} (0.80)^{10} (0.20)^0$$
$$\approx 0.376$$

The power of the test to detect an increase in foul-shot percentage from 60% to 80% is about 37.6%.

d) The power of the test to detect improvement in foul-shooting can be increased by increasing the number of shots, or by keeping the number of shots at 10 but increasing the level of significance by declaring that 8, 9, or 10 shots made will convince the coach that the player has improved. In other words, the coach can increase the power of the test by lowering the standard of proof.

18. Pottery.

H_0 : The new clay is no better than the old, and breaks 40% of the time. ($p = 0.40$)
H_A : The new clay is better than the old, and breaks less than 40% of the time. ($p < 0.40$)

a) The fired pieces can be considered Bernoulli trials. There are only two possible outcomes, broken and unbroken. The probability of breaking is constant at $p = 0.40$. It is reasonable to think that the pieces break independently of each other. Use $Binom(10, 0.40)$.

Let X = the number of broken pieces out of $n = 10$.

$P(\text{at most one breaks}) = P(X \le 1)$

$$= P(X = 0) + P(X = 1)$$
$$= {}_{10}C_0(0.40)^0(0.60)^{10} + {}_{10}C_1(0.40)^1(0.60)^9$$
$$\approx 0.0464$$

b) The artist made a Type I error.

c) The probability Type II error can be calculated for specific values of the new probability of success. Each true value of p has a Type II error calculation associated with it. In this case, we are finding the probability of Type II error if the pieces break only 20% of the time instead of 40% of the time. She won't notice that the clay is better is 2 or more pieces break. Use $Binom(10, 0.80)$.

Let X = the number of broken pieces out of $n = 10$.

$P(\text{at least 2 break}) = P(X \ge 2)$

$$= P(X = 2) + \cdots + P(X = 10)$$
$$= {}_{10}C_2(0.20)^2(0.80)^8 + \cdots + {}_{10}C_{10}(0.20)^{10}(0.80)^0$$
$$\approx 0.6242$$

The probability that she makes a Type II error (not noticing that the clay is better) is approximately 0.6242.

d) The power of the test to detect improvement in the clay can be increased by increasing the number of pieces fired, or by keeping the number of pieces at 10 but increasing the level of significance by declaring that 0, 1, or 2 broken pieces will convince the artist that the player has improved. In other words, the artist can improve the power by lowering her standard of proof.

19. Survey.

Only the Laborers data may be used to construct a confidence interval.

The Laborers data are okay, since $n = 300$ is less than 10% of 6235 employees, and $n\hat{p} = (300)(0.09) = 27$ and $nq = (300)(0.91) = 273$ are both greater than 10, so the sample is large enough. (We are assuming that the samples are random and free from biases.)

The Clerical data are not okay, since $n = 200$ is greater than 10% of the population of 1520 employees.

The Management data are not okay, since $n\hat{p} = (25)(0.08) = 2$, which is not greater than 10. The Success/Failure condition is not satisfied.

20. Fire safety.

Only the Single-family home data may be used to construct a confidence interval.

The Single-family home data are okay, since $n = 200$ is less than 10% of 7742 buildings, and $n\hat{p} = (200)(0.07) = 14$ and $nq = (200)(0.93) = 186$ are both greater than 10, so the sample is large enough. (Assuming that the samples are random and free from biases.)

The Apartment building data are not okay, since $n\hat{p} = (20)(0.10) = 2$, which is not greater than 10. The Success/Failure condition is not satisfied.

The Commercial data are not okay, since $n = 70$ is greater than 10% of the population of 407 buildings.

21. Little league.

a) **Plausible independence:** It is reasonable to think that one pitcher's condition would not affect the conditions of other pitchers.
Randomization condition: The study is not identified at random. At the very least, let's hope that the sample is representative of all young pitchers.
10% condition: 298 young pitchers are less than 10% of all young pitchers.
Success/Failure condition: $n\hat{p} = (0.26)(298) = 77$ and $nq = (0.74)(298) = 221$ are both greater than 10, so the sample is large enough.

Since the conditions are met, we can use a one-proportion z-interval to estimate the proportion of young pitchers who complained of elbow pain.

$$\hat{p} \pm z^* \sqrt{\frac{\hat{p}\hat{q}}{n}} = (0.26) \pm 1.645 \sqrt{\frac{(0.26)(0.74)}{298}} = (21.8\%, 30.2\%)$$

We are 90% confident that between 21.8% and 30.2% of young pitchers complained of elbow pain.

b) The coaches claim that "only" 1 kid in 5 is at risk of arm injury is not consistent with interval, since 20% is below the lower end of the interval.

22. News sources.

H_0: The proportion of people that get news from the Internet is 25%. ($p = 0.25$)
H_A: The proportion of people that get news from the Internet is greater than 25%. ($p > 0.25$)

To get an idea of the level of confidence associated with this confidence interval, imagine working for Pew Research. You have 1593 randomly selected people, but you have no idea how they are going to respond, but you want a margin of error of 3%. Use the most conservative estimates you can for $\hat{p}$ and $\hat{q}$, namely 0.5. We can solve for $z^* \approx 2.39$, and estimate that this confidence interval has at least 98% confidence.

$$ME = z^* \sqrt{\frac{\hat{p}\hat{q}}{n}}$$

$$0.03 = z^* \sqrt{\frac{(0.5)(0.5)}{1593}}$$

$$z^* = \frac{0.03}{\sqrt{\frac{(0.5)(0.5)}{1593}}}$$

$$z^* \approx 2.39$$

We are 98% confidence that the proportion of people who get news from the Internet is 33% $\pm$ 3%, or 30% to 36%. This interval provides us with a reason to reject the null hypothesis, since 25% is not in the interval. There is strong evidence that the proportion of people who get news from the Internet is greater than 25%.

Chapter 22 – Comparing Two Proportions

1. **Arthritis.**

    a) **Randomization condition:** Americans age 65 and older were selected randomly.
    10% condition: 1012 men and 1062 women are less than 10% of all men and women.
    Independent samples condition: The sample of men and the sample of women were drawn independently of each other.
    Success/Failure condition: $n\hat{p}$ (men) = 411, $n\hat{q}$ (men) = 601, $n\hat{p}$ (women) = 535, and $n\hat{q}$ (women) = 527 are all greater than 10, so the samples are both large enough.

    Since the conditions have been satisfied, we will find a two-proportion z-interval.

    b) $(\hat{p}_F - \hat{p}_M) \pm z^* \sqrt{\dfrac{\hat{p}_F \hat{q}_F}{n_F} + \dfrac{\hat{p}_M \hat{q}_M}{n_M}} = \left(\dfrac{535}{1062} - \dfrac{411}{1012}\right) \pm 1.960 \sqrt{\dfrac{\left(\frac{535}{1062}\right)\left(\frac{527}{1062}\right)}{1062} + \dfrac{\left(\frac{411}{1012}\right)\left(\frac{601}{1012}\right)}{1012}} = (0.055, 0.140)$

    c) We are 95% confident that the proportion of American women age 65 and older who suffer from arthritis is between 5.5% and 14.0% higher than the proportion of American men the same age who suffer from arthritis.

    d) Since the interval for the difference in proportions of arthritis sufferers does not contain 0, there is strong evidence that arthritis is more likely to afflict women than men.

2. **Graduation.**

    a) **Randomization condition:** Assume that the samples are representative of all recent graduates.
    10% condition: Although large, the samples are less than 10% of all graduates.
    Independent samples condition: The sample of men and the sample of women were drawn independently of each other.
    Success/Failure condition: The samples are very large, certainly large enough for the methods of inference to be used.

    Since the conditions have been satisfied, we will find a two-proportion z-interval.

    b) $(\hat{p}_F - \hat{p}_M) \pm z^* \sqrt{\dfrac{\hat{p}_F \hat{q}_F}{n_F} + \dfrac{\hat{p}_M \hat{q}_M}{n_M}}$

    $= (0.881 - 0.849) \pm 1.960 \sqrt{\dfrac{(0.881)(0.119)}{12,678} + \dfrac{(0.849)(0.151)}{12,460}} = (0.024, 0.040)$

    c) We are 95% confident that the proportion of 24-year-old American women who have graduated from high school is between 2.4% and 4.0% higher than the proportion of American men the same age who have graduated from high school.

    d) Since the interval for the difference in proportions of high school graduates does not contain 0, there is strong evidence that women are more likely than men to complete high school.

3. **Pets.**

   a) $SE(\hat{p}_{Herb} - \hat{p}_{None}) = \sqrt{\dfrac{\hat{p}_{Herb}\hat{q}_{Herb}}{n_{Herb}} + \dfrac{\hat{p}_{None}\hat{q}_{None}}{n_{None}}} = \sqrt{\dfrac{\left(\frac{473}{827}\right)\left(\frac{354}{827}\right)}{827} + \dfrac{\left(\frac{19}{130}\right)\left(\frac{111}{130}\right)}{130}} = 0.035$

   b) Randomization condition: Assume that the dogs studied were representative of all dogs.
   10% condition: 827 dogs from homes with herbicide used regularly and 130 dogs from homes with no herbicide used are less than 10% of all dogs.
   Independent samples condition: The samples were drawn independently of each other.
   Success/Failure condition: $n\hat{p}$(herb) = 473, $n\hat{q}$(herb) = 354, $n\hat{p}$(none) = 19, and $n\hat{q}$(none) = 111 are all greater than 10, so the samples are both large enough.

   Since the conditions have been satisfied, we will find a two-proportion z-interval.

   $$\left(\hat{p}_{Herb} - \hat{p}_{None}\right) \pm z^{*}\sqrt{\dfrac{\hat{p}_{Herb}\hat{q}_{Herb}}{n_{Herb}} + \dfrac{\hat{p}_{None}\hat{q}_{None}}{n_{None}}}$$

   $$= \left(\dfrac{473}{827} - \dfrac{19}{130}\right) \pm 1.960\sqrt{\dfrac{\left(\frac{473}{827}\right)\left(\frac{354}{827}\right)}{827} + \dfrac{\left(\frac{19}{130}\right)\left(\frac{111}{130}\right)}{130}} = (0.356, 0.495)$$

   c) We are 95% confident that the proportion of pets with a malignant lymphoma in homes where herbicides are used is between 35.6% and 49.5% higher than the proportion of pets with lymphoma in homes where no pesticides are used.

4. **Carpal Tunnel.**

   a) $SE(\hat{p}_{Surg} - \hat{p}_{Splint}) = \sqrt{\dfrac{\hat{p}_{Surg}\hat{q}_{Surg}}{n_{Surg}} + \dfrac{\hat{p}_{Splint}\hat{q}_{Splint}}{n_{Splint}}} = \sqrt{\dfrac{(0.80)(0.20)}{88} + \dfrac{(0.54)(0.46)}{88}} = 0.068$

   b) Randomization condition: It's not clear whether or not this study was an experiment. If so, assume that the subjects were randomly allocated to treatment groups. If not, assume that the subjects are representative of all carpal tunnel sufferers.
   10% condition: 88 subjects in each group are less than 10% of all carpal tunnel sufferers.
   Independent samples condition: The improvement rates of the two groups are not related.
   Success/Failure condition: $n\hat{p}$(surg) = (88)(0.80) = 70, $n\hat{q}$(surg) = (88)(0.20) = 18, $n\hat{p}$(splint) = (88)(0.54) = 48, and $n\hat{q}$(splint) = (88)(0.46) = 40 are all greater than 10, so the samples are both large enough.

   Since the conditions have been satisfied, we will find a two-proportion z-interval.

   $$\left(\hat{p}_{Surg} - \hat{p}_{Splint}\right) \pm z^{*}\sqrt{\dfrac{\hat{p}_{Surg}\hat{q}_{Surg}}{n_{Surg}} + \dfrac{\hat{p}_{Splint}\hat{q}_{Splint}}{n_{Splint}}}$$

   $$= (0.80 - 0.54) \pm 1.960\sqrt{\dfrac{(0.80)(0.20)}{88} + \dfrac{(0.54)(0.46)}{88}} = (0.126, 0.394)$$

   c) We are 95% confident that the proportion of patients who show improvement in carpal tunnel syndrome with surgery is between 12.6% and 39.4% higher than the proportion who show improvement with wrist splints.

5. Prostate cancer.

a) This was an experiment. Men were randomly assigned to imposed treatments. They were assigned to either have prostate surgery or assigned to not have prostate surgery.

b) **Randomization condition:** The men were randomly assigned to the two treatment groups.
10% condition: 347 men who had surgery and 348 men who did not have surgery are both less than 10% of all men.
Independent samples condition: The groups were assigned randomly, so the groups are not related.
Success/Failure condition: $n\hat{p}$(surg) = 16, $n\hat{q}$(surg) = 331, $n\hat{p}$(none) = 31, and $n\hat{q}$(splint) = 317 are all greater than 10, so the samples are both large enough.

Since the conditions have been satisfied, we will find a two-proportion z-interval.

$$\left(\hat{p}_{None} - \hat{p}_{Surg}\right) \pm z^* \sqrt{\frac{\hat{p}_{None}\hat{q}_{None}}{n_{None}} + \frac{\hat{p}_{Surg}\hat{q}_{Surg}}{n_{Surg}}}$$

$$= \left(\tfrac{31}{348} - \tfrac{16}{347}\right) \pm 1.960 \sqrt{\frac{\left(\tfrac{31}{348}\right)\left(\tfrac{317}{348}\right)}{348} + \frac{\left(\tfrac{16}{347}\right)\left(\tfrac{331}{347}\right)}{347}} = (0.006, 0.080)$$

We are 95% confident that the proportion of patients who die from prostate cancer after having no surgery is between 0.60% and 8.0% higher than the proportion of patients who die after having surgery.

c) Since 0 is not contained in the interval, there is evidence that surgery may be effective in preventing death from prostate cancer.

6. Race and smoking.

a) **Randomization condition:** Assume that the survey was conducted randomly.
10% condition: 550 white adults and 550 black adults are both less than 10% of all adults.
Independent samples condition: The samples are independent of one another.
Success/Failure condition: $n\hat{p}$(white) = (550)(0.248) = 136, $n\hat{q}$(white) = (550)(0.752) = 414, $n\hat{p}$(black) = (550)(0.257) = 141, and $n\hat{q}$(black) = (550)(0.743) = 409 are all greater than 10, so the samples are both large enough.

Since the conditions have been satisfied, we will find a two-proportion z-interval.

$$\left(\hat{p}_{Black} - \hat{p}_{White}\right) \pm z^* \sqrt{\frac{\hat{p}_{Black}\hat{q}_{Black}}{n_{Black}} + \frac{\hat{p}_{White}\hat{q}_{White}}{n_{White}}}$$

$$= (0.257 - 0.248) \pm 1.645 \sqrt{\frac{(0.257)(0.743)}{550} + \frac{(0.248)(0.752)}{550}} = (-0.034, 0.052)$$

We are 90% confident that the proportion of black smokers is between 3.4% lower and 5.2% higher than the proportion of white smokers.

b) H_0 : The proportion of black smokers is the same as the proportion of white smokers.
$$\left(p_{Black} = p_{White} \text{ or } p_{Black} - p_{White} = 0\right)$$
H_A : The proportion of black smokers is different from the proportion of white smokers.
$$\left(p_{Black} \neq p_{White} \text{ or } p_{Black} - p_{White} \neq 0\right)$$

Since 0 is contained within the confidence interval, we fail to reject the null hypothesis. There is no evidence of a race-based difference in smoking percentages.

7. **Politics.**

   a) The margin of error is larger for the difference in proportions of support because differences have larger standard errors than single samples.

   b) The confidence interval for difference in support was 2% ± 4%, or (–2%, 6%). Since the interval contains 0, there is no evidence of a difference in the proportion of support for antiterrorist legislation.

8. **War.**

   a) The poll estimated that the difference in support between Republicans and Democrats was 75% – 30%, or 45% higher for the Republicans.

   b) The margin of error for the difference would be greater than 3%, the margin of error for the single sample. Differences have standard errors that are larger than standard errors for single samples.

9. **Teen smoking, part I.**

   a) This is a prospective observational study.

   b) H_0: The proportion of teen smokers among the group whose parents disapprove of smoking is the same as the proportion of teen smokers among the group whose parents are lenient about smoking. $\left(p_{Dis} = p_{Len} \text{ or } p_{Dis} - p_{Len} = 0\right)$

   H_A : The proportion of teen smokers among the group whose parents disapprove of smoking is lower than the proportion of teen smokers among the group whose parents are lenient about smoking. $\left(p_{Dis} < p_{Len} \text{ or } p_{Dis} - p_{Len} < 0\right)$

   c) **Randomization condition:** Assume that the teens surveyed are representative of all teens. **10% condition:** 284 and 41 are both less than 10% of all teens.
   Independent samples condition: The groups were surveyed independently.
   Success/Failure condition: $n\hat{p}$ (disapprove) = 54, $n\hat{q}$ (disapprove) = 230, $n\hat{p}$ (lenient) = 11, and $n\hat{q}$ (lenient) = 30 are all greater than 10, so the samples are both large enough.

   Since the conditions have been satisfied, we will model the sampling distribution of the difference in proportion with a Normal model with mean 0 and standard deviation

   estimated by $SE_{pooled}\left(\hat{p}_{Dis} - \hat{p}_{Len}\right) = \sqrt{\dfrac{\hat{p}_{pooled}\hat{q}_{pooled}}{n_{Dis}} + \dfrac{\hat{p}_{pooled}\hat{q}_{pooled}}{n_{Len}}} = \sqrt{\dfrac{\left(\frac{65}{325}\right)\left(\frac{260}{325}\right)}{284} + \dfrac{\left(\frac{65}{325}\right)\left(\frac{260}{325}\right)}{41}} = 0.0668.$

d) The observed difference between the proportions is 0.190 – 0.268 = – 0.078.

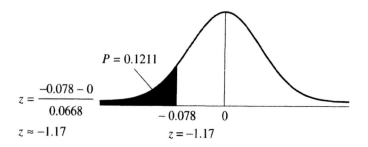

$$P = 0.1211$$

Since the _P_-value = 0.1211 is high, we fail to reject the null hypothesis. There is little evidence to suggest that parental attitudes influence teens' decisions to smoke.

$$z = \frac{-0.078 - 0}{0.0668}$$

$$z \approx -1.17$$

e) If there is no difference in the proportions, there is about a 12% chance of seeing the observed difference or larger by natural sampling variation.

f) If teens' decisions about smoking _are_ influenced, we have committed a Type II error.

10. Depression.

a) This is a prospective observational study.

b) H_0: The proportion of cardiac patients without depression who died within the 4 years is the same as the proportion of cardiac patients with depression who died during the same time period. $\left(p_{None} = p_{Dep} \text{ or } p_{None} - p_{Dep} = 0\right)$

H_A: The proportion of cardiac patients without depression who died within the 4 years is the less than the proportion of cardiac patients with depression who died during the same time period. $\left(p_{None} < p_{Dep} \text{ or } p_{None} - p_{Dep} < 0\right)$

c) **Randomization condition:** Assume that the cardiac patients followed by the study are representative of all cardiac patients.
10% condition: 361 and 89 are both less than 10% of all teens.
Independent samples condition: The groups are not associated.
Success/Failure condition: $n\hat{p}$ (no depression) = 67, $n\hat{q}$ (no depression) = 294, $n\hat{p}$ (depression) = 26, and $n\hat{q}$ (depression) = 63 are all greater than 10, so the samples are both large enough.

Since the conditions have been satisfied, we will model the sampling distribution of the difference in proportion with a Normal model with mean 0 and standard deviation

$$\text{estimated by } SE_{pooled}\left(\hat{p}_{None} - \hat{p}_{Dep}\right) = \sqrt{\frac{\hat{p}_{pooled}\hat{q}_{pooled}}{n_{None}} + \frac{\hat{p}_{pooled}\hat{q}_{pooled}}{n_{Dep}}} = \sqrt{\frac{\left(\frac{93}{450}\right)\left(\frac{357}{450}\right)}{361} + \frac{\left(\frac{93}{450}\right)\left(\frac{357}{450}\right)}{89}} \approx 0.0479.$$

d) The observed difference between the proportions is:
0.1856 – 0.2921 = – 0.1065.

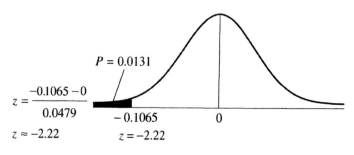

$$P = 0.0131$$

Since the _P_-value = 0.0131 is low, we reject the null hypothesis. There is strong evidence to suggest that the proportion of non-depressed cardiac patients who die within 4 years is less than the proportion of depressed cardiac patients who die within 4 years.

$$z = \frac{-0.1065 - 0}{0.0479}$$

$$z \approx -2.22$$

e) If there is no difference in the proportions, we will see an observed difference this large or larger only about 1.3% of the time by natural sampling variation.

f) If cardiac patients without depression don't actually have a lower proportion of deaths in 4 years than cardiac patients with depression, then we have committed a Type I error.

11. Teen smoking, part II.

a) Since the conditions have already been satisfied in Exercise 9, we will find a two-proportion z-interval.

$$\left(\hat{p}_{Dis} - \hat{p}_{Len}\right) \pm z^* \sqrt{\frac{\hat{p}_{Dis}\hat{q}_{Dis}}{n_{Dis}} + \frac{\hat{p}_{Len}\hat{q}_{Len}}{n_{Len}}}$$

$$= \left(\tfrac{54}{284} - \tfrac{11}{41}\right) \pm 1.960 \sqrt{\frac{\left(\tfrac{54}{284}\right)\left(\tfrac{230}{284}\right)}{284} + \frac{\left(\tfrac{11}{41}\right)\left(\tfrac{30}{41}\right)}{41}} = (-0.065, 0.221)$$

b) We are 95% confident that the proportion of teens whose parents disapprove of smoking who will eventually smoke is between 6.5% less and 22.1% more than for teens with parents who are lenient about smoking.

c) We expect 95% of random samples of this size to produce intervals that contain the true difference between the proportions.

12. Depression revisited.

a) Since the conditions have already been satisfied in Exercise 10, we will find a two-proportion z-interval.

$$\left(\hat{p}_{None} - \hat{p}_{Dep}\right) \pm z^* \sqrt{\frac{\hat{p}_{None}\hat{q}_{None}}{n_{None}} + \frac{\hat{p}_{Dep}\hat{q}_{Dep}}{n_{Dep}}}$$

$$= \left(\tfrac{67}{361} - \tfrac{26}{89}\right) \pm 1.960 \sqrt{\frac{\left(\tfrac{67}{361}\right)\left(\tfrac{294}{361}\right)}{361} + \frac{\left(\tfrac{26}{89}\right)\left(\tfrac{63}{89}\right)}{89}} = (0.004, 0.209)$$

b) We are 95% confident that the proportion of cardiac disease patients who die within 4 years is between 0.4% and 20.9% higher for depressed patients than for non-depressed patients.

c) We expect 95% of random samples of this size to produce intervals that contain the true difference between the proportions.

13. Pregnancy.

a) H_0: The proportion of live births is the same for women under the age of 38 as it is for women 38 or older. $\left(p_{<38} = p_{\geq 38} \text{ or } p_{<38} - p_{\geq 38} = 0\right)$

H_A: The proportion of live births is different for women under the age of 38 than for women 38 or older. $\left(p_{<38} \neq p_{\geq 38} \text{ or } p_{<38} - p_{\geq 38} \neq 0\right)$

Randomization condition: Assume that the women studied are representative of all women.
10% condition: 157 and 89 are both less than 10% of all women.
Independent samples condition: The groups are not associated.

Success/Failure condition: $n\hat{p}$ (under 38) = 42, $n\hat{q}$ (under 38) = 115, $n\hat{p}$ (38 and over) = 7, and $n\hat{q}$ (38 and over) = 82 are not all greater than 10, since the observed number of live births is only 7. However, if we check the pooled value, $n\hat{p}_{pooled}$ (38 and over) = (89)(0.1992) = 18. All of the samples are large enough.

Since the conditions have been satisfied, we will model the sampling distribution of the difference in proportion with a Normal model with mean 0 and standard deviation

$$estimated\ by\ SE_{pooled}\left(\hat{p}_{<38} - \hat{p}_{\geq38}\right) = \sqrt{\frac{\hat{p}_{pooled}\hat{q}_{pooled}}{n_{<38}} + \frac{\hat{p}_{pooled}\hat{q}_{pooled}}{n_{\geq38}}} = \sqrt{\frac{\left(\frac{49}{246}\right)\left(\frac{197}{246}\right)}{157} + \frac{\left(\frac{49}{246}\right)\left(\frac{197}{246}\right)}{89}} \approx 0.0530.$$

The observed difference between the proportions is:
0.2675 – 0.0787 = 0.1888.

Since the *P*-value = 0.0004 is low, we reject the null hypothesis. There is strong evidence to suggest a difference in the proportion of live births for women under 38 and women 38 and over at this clinic. In fact, the evidence suggests that women under 38 have a higher proportion of live births.

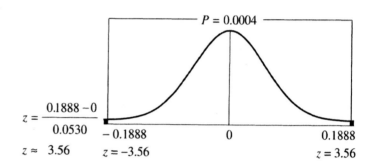

$z = \dfrac{0.1888 - 0}{0.0530}$

$z \approx 3.56$

b) $\left(\hat{p}_{<38} - \hat{p}_{\geq38}\right) \pm z^* \sqrt{\dfrac{\hat{p}_{<38}\hat{q}_{<38}}{n_{<38}} + \dfrac{\hat{p}_{\geq38}\hat{q}_{\geq38}}{n_{\geq38}}}$

$= \left(\frac{42}{157} - \frac{7}{89}\right) \pm 1.960 \sqrt{\dfrac{\left(\frac{42}{157}\right)\left(\frac{115}{157}\right)}{157} + \dfrac{\left(\frac{7}{89}\right)\left(\frac{82}{89}\right)}{89}} = (0.100, 0.278)$

We are 95% confident that the proportion of live births for patients at this clinic is between 10.0% and 27.8% higher for women under 38 than for women 38 and over. However, the Success/Failure condition is not met for the older women, so we should be cautious when using this interval. (The expected number of successes from the pooled proportion cannot be used for a condition for a confidence interval. It's based upon an assumption that the proportions are the same. We don't make that assumption in a confidence interval. In fact, we are implicitly assuming a *difference*, by finding an interval for the difference in proportion.)

14. Suicide.

a) H_0: The proportion of attempted suicides is the same for adopted children as it is for children who were not adopted. $\left(p_A = p_N \text{ or } p_A - p_N = 0\right)$

H_A: The proportion of attempted suicides is different for adopted children than it is for children who were not adopted. $\left(p_A \neq p_N \text{ or } p_A - p_N \neq 0\right)$

Randomization condition: Assume that the teens studied are representative of all teens.
10% condition: 214 and 6363 are both less than 10% of all teens.
Independent samples condition: The groups are not associated.

Success/Failure condition: $n\hat{p}$ (adopted) = 16, $n\hat{q}$ (adopted) = 198, $n\hat{p}$ (not adopted) = 197, and $n\hat{q}$ (not adopted) = 6166 are all greater than 10, so both samples are large enough.

Since the conditions have been satisfied, we will model the sampling distribution of the difference in proportion with a Normal model with mean 0 and standard deviation

estimated by $SE_{\text{pooled}}(\hat{p}_A - \hat{p}_N) = \sqrt{\dfrac{\hat{p}_{\text{pooled}}\hat{q}_{\text{pooled}}}{n_A} + \dfrac{\hat{p}_{\text{pooled}}\hat{q}_{\text{pooled}}}{n_N}} = \sqrt{\dfrac{\left(\frac{213}{6577}\right)\left(\frac{6364}{6577}\right)}{214} + \dfrac{\left(\frac{213}{6577}\right)\left(\frac{6364}{6577}\right)}{6363}} \approx 0.0123.$

The observed difference between the proportions is:
0.07477 – 0.03096 = 0.04381.

Since the *P*-value = 0.0004 is low, we reject the null hypothesis. There is strong evidence of a difference in the proportion of attempted suicides for adopted children and children who were not adopted. In fact, the evidence suggests adopted children have a greater proportion of attempted suicides.

$z = \dfrac{-0.04381 - 0}{0.01230}$

$z \approx 3.56$

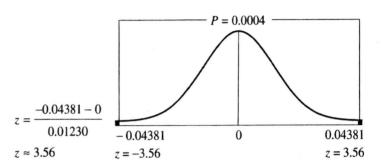

P = 0.0004

– 0.04381   0   0.04381
$z = -3.56$ $z = 3.56$

b) $(\hat{p}_A - \hat{p}_N) \pm z^* \sqrt{\dfrac{\hat{p}_A \hat{q}_A}{n_A} + \dfrac{\hat{p}_N \hat{q}_N}{n_N}}$

$= \left(\dfrac{16}{214} - \dfrac{197}{6363}\right) \pm 1.960 \sqrt{\dfrac{\left(\frac{16}{214}\right)\left(\frac{198}{214}\right)}{214} + \dfrac{\left(\frac{197}{6363}\right)\left(\frac{6166}{6363}\right)}{6363}} = (0.008, 0.079)$

We are 95% confident that the proportion of attempted suicides is between 0.8% and 7.9% higher for adopted children than for children who were not adopted.

15. Politics and sex.

H_0: The proportion of voters in support of the candidate is the same before and after news of his extramarital affair got out. $(p_B = p_A$ or $p_B - p_A = 0)$

H_A: The proportion of voters in support of the candidate has decreased after news of his extramarital affair got out. $(p_B > p_A$ or $p_B - p_A > 0)$

Randomization condition: Voters were randomly selected.
10% condition: 630 and 1010 are both less than 10% of all voters.
Independent samples condition: Since the samples were random, the groups are independent.
Success/Failure condition: $n\hat{p}$ (before) = (630)(0.54) = 340, $n\hat{q}$ (before) = (630)(0.46) = 290, $n\hat{p}$ (after) = (1010)(0.51) = 515, and $n\hat{q}$ (after) = (1010)(0.49) = 505 are all greater than 10, so both samples are large enough.

Since the conditions have been satisfied, we will model the sampling distribution of the difference in proportion with a Normal model with mean 0 and standard deviation estimated by:

$$SE_{pooled}(\hat{p}_B - \hat{p}_A) = \sqrt{\frac{\hat{p}_{pooled}\hat{q}_{pooled}}{n_B} + \frac{\hat{p}_{pooled}\hat{q}_{pooled}}{n_A}} = \sqrt{\frac{(0.5215)(0.4785)}{630} + \frac{(0.5215)(0.4785)}{1010}} \approx 0.02536.$$

The observed difference between the proportions is:
0.54 – 0.51 = 0.03.

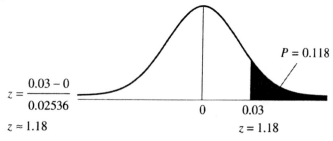

Since the *P*-value = 0.118 is fairly high, we fail to reject the null hypothesis. There is little evidence of a decrease in the proportion of voters in support of the candidate after the news of his extramarital affair got out.

$$z = \frac{0.03 - 0}{0.02536}$$

$$z \approx 1.18$$

P = 0.118

0   0.03

z = 1.18

16. Retirement.

H_0: The proportion of those who anticipate having enough money to live comfortably in retirement is the same for men and women. $(p_M = p_W \text{ or } p_M - p_W = 0)$

H_A: The proportion of those who anticipate having enough money to live comfortably in retirement is different for men and women. $(p_M \neq p_W \text{ or } p_M - p_W \neq 0)$

Randomization condition: Assume that the people surveyed are representative of all people.
10% condition: 250 and 250 are both less than 10% of all people.
Independent samples condition: The groups are independent.
Success/Failure condition: $n\hat{p}$ (men) = (250)(0.27) = 67.5, $n\hat{q}$ (men) = (250)(0.73) = 182.5, $n\hat{p}$ (women) = (250)(0.18) = 45, and $n\hat{q}$ (women) = (250)(0.82) = 205 are all greater than 10, so both samples are large enough.

Since the conditions have been satisfied, we will model the sampling distribution of the difference in proportion with a Normal model with mean 0 and standard deviation estimated by:

$$SE_{pooled}(\hat{p}_M - \hat{p}_W) = \sqrt{\frac{\hat{p}_{pooled}\hat{q}_{pooled}}{n_M} + \frac{\hat{p}_{pooled}\hat{q}_{pooled}}{n_W}} = \sqrt{\frac{(0.225)(0.775)}{250} + \frac{(0.225)(0.775)}{250}} \approx 0.03735.$$

The observed difference between the proportions is:
0.27 – 0.18 = 0.09.

Since the *P*-value = 0.0160 is low, we reject the null hypothesis. There is strong evidence of a difference in the proportion of those who anticipated having enough money to live comfortably in retirement for men and women. In fact, the evidence suggests that a greater proportion of men feel comfortable about retirement.

P = 0.0160

$$z = \frac{0.09 - 0}{0.03735}$$

$$z \approx 2.41$$

– 0.09   0   0.09

z = –2.41   *z* = 2.41

17. Twins.

a) H_0: The proportion of multiple births is the same for white women and black women. $\left(p_W = p_B \text{ or } p_W - p_B = 0\right)$

H_A: The proportion of multiple births is different for white women and black women. $\left(p_W \neq p_B \text{ or } p_W - p_B \neq 0\right)$

Randomization condition: Assume that these women are representative of all women.
10% condition: 3132 and 606 are both less than 10% of all people.
Independent samples condition: The groups are independent.
Success/Failure condition: $n\hat{p}$ (white) = 94, $n\hat{q}$ (white) = 3038, $n\hat{p}$ (black) = 20, and $n\hat{q}$ (black) = 586 are all greater than 10, so both samples are large enough.

Since the conditions have been satisfied, we will model the sampling distribution of the difference in proportion with a Normal model with mean 0 and standard deviation estimated by:

$$SE_{pooled}\left(\hat{p}_W - \hat{p}_B\right) = \sqrt{\frac{\hat{p}_{pooled}\hat{q}_{pooled}}{n_W} + \frac{\hat{p}_{pooled}\hat{q}_{pooled}}{n_B}} = \sqrt{\frac{\left(\frac{114}{3738}\right)\left(\frac{3624}{3738}\right)}{3132} + \frac{\left(\frac{114}{3738}\right)\left(\frac{3624}{3738}\right)}{606}} \approx 0.007631.$$

The observed difference between the proportions is: $0.030 - 0.033 = -0.003$.

Since the *P*-value = 0.6951 is high, we fail to reject the null hypothesis. There is no evidence of a difference between the proportions of multiple births for white women and black women.

$z = \dfrac{0.003 - 0}{0.07631}$

$z \approx -0.39$

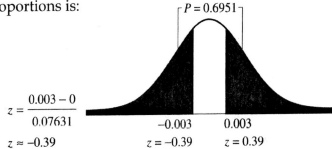

b) If there is actually a difference between the proportions of multiple births for white women and black women, then we have committed a Type II error.

18. Shopping.

a) H_0: The proportion of men who have purchased books online is the same as the proportion of women who have purchased books online. $\left(p_M = p_W \text{ or } p_M - p_W = 0\right)$

H_A: The proportion of men who have purchased books online is greater than the proportion of women who have purchased books online. $\left(p_M > p_W \text{ or } p_M - p_W > 0\right)$

Randomization condition: The men and women were chosen randomly.
10% condition: 222 and 208 are both less than 10% of all people.
Independent samples condition: The groups were chosen independently.
Success/Failure condition: $n\hat{p}$ (men) = (222)(0.21) = 47, $n\hat{q}$ (men) = (222)(0.79) = 175, $n\hat{p}$ (women) = (208)(0.18) = 37, and $n\hat{q}$ (women) = (208)(0.82) = 171 are all greater than 10, so both samples are large enough.

Since the conditions have been satisfied, we will model the sampling distribution of the difference in proportion with a Normal model with mean 0 and standard deviation estimated by:

$$SE_{pooled}(\hat{p}_M - \hat{p}_W) = \sqrt{\frac{\hat{p}_{pooled}\hat{q}_{pooled}}{n_M} + \frac{\hat{p}_{pooled}\hat{q}_{pooled}}{n_W}} = \sqrt{\frac{(0.1955)(8045)}{222} + \frac{(0.1955)(8045)}{208}} \approx 0.03827.$$

The observed difference between the proportions is:
0.21 – 0.18 = 0.03.

Since the *P*-value = 0.2166 is high, we fail to reject the null hypothesis. There is no evidence that the proportion of men who have purchased books online is greater than the proportion of women who have purchased books online.

$z = \dfrac{0.03 - 0}{0.03827}$

$z \approx 0.78$

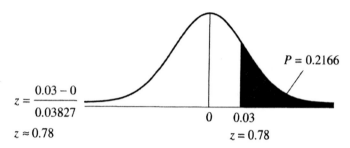

b) If there is actually a difference between the proportions of men who have purchased books online and the proportion of women who have purchased books online, then we have committed a Type II error.

19. Mammograms.

a) H_0: The proportion of deaths from breast cancer is the same for women who never had a mammogram as for women who had mammograms. $(p_N = p_M$ or $p_N - p_M = 0)$

H_A: The proportion of deaths from breast cancer is greater for women who never had a mammogram than for women who had mammograms. $(p_N > p_M$ or $p_N - p_M > 0)$

Randomization condition: Assume that the women are representative of all women.
10% condition: 30,565 and 30,131 are both less than 10% of all women.
Independent samples condition: The groups were chosen independently.
Success/Failure condition: $n\hat{p}$(never) = 196, $n\hat{q}$(never) = 30,369, $n\hat{p}$(mammogram) = 153, and $n\hat{q}$(mammogram) = 29,978 are all greater than 10, so both samples are large enough.

Since the conditions have been satisfied, we will model the sampling distribution of the difference in proportion with a Normal model with mean 0 and standard deviation estimated by:

$$SE_{pooled}(\hat{p}_N - \hat{p}_M) = \sqrt{\frac{\hat{p}_{pooled}\hat{q}_{pooled}}{n_N} + \frac{\hat{p}_{pooled}\hat{q}_{pooled}}{n_M}} = \sqrt{\frac{\left(\frac{349}{60,696}\right)\left(\frac{60,347}{60,696}\right)}{30,565} + \frac{\left(\frac{349}{60,696}\right)\left(\frac{60,347}{60,696}\right)}{30,131}} \approx 0.0006138.$$

The observed difference between the proportions is: 0.006413 – 0.005078 = 0.001335.

Since the *P*-value = 0.0148 is low, we reject the null hypothesis. There is strong evidence that the proportion of breast cancer deaths for women who have never had a mammogram is greater than the proportion of deaths from breast cancer for women who underwent screening by mammogram.

$$z = \frac{0.001335 - 0}{0.0006198}$$

$$z \approx 2.17$$

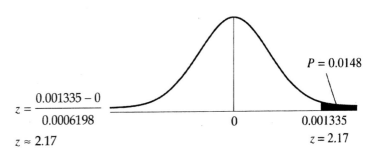

b) If there is actually no difference between the proportions of deaths from breast cancer for women who have and have not had mammograms, we have committed a Type I error.

20. Mammograms redux.

a) H_0: The proportion of deaths from breast cancer is the same for women who never had a mammogram as for women who had mammograms. $(p_N = p_M$ or $p_N - p_M = 0)$

H_A: The proportion of deaths from breast cancer is greater for women who never had a mammogram than for women who had mammograms. $(p_N > p_M$ or $p_N - p_M > 0)$

Randomization condition: Assume that the women are representative of all women.
10% condition: 21,195 and 21,088 are both less than 10% of all women.
Independent samples condition: The groups were chosen independently.
Success/Failure condition: $n\hat{p}$ (never) = 66, $n\hat{q}$ (never) = 21,129, $n\hat{p}$ (mammogram) = 63, and $n\hat{q}$ (mammogram) = 21,025 are all greater than 10, so both samples are large enough.

Since the conditions have been satisfied, we will model the sampling distribution of the difference in proportion with a Normal model with mean 0 and standard deviation estimated by:

$$SE_{pooled}\left(\hat{p}_N - \hat{p}_M\right) = \sqrt{\frac{\hat{p}_{pooled}\hat{q}_{pooled}}{n_N} + \frac{\hat{p}_{pooled}\hat{q}_{pooled}}{n_M}} = \sqrt{\frac{\left(\frac{129}{42,283}\right)\left(\frac{42,154}{42,283}\right)}{21,195} + \frac{\left(\frac{129}{42,283}\right)\left(\frac{42,154}{42,283}\right)}{21,088}} \approx 0.000536.$$

The observed difference between the proportions is: 0.003114 – 0.002987 = 0.000127.

Since the *P*-value = 0.4068 is high, we fail to reject the null hypothesis. There is no evidence that the proportion of breast cancer deaths for women who have never had a mammogram is greater than the proportion of deaths from breast cancer for women who underwent screening by mammogram.

$$z = \frac{0.000127 - 0}{0.000536}$$

$$z \approx 0.24$$

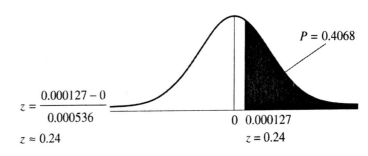

b) If the proportion of deaths from breast cancer for women who have not had mammograms is actually greater than the proportion of deaths from breast cancer for women who have had mammograms, we have committed a Type II error.

21. Pain.

a) **Randomization condition:** The patients were randomly selected AND randomly assigned to treatment groups. If that's not random enough for you, I don't know what is!
10% condition: 112 and 108 are both less than 10% of all people with joint pain.
Success/Failure condition: $n\hat{p}(A) = 84$, $n\hat{q}(A) = 28$, $n\hat{p}(B) = 66$, and $n\hat{q}(B) = 42$ are all greater than 10, so both samples are large enough.

Since the conditions are met, we can use a one-proportion z-interval to estimate the percentage of patients who may get relief from medication A.

$$\hat{p} \pm z^* \sqrt{\frac{\hat{p}\hat{q}}{n}} = \left(\frac{84}{112}\right) \pm 1.960 \sqrt{\frac{\left(\frac{84}{112}\right)\left(\frac{28}{112}\right)}{112}} = (67.0\%, 83.0\%)$$

We are 95% confident that between 67.0% and 83.0% of patients with joint pain will find medication A to be effective.

b) Since the conditions were met in part a, we can use a one-proportion z-interval to estimate the percentage of patients who may get relief from medication B.

$$\hat{p} \pm z^* \sqrt{\frac{\hat{p}\hat{q}}{n}} = \left(\frac{66}{108}\right) \pm 1.960 \sqrt{\frac{\left(\frac{66}{108}\right)\left(\frac{42}{108}\right)}{108}} = (51.9\%, 70.3\%)$$

We are 95% confident that between 51.9% and 70.3% of patients with joint pain will find medication B to be effective.

c) The 95% confidence intervals overlap, which might lead one to believe that there is no evidence of a difference in the proportions of people who find each medication effective. However, if one was lead to believe that, one should proceed to part…

d) Most of the conditions were checked in part a. We only have one more to check:
Independent samples condition: The groups were assigned randomly, so there is no reason to believe there is a relationship between them.

Since the conditions have been satisfied, we will find a two-proportion z-interval.

$$\left(\hat{p}_A - \hat{p}_B\right) \pm z^* \sqrt{\frac{\hat{p}_A \hat{q}_A}{n_A} + \frac{\hat{p}_B \hat{q}_B}{n_B}}$$

$$= \left(\tfrac{84}{112} - \tfrac{66}{108}\right) \pm 1.960 \sqrt{\frac{\left(\frac{84}{112}\right)\left(\frac{28}{112}\right)}{112} + \frac{\left(\frac{66}{108}\right)\left(\frac{42}{108}\right)}{112}} = (0.017, 0.261)$$

We are 95% confident that the proportion of patients with joint pain who will find medication A effective is between 1.70% and 26.1% higher than the proportion of patients who will find medication B effective.

e) The interval does not contain zero. There is evidence that medication A is more effective than medication B.

f) The two-proportion method is the proper method. By attempting to use two, separate, confidence intervals, you are adding standard deviations when looking for a difference in proportions. We know from our previous studies that *variances* add when finding the standard deviation of a difference. The two-proportion method does this.

22. Gender gap.

a) Randomization condition: The poll was probably random, although not specifically stated.
10% condition: 473 and 522 are both less than 10% of all voters.
Success/Failure condition: $n\hat{p}$(men) = 246, $n\hat{q}$(men) = 227, $n\hat{p}$(women) = 235, and $n\hat{q}$(women) = 287 are all greater than 10, so both samples are large enough.

Since the conditions are met, we can use a one-proportion z-interval to estimate the percentage of men who may vote for the candidate.

$$\hat{p} \pm z^* \sqrt{\frac{\hat{p}\hat{q}}{n}} = (0.52) \pm 1.960 \sqrt{\frac{(0.52)(0.48)}{473}} = (47.5\%, 56.5\%)$$

We are 95% confident that between 47.5% and 56.5% of men may vote for the candidate.

b) Since the conditions were met in part a, we can use a one-proportion z-interval to estimate the percentage of women who may vote for the candidate.

$$\hat{p} \pm z^* \sqrt{\frac{\hat{p}\hat{q}}{n}} = (0.45) \pm 1.960 \sqrt{\frac{(0.45)(0.55)}{522}} = (40.7\%, 49.3\%)$$

We are 95% confident that between 40.7% and 49.3% of women may vote for the candidate.

c) The 95% confidence intervals overlap, which might make you think that there is no evidence of a difference in the proportions of men and women who may vote for the candidate. However, if you think that, don't delay! Move on to part...

d) Most of the conditions were checked in part a. We only have one more to check:
Independent samples condition: There is no reason to believe that the samples of men and women influence each other in any way.

Since the conditions have been satisfied, we will find a two-proportion z-interval.

$$(\hat{p}_M - \hat{p}_W) \pm z^* \sqrt{\frac{\hat{p}_M\hat{q}_M}{n_M} + \frac{\hat{p}_W\hat{q}_W}{n_W}}$$

$$= (0.52 - 0.45) \pm 1.960 \sqrt{\frac{(0.52)(0.48)}{473} + \frac{(0.45)(0.55)}{522}} = (0.008, 0.132)$$

We are 95% confident that the proportion of men who may vote for the candidate is between 0.8% and 13.2% higher than the proportion of women who may vote for the candidate.

e) The interval does not contain zero. There is evidence that the proportion of men may vote for the candidate is greater than the proportion of women who may vote for the candidate.

f) The two-proportion method is the proper method. By attempting to use two, separate, confidence intervals, you are adding standard deviations when looking for a difference in proportions. We know from our previous studies that *variances* add when finding the standard deviation of a difference. The two-proportion method does this.

Review of Part V – From the Data at Hand to the World at Large

1. **Herbal cancer.**

   H_0: The cancer rate for those taking the herb is the same as the cancer rate for those not taking the herb. $\left(p_{Herb} = p_{Not} \text{ or } p_{Herb} - p_{Not} = 0\right)$

   H_A: The cancer rate for those taking the herb is higher than the cancer rate for those not taking the herb. $\left(p_{Herb} > p_{Not} \text{ or } p_{Herb} - p_{Not} > 0\right)$

2. **Colorblind.**

   a) **10% condition:** The 325 male students are probably representative of all males, and 325 students are less than 10% of the population of males.
   Success/Failure condition: $np = (325)(0.08) = 26$ and $nq = (325)(0.92) = 299$ are both greater than 10, so the sample is large enough.

   Since the conditions have been satisfied, a Normal model can be used to model the sampling distribution of the proportion of colorblind men among 325 students.

   b) $\mu_{\hat{p}} = p = 0.08$

   $$\sigma(\hat{p}) = \sqrt{\frac{pq}{n}} = \sqrt{\frac{(0.08)(0.92)}{325}} \approx 0.015$$

   c)

   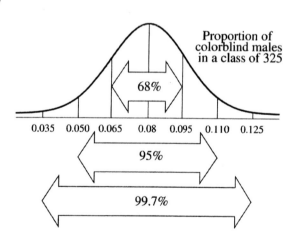

   d) According to the Normal model, we expect about 68% of classes with 325 males to have between 6.5% and 9.5% colorblind males. We expect about 95% of such classes to have between 5% and 11% colorblind males. About 99.7% of such classes are expected to have between 3.5% and 12.5% colorblind males.

3. **Birth days.**

   a) If births are distributed uniformly across all days, we expect the number of births on each day to be $np = (72)(\frac{1}{7}) \approx 10.29$.

   b) **10% condition:** The 72 births are likely to be representative of all births at the hospital with regards to day of birth, and 72 births are less than 10% of the births.
   Success/Failure condition: The expected number of births on a particular day of the week is $np = (72)(\frac{1}{7}) \approx 10.29$ and the expected number of births not on that particular day is $nq = (72)(\frac{6}{7}) \approx 61.71$. These are both greater than 10, so the sample is large enough.

Since the conditions have been satisfied, a Normal model can be used to model the sampling distribution of the proportion of 72 births that occur on a given day of the week.

$$\mu_{\hat{p}} = p = \tfrac{1}{7} \approx 0.1429$$

$$\sigma(\hat{p}) = \sqrt{\frac{pq}{n}} = \sqrt{\frac{(\frac{1}{7})(\frac{6}{7})}{72}} \approx 0.04124$$

There were 7 births on Mondays, so $\hat{p} = \dfrac{7}{72} \approx 0.09722$. This is only about a 1.11 standard deviations below the expected proportion, so there's no evidence that this is an unusual occurrence.

c) The 17 births on Tuesdays represent an unusual occurrence. For Tuesdays, $\hat{p} = \dfrac{17}{72} \approx 0.2361$, which is about 2.26 standard deviations above the expected proportion of births. There is evidence to suggest that the proportion of births on Tuesdays is higher than expected, if births are distributed uniformly across days.

d) Some births are scheduled for the convenience of the doctor and/or the mother.

4. Polling.

a) No, the number of votes would not always be the same. We expect a certain amount of variability when sampling.

b) This is NOT a problem about confidence intervals. We already know the true proportion of voters who voted for Clinton. This problem deals with the sampling distribution of that proportion.

We would expect 95% of our sample proportions of Clinton voters to be within 1.960 standard deviations of the true proportion of Clinton voters, 43%.

$$\sigma(\hat{p}_C) = \sqrt{\frac{p_C q_C}{n}} = \sqrt{\frac{(0.43)(0.57)}{100}} \approx 4.95\%$$

So, we expect 95% of our sample proportions to be within 1.960(4.95%) = 9.7% of 43%, or between 33.3% and 52.7%.

c) This is NOT a problem about confidence intervals. We already know the true proportion of voters who voted for Perot. This problem deals with the sampling distribution of that proportion.

We would expect 95% of our sample proportions of Perot voters to be within 1.960 standard deviations of the true proportion of Perot voters, 19%.

$$\sigma(\hat{p}_P) = \sqrt{\frac{p_P q_P}{n}} = \sqrt{\frac{(0.19)(0.81)}{100}} \approx 3.92\%$$

So, we expect 95% of our sample proportions to be within 1.960(3.92%) = 7.7% of 19%, or between 11.3% and 26.7%.

d) The sample proportion of Perot voters is expected to vary less than the sample proportion of Bush voters. Proportions farther away from 50% have smaller standard errors. Calculate the standard deviation of the proportion of Bush voters to convince yourself!

5. Leaky gas tanks.

  a) H_0: The proportion of leaky gas tanks is 40%. $(p = 0.40)$
  H_A: The proportion of leaky gas tanks is less than 40%. $(p < 0.40)$

  b) **Randomization condition:** A random sample of 27 service stations in California was taken.
  10% condition: 27 service stations are less than 10% of all service stations in California.
  Success/Failure condition: $np = (27)(0.40) = 10.8$ and $nq = (27)(0.60) = 16.2$ are both greater than 10, so the sample is large enough.

  c) Since the conditions have been satisfied, a Normal model can be used to model the sampling distribution of the proportion, with $\mu_{\hat{p}} = p = 0.40$ and

  $\sigma(\hat{p}) = \sqrt{\dfrac{pq}{n}} = \sqrt{\dfrac{(0.40)(0.60)}{27}} \approx 0.09428.$ We can perform a one-proportion z-test.

  The observed proportion of leaky gas tanks is $\hat{p} = \dfrac{7}{27} \approx 0.2593.$

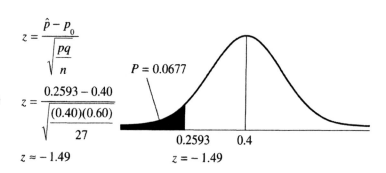

$$z = \frac{\hat{p} - p_0}{\sqrt{\dfrac{pq}{n}}}$$

$$z = \frac{0.2593 - 0.40}{\sqrt{\dfrac{(0.40)(0.60)}{27}}}$$

$z \approx -1.49$

$P = 0.0677$

$0.2593 \quad 0.4$

$z = -1.49$

  d) Since the *P*-value = 0.0677 is relatively high, we fail to reject the null hypothesis. There is little evidence that the proportion of leaky gas tanks is less than 40%. The new program doesn't appear to be effective in decreasing the proportion of leaky gas tanks.

  e) If the program actually works, we haven't done anything *wrong*. Our methods are correct. Statistically speaking, we have committed a Type II error.

  f) In order to decrease the probability of making this type of error, we could lower our standards of proof, by raising the level of significance. This will increase the power of the test to detect a decrease in the proportion of leaky gas tanks. Another way to decrease the probability that we make a Type II error is to sample more service stations. This will decrease the variation in the sample proportion, making our results more reliable.

  g) Increasing the level of significance is advantageous, since it decreases the probability of making a Type II error, and increases the power of the test. However, it also increases the probability that a Type I error is made, in this case, thinking that the program is effective when it really is not effective.
  Increasing the sample size decreases the probability of making a Type II error and increases power, but can be costly and time-consuming.

6. Surgery and germs.

  a) Lister imposed a treatment, the use of carbolic acid as a disinfectant. This is an experiment.

b) H_0: The survival rate when carbolic acid is used is the same as the survival rate when carbolic acid is not used. $\left(p_C = p_N \text{ or } p_C - p_N = 0\right)$

H_A: The survival rate when carbolic acid is used is greater than the survival rate when carbolic acid is not used. $\left(p_C > p_N \text{ or } p_C - p_N > 0\right)$

Randomization condition: There is no mention of random assignment. Assume that the two groups of patients were similar, and amputations took place under similar conditions, with the use of carbolic acid being the only variable.
10% condition: 40 and 35 are both less than 10% of all possible amputations.
Independent samples condition: It is reasonable to think that the groups were not related in any way.
Success/Failure condition: $n\hat{p}$ (carbolic acid) = 34, $n\hat{q}$ (carbolic acid) = 6, $n\hat{p}$ (none) = 19, and $n\hat{q}$ (none) = 16. The number of patients who died in the carbolic acid group is only 6, but the expected number of deaths using the pooled proportion, $n\hat{q}_{pooled} = (40)(\frac{22}{75}) = 11.7$, so the samples are both large enough.

Since the conditions have been satisfied, we will perform a two-proportion z-test. We will model the sampling distribution of the difference in proportion with a Normal model with mean 0 and standard deviation estimated by:

$$SE_{pooled}\left(\hat{p}_C - \hat{p}_N\right) = \sqrt{\frac{\hat{p}_{pooled}\hat{q}_{pooled}}{n_C} + \frac{\hat{p}_{pooled}\hat{q}_{pooled}}{n_N}} = \sqrt{\frac{\left(\frac{53}{75}\right)\left(\frac{22}{75}\right)}{40} + \frac{\left(\frac{53}{75}\right)\left(\frac{22}{75}\right)}{35}} \approx 0.1054.$$

The observed difference between the proportions is:
$0.85 - 0.5429 = 0.3071$.

Since the *P*-value = 0.0018 is low, we reject the null hypothesis. There is strong evidence that the survival rate is higher when carbolic acid is used to disinfect the operating room than when carbolic acid is not used.

$z = \dfrac{0.3071 - 0}{0.1054}$

$z \approx 2.91$

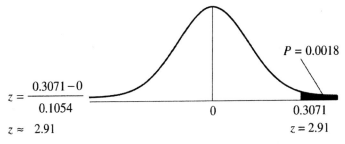

$P = 0.0018$

0

0.3071

$z = 2.91$

c) We don't know whether or not patients were randomly assigned to treatments, and we don't know whether or not blinding was used.

7. Scrabble.

a) The researcher believes that the true proportion of As is within 10% of the estimated 54%, namely, between 44% and 64%.

b) A large margin of error is usually associated with a small sample, but the sample consisted of "many" hands. The margin of error is large because the standard error of the sample is large. This occurs because the true proportion of As in a hand is close to 50%, the most difficult proportion to predict.

c) This provides no evidence that the simulation is faulty. The true proportion of As is contained in the confidence interval. The researcher's results are consistent with 63% As.

8. Dice.

Die rolls are truly independent, and the distribution of the outcomes of die rolls is not skewed (it's uniform). According to the CLT, the sampling distribution of $\bar{y}$, the average for 10 die rolls, can be approximated by a Normal model, with $\mu_{\bar{y}} = 3.5$ and standard deviation $\sigma(\bar{y}) = \dfrac{1.7}{\sqrt{10}} \approx 0.538$, even though 10 rolls is a fairly small sample.

According to the Normal model, the probability that the average of 10 die rolls is between 3 and 4 (and therefore the probability of the sum of 10 die rolls is between 30 and 40) is approximately 0.647.

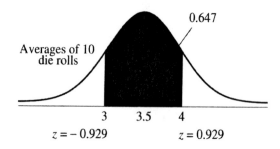

9. News sources.

a) The Pew Research Foundation believes that the true proportion of people who obtain news from the Internet is between 30% and 36%.

b) The smaller sample size in the limited sample would result in a larger standard error. This would make the margin of error larger, as well.

c) $\hat{p} \pm z^* \sqrt{\dfrac{\hat{p}\hat{q}}{n}} = (0.45) \pm 1.960 \sqrt{\dfrac{(0.45)(0.55)}{239}} = (38.7\%, 51.3\%)$

We are 95% confident that between 38.7% and 51.3% of active traders rely on the Internet for investment information.

d) The sample of 239 active traders is smaller than either of the earlier samples. This results in a larger margin of error.

10. Death penalty.

a) **Plausible Independence:** There is no reason to believe that one randomly selected American adult's response will affect another's.
Randomization condition: Gallup randomly selected 537 American adults.
10% condition: 537 results is less than 10% of all American adults.
Success/Failure condition: $n\hat{p} = (537)(0.52) = 279$ and $n\hat{q} = (537)(0.48) = 258$ are both greater than 10, so the sample is large enough.

Since the conditions are met, we can use a one-proportion z-interval to estimate the percentage of American adults who favor the death penalty.

$\hat{p} \pm z^* \sqrt{\dfrac{\hat{p}\hat{q}}{n}} = (0.52) \pm 1.960 \sqrt{\dfrac{(0.52)(0.48)}{537}} = (47.8\%, 56.2\%)$

We are 95% confident that between 47.8% and 56.2% of American adults favor the death penalty.

b) Since the interval extends below 50%, it is not clear that the death penalty has majority support.

c)

$$ME = z^* \sqrt{\frac{\hat{p}\hat{q}}{n}}$$

$$0.02 = 2.326 \sqrt{\frac{(0.50)(0.50)}{n}}$$

$$n = \frac{(2.326)^2 (0.50)(0.50)}{(0.02)^2}$$

$$n \approx 3382 \text{ people}$$

We do not know the true proportion of American adults in favor of the death penalty, so use $\hat{p} = \hat{q} = 0.50$, for the most cautious estimate. In order to determine the proportion of American adults in favor of the death penalty to within 2% with 98% confidence, we would have to sample at least 3382 people.

11. Bimodal.

a) The *sample's* distribution (NOT the *sampling* distribution), is expected to look more and more like the distribution of the population, in this case, bimodal.

b) The expected value of the sample's mean is expected to be μ, the population mean, regardless of sample size.

c) The variability of the sample mean, $\sigma(\bar{y})$, is $\frac{\sigma}{\sqrt{n}}$, the population standard deviation divided by the square root of the sample size, regardless of the sample size.

d) As the sample size increases, the sampling distribution model becomes closer and closer to a Normal model.

12. Vitamin D.

a) Certainly, the 1546 women are less than 10% of all African-American women, and $n\hat{p} = (1546)(0.42) = 649$ and $n\hat{q} = (1546)(0.58) = 897$ are both greater than 10, so the sample is large enough. We would like to know that the sample is random. This would help assure us that these women were chosen independently.

b) $\hat{p} \pm z^* \sqrt{\frac{\hat{p}\hat{q}}{n}} = (0.42) \pm 1.960 \sqrt{\frac{(0.42)(0.58)}{1546}} = (39.5\%, 44.5\%)$.

c) We are 95% confident that between 39.5% and 44.5% of African-American women have a vitamin D deficiency.

d) 95% of all random samples of this size will produce intervals that contain the true proportion of African-American women who have a vitamin D deficiency.

13. Archery.

a) $\mu_{\hat{p}} = p = 0.80$

$$\sigma(\hat{p}) = \sqrt{\frac{pq}{n}} = \sqrt{\frac{(0.80)(0.20)}{200}} \approx 0.028$$

b) $np = (200)(0.80) = 160$ and $nq = (200)(0.20) = 40$ are both greater than 10, so the Normal model is appropriate.

c) The Normal model of the sampling distribution of the proportion of bull's-eyes she makes out of 200 is at the right.

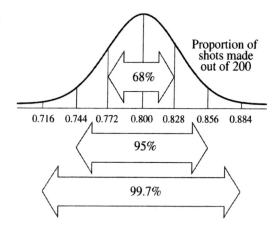

Approximately 68% of the time, we expect her to hit the bull's-eye on between 77.2% and 82.8% of her shots. Approximately 95% of the time, we expect her to hit the bull's-eye on between 74.4% and 85.6% of her shots. Approximately 99.7% of the time, we expect her to hit the bull's-eye on between 71.6% and 88.4% of her shots.

d) According to the Normal model, the probability that she hits the bull's-eye in at least 85% of her 200 shots is approximately 0.039.

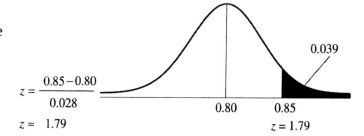

$$z = \frac{0.85 - 0.80}{0.028}$$

$$z \approx 1.79$$

14. Free throws.

a) **Randomization condition:** Assume that these free throws are representative of Tim Duncan's free throw ability in these seasons.
10% condition: 662 and 568 are less than 10% of all possible free throws.
Independent samples condition: The data were collected in two different seasons.
Success/Failure condition: $n\hat{p}('00\text{-}'01) = 409$, $n\hat{q}('00\text{-}'01) = 253$, $n\hat{p}('01\text{-}'02) = 460$, and $n\hat{q}('01\text{-}'02) = 108$ are all greater than 10, so the samples are both large enough.

Since the conditions have been satisfied, we will find a two-proportion z-interval.

$$\left(\hat{p}_{2002} - \hat{p}_{2001}\right) \pm z^* \sqrt{\frac{\hat{p}_{2002}\hat{q}_{2002}}{n_{2002}} + \frac{\hat{p}_{2001}\hat{q}_{2001}}{n_{2001}}} = \left(\tfrac{460}{568} - \tfrac{409}{662}\right) \pm 1.960 \sqrt{\frac{\left(\tfrac{460}{568}\right)\left(\tfrac{108}{568}\right)}{568} + \frac{\left(\tfrac{409}{662}\right)\left(\tfrac{253}{662}\right)}{662}} = (0.143, 0.241)$$

We are 95% confident that Tim Duncan's free throw percentage increased between 14.3% and 24.1% from the 2000-2001 season to the 2001-2002 season.

b) Since the interval for the difference in percentage of free throws made is well above 0, there is strong evidence that Tim Duncan has become a better free throw shooter.

15. Twins.

H_0: The proportion of preterm twin births in 1990 is the same as the proportion of preterm twin births in 2000. $\left(p_{1990} = p_{2000} \text{ or } p_{1990} - p_{2000} = 0\right)$

H_A: The proportion of preterm twin births in 1990 is the less than the proportion of preterm twin births in 2000. $\left(p_{1990} < p_{2000} \text{ or } p_{1990} - p_{2000} < 0\right)$

Randomization condition: Assume that these births are representative of all twin births.
10% condition: 43 and 48 are both less than 10% of all twin births.
Independent samples condition: The samples are from different years, so they are unlikely to be related.
Success/Failure condition: $n\hat{p}(1990) = 20$, $n\hat{q}(1990) = 23$, $n\hat{p}(2000) = 26$, and $n\hat{q}(2000) = 22$ are all greater than 10, so both samples are large enough.

Since the conditions have been satisfied, we will perform a two-proportion z-test. We will model the sampling distribution of the difference in proportion with a Normal model with mean 0 and standard deviation estimated by:

$$SE_{pooled}(\hat{p}_{1990} - \hat{p}_{2000}) = \sqrt{\frac{\hat{p}_{pooled}\hat{q}_{pooled}}{n_{1900}} + \frac{\hat{p}_{pooled}\hat{q}_{pooled}}{n_{2000}}} = \sqrt{\frac{\left(\frac{46}{91}\right)\left(\frac{45}{91}\right)}{43} + \frac{\left(\frac{46}{91}\right)\left(\frac{45}{91}\right)}{48}} \approx 0.1050.$$

The observed difference between the proportions is:
$0.4651 - 0.5417 = -0.0766$

Since the *P*-value = 0.2329 is high, we fail to reject the null hypothesis. There is no evidence of an increase in the proportion of preterm twin births from 1990 to 2000, at least not at this large city hospital.

$z = \dfrac{-0.0766 - 0}{0.1050}$

$z \approx -0.73$

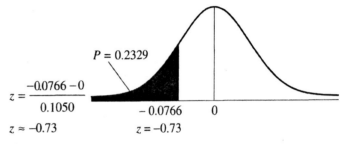

$P = 0.2329$

$-0.0766 \qquad 0$

$z = -0.73$

16. Eclampsia.

a) **Randomization condition:** Although not specifically stated, these results are from a large-scale experiment, which was undoubtedly properly randomized.
   10% condition: 4999 and 4993 are less than 10% of all pregnant women.
   Independent samples condition: Subjects were randomly assigned to the treatments.
   Success/Failure condition: $n\hat{p}(\text{mag. sulf.}) = 1201$, $n\hat{q}(\text{mag. sulf.}) = 3798$, $n\hat{p}(\text{placebo}) = 228$, and $n\hat{q}(\text{placebo}) = 4765$ are all greater than 10, so both samples are large enough.

Since the conditions have been satisfied, we will find a two-proportion z-interval.

$$(\hat{p}_{MS} - \hat{p}_N) \pm z^* \sqrt{\frac{\hat{p}_{MS}\hat{q}_{MS}}{n_{MS}} + \frac{\hat{p}_N\hat{q}_N}{n_N}} = \left(\frac{1201}{4999} - \frac{228}{4993}\right) \pm 1.960 \sqrt{\frac{\left(\frac{1201}{4999}\right)\left(\frac{3798}{4999}\right)}{4999} + \frac{\left(\frac{228}{4993}\right)\left(\frac{4765}{4993}\right)}{4993}} = (0.181, 0.208)$$

We are 95% confident that the proportion of pregnant women who will experience side effects while taking magnesium sulfide will be between 18.1% and 20.8% higher than the proportion of women that will experience side effects while not taking magnesium sulfide.

b) H_0: The proportion of pregnant women who will develop eclampsia is the same for women taking magnesium sulfide as it is for women not taking magnesium sulfide.
   $$(p_{MS} = p_N \text{ or } p_{MS} - p_N = 0)$$

   H_A: The proportion of pregnant women who will develop eclampsia is lower for women taking magnesium sulfide than for women not taking magnesium sulfide.
   $$(p_{MS} < p_N \text{ or } p_{MS} - p_N < 0)$$

Success/Failure condition: $n\hat{p}$(mag. sulf.) = 40, $n\hat{q}$(mag. sulf.) = 4959, $n\hat{p}$(placebo) = 96, and $n\hat{q}$(placebo) = 4897 are all greater than 10, so both samples are large enough.

Since the conditions have been satisfied (some in part a), we will perform a two-proportion z-test. We will model the sampling distribution of the difference in proportion with a Normal model with mean 0 and standard deviation estimated by:

$$SE_{pooled}\left(\hat{p}_{MS} - \hat{p}_{N}\right) = \sqrt{\frac{\hat{p}_{pooled}\hat{q}_{pooled}}{n_{MS}} + \frac{\hat{p}_{pooled}\hat{q}_{pooled}}{n_{N}}} = \sqrt{\frac{\left(\frac{136}{9992}\right)\left(\frac{9856}{9992}\right)}{4999} + \frac{\left(\frac{136}{9992}\right)\left(\frac{9856}{9992}\right)}{4993}} \approx 0.002318.$$

The observed difference between the proportions is 0.00800 – 0.01923 = – 0.01123, which is approximately 4.84 standard errors below the expected difference in proportion of 0.

Since the P-value = 6.4×10^{-7} is very low, we reject the null hypothesis. There is strong evidence that the proportion of pregnant women who develop eclampsia will be lower for women taking magnesium sulfide than for those not taking magnesium sulfide.

17. Eclampsia.

a) H_0: The proportion of pregnant women who die after developing eclampsia is the same for women taking magnesium sulfide as it is for women not taking magnesium sulfide. $\left(p_{MS} = p_N \text{ or } p_{MS} - p_N = 0\right)$

H_A: The proportion of pregnant women who die after developing eclampsia is lower for women taking magnesium sulfide than for women not taking magnesium sulfide. $\left(p_{MS} < p_N \text{ or } p_{MS} - p_N < 0\right)$

b) **Randomization condition:** Although not specifically stated, these results are from a large-scale experiment, which was undoubtedly properly randomized.
10% condition: 40 and 96 are less than 10% of all pregnant women.
Independent samples condition: Subjects were randomly assigned to the treatments.
Success/Failure condition: $n\hat{p}$(mag. sulf.) = 11, $n\hat{q}$(mag. sulf.) = 29, $n\hat{p}$(placebo) = 20, and $n\hat{q}$(placebo) = 76 are all greater than 10, so both samples are large enough.

Since the conditions have been satisfied, we will perform a two-proportion z-test. We will model the sampling distribution of the difference in proportion with a Normal model with mean 0 and standard deviation estimated by:

$$SE_{pooled}\left(\hat{p}_{MS} - \hat{p}_{N}\right) = \sqrt{\frac{\hat{p}_{pooled}\hat{q}_{pooled}}{n_{MS}} + \frac{\hat{p}_{pooled}\hat{q}_{pooled}}{n_{N}}} = \sqrt{\frac{\left(\frac{31}{136}\right)\left(\frac{105}{136}\right)}{40} + \frac{\left(\frac{31}{136}\right)\left(\frac{105}{136}\right)}{96}} \approx 0.07895.$$

c) The observed difference between the proportions is:
0.275 – 0.2083 = 0.0667

Since the P-value = 0.8008 is high, we fail to reject the null hypothesis. There is no evidence that the proportion of women who may die after developing eclampsia is lower for women taking magnesium sulfide than for women who are not taking the drug.

$z = \dfrac{0.0667 - 0}{0.07895}$

$z \approx 0.84$

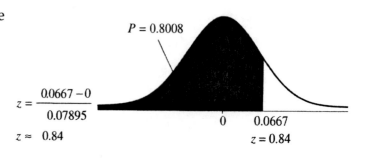

d) There is not sufficient evidence to conclude that magnesium sulfide is effective in preventing death when eclampsia develops.

e) If magnesium sulfide is effective in preventing death when eclampsia develops, then we have made a Type II error.

f) To increase the power of the test to detect a decrease in death rate due to magnesium sulfide, we could increase the sample size or increase the level of significance.

g) Increasing the sample size lowers variation in the sampling distribution, but may be costly. The sample size is already quite large. Increasing the level of significance increases power by increasing the likelihood of rejecting the null hypothesis, but increases the chance of making a Type I error, namely thinking that magnesium sulfide is effective when it is not.

18. Eggs.

a) According to the Normal model, approximately 33.7% of these eggs weigh more than 62 grams.

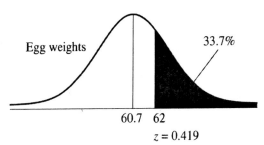

b) **Random sampling condition:** The dozen eggs are selected randomly.
10% condition: The dozen eggs are less than 10% of all eggs.

The mean egg weight is $\mu = 60.7$ grams, with standard deviation $\sigma = 3.1$ grams. Since the distribution of egg weights is Normal, we can model the sampling distribution of the mean egg weight of a dozen eggs with a Normal model, with $\mu_{\bar{y}} = 60.7$ grams and standard deviation $\sigma(\bar{y}) = \dfrac{3.1}{\sqrt{12}} \approx 0.895$ grams.

According to the Normal model, the probability that a randomly selected dozen eggs have a mean greater than 62 grams is approximately 0.073.

$$z = \dfrac{62 - 60.7}{\dfrac{3.1}{\sqrt{12}}}$$

$$z \approx 1.453$$

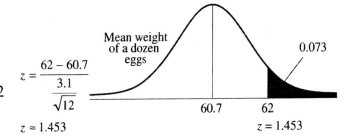

c) The average weight of a dozen eggs can be modeled by $N(60.7, 0.895)$, so the total weight of a dozen eggs can be modeled by $N(728.4, 10.74)$.

Approximately 68% of the cartons of a dozen eggs would weigh between 717.1 and 739.1 grams. Approximately 95% of the cartons would weigh between 706.9 and 749.9 grams. Approximately 99.7% of the cartons would weigh between 696.2 and 760.6 grams.

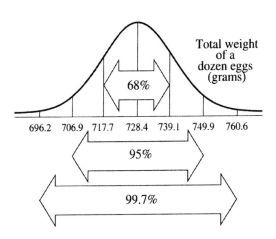

19. Polling disclaimer.

a) It is not clear what specific question the pollster asked. Otherwise, they did a great job of identifying the W's.

b) A sample that was stratified by age, sex, region, and education was used.

c) The margin of error was 4%.

d) Since "no more than 1 time in 20 should chance variations in the sample cause the results to vary by more than 4 percentage points", the confidence level is 19/20 = 95%.

e) The subgroups had smaller sample sizes than the larger group. The standard errors in these subgroups were larger as a result, and this caused the margins of error to be larger.

f) They cautioned readers about response bias due to wording and order of the questions.

20. Enough eggs.

$$ME = z^* \sqrt{\frac{\hat{p}\hat{q}}{n}}$$

$$0.02 = 1.960 \sqrt{\frac{(0.75)(0.25)}{n}}$$

$$n = \frac{(1.960)^2 (0.75)(0.25)}{(0.02)^2}$$

$$n \approx 1801 \text{ eggs}$$

ISA Babcock needs to collect data on about 1800 hens in order to advertise the production rate for the B300 Layer with 95% confidence with a margin of error of $\pm 2\%$.

21. Teen deaths.

a) H_0 : The percentage of fatal accidents involving teenage girls is 14.3%, the same as the overall percentage of fatal accidents involving teens . ($p = 0.143$)
H_A : The percentage of fatal accidents involving teenage girls is lower than 14.3%, the overall percentage of fatal accidents involving teens . ($p < 0.143$)

Plausible independence: It is reasonable to think that accidents occur independently.
Randomization condition: Assume that the 388 accidents observed are representative of all accidents.
10% condition: The sample of 388 accidents is less than 10% of all accidents.
Success/Failure condition: $np = (388)(0.143) = 55.484$ and $nq = (388)(0.857) = 332.516$ are both greater than 10, so the sample is large enough.

The conditions have been satisfied, so a Normal model can be used to model the sampling distribution of the proportion, with $\mu_{\hat{p}} = p = 0.143$ and

$$\sigma(\hat{p}) = \sqrt{\frac{pq}{n}} = \sqrt{\frac{(0.143)(0.857)}{388}} \approx 0.01777 .$$

We can perform a one-proportion z-test. The observed proportion of fatal accidents involving teen girls is $\hat{p} = \dfrac{44}{388} \approx 0.1134$.

Since the *P*-value = 0.0479 is low, we reject the null hypothesis. There is some evidence that the proportion of fatal accidents involving teen girls is less than the overall proportion of fatal accidents involving teens.

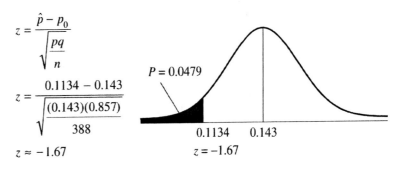

$$z = \frac{\hat{p} - p_0}{\sqrt{\dfrac{pq}{n}}}$$

$$z = \frac{0.1134 - 0.143}{\sqrt{\dfrac{(0.143)(0.857)}{388}}}$$

$$z \approx -1.67$$

b) If the proportion of fatal accidents involving teenage girls is really 14.3%, we expect to see the observed proportion, 11.34%, in about 4.79% of samples of size 388 simply due to sampling variation.

22. Perfect pitch.

a) H_0: The proportion of Asian students with perfect pitch is the same as the proportion of non-Asians with perfect pitch. $\left(p_A = p_N \text{ or } p_A - p_N = 0\right)$

H_A: The proportion of Asian students with perfect pitch is the different than the proportion of non-Asians with perfect pitch. $\left(p_A \neq p_N \text{ or } p_A - p_N \neq 0\right)$

b) Since $P < 0.0001$, which is very low, we reject the null hypothesis. There is strong evidence of a difference in the proportion of Asians with perfect pitch and the proportion of non-Asians with perfect pitch. There is evidence that Asians are more likely to have perfect pitch.

c) If there is no difference in the proportion of students with perfect pitch, we would expect the observed difference of 25% to be seen simply due to sampling variation in less than 1 out of every 10,000 samples of 2700 students.

d) The data do not prove anything about genetic differences causing differences in perfect pitch. Asians are merely more likely to have perfect pitch. There may be lurking variables other than genetics that are causing the higher rate of perfect pitch.

23. Largemouth bass.

a) One would expect many small fish, with a few large fish.

b) We cannot determine the probability that a largemouth bass caught from the lake weighs over 3 pounds because we don't know the exact shape of the distribution. We know that it is NOT Normal.

c) It would be quite risky to attempt to determine whether or not the mean weight of 5 fish was over 3 pounds. With a skewed distribution, a sample of size 5 is not large enough for the Central Limit Theorem to guarantee that a Normal model is appropriate to describe the distribution of the mean.

d) A sample of 60 randomly selected fish is large enough for the Central Limit Theorem to guarantee that a Normal model is appropriate to describe the sampling distribution of the mean, as long as 60 fish is less than 10% of the population of all the fish in the lake.

The mean weight is $\mu = 3.5$ pounds, with standard deviation $\sigma = 2.2$ pounds. Since the sample size is sufficiently large, we can model the sampling distribution of the mean weight of 60 fish with a Normal model, with $\mu_{\bar{y}} = 3.5$ pounds and standard deviation

$$\sigma(\bar{y}) = \frac{2.2}{\sqrt{60}} \approx 0.284 \text{ pounds}.$$

According to the Normal model, the probability that 60 randomly selected fish average more than 3 pounds is approximately 0.961.

$$z = \frac{3 - 3.5}{\frac{2.2}{\sqrt{60}}}$$

$$z \approx -1.76$$

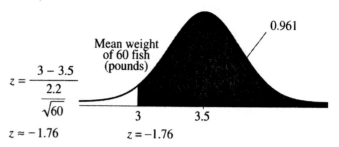

$$z = -1.76$$

24. Cheating.

a) **Plausible independence:** There is no reason to believe that students selected at random would influence each others responses.
   Randomization condition: The 4500 students were selected randomly.
   10% condition: 4500 students is less than 10% of all students.
   Success/Failure condition: $n\hat{p} = (4500)(0.74) = 3330$ and $n\hat{q} = (4500)(0.26) = 1170$ are both greater than 10, so the sample is large enough.

   Since the conditions are met, we can use a one-proportion z-interval to estimate the percentage of students who have cheated at least once.

   $$\hat{p} \pm z^* \sqrt{\frac{\hat{p}\hat{q}}{n}} = (0.74) \pm 1.645 \sqrt{\frac{(0.74)(0.26)}{4500}} = (72.9\%, 75.1\%)$$

b) We are 90% confident that between 72.9% and 75.1% of high school students have cheated at least once.

c) About 90% of random samples of size 4500 will produce intervals that contain the true proportion of high school students who have cheated at least once.

d) A 95% confidence interval would be wider. Greater confidence requires a larger margin of error.

25. Language.

a) **10% condition:** The 60 people selected randomly represent less than 10% of all people.
   Success/Failure condition: $np = (60)(0.80) = 48$ and $nq = (60)(0.20) = 12$ are both greater than 10.

   Therefore, the sampling distribution model for the proportion of 60 randomly selected people who have left-brain language control is Normal, with $\mu_{\hat{p}} = p = 0.80$ and standard deviation $\sigma(\hat{p}) = \sqrt{\frac{pq}{n}} = \sqrt{\frac{(0.80)(0.20)}{60}} \approx 0.0516.$

b) According to the Normal model, the probability that over 75% of these 60 people have left-brain language control is approximately 0.894.

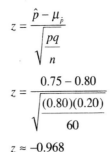

$$z = \frac{\hat{p} - \mu_{\hat{p}}}{\sqrt{\dfrac{pq}{n}}}$$

$$z = \frac{0.75 - 0.80}{\sqrt{\dfrac{(0.80)(0.20)}{60}}}$$

$$z \approx -0.968$$

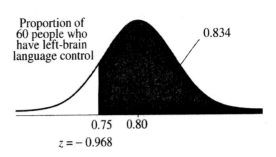

Proportion of 60 people who have left-brain language control

0.834

0.75 0.80

$z = -0.968$

c) If the sample had consisted of 100 people, the probability would have been higher. A larger sample results in a smaller standard deviation for the sample proportion.

d) Answers may vary. Let's consider three standard deviations below the expected proportion to be "almost certain". It would take a sample of (exactly!) 576 people to make sure that 75% would be 3 standard deviations below the expected percentage of people with left-brain language control.

Using round numbers for n instead of z, about 500 people in the sample would make the probability of choosing a sample with at least 75% of the people having left-brain language control is a whopping 0.997.

$$z = \frac{\hat{p} - \mu_{\hat{p}}}{\sqrt{\dfrac{pq}{n}}}$$

$$-3 = \frac{0.75 - 0.80}{\sqrt{\dfrac{(0.80)(0.20)}{n}}}$$

$$n = \frac{(-3)^2(0.80)(0.20)}{(0.75 - 0.80)^2} = 576$$

It all depends on what "almost certain" means to you.

26. Cigarettes.

a) H$_0$: 30% of high school students smoke. ($p = 0.30$)
H$_A$: Less than 30% of high school students smoke. ($p < 0.30$)

b) Randomization condition: The CDC randomly sampled 10,204 high school students.
10% condition: The sample of 10,204 students is less than 10% of all high school students.
Success/Failure condition: $np = (10,204)(0.30) = 3061.2$ and $nq = (10,204)(0.70) = 7142.8$ are both greater than 10, so the sample is large enough. (And then some!)

The conditions have been satisfied, so a Normal model can be used to model the sampling distribution of the proportion, with $\mu_{\hat{p}} = p = 0.30$ and $\sigma(\hat{p}) = \sqrt{\dfrac{pq}{n}} = \sqrt{\dfrac{(0.30)(0.70)}{10,204}} \approx 0.00454$.

c) We can perform a one-proportion z-test. The observed proportion of high school students who smoke is $\hat{p} = 0.285$. This proportion is about 3.31 standard deviations below the hypothesized proportion of smokers.

The P-value of this test is 0.00047.

d) If the proportion of students who smoke is actually 30%, the probability that a sample of this size would have a sample proportion of 28.5% or lower is 0.00047.

e) Since the P-value = 0.00047 is low, we reject the null hypothesis. There is strong evidence that less than 30% of high school students smoked in 2001. The goal is on track.

f) If the conclusion is incorrect, a Type I error has been made.

27. Crohn's disease.

a) **Plausible independence:** It is reasonable to think that the patients would respond to infliximab independently of each other.
Randomization condition: Assume that the 573 patients are representative of all Crohn's disease sufferers.
10% condition: 573 patients are less than 10% of all sufferers of Crohn's disease.
Success/Failure condition: $n\hat{p} = 335$ and $n\hat{q} = 238$ are both greater than 10, so the sample is large enough.

Since the conditions are met, we can use a one-proportion z-interval to estimate the percentage of Crohn's disease sufferers who respond positively to infliximab.

$$\hat{p} \pm z^* \sqrt{\frac{\hat{p}\hat{q}}{n}} = \left(\frac{335}{573}\right) \pm 1.960\sqrt{\frac{\left(\frac{335}{573}\right)\left(\frac{238}{573}\right)}{573}} = (54.5\%, 62.5\%)$$

b) We are 95% confident that between 54.5% and 62.5% of Crohn's disease sufferers would respond positively to infliximab.

c) 95% of random samples of size 573 will produce intervals that contain the true proportion of Crohn's disease sufferers who respond positively to infliximab.

28. Teen smoking.

10% condition: The freshman class is less than 10% of all teenagers. Assume that the freshman class is representative of all teenagers. This may not be a reasonable assumption. There are many interlocking relationships between smoking, socioeconomic status, and college attendance. This class may not be representative of all teens with regards to smoking simply because they are in college. Be cautious with your conclusions!
Success/Failure condition: $np = (522)(0.30) = 156.6$ and $nq = (522)(0.70) = 365.4$ are both greater than 10.

Therefore, the sampling distribution model for the proportion of 522 students who smoke is Normal, with $\mu_{\hat{p}} = p = 0.30$, and standard deviation $\sigma(\hat{p}) = \sqrt{\frac{pq}{n}} = \sqrt{\frac{(0.30)(0.70)}{522}} \approx 0.0201$.

40% is about 4.99 standard deviations below the expected proportion of smokers. According to the Normal model, the probability that more than 40% of these students smoke is very small, about 3.1×10^{-7}. It is very unlikely that more than 40% the freshman class smokes.

29. Alcohol abuse.

$$ME = z^* \sqrt{\frac{\hat{p}\hat{q}}{n}}$$

$$0.04 = 1.645 \sqrt{\frac{(0.5)(0.5)}{n}}$$

$$n = \frac{(1.645)^2 (0.5)(0.5)}{(0.04)^2}$$

$$n \approx 423$$

The university will have to sample at least 423 students in order to estimate the proportion of students who have been drunk with in the past week to within ± 4%, with 90% confidence.

30. Errors.

a) Since a treatment (the additive) is imposed, this is an experiment.

b) The company is only interested in a decrease in the percentage of cars needing repairs, so they will perform a one-sided test.

c) The independent laboratory will make a Type I error if they decide that the additive reduces the number of repairs, when it actually makes no difference in the number of repairs.

d) The independent laboratory will make a Type II error if they decide that the additive makes no difference in the number of repairs, when it actually reduces the number of repairs.

e) The additive manufacturer would consider a Type II error more serious. The lab claims that the manufacturer's product doesn't work, and it actually does.

f) Since this was a controlled experiment, the company can conclude that the additive is the reason that the cabs are running better. They should be cautious recommending it for all cars. There is evidence that the additive works well for cabs, which get heavy use. It might not be effective in cars with a different pattern of use than cabs.

31. Preemies.

a) **Randomization condition:** Assume that these kids are representative of all kids.
10% condition: 242 and 233 are less than 10% of all kids.
Independent samples condition: The groups are independent.
Success/Failure condition: $n\hat{p}$ (preemies) $= (242)(0.74) = 179$, $n\hat{q}$ (preemies) $= (242)(0.26) = 63$, $n\hat{p}$ (normal weight) $= (233)(0.83) = 193$, and $n\hat{q}$ (normal weight) $= 40$ are all greater than 10, so the samples are both large enough.

Since the conditions have been satisfied, we will find a two-proportion z-interval.

$$\left(\hat{p}_N - \hat{p}_P\right) \pm z^* \sqrt{\frac{\hat{p}_N \hat{q}_N}{n_N} + \frac{\hat{p}_P \hat{q}_P}{n_P}} = (0.83 - 0.74) \pm 1.960 \sqrt{\frac{(0.83)(0.17)}{233} + \frac{(0.74)(0.26)}{242}} = (0.017, 0.163)$$

We are 95% confident that between 1.7% and 16.3% more normal birth-weight children graduated from high school than children who were born premature.

b) Since the interval for the difference in percentage of high school graduates is above 0, there is evidence normal birth-weight children graduate from high school at a greater rate than premature children.

c) If preemies do not have a lower high school graduation rate than normal birth-weight children, then we made a Type I error. We rejected the null hypothesis of "no difference" when we shouldn't have.

32. Safety.

a) $\hat{p} \pm z^* \sqrt{\dfrac{\hat{p}\hat{q}}{n}} = (0.14) \pm 1.960 \sqrt{\dfrac{(0.14)(0.86)}{814}} = (11.6\%, 16.4\%)$

We are 95% confident that between 11.6% and 16.4% of Texas children wear helmets when biking, roller skating, or skateboarding.

b) These data might not be a random sample.

c)

$$ME = z^* \sqrt{\dfrac{\hat{p}\hat{q}}{n}}$$

$$0.04 = 2.326 \sqrt{\dfrac{(0.14)(0.86)}{n}}$$

$$n = \dfrac{(2.326)^2 (0.14)(0.86)}{(0.04)^2}$$

$$n \approx 408$$

If we use the 14% estimate obtained from the first study, the researchers will need to observe at least 408 kids in order to estimate the proportion of kids who wear helmets to within 4%, with 98% confidence.

(If you use a more cautious approach, estimating that 50% of kids wear helmets, you need a whopping 846 observations. Are you beginning to see why pilot studies are conducted?)

33. Fried PCs.

a) H_0: The computer is undamaged.
H_A: The computer is damaged.

b) The biggest advantage is that all of the damaged computers will be detected, since, historically, damaged computers never pass all the tests. The disadvantage is that only 80% of undamaged computers pass all the tests. The engineers will be classifying 20% of the undamaged computers as damaged.

c) In this example, a Type I error is rejecting an undamaged computer. To allow this to happen only 5% of the time, the engineers would reject any computer that failed 3 or more tests, since 95% of the undamaged computers fail two or fewer tests.

d) The power of the test in part c is 20%, since only 20% of the damaged machines fail 3 or more tests.

e) By declaring computers "damaged" if the fail 2 or more tests, the engineers will be rejecting only 7% of undamaged computers. From 5% to 7% is an increase of 2% in α. Since 90% of the damaged computers fail 2 or more tests, the power of the test is now 90%, a substantial increase.

34. Power.

a) Power will increase, since the variability in the sampling distribution will decrease. We are more certain of all our decisions when there is less variability.

b) Power will decrease, since we are rejecting the null hypothesis less often.

Chapter 23 – Inferences About Means

1. ***t*-models, part I.**

   a) 1.74   **b)** 2.37   **c)** 0.0524   **d)** 0.0889

2. ***t*-models, part II.**

   a) 2.37   **b)** 2.36   **c)** 0.9829   **d)** 0.0381

3. ***t*-models, part III.**

   As the number of degrees of freedom increases, the shape and center of *t*-models do not change. The spread of *t*-models decreases as the number of degrees of freedom increases.

4. ***t*-models, part IV (last one!).**

   As the number of degrees of freedom increases, the critical value of *t* for a 95% confidence interval gets smaller, approaching approximately 1.960, the critical value of *z* for a 95% confidence interval.

5. **Cattle.**

   a) Not correct. A confidence interval is for the mean weight gain of the population of all cows. It says nothing about individual cows. This interpretation also appears to imply that there is something special about the interval that was generated, when this interval is actually one of many that could have been generated, depending on the cows that were chosen for the sample.

   b) Not correct. A confidence interval is for the mean weight gain of the population of all cows, not individual cows.

   c) Not correct. We don't need a confidence interval about the average weight gain for cows in this study. We are certain that the mean weight gain of the cows in this study is 56 pounds. Confidence intervals are for the mean weight gain of the population of all cows.

   d) Not correct. This statement implies that the average weight gain varies. It doesn't. We just don't know what it is, and we are trying to find it. The average weight gain is either between 45 and 67 pounds, or it isn't.

   e) Not correct. This statement implies that there is something special about our interval, when this interval is actually one of many that could have been generated, depending on the cows that were chosen for the sample. The correct interpretation is that 95% of samples of this size will produce an interval that will contain the mean weight gain of the population of all cows.

6. **Teachers.**

   a) Not correct. Actually, 9 out of 10 samples will produce intervals that will contain the mean salary for Nevada teachers. Different samples are expected to produce different intervals.

   b) Correct! This is the one!

   c) Not correct. A confidence interval is about the mean salary of the population of Nevada teachers, not the salaries of individual teachers.

d) Not correct. A confidence interval is about the mean salary of the population of Nevada teachers and doesn't tell us about the sample, nor does it tell us anything about individual salaries.

e) Not correct. The population is teachers' salaries in Nevada, not the entire United States.

7. Pulse rates.

a) We are 95% confident that the mean pulse rate of adults is between 70.9 and 74.5 beats per minute.

b) The width of the interval is about 74.5 – 70.9 = 3.6 beats per minute. The margin of error is half of that, about 1.8 beats per minute.

c) The margin of error would have been larger. More confidence requires a larger critical value of t, which increases the margin of error.

8. Crawling.

a) We are 95% confident that the mean age at which babies begin to crawl is between 29.2 and 31.8 weeks.

b) The width of the interval is about 31.8 – 29.2 = 2.6 weeks. The margin of error is half of that, about 1.3 weeks.

c) The margin of error would have been smaller. Less confidence requires a smaller critical value of t, which decreases the margin of error.

9. Normal temperature.

a) **Randomization condition:** The adults were randomly selected.
10% condition: 52 adults are less than 10% of all adults.
Nearly Normal condition: The sample of 52 adults is large, and the histogram shows no serious skewness, outliers, or multiple modes.

The people in the sample had a mean temperature of 98.2846° and a standard deviation in temperature of 0.682379°. Since the conditions are satisfied, the sampling distribution of the mean can be modeled by a Student's t model, with 52 – 1 = 51 degrees of freedom. We will use a one-sample t-interval with 98% confidence for the mean body temperature. (By hand, use $t_{50}^* \approx 2.403$ from the table.)

b) $\bar{y} \pm t_{n-1}^* \left(\dfrac{s}{\sqrt{n}} \right) = 98.2846 \pm t_{51}^* \left(\dfrac{0.682379}{\sqrt{52}} \right) \approx (98.06, 98.51)$

c) We are 98% confident that the mean body temperature for adults is between 98.06°F and 98.51°F. (If you calculated the interval by hand, using $t_{50}^* \approx 2.403$ from the table, your interval may be slightly different than intervals calculated using technology. With the rounding used here, they are identical. Even if they aren't, it's not a big deal.)

d) 98% of all random samples of size 52 will produce intervals that contain the true mean body temperature of adults.

e) Since the interval is completely below the body temperature of 98.6°F, there is strong evidence that the true mean body temperature of adults is lower than 98.6°F.

10. Parking.

a) **Randomization condition:** The weekdays were not randomly selected. We will assume that the weekdays in our sample are representative of all weekdays.
10% condition: 44 weekdays are less than 10% of all weekdays.
Nearly Normal condition: We don't have the actual data, but since the sample of 44 weekdays is fairly large it is okay to proceed.

The weekdays in the sample had a mean revenue of $126 and a standard deviation in revenue of $15. The sampling distribution of the mean can be modeled by a Student's t model, with $44 - 1 = 43$ degrees of freedom. We will use a one-sample t-interval with 90% confidence for the mean daily income of the parking garage. (By hand, use $t^*_{40} \approx 1.684$ from the table.)

b) $\bar{y} \pm t^*_{n-1}\left(\dfrac{s}{\sqrt{n}}\right) = 126 \pm t^*_{43}\left(\dfrac{15}{\sqrt{44}}\right) \approx (122.2, 129.8)$

c) We are 90% confident that the mean daily income of the parking garage is between $122.20 and $129.80. (If you calculated the interval by hand, using $t^*_{40} \approx 1.684$ from the table, your interval will be (122.19, 129.81), ever so slightly wider from the interval calculated using technology. This is not a big deal.)

d) 90% of all random samples of size 44 will produce intervals that contain the true mean daily income of the parking garage.

e) Since the interval is completely below the $130 predicted by the consultant, there is evidence that the average daily parking revenue is lower than $130.

11. Normal temperatures, part II.

a) The 90% confidence interval would be narrower than the 98% confidence interval. We can be more precise with our interval when we are less confident.

b) The 98% confidence interval has a greater chance of containing the true mean body temperature of adults than the 90% confidence interval, but the 98% confidence interval is less precise (wider) than the 90% confidence interval.

c) The 98% confidence interval would be narrower if the sample size were increased from 52 people to 500 people. The smaller standard error would result in a smaller margin of error.

d) Our sample of 52 people gave us a 98% confidence interval with a margin of error of $(98.51 - 98.05)/2 = 0.225°F$. In order to get a margin of error of 0.1, less than half of that, we need a sample over 4 times as large. It should be safe to use $t^*_{100} \approx 2.364$ from the table, since the sample will need to be larger than 101. Or we could use $z^* \approx 2.326$, since we expect the sample to be large. The important thing is to pick *something*, and go with it. This is just an *estimate*! We need a sample of about 252 people in order to estimate the mean body temperature of adults to within 0.1°F.

$ME = t^*_{n-1}\left(\dfrac{s}{\sqrt{n}}\right)$

$0.1 = 2.326\left(\dfrac{0.682379}{\sqrt{n}}\right)$

$n = \dfrac{(2.326)^2(0.682379)^2}{(0.1)^2}$

$n \approx 252$

12. Parking II.

a) The 95% confidence interval would be wider than the 90% confidence interval. We can be more confident that our interval contains the mean parking revenue when we are less precise. This would be better for the city because the 95% confidence interval is more likely to contain the true mean parking revenue.

b) The 95% confidence interval is wider than the 90% confidence interval, and therefore less precise. It would be difficult for budget planners to use this wider interval, since they need precise figures for the budget.

c) By collecting a larger sample of parking revenue on weekdays, they could create a more precise interval without sacrificing confidence.

d) The confidence interval that was calculated in Exercise 10 won't help us to estimate the sample size. That interval was for 90% confidence. Now we want 95% confidence. A quick estimate with a critical value of $z^* = 2$ (from the 68-95-99.7 rule) gives us a sample size of 100, which will probably work fine. Let's be a bit more precise, just for fun! Conservatively, let's choose t^* with fewer degrees of freedom, which will give us a wider interval. From the table, the next available number of degrees of freedom is $t^*_{80} \approx 1.990$, not much different than the estimate of 2 that was used before. If we substitute 1.990 for t^*, we can estimate a sample size of about 99. But try not to lose sight of the fact that these are just *estimates*. Why not play it a bit safe? Use $n = 100$.

$$ME = t^*_{n-1}\left(\frac{s}{\sqrt{n}}\right)$$

$$3 = 2\left(\frac{15}{\sqrt{n}}\right)$$

$$n = \frac{(2)^2(15)^2}{(3)^2}$$

$$n = 100$$

13. Hot dogs.

a) $\bar{y} \pm t^*_{n-1}\left(\dfrac{s}{\sqrt{n}}\right) = 310 \pm t^*_{39}\left(\dfrac{36}{\sqrt{40}}\right) \approx (298.5,\ 321.5)$

b) We have assumed that the hot dog weights are independent and that the distribution of the population of hot dog weights is Normal. The conditions to check are:
Randomization condition: We don't know that the hot dogs were sampled at random, but it is reasonable to think that the hot dogs are representative of hot dogs of this type.
10% condition: 40 hot dogs are less than 10% of all hot dogs.
Nearly Normal condition: We don't have the actual data, but since the sample of 40 hot dogs is fairly large, it is okay to proceed.

c) We are 95% confident that the mean sodium content in this type of "reduced sodium" hot dogs is between 298.5 and 321.5 mg.

14. Speed of Light.

a) $\bar{y} \pm t^*_{n-1}\left(\dfrac{s}{\sqrt{n}}\right) = 756.22 \pm t^*_{22}\left(\dfrac{107.12}{\sqrt{23}}\right) \approx (709.9,\ 802.5)$

b) We are 95% confident that the speed of light is between 299,709.9 and 299,802.5 km/sec.

c) We have assumed that the measurements are independent of each other and that the distribution of the population of all possible measurements is Normal. The assumption of independence seems reasonable, but it might be a good idea to look at a display of the measurements made by Michelson to verify that the Nearly Normal Condition is satisfied.

15. Second dog.

a) This larger sample should produce a more accurate estimate of the mean sodium content in the hot dogs. A larger sample has a smaller standard error, which results in a smaller margin of error.

b) $SE(\bar{y}) = \left(\dfrac{s}{\sqrt{n}}\right) = \left(\dfrac{32}{\sqrt{60}}\right) \approx 4.1$ mg sodium.

c) **Randomization condition:** We don't know that the hot dogs were sampled at random, but it is reasonable to think that the hot dogs are representative of hot dogs of this type.
10% condition: 60 hot dogs are less than 10% of all hot dogs.
Nearly Normal condition: We don't have the actual data, but since the sample of 60 hot dogs is large it is okay to proceed.

The hot dogs in the sample had a mean sodium content of 318 mg, and a standard deviation in sodium content of 32 mg. Since the conditions have been satisfied, construct a one-sample t-interval, with $60 - 1 = 59$ degrees of freedom, at 95% confidence.

$$\bar{y} \pm t^*_{n-1}\left(\frac{s}{\sqrt{n}}\right) = 318 \pm t^*_{59}\left(\frac{32}{\sqrt{60}}\right) \approx (309.7,\ 326.3)$$

We are 95% confident that the mean sodium content of this type of hot dog is between 309.7 and 326.3 mg.

d) If a "reduced sodium" hot dog has to have at least 30% less sodium than a hot dog containing an average of 465 mg of sodium, then it must have less than $0.70(465) = 325.5$ mg of sodium. Since this value is contained in our 95% confidence interval, there is little evidence to suggest that this type of hot dog can be labeled as "reduced sodium".

16. Better light.

a) $SE(\bar{y}) = \left(\dfrac{s}{\sqrt{n}}\right) = \left(\dfrac{79.0}{\sqrt{100}}\right) = 7.9$ km/sec.

b) The interval should be narrower. There are three reasons for this: the larger sample size results in a smaller standard error (reducing the margin of error), the larger sample size results in a greater number of degrees of freedom (decreasing the value of t^*, reducing the margin of error), and the smaller standard deviation in measurements results in a smaller standard error (reducing the margin of error). Additionally, the interval will have a different center, since the sample mean is different.

c) We must assume that the measurements are independent of one another. Since the sample size is large, the Nearly Normal Condition is overridden, but it would still be nice to look at a graphical display of the measurements. A one-sample t-interval for the speed of light can be constructed, with $100 - 1 = 99$ degrees of freedom, at 95% confidence.

$$\bar{y} \pm t_{n-1}^* \left(\frac{s}{\sqrt{n}} \right) = 852.4 \pm t_{99}^* \left(\frac{79.0}{\sqrt{100}} \right) \approx (836.72, 868.08)$$

We are 95% confident that the speed of light is between 299,836.72 and 299,868.08 km/sec.

Since the interval for the new method does not contain the true speed of light as reported by Stigler, 299,710.5 km/sec., there is no evidence to support the accuracy of Michelson's new methods.

The interval for Michelson's old method (from Exercise 14) does contain the true speed of light as reported by Stigler. There is some evidence that Michelson's previous measurement technique was a good one, if not very precise.

17. TV Safety.

a) The inspectors are performing an upper-tail test. They need to prove that the stands will support 500 pounds (or more) easily.

b) The inspectors commit a Type I error if they certify the stands as safe, when they are not.

c) The inspectors commit a Type II error if they decide the stands are not safe, when they are.

18. Catheters.

a) Quality control personnel are conducting a two-sided test. If the catheters are too big, they won't fit through the vein. If they are too small, the examination apparatus may not fit through the catheter.

b) The quality control personnel commit a Type I error if catheters are rejected, when in fact the diameters are fine. The manufacturing process is stopped needlessly.

c) The quality control personnel commit a Type II error if catheters are being produced that do not meet the specifications, and this goes unnoticed. Defective catheters are being produced and sold.

19. TV safety revisited.

a) The value of α should be decreased. This means a smaller chance of declaring the stands safe under the null hypothesis that they are not safe.

b) The power of the test is the probability of correctly detecting that the stands can safely hold over 500 pounds.

c) 1) The company could redesign the stands so that their strength is more consistent, as measured by the standard deviation. Redesigning the manufacturing process is likely to be quite costly.
2) The company could increase the number of stands tested. This costs them both time to perform the test and money to pay the quality control personnel.
3) The company could increase α, effectively lowering their standards for what is required to certify the stands "safe". This is a big risk, since there is a greater chance of Type I error, namely allowing unsafe stands to be sold.
4) The company could make the stands stronger, increasing the mean amount of weight that the stands can safely hold. This type of redesign is expensive.

20. Catheters again.

a) If the level of significance is lowered to $\alpha = 0.01$, the probability of Type II error will increase. Lowering α will lower the probability of incorrectly rejecting the null hypothesis when it's true, which will increase the probability of incorrectly *failing* to reject the null hypothesis when it is *false*.

b) The power of this test is the probability of correctly detecting deviations from 2 mm in diameter.

c) The power of the test will increase as the actual mean diameter gets farther and farther away from 2 mm. Larger deviations from what is expected are easier to detect than small deviations.

d) In order to increase the power of the test, the company could increase the sample size, reducing the standard error of the sampling distribution model. They could also increase the value of α, requiring a lower standard of proof to identify a faulty manufacturing process.

21. Marriage.

a) H_0: The mean age at which American men first marry is 23.3 years. $(\mu = 23.3)$
   H_A: The mean age at which American men first marry is greater than 23.3 years. $(\mu > 23.3)$

b) **Randomization condition:** The 40 men were selected randomly.
   10% condition: 40 men are less than 10% of all recently married men.
   Nearly Normal condition: The population of ages of men at first marriage is likely to be skewed to the right. It is much more likely that there are men who marry for the first time at an older age than at an age that is very young. We should examine the distribution of the sample to check for serious skewness and outliers, but with a large sample of 40 men, it should be safe to proceed.

c) Since the conditions for inference are satisfied, we can model the sampling distribution of the mean age of men at first marriage with $N\left(23.3, \dfrac{\sigma}{\sqrt{n}}\right)$. Since we do not know σ, the standard deviation of the population, $\sigma(\bar{y})$ will be estimated by $SE(\bar{y}) = \dfrac{s}{\sqrt{n}}$, and we will use a Student's t model, with $40 - 1 = 39$ degrees of freedom, $t_{39}\left(23.3, \dfrac{s}{\sqrt{40}}\right)$.

d) The mean age at first marriage in the sample was 24.2 years, with a standard deviation in age of 5.3 years. Use a one-sample t-test, modeling the sampling distribution of $\bar{y}$ with $t_{39}\left(23.3, \dfrac{5.3}{\sqrt{40}}\right)$.
   The *P*-value is 0.1447.

$$t = \frac{\bar{y} - \mu_0}{SE(\bar{y})}$$

$$t = \frac{24.2 - 23.3}{\dfrac{5.3}{\sqrt{40}}}$$

$$t \approx 1.07$$

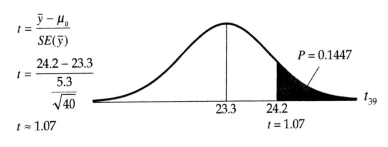

$P = 0.1447$

23.3   24.2
$t = 1.07$

t_{39}

e) If the mean age at first marriage is still 23.3 years, there is a 14.5% chance of getting a sample mean of 24.2 years or older simply from natural sampling variation.

f) Since the *P*-value = 0.1447 is high, we fail to reject the null hypothesis. There is no evidence to suggest that the mean age of men at first marriage has changed from 23.3 years, the mean in 1960.

22. Fuel economy.

a) H_0: The mean mileage of the cars in the fleet is 26 mpg. $(\mu = 26)$

H_A: The mean mileage of the cars in the fleet is less than 26 mpg. $(\mu < 26)$

b) **Randomization condition:** The 50 trips were selected randomly.
10% condition: 50 trips are less than 10% of all trips.
Nearly Normal condition: We don't have the actual data, so we cannot look at the distribution of the data, but the sample is large, so we can proceed.

c) Since the conditions for inference are satisfied, we can model the sampling distribution of the mean mileage of cars in the fleet with $N\left(26, \dfrac{\sigma}{\sqrt{n}}\right)$. Since we do not know σ, the standard deviation of the population, $\sigma(\bar{y})$ will be estimated by $SE(\bar{y}) = \dfrac{s}{\sqrt{n}}$, and we will use a Student's *t* model, with 50 – 1 = 49 degrees of freedom, $t_{49}\left(26, \dfrac{s}{\sqrt{50}}\right)$.

d) The trips in the sample had a mean mileage of 25.02 mpg, with a standard deviation of 4.83 mpg. Use a one-sample *t*-test, modeling the sampling distribution of $\bar{y}$ with $t_{49}\left(26, \dfrac{4.83}{\sqrt{50}}\right)$.
The *P*-value is 0.0789.

$t = \dfrac{\bar{y} - \mu_0}{SE(\bar{y})}$

$t = \dfrac{25.02 - 26}{\dfrac{4.83}{\sqrt{50}}}$

$t \approx -1.43$

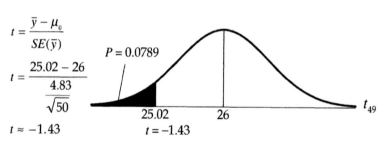

$P = 0.0789$

25.02   26 t_{49}

$t = -1.43$

e) If the mean mileage of cars in the fleet is 26 mpg, the chance that a sample mean of a sample of size 50 is 25.02 mpg or less simply due to sampling error is 7.9%.

f) Since the *P*-value = 0.0789 is fairly high, we fail to reject the null hypothesis. There is little evidence to suggest that the mean mileage of cars in the fleet is less than 26 mpg.

23. Ruffles.

a) **Randomization condition:** The 6 bags were not selected at random, but it is reasonable to think that these bags are representative of all bags of chips.
10% condition: 6 bags are less than 10% of all bags of chips.
Nearly Normal condition: The histogram of the weights of chips in the sample is nearly normal.

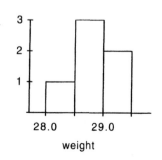

weight

b) $\bar{y} \approx 28.78$ grams, $s \approx 0.40$ grams

c) Since the conditions for inference have been satisfied, use a one-sample *t*-interval, with 6 – 1 = 5 degrees of freedom, at 95% confidence.

$$\bar{y} \pm t^*_{n-1}\left(\frac{s}{\sqrt{n}}\right) = 28.78 \pm t^*_5\left(\frac{0.40}{\sqrt{6}}\right) \approx (28.36, 29.21)$$

d) We are 95% confident that the mean weight of the contents of Ruffles bags is between 28.36 and 29.21 grams.

e) Since the interval is above the stated weight of 28.3 grams, there is evidence that the company is filling the bags to more than the stated weight, on average.

24. Doritos.

a) **Randomization condition:** The 6 bags were not selected at random, but it is reasonable to think that these bags are representative of all bags of chips.
10% condition: 6 bags are less than 10% of all bags of chips.
Nearly Normal condition: The Normal probability plot is reasonably straight. Although the histogram of the weights of chips in the sample is not symmetric, any apparent "skewness" is the result of a single bag of chips. It is safe to proceed.

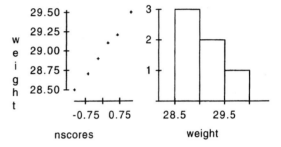

b) $\bar{y} \approx 28.98$ grams, $s \approx 0.36$ grams

c) Since the conditions for inference have been satisfied, use a one-sample *t*-interval, with 6 – 1 = 5 degrees of freedom, at 95% confidence.

$$\bar{y} \pm t^*_{n-1}\left(\frac{s}{\sqrt{n}}\right) = 28.98 \pm t^*_5\left(\frac{0.36}{\sqrt{6}}\right) \approx (28.61, 29.36)$$

d) We are 95% confident that the mean weight of the contents of Doritos bags is between 28.61 and 29.36 grams.

e) Since the interval is above the stated weight of 28.3 grams, there is evidence that the company is filling the bags to more than the stated weight, on average.

25. Cars.

H$_0$: The mean weight of cars currently licensed in the U.S. is 3000 pounds. ($\mu = 3000$)

H$_A$: The mean weight of cars currently licensed in the U.S. is not 3000 pounds. ($\mu \neq 3000$)

Randomization condition: The 91 cars in the sample were randomly selected.
10% condition: 91 cars are less than 10% of all cars.
Nearly Normal condition: We don't have the actual data, so we cannot look at a graphical display, but since the sample is large, it is safe to proceed.

The cars in the sample had a mean weight of 2919 pounds and a standard deviation in weight of 531.5 pounds. Since the conditions for inference are satisfied, we can model the sampling distribution of the mean weight of cars currently licensed in the U.S. with a

Student's t model, with $91 - 1 = 90$ degrees of freedom, $t_{90}\left(3000, \dfrac{531.5}{\sqrt{91}}\right)$.

We will perform a one-sample t-test.

Since the P-value $= 0.1495$ is high, we fail to reject the null hypothesis. There is little evidence to suggest that the mean weight of cars in the U.S. is different than 3000 pounds.

$$t = \frac{\bar{y} - \mu_0}{SE(\bar{y})}$$

$$t = \frac{2919 - 3000}{\dfrac{531.5}{\sqrt{91}}}$$

$$t \approx -1.45$$

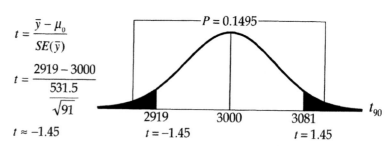

26. Portable phones.

H_0: The mean range of this type of portable phone is 150 feet. $(\mu = 150)$

H_A: The mean range of this type of portable phone is not 150 feet. $(\mu \neq 150)$

Randomization condition: The 44 phones in the sample were randomly selected.
10% condition: 44 phones are less than 10% of all cars.
Nearly Normal condition: We don't have the actual data, so we cannot look at a graphical display, but since the sample is fairly large, it is safe to proceed.

The phones in the sample had a mean range of 142 feet and a standard deviation of 12 feet. Since the conditions for inference are satisfied, we can model the sampling distribution of the mean range of this type of portable phone with a Student's t model, with

$44 - 1 = 43$ degrees of freedom, $t_{43}\left(150, \dfrac{12}{\sqrt{44}}\right)$. We will perform a one-sample t-test.

Since the P-value $= 0.0000654$ is very low, we reject the null hypothesis. There is strong evidence that the mean range of this type of phone is not 150 feet. Our evidence suggests that the mean range is actually less than 150 feet.

$$t = \frac{\bar{y} - \mu_0}{SE(\bar{y})}$$

$$t = \frac{142 - 150}{\dfrac{12}{\sqrt{44}}}$$

$$t \approx -4.42$$

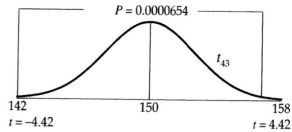

27. Chips ahoy.

a) **Randomization condition:** The bags of cookies were randomly selected.
10% condition: 16 bags are less than 10% of all bags
Nearly Normal condition: The Normal probability plot is reasonably straight, and the histogram of the number of chips per bag is unimodal and symmetric.

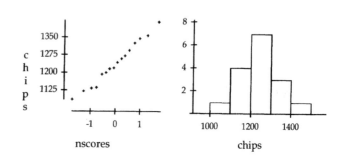

b) The bags in the sample had with a mean number of chips or 1238.19, and a standard deviation of 94.282 chips. Since the conditions for inference have been satisfied, use a one-sample *t*-interval, with 16 – 1 = 15 degrees of freedom, at 95% confidence.

$$\bar{y} \pm t^*_{n-1}\left(\frac{s}{\sqrt{n}}\right) = 1238.19 \pm t^*_{15}\left(\frac{94.282}{\sqrt{16}}\right) \approx (1187.9, 1288.4)$$

We are 95% confident that the mean number of chips in an 18-ounce bag of Chips Ahoy cookies is between 1187.9 and 1288.4.

c) H_0: The mean number of chips per bag is 1000. $(\mu = 1000)$

H_A: The mean number of chips per bag is greater than 1000. $(\mu > 1000)$

Since the confidence interval is well above 1000, there is strong evidence that the mean number of chips per bag is well above 1000.

However, since the "1000 Chip Challenge" is about individual bags, not means, the claim made by Nabisco may not be true. If the mean was around 1188 chips, the low end of our confidence interval, and the standard deviation of the population was about 94 chips, our best estimate obtained from our sample, a bag containing 1000 chips would be about 2 standard deviations below the mean. This is not likely to happen, but not an outrageous occurrence. These data do not provide evidence that the "1000 Chip Challenge" is true.

28. Yogurt.

a) **Randomization condition:** The brands of vanilla yogurt may not be a random sample, but they are probably representative of all brands of yogurt.
10% condition: 14 brands of vanilla yogurt may not be less than 10% of all yogurt brands. Are there 140 brands of vanilla yogurt available?
Independence assumption: The Randomization Condition and the 10% Condition are designed to check the reasonableness of the assumption of independence. We had some trouble verifying these conditions. But is the calorie content per serving of one brand of yogurt likely to be associated with that of another brand? Probably not. It's okay to proceed.
Nearly Normal condition: The Normal probability plot is reasonably straight, and the histogram of the number of calories per serving is unimodal and symmetric.

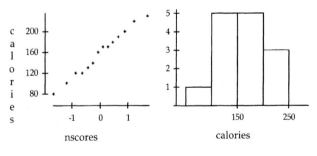

b) The brands in the sample had a mean calorie content of 157.857 calories, and a standard deviation of 44.7521 calories. Since the conditions for inference have been satisfied, use a one-sample *t*-interval, with 14 – 1 = 13 degrees of freedom, at 95% confidence.

$$\bar{y} \pm t^*_{n-1}\left(\frac{s}{\sqrt{n}}\right) = 157.857 \pm t^*_{13}\left(\frac{44.7521}{\sqrt{14}}\right) \approx (132.0, 183.7)$$

c) We are 95% confident that the mean calorie content in a serving of vanilla yogurt is between 132.0 and 183.7 calories. There is evidence that the estimate of 120 calories made in the diet guide is too low. The 95% confidence interval is well above 120 calories.

29. Maze.

a) **Independence assumption:** It is reasonable to think that the rats' times will be independent, as long as the times are for different rats.

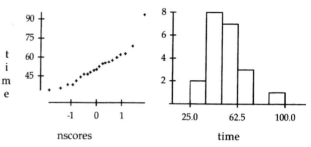

Nearly Normal condition: There is an outlier in both the Normal probability plot and the histogram that should probably be eliminated before continuing the test. One rat took a long time to complete the maze.

b) H_0: The mean time for rats to complete this maze is 60 seconds. $(\mu = 60)$

H_A: The mean time for rats to complete this maze is not 60 seconds. $(\mu \neq 60)$

The rats in the sample finished the maze with a mean time of 52.21 seconds and a standard deviation in times of 13.5646 seconds. Since the conditions for inference are satisfied, we can model the sampling distribution of the mean time in which rats complete the maze with a Student's t model, with

$21 - 1 = 20$ degrees of freedom, $t_{20}\left(60, \dfrac{13.5646}{\sqrt{21}}\right)$. We will perform a one-sample t-test.

Since the P-value = 0.0160 is low, we reject the null hypothesis. There is evidence that the mean time required for rats to finish the maze is not 60 seconds. Our evidence suggests that the mean time is actually less than 60 seconds.

$$t = \frac{\bar{y} - \mu_0}{SE(\bar{y})}$$

$$t = \frac{52.21 - 60}{\dfrac{13.56}{\sqrt{21}}}$$

$$t \approx -2.63$$

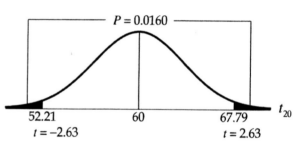

c) Without the outlier, the rats in the sample finished the maze with a mean time of 50.13 seconds and standard deviation in times of 9.90 seconds. Since the conditions for inference are satisfied, we can model the sampling distribution of the mean time in which rats complete the maze with a Student's t model, with $20 - 1 = 19$ degrees of freedom,

$t_{19}\left(60, \dfrac{9.90407}{\sqrt{20}}\right)$. We will perform a one-sample t-test.

This test results in a value of $t = -4.46$, and a two-sided P-value = 0.0003. Since the P-value is low, we reject the null hypothesis. There is evidence that the mean time required for rats to finish the maze is not 60 seconds. Our evidence suggests that the mean time is actually less than 60 seconds.

d) According to both tests, there is evidence that the mean time required for rats to complete the maze is different than 60 seconds. The maze does not meet the "one-minute average" requirement. It should be noted that the test without the outlier is the appropriate test. The one slow rat made the mean time required seem much higher than it probably was.

30. Braking.

H_0: The mean braking distance is 125 feet. The tread pattern works adequately. $(\mu = 125)$

H_A: The mean braking distance is greater than 125 feet, and the new tread pattern should not be used. $(\mu > 125)$

Independence assumption: It is reasonable to think that the braking distances on the test track are independent of each other.
Nearly Normal condition: The braking distance of 102 feet is an outlier. After it is removed, the Normal probability plot is reasonably straight, and the histogram of braking distances unimodal and symmetric.

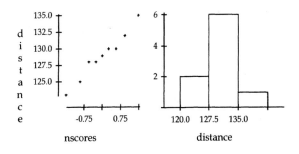

The braking distances in the sample had a mean of 128.889 feet, and a standard deviation of 3.55121 feet. Since the conditions for inference are satisfied, we can model the sampling distribution of the mean braking distance with a Student's t model, with $9 - 1 = 8$ degrees of freedom, $t_9\left(125, \dfrac{3.55121}{\sqrt{9}}\right)$. We will perform a one-sample t-test.

Since the *P*-value = 0.0056 is low, we reject the null hypothesis. There is strong evidence that the mean braking distance of cars with these tires is greater than 125 feet. The new tread pattern should not be adopted.

$$t = \frac{\bar{y} - \mu_0}{SE(\bar{y})}$$

$$t = \frac{128.889 - 125}{\dfrac{3.55121}{\sqrt{9}}}$$

$$t \approx 3.29$$

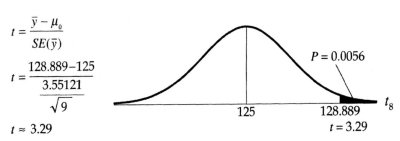

Chapter 24 – Comparing Means

1. **Learning math.**

   a) The margin of error of this confidence interval is $(11.427 - 5.573)/2 = 2.927$ points.

   b) The margin of error for a 98% confidence interval would have been larger. The critical value of t^* is larger for higher confidence levels. We need a wider interval to increase the likelihood that we catch the true mean difference in test scores within our interval. In other words, greater confidence comes at the expense of precision.

   c) We are 95% confident that the mean score for the CPMP math students will be between 5.573 and 11.427 points higher on this assessment than the mean score of the traditional students.

   d) Since the entire interval is above 0, there is strong evidence that students who learn with CPMP will have higher mean scores is algebra than those in traditional programs.

2. **Stereograms.**

   a) We are 90% confident that the mean time required to "fuse" the image for people who receive no information or verbal information only will be between 0.55 and 5.47 seconds longer than the mean time required to "fuse" the image for people who receive both verbal and visual information.

   b) Since the entire interval is above 0, there is evidence that viewing the picture of the image helps people "see" the 3D image.

   c) The margin of error for this interval is $(5.47 - 0.55)/2 = 2.46$ seconds.

   d) 90% of all random samples of this size will produce intervals that will contain the true value of the mean difference between the times of the two groups.

   e) A 99% confidence interval would be wider. The critical value of t^* is larger for higher confidence levels. We need a wider interval to increase the likelihood that we catch the true mean difference in test scores within our interval. In other words, greater confidence comes at the expense of precision.

   f) The conclusion reached may very well change. A wider interval may contain the mean difference of 0, failing to provide evidence of a difference in mean times.

3. **CPMP, again.**

   a) H_0: The mean score of CPMP students is the same as the mean score of traditional students. $\left(\mu_C = \mu_T \text{ or } \mu_C - \mu_T = 0 \right)$

   H_A: The mean score of CPMP students is different from the mean score of traditional students. $\left(\mu_C \neq \mu_T \text{ or } \mu_C - \mu_T \neq 0 \right)$

b) **Independent groups assumption:** Scores of students from different classes should be independent.
Randomization condition: Although not specifically stated, classes in this experiment were probably randomly assigned to either CPMP or traditional curricula.
10% condition: 312 and 265 are less than 10% of all students.
Nearly Normal condition: We don't have the actual data, so we can't check the distribution of the sample. However, the samples are large. The Central Limit Theorem allows us to proceed.

Since the conditions are satisfied, we can use a two-sample *t*-test with 583 degrees of freedom (from the computer).

c) If the mean scores for the CPMP and traditional students are really equal, there is less than a 1 in 10,000 chance of seeing a difference as large or larger than the observed difference just from natural sampling variation.

d) Since the *P*-value < 0.0001, reject the null hypothesis. There is strong evidence that the CPMP students have a different mean score than the traditional students. The evidence suggests that the CPMP students have a higher mean score.

4. **CPMP and word problems.**

H_0: The mean score of CPMP students is the same as the mean score of traditional students. $\left(\mu_C = \mu_T \text{ or } \mu_C - \mu_T = 0\right)$

H_A: The mean score of CPMP students is different from the mean score of traditional students. $\left(\mu_C \neq \mu_T \text{ or } \mu_C - \mu_T \neq 0\right)$

Independent groups assumption: Scores of students from different classes should be independent.
Randomization condition: Although not specifically stated, classes in this experiment were probably randomly assigned to either CPMP or traditional curricula.
10% condition: 320 and 273 are less than 10% of all students.
Nearly Normal condition: We don't have the actual data, so we can't check the distribution of the sample. However, the samples are large. The Central Limit Theorem allows us to proceed.

Since the conditions are satisfied, it is appropriate to model the sampling distribution of the difference in means with a Student's *t*-model, with 590.05 degrees of freedom (from the approximation formula).

We will perform a two-sample *t*-test. The sampling distribution model has mean 0, with standard error: $SE(\bar{y}_C - \bar{y}_T) = \sqrt{\dfrac{32.1^2}{320} + \dfrac{28.5^2}{273}} \approx 2.489.$

The observed difference between the mean scores is 57.4 – 53.9 = 3.5.

Since the *P*-value = 0.1602, we fail to reject the null hypothesis. There is no evidence that the CPMP students have a different mean score on the word problems test than the traditional students.

$$t = \frac{(\bar{y}_c - \bar{y}_\tau) - (0)}{SE(\bar{y}_c - \bar{y}_\tau)}$$

$$t \approx \frac{3.5}{2.489}$$

$$t \approx 1.406$$

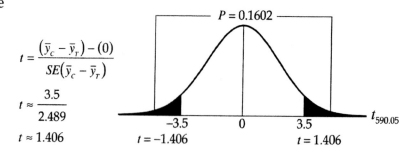

5. **Commuting.**

   a) **Independent groups assumption:** Since the choice of route was determined at random, the commuting times for Route A are independent of the commuting times for Route B.
   Randomization condition: The man randomly determined which route he would travel on each day.
   10% condition: 20 days on each route represents less than 10% of all possible days.
   Nearly Normal condition: The histograms of travel times for the routes are roughly unimodal and symmetric. (Given)

   Since the conditions are satisfied, it is appropriate to model the sampling distribution of the difference in means with a Student's *t*-model, with 33.1 degrees of freedom (from the approximation formula). We will construct a two-sample *t*-interval, with 95% confidence.

   $$(\bar{y}_B - \bar{y}_A) \pm t^*_{df} \sqrt{\frac{s_B^2}{n_B} + \frac{s_A^2}{n_A}} = (43 - 40) \pm t^*_{33.1} \sqrt{\frac{2^2}{20} + \frac{3^2}{20}} \approx (1.36, 4.64)$$

   We are 95% confident that Route B has a mean commuting time between 1.36 and 4.64 minutes longer than the mean commuting time of Route A.

   b) Since 5 minutes is beyond the high end of the interval, there is no evidence that the Route B is an average of 5 minutes longer than Route A. It appears that the old-timer may be exaggerating the average difference in commuting time.

6. **Pulse rates.**

   a) The boxplots suggest that the mean pulse rates for men and women are roughly equal, but that females' pulse rates are more variable.

   b) **Independent groups assumption:** There is no reason to believe that the pulse rates for men and women are related.
   Randomization condition: There is no mention of randomness, but we can assume that the researcher chose a representative sample of men and women with regards to pulse rate.
   10% condition: 28 and 24 are less than 10% of all men and women.
   Nearly Normal condition: The boxplots are reasonably symmetric. Let's hope the distributions of the samples are unimodal, too.

   The conditions for inference are satisfied, so we can analyze these data using the methods discussed in this chapter.

c) Since the conditions are satisfied, it is appropriate to model the sampling distribution of the difference in means with a Student's t-model, with 40.2 degrees of freedom (from the approximation formula). We will construct a two-sample t-interval, with 90% confidence.

$$(\bar{y}_M - \bar{y}_F) \pm t^*_{df}\sqrt{\frac{s_M^2}{n_M} + \frac{s_F^2}{n_F}} = (72.75 - 72.625) \pm t^*_{40.2}\sqrt{\frac{5.37225^2}{28} + \frac{7.69987^2}{24}} \approx (-3.025, 3.275)$$

We are 90% confident that the mean pulse rate for men is between 3.025 points lower and 3.275 points higher than the mean pulse rate for women.

d) Since 0 is in the interval, there is no evidence of a difference in mean pulse rate for men and women. This confirms our answer to part a.

7. **Cereal.**

Independent groups assumption: The percentage of sugar in the children's cereals is unrelated to the percentage of sugar in adult's cereals.
Randomization condition: It is reasonable to assume that the cereals are representative of all children's cereals and adult cereals, in regard to sugar content.
10% condition: 19 and 28 are less than 10% of all cereals.
Nearly Normal condition: The histogram of adult cereal sugar content is skewed to the right, but the sample sizes are of reasonable size. The Central Limit Theorem allows us to proceed.

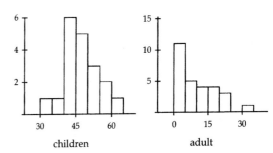

Since the conditions are satisfied, it is appropriate to model the sampling distribution of the difference in means with a Student's t-model, with 42 degrees of freedom (from the approximation formula). We will construct a two-sample t-interval, with 95% confidence.

$$(\bar{y}_C - \bar{y}_A) \pm t^*_{df}\sqrt{\frac{s_C^2}{n_C} + \frac{s_A^2}{n_A}} = (46.8 - 10.1536) \pm t^*_{42}\sqrt{\frac{6.41838^2}{19} + \frac{7.61239^2}{28}} \approx (32.49, 40.80)$$

We are 95% confident that children's cereals have a mean sugar content that is between 32.49% and 40.80% higher than the mean sugar content of adult cereals.

8. **Egyptians.**

a) **Independent groups assumption:** The skull breadth of Egyptians in 4000 B.C.E is independent of the skull breadth of Egyptians almost 4 millennia later!
Randomization condition: It is reasonable to assume that the skulls measured have skull breadths that are representative of all Egyptians of the time.
10% condition: 30 and 30 are less than 10% of all Egyptians of the times.
Nearly Normal condition: The histograms of skull breadths are both unimodal and symmetric.

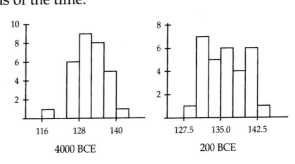

b) Since the conditions are satisfied, it is appropriate to model the sampling distribution of the difference in means with a Student's t-model, with 54 degrees of freedom (from the approximation formula). We will construct a two-sample t-interval, with 95% confidence.

$$(\bar{y}_{200} - \bar{y}_{4K}) \pm t^*_{df}\sqrt{\frac{s^2_{200}}{n_{200}} + \frac{s^2_{4K}}{n_{4K}}} = (135.633 - 131.367) \pm t^*_{54}\sqrt{\frac{4.03846^2}{30} + \frac{5.12925^2}{30}} \approx (1.88, 6.66)$$

We are 95% confident that Egyptian males in 200 B.C.E. had a mean skull breadth between 1.88 and 6.66 mm larger than the mean skull breadth of Egyptian males in 4000 B.C.E.

c) Since the interval is completely above 0, there is evidence that the mean breadth of males' skulls has changed over this time period. The evidence suggests that the mean skull breadth has increased.

9. Reading.

H_0: The mean reading comprehension score of students who learn by the new method is the same as the mean score of students who learn by traditional methods.
$$(\mu_N = \mu_T \text{ or } \mu_N - \mu_T = 0)$$

H_A: The mean reading comprehension score of students who learn by the new method is greater than the mean score of students who learn by traditional methods.
$$(\mu_N > \mu_T \text{ or } \mu_N - \mu_T > 0)$$

Independent groups assumption: Student scores in one group should not have an impact on the scores of students in the other group.
Randomization condition: Students were randomly assigned to classes.
10% condition: 18 and 20 are less than 10% of all students.
Nearly Normal condition: The histograms of the scores are unimodal and symmetric.

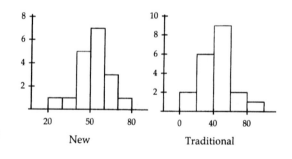

Since the conditions are satisfied, it is appropriate to model the sampling distribution of the difference in means with a Student's t-model, with 33 degrees of freedom (from the approximation formula). We will perform a two-sample t-test. We know:

$$\begin{array}{ll} \bar{y}_N = 51.7222 & \bar{y}_T = 41.8 \\ s_N = 11.7062 & s_T = 17.4495 \\ n_N = 18 & n_T = 20 \end{array}$$

The sampling distribution model has mean 0, with standard error:

$$SE(\bar{y}_N - \bar{y}_T) = \sqrt{\frac{11.7062^2}{18} + \frac{17.4495^2}{20}} \approx 4.779.$$

The observed difference between the mean scores is $51.7222 - 41.8 \approx 9.922$.

Since the *P*-value = 0.0228 is low, we reject the null hypothesis. There is evidence that the students taught using the new activities have a higher mean score on the reading comprehension test than the students taught using traditional methods.

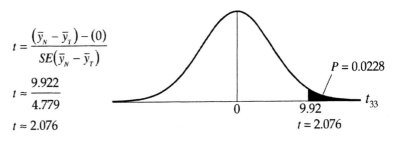

$$t = \frac{(\bar{y}_N - \bar{y}_T) - (0)}{SE(\bar{y}_N - \bar{y}_T)}$$

$$t \approx \frac{9.922}{4.779}$$

$$t \approx 2.076$$

10. Streams.

a) H_0: Streams with limestone substrates and streams with shale substrates have the same mean pH level. $(\mu_L = \mu_S \text{ or } \mu_L - \mu_S = 0)$

H_A: Streams with limestone substrates and streams with shale substrates have different mean pH levels. $(\mu_L \neq \mu_S \text{ or } \mu_L - \mu_S \neq 0)$

b) **Independent groups assumption:** pH levels from the two types of streams are independent.
Independence assumption: Since we don't know if the streams were chosen randomly, assume that the pH level of one stream does not affect the pH of another stream. This seems reasonable.
Nearly Normal condition: The boxplots provided show that the pH levels of the streams may be skewed (since the median is either the upper or lower quartile for the shale streams and the lower whisker of the limestone streams is stretched out), and there are outliers. However, since there are 133 degrees of freedom, we know that the sample sizes are large. It should be safe to proceed.

c) Since the *P*-value ≤ 0.0001 is low, we reject the null hypothesis. There is strong evidence that the streams with limestone substrates have mean pH levels different than those of streams with shale substrates. The limestone streams are less acidic on average.

11. Hurricanes.

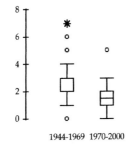
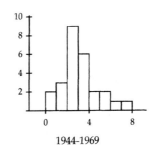
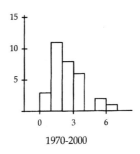

The histograms of the number of hurricanes per year for the two time periods are both skewed, but that's not a problem. The sample sizes are large enough for the Central Limit Theorem to allow use to use our methods of inference. However, the outliers in the data prevent us from using inference. The Central Limit Theorem can't help us out here. It's probably best not to proceed with a test.

12. Memory.

a) If the mean memory scores for people taking gingko biloba and people not taking it are the same, there is a 93.74% chance of seeing a difference in mean memory score this large of larger simply from natural sampling variability.

b) Since the *P*-value is so high, there is no evidence that the mean memory test score for gingko biloba users is higher than the mean memory test score for non-users.

c) Proponents of gingko biloba would insist that we had made a Type II error, incorrectly failing to reject a null hypothesis that is false.

13. Baseball.

a)

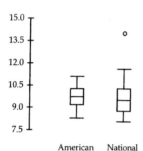 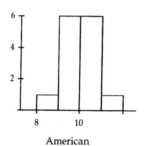 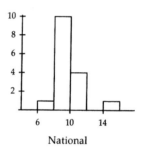

The distributions are Nearly Normal, except for an outlier in the National League.

b) $\bar{y} \pm t^*_{n-1}\left(\dfrac{s}{\sqrt{n}}\right) = 9.80714 \pm t^*_{13}\left(\dfrac{0.805237}{\sqrt{14}}\right) \approx (9.34,\ 10.27)$

We are 95% confident that the mean number of runs scored per game in American League stadiums is between 9.34 and 10.27.

c) The data for the National League stadiums contains an outlier. In order to use the methods of inference, the outlier would have to be removed, and mentioned separately.

d) The average of 14 runs scored per game in Coors Field is unusual. It is almost 2 IQRs above quartile three, more than enough to classify as an outlier. Looking at the boxplot, it just seems out of place.

e) If you attempt to use two confidence intervals to assess a difference in means, you are actually adding standard deviations. But it's the variances that add, not the standard deviations. The two-sample difference of means procedure takes this into account.

14. Handy.

a) Males: $\bar{y}_M \pm t^*_{n-1}\left(\dfrac{s}{\sqrt{n}}\right) = 19.39 \pm t^*_{49}\left(\dfrac{2.52}{\sqrt{50}}\right) \approx (18.67,\ 20.11)$

We are 95% confident that males can place between 18.67 and 20.11 pegs on average.

Females: $\bar{y}_F \pm t^*_{n-1}\left(\dfrac{s}{\sqrt{n}}\right) = 17.91 \pm t^*_{49}\left(\dfrac{3.39}{\sqrt{50}}\right) \approx (16.95,\ 18.87)$

We are 95% confident that females can place between 16.95 and 18.87 pegs on average.

b) It may appear to suggest that there is no difference in the mean number of pegs placed by males and females, but a two-sample *t*-interval should be constructed to assess the difference in mean number of pegs placed.

c) $(\bar{y}_M - \bar{y}_F) \pm t^*_{df} \sqrt{\dfrac{s_M^2}{n_M} + \dfrac{s_F^2}{n_F}} = (19.39 - 17.91) \pm t^*_{90.49} \sqrt{\dfrac{2.52^2}{50} + \dfrac{3.39^2}{50}} \approx (0.29,\, 2.67)$

d) We are 95% confident that the mean number of pegs placed by males is between 0.29 and 2.67 pegs higher than the mean number of pegs placed by females.

e) The two-sample *t*-interval is the correct procedure.

f) If you attempt to use two confidence intervals to assess a difference in means, you are actually adding standard deviations. But it's the variances that add, not the standard deviations. The two-sample difference of means procedure takes this into account.

15. Double header.

a) $(\bar{y}_A - \bar{y}_N) \pm t^*_{df} \sqrt{\dfrac{s_A^2}{n_A} + \dfrac{s_N^2}{n_N}} = (9.80714 - 9.58125) \pm t^*_{23} \sqrt{\dfrac{0.805237^2}{14} + \dfrac{1.54929^2}{16}} \approx (-0.69,\, 1.14)$

b) We are 95% confident that the mean number of runs scored in American League stadiums is between 0.69 runs lower and 1.14 runs higher than the mean number of runs scored in National League stadiums.

c) Since the interval contains 0, there is no evidence of a difference in the mean number of runs scored per game in the stadiums of the two leagues.

d) $(\bar{y}_A - \bar{y}_N) \pm t^*_{df} \sqrt{\dfrac{s_A^2}{n_A} + \dfrac{s_N^2}{n_N}} = (9.80714 - 9.28667) \pm t^*_{26} \sqrt{\dfrac{0.805237^2}{14} + \dfrac{1.04120^2}{15}} \approx (-0.19,\, 1.23)$

With the mean number of runs scored at Coors Field removed, we are 95% confident that the mean number of runs scored in American League stadiums is between 0.19 runs lower and 1.23 runs higher than the mean number of runs scored in National League stadiums. Since 0 is still in the interval, there is no evidence of a difference in the mean number of runs scored per game in the stadiums of the two leagues.

16. Hard water.

a) H$_0$: The mean mortality rate is the same for towns North and South of Derby.
$$\left(\mu_N = \mu_S \ \text{ or } \ \mu_N - \mu_S = 0\right)$$

H$_A$: The mean mortality rate is different for towns North and South of Derby.
$$\left(\mu_N \neq \mu_S \ \text{ or } \ \mu_N - \mu_S \neq 0\right)$$

Independent groups assumption: The towns were sampled independently.
Independence assumption: Assume that the mortality rates are in each town are independent of the mortality rates in the others.
Nearly Normal condition: We don't have the actual data, so we can't look at histograms of the distributions, but the samples are fairly large. It should be okay to proceed.

Since the conditions are satisfied, it is appropriate to model the sampling distribution of the difference in means with a Student's *t*-model, with 53.49 degrees of freedom (from the approximation formula). We will perform a two-sample *t*-test.

The sampling distribution model has mean 0, with standard error:

$$SE(\bar{y}_N - \bar{y}_S) = \sqrt{\frac{138.470^2}{34} + \frac{151.114^2}{27}} \approx 37.546.$$

The observed difference between the mean scores is
1631.59 – 1388.85 = 242.74.

Since the *P*-value = 3.2×10^{-8} is low, we reject the null hypothesis. There is strong evidence that the mean mortality rate different for towns north and south of Derby. There is evidence that the mortality rate north of Derby is higher.

$$t = \frac{(\bar{y}_N - \bar{y}_S) - (0)}{SE(\bar{y}_N - \bar{y}_S)}$$

$$t \approx \frac{242.74}{37.546}$$

$$t \approx 6.47$$

b) Since there is an outlier in the data north of Derby, the conditions for inference are not satisfied, and it is risky to use the two-sample *t*-test. The outlier should be removed, and the test should be performed again. Without the actual data, we are not able to do this. The test without the outlier would *probably* help us reach the same conclusion, but there is no way to be sure.

17. Job satisfaction.

A two-sample *t*-procedure is not appropriate for these data, because the two groups are not independent. They are before and after satisfaction scores for the same workers. Workers that have high levels of job satisfaction before the exercise program is implemented may tend to have higher levels of job satisfaction than other workers after the program as well.

18. Summer school.

A two-sample *t*-procedure is not appropriate for these data, because the two groups are not independent. They are before and after scores for the same students. Students with high scores before summer school may tend to have higher scores after summer school as well.

19. Sex and violence.

a) Since the *P*-value = 0.136 is high, we fail to reject the null hypothesis. There is no evidence of a difference in the mean number of brands recalled by viewers of sexual content and viewers of violent content.

b) H_0: The mean number of brands recalled is the same for viewers of sexual content and viewers of neutral content. $(\mu_S = \mu_N$ or $\mu_S - \mu_N = 0)$

H_A: The mean number of brands recalled is different for viewers of sexual content and viewers of neutral content. $(\mu_S \neq \mu_N$ or $\mu_S - \mu_N \neq 0)$

Independent groups assumption: Recall of one group should not affect recall of another.
Randomization condition: Subjects were randomly assigned to groups.
10% condition: 108 and 108 are less than 10% of all TV viewers.
Nearly Normal condition: The samples are large.

Since the conditions are satisfied, it is appropriate to model the sampling distribution of the difference in means with a Student's *t*-model, with 214 degrees of freedom (from the approximation formula). We will perform a two-sample *t*-test.

The sampling distribution model has mean 0, with standard error:

$$SE(\bar{y}_S - \bar{y}_N) = \sqrt{\frac{1.76^2}{108} + \frac{1.77^2}{108}} \approx 0.24.$$

The observed difference between the mean scores is 1.71 – 3.17 = – 1.46.

Since the *P*-value = 5.5×10^{-9} is low, we reject the null hypothesis. There is strong evidence that the mean number of brand names recalled is different for viewers of sexual content and viewers of neutral content. The evidence suggests that viewers of neutral ads remember more brand names on average than viewers of sexual content.

$$t = \frac{(\bar{y}_S - \bar{y}_N) - (0)}{SE(\bar{y}_S - \bar{y}_N)}$$

$$t \approx \frac{-1.46}{0.24}$$

$$t \approx -6.08$$

20. Ad campaign.

a) We are 95% confident that the mean number of ads remembered by viewers of shows with violent content will be between 1.6 and 0.6 lower than the mean number of brand names remembered by viewers of shows with neutral content.

b) If they want viewers to remember their brand names, they should consider advertising on shows with neutral content, as opposed to shows with violent content.

21. Sex and violence II.

a) H_0: The mean number of brands recalled is the same for viewers of violent content and viewers of neutral content. $(\mu_V = \mu_N$ or $\mu_V - \mu_N = 0)$

H_A: The mean number of brands recalled is different for viewers of violent content and viewers of neutral content. $(\mu_V \neq \mu_N$ or $\mu_V - \mu_N \neq 0)$

Independent groups assumption: Recall of one group should not affect recall of another.
Randomization condition: Subjects were randomly assigned to groups.
10% condition: 101 and 103 are less than 10% of all TV viewers.
Nearly Normal condition: The samples are large.

Since the conditions are satisfied, it is appropriate to model the sampling distribution of the difference in means with a Student's *t*-model, with 201.96 degrees of freedom (from the approximation formula). We will perform a two-sample *t*-test.

The sampling distribution model has mean 0, with standard error:

$$SE(\bar{y}_V - \bar{y}_N) = \sqrt{\frac{1.61^2}{101} + \frac{1.62^2}{103}} \approx 0.226.$$

The observed difference between the mean scores is 3.02 – 4.65 = – 1.63.

$$t = \frac{(\bar{y}_V - \bar{y}_N) - (0)}{SE(\bar{y}_V - \bar{y}_N)}$$

$$t \approx \frac{-1.63}{0.226}$$

$$t \approx -7.21$$

Since the *P*-value = 1.1×10^{-11} is low, we reject the null hypothesis. There is strong evidence that the mean number of brand names recalled is different for viewers of violent content and viewers of neutral content. The evidence suggests that viewers of neutral ads remember more brand names on average than viewers of violent content.

b) $(\bar{y}_N - \bar{y}_S) \pm t_{df}^* \sqrt{\dfrac{s_N^2}{n_N} + \dfrac{s_S^2}{n_S}} = (4.65 - 2.72) \pm t_{204.8}^* \sqrt{\dfrac{1.62^2}{103} + \dfrac{1.85^2}{106}} \approx (1.456, 2.404)$

We are 95% confident that the mean number of brand names recalled 24 hours later is between 1.46 and 2.40 higher for viewers of shows with neutral content than for viewers of shows with sexual content.

22. Ad recall.

a) He might attempt to conclude that the mean number of brand names recalled is greater after 24 hours.

b) The groups are not independent. They are the same people, asked at two different time periods.

c) A person with high recall right after the show might tend to have high recall 24 hours later as well. Also, the first interview may have helped the people to remember the brand names for a longer period of time than they would have otherwise.

d) Randomly assign half of the group watching that type of content to be interviewed immediately after watching, and assign the other half to be interviewed 24 hours later.

23. Lower scores?

a) Assuming that the conditions for inference were met by the NAEP, a 95% confidence interval for the difference in mean score is:

$(\bar{y}_{1996} - \bar{y}_{2000}) \pm t_{df}^* \sqrt{\dfrac{s_{1996}^2}{n_{1996}} + \dfrac{s_{2000}^2}{n_{2000}}} = (150 - 147) \pm 1.960(1.22) \approx (0.61, 5.39)$

Since the samples sizes are very large, it should be safe to use $z^* = 1.960$ for the critical value of t. We are 95% confident that the mean score in 2000 was between 0.61 and 5.39 points lower than the mean score in 1996. Since 0 is not contained in the interval, this provides evidence that the mean score has decreased from 1996 to 2000.

b) Both sample sizes are very large, which will make the standard errors of these samples very small. They are both likely to be very accurate. The difference in sample size shouldn't make you any more certain or any less certain.

However, these results are completely dependent upon whether or not the conditions for inference were met. If, by sampling more students, the NAEP sampled from a different population, then the two years are incomparable.

24. The Internet.

a) The differences that were observed between the group of students with Internet access and those without were too great to be attributed to natural sampling variation.

b) The researchers have incorrectly rejected their null hypothesis of no difference between the groups, committing a Type I error.

c) There is evidence of an association between Internet access and mean science score, but this does not prove that access to the Internet causes higher scores. There may be other variables involved, such as socioeconomic status or the education level of the parents. We would need results from a controlled experiment to determine cause and effect.

25. Statistics journals.

Independent groups assumption: These were articles submitted for publication in two different journals.
Randomization condition: It is necessary to assume that these articles are representative of all articles of this type with respect to publication delay.
10% condition: 288 and 209 are less than 10% of all articles submitted for publication to these journals.
Nearly Normal condition: The samples are large, so as long as there are no outliers in the data, it is okay to proceed.

Since the conditions are satisfied, it is appropriate to model the sampling distribution of the difference in means with a Student's t-model, with 338.30 degrees of freedom (from the approximation formula). We will construct a two-sample t-interval, with 90% confidence.

$$(\bar{y}_{Ap} - \bar{y}_{Am}) \pm t^*_{df}\sqrt{\frac{s^2_{Ap}}{n_{Ap}} + \frac{s^2_{Am}}{n_{Am}}} = (31 - 21) \pm t^*_{338.30}\sqrt{\frac{12^2}{209} + \frac{8^2}{288}} \approx (8.43, 11.57)$$

We are 90% confident that the mean number of months by which publication is delayed is between 8.43 and 11.57 months greater for *Applied Statistics* than it is for *The American Statistician*.

26. Music and memory.

a) H_0: The mean memory test score is the same for those who listen to Mozart as it is for those who listen to rap music. $\left(\mu_M = \mu_R \text{ or } \mu_M - \mu_R = 0\right)$

H_A: The mean memory test score is greater for those who listen to Mozart than it is for those who listen to rap music. $\left(\mu_M > \mu_R \text{ or } \mu_M - \mu_R > 0\right)$

Independent groups assumption: The groups are not related in regards to memory score.
Randomization condition: Subjects were randomly assigned to groups.
10% condition: 20 and 29 are less than 10% of all people.
Nearly Normal condition: We don't have the actual data. We will assume that the distributions of the populations of memory test scores are Normal.

Since the conditions are satisfied, it is appropriate to model the sampling distribution of the difference in means with a Student's t-model, with 45.88 degrees of freedom (from the approximation formula). We will perform a two-sample t-test.

The sampling distribution model has mean 0, with standard error:

$$SE(\bar{y}_M - \bar{y}_R) = \sqrt{\frac{3.19^2}{20} + \frac{3.99^2}{29}} \approx 1.0285.$$

The observed difference between the mean number of objects remembered is 10.0 – 10.72 = – 0.72.

$$t = \frac{(\bar{y}_M - \bar{y}_R) - (0)}{SE(\bar{y}_M - \bar{y}_R)}$$

$$t \approx \frac{-0.72}{1.0285}$$

$$t \approx -0.70$$

Since the P-value = 0.7563 is high, we fail to reject the null hypothesis. There is no evidence that the mean number of objects remembered by those who listen to Mozart is higher than the mean number of objects remembered by those who listen to rap music.

b) $(\bar{y}_M - \bar{y}_N) \pm t^*_{df}\sqrt{\frac{s_M^2}{n_M} + \frac{s_N^2}{n_N}} = (10.0 - 12.77) \pm t^*_{19.09}\sqrt{\frac{3.19^2}{20} + \frac{4.73^2}{13}} \approx (-5.351, -0.189)$

We are 90% confident that the mean number of objects remembered by those who listen to Mozart is between 0.189 and 5.352 objects lower than the mean of those who listened to no music.

27. Mozart.

a) H_0: The mean memory test score is the same for those who listen to rap as it is for those who listen to no music. $(\mu_R = \mu_N$ or $\mu_R - \mu_N = 0)$

H_A: The mean memory test score is lower for those who listen to rap than it is for those who listen to no music. $(\mu_R < \mu_N$ or $\mu_R - \mu_N < 0)$

Independent groups assumption: The groups are not related in regards to memory score.
Randomization condition: Subjects were randomly assigned to groups.
10% condition: 29 and 13 are less than 10% of all people.
Nearly Normal condition: We don't have the actual data. We will assume that the distributions of the populations of memory test scores are Normal.

Since the conditions are satisfied, it is appropriate to model the sampling distribution of the difference in means with a Student's t-model, with 20.00 degrees of freedom (from the approximation formula). We will perform a two-sample t-test.

The sampling distribution model has mean 0, with standard error:

$$SE(\bar{y}_R - \bar{y}_N) = \sqrt{\frac{3.99^2}{29} + \frac{4.73^2}{13}} \approx 1.5066.$$

The observed difference between the mean number of objects remembered is 10.72 – 12.77 = – 2.05.

$$t = \frac{(\bar{y}_R - \bar{y}_N) - (0)}{SE(\bar{y}_R - \bar{y}_N)}$$

$$t \approx \frac{-2.05}{1.5066}$$

$$t \approx -1.36$$

Since the P-value = 0.0944 is high, we fail to reject the null hypothesis. There is little evidence that the mean number of objects remembered by those who listen to rap is lower than the mean number of objects remembered by those who listen to no music.

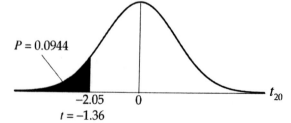

b) We did not conclude that there was a difference in the number of items remembered.

28. Cuckoos.

In order to determine whether the mean length of cuckoo eggs is the same for different species, we will conduct three hypothesis tests.

Independent groups assumption: The eggs were collected from the nests of three different species of bird.
Randomization condition: Assume that the eggs are representative of all cuckoo eggs laid in the nest of the particular species of bird.
10% condition: 14, 16, and 15 are less than 10% of all cuckoo eggs.
Nearly Normal condition: The histograms of the distribution of the lengths of cuckoo eggs found in sparrow and robin nests are unimodal and symmetric. The histogram of the distribution of the lengths of cuckoo eggs found in wagtail nests is uniform, but since there are no outliers and the sample size is not too small, it should be safe to proceed.

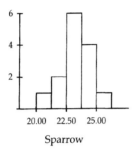
Sparrow

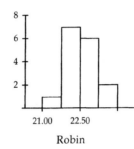
Robin

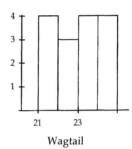
Wagtail

1) H_0: The mean length of cuckoo eggs is the same whether the foster parents are sparrows or robins. $(\mu_S = \mu_R$ or $\mu_S - \mu_R = 0)$

H_A: The mean length of cuckoo eggs is different, depending on whether the foster parents are sparrows or robins. $(\mu_S \neq \mu_R$ or $\mu_S - \mu_R \neq 0)$

Since the conditions are satisfied, it is appropriate to model the sampling distribution of the difference in means with a Student's *t*-model, with 21.60 degrees of freedom (from the approximation formula). We will perform a two-sample *t*-test.

The sampling distribution model has mean 0, with standard error:

$$SE(\bar{y}_S - \bar{y}_R) = \sqrt{\frac{1.06874^2}{14} + \frac{0.68452^2}{16}} \approx 0.3330$$

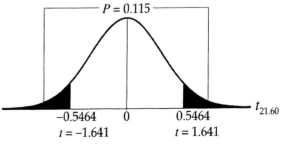

The observed difference between the mean length of the cuckoo eggs is
23.1214 – 22.575 = 0.5464.

Since the *P*-value = 0.115 is high, we fail to reject the null

$$t = \frac{(\bar{y}_S - \bar{y}_R) - (0)}{SE(\bar{y}_S - \bar{y}_R)}$$

$$t \approx \frac{0.5464}{0.3330}$$

$$t \approx 1.641$$

hypothesis. There is little evidence that the mean length of cuckoo eggs is different when the foster parents are sparrows than when they are robins.

2) H_0: The mean length of cuckoo eggs is the same whether the foster parents are sparrows or wagtails. $\left(\mu_s = \mu_w \text{ or } \mu_s - \mu_w = 0 \right)$

   H_A: The mean length of cuckoo eggs is different, depending on whether the foster parents are sparrows or wagtails. $\left(\mu_s \neq \mu_w \text{ or } \mu_s - \mu_w \neq 0 \right)$

   This test is virtually identical in mechanics to the first test. We know:

$\bar{y}_s = 23.1214$	$\bar{y}_w = 22.9033$	$t = 0.549$
$s_s = 1.06874$	$s_w = 1.06762$	$df = 26.86$
$n_s = 14$	$n_w = 15$	$P-Value = 0.587$

   Since the *P*-value = 0.587 is high, we fail to reject the null hypothesis. There is little evidence that the mean length of cuckoos eggs is different when the foster parents are sparrows than when they are wagtails.

3) H_0: The mean length of cuckoo eggs is the same whether the foster parents are robins or wagtails. $\left(\mu_R = \mu_w \text{ or } \mu_R - \mu_w = 0 \right)$

   H_A: The mean length of cuckoo eggs is different, depending on whether the foster parents are robins or wagtails. $\left(\mu_R \neq \mu_w \text{ or } \mu_R - \mu_w \neq 0 \right)$

   This test is virtually identical in mechanics to the first test. We know:

$\bar{y}_R = 22.575$	$\bar{y}_w = 22.9033$	$t = -1.012$
$s_R = 0.68452$	$s_w = 1.06762$	$df = 23.60$
$n_R = 16$	$n_w = 15$	$P-Value = 0.322$

   Since the *P*-value = 0.322 is high, we fail to reject the null hypothesis. There is little evidence that the mean length of cuckoo eggs is different when the foster parents are robins than when they are wagtails.

   There is no evidence to suggest a difference in mean length of cuckoo eggs that are laid in the nests of different foster parents. In general, we should be wary of doing three *t*-tests on the same data. Our Type I error is not the same for doing three tests as it is for one test. However, because none of the tests showed significant differences, this is less of a concern here.

Chapter 25 – Paired Samples and Blocks

1. More eggs?

a) Randomly assign 50 hens to each of the two kinds of feed. Compare the mean egg production of the two groups at the end of one month.

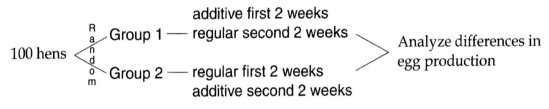

b) Randomly divide the 100 hens into two groups of 50 hens each. Feed the hens in the first group the regular feed for two weeks, then switch to the additive for 2 weeks. Feed the hens in the second group the additive for two weeks, then switch to the regular feed for two weeks. Subtract each hen's "regular" egg production from her "additive" egg production, and analyze the mean difference in egg production.

additive first 2 weeks
Group 1 — regular second 2 weeks
100 hens Random
Group 2 — regular first 2 weeks
additive second 2 weeks
Analyze differences in egg production

c) The matched pairs design in part b is the stronger design. Hens vary in their egg production regardless of feed. This design controls for that variability by matching the hens with themselves.

2. MTV.

a) Randomly assign half of the volunteers to do the puzzles in a quiet room, and assign the other half to do the puzzles with MTV on. Compare the mean time of the group in the quiet room to the mean time of the group watching MTV.

Volunteers Random
Group 1 — Quiet room
Group 2 — MTV
Compare mean crossword times

b) Randomly assign half of the volunteers to do a puzzle in a quiet room, and assign the other half to do the puzzles with MTV on. Then have each do a puzzle under the other condition. Subtract each volunteer's "quiet" time from his or her "MTV" time, and analyze the mean difference in times.

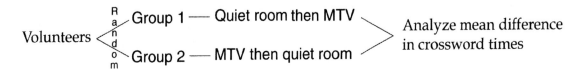

c) The matched pairs design in part b is the stronger design. People vary in their ability to do crossword puzzles. This design controls for that variability by matching the volunteers with themselves.

3. Women.

a) The paired t-test is appropriate. The labor force participation rate for two different years was paired by city.

b) Since the P-value = 0.0244, there is evidence of a difference in the average labor force participation rate for women between 1968 and 1972. The evidence suggests an increase in the participation rate for women.

4. BST.

a) **Paired data assumption:** We are testing the same cows, before and after injections of BST. **Randomization condition:** These cows are likely to be representative of all Ayrshires. **10% condition:** 60 cows are less than 10% of all cows.
Normal population assumption: We don't have the list of individual differences, so we can't look at a histogram. The sample is large, so we may proceed.

Since the conditions are satisfied, the sampling distribution of the difference can be modeled with a Student's t-model with $60 - 1 = 59$ degrees of freedom. We will find a paired t-interval, with 95% confidence.

b) $\bar{d} \pm t^*_{n-1}\left(\dfrac{s_d}{\sqrt{n}}\right) = 14 \pm t^*_{59}\left(\dfrac{5.2}{\sqrt{60}}\right) \approx (12.66,\ 15.34)$

c) We are 95% confident that the mean increase in daily milk production for Ayshire cows after BST injection is between 12.66 and 15.34 pounds.

d) 25% of 47 pounds is 11.75 pounds. According to the interval generated in part b, the average increase in milk production is more than this, so the farmer can justify the extra expense for BST.

5. Rain.

a) The two-sample t-test is appropriate for these data. The seeded and unseeded clouds are not paired in any way. They are independent.

b) Since the P-value = 0.0538, there is some evidence that the mean rainfall from seeded clouds is greater than the mean rainfall from unseeded clouds.

6. BST II

Although the data from each herd of cows are paired, we are asked to compare the paired differences from each herd. The herds are independent, so we will use a two-sample t-test.

H_0: The mean increase in milk production due to BST is the same for both breeds.
$\left(\mu_{dA} = \mu_{dJ} \text{ or } \mu_{dA} - \mu_{dJ} = 0\right)$

H_A: The mean increase in milk production due to BST is different for the two breeds.
$\left(\mu_{dA} \neq \mu_{dJ} \text{ or } \mu_{dA} - \mu_{dJ} \neq 0\right)$

Independent groups assumption: The cows are from different herds.
Randomization condition: Assume that the cows are representative of their breeds.
10% condition: 60 Ayrshires and 52 Jerseys are less than 10% of these breeds.
Nearly Normal condition: We don't have the actual data, so we can't check the distribution of the two sets of differences. However, the samples are large. The Central Limit Theorem allows us to proceed.

Since the conditions are satisfied, it is appropriate to model the sampling distribution of the difference in means (actually, the difference in mean differences!) with a Student's *t*-model, with 109.55 degrees of freedom (from the approximation formula).

We will perform a two-sample *t*-test. The sampling distribution model has mean 0, with

standard error: $SE(\bar{d}_A - \bar{d}_J) = \sqrt{\dfrac{5.2^2}{60} + \dfrac{4.8^2}{52}} \approx 0.945$.

The observed difference between the mean differences is 14 – 9 = 5.

$t = \dfrac{(\bar{d}_A - \bar{d}_J) - (0)}{SE(\bar{d}_A - \bar{d}_J)}$

Since the *P*-value = 6.4×10^{-7} is very small, we reject the null hypothesis. There is strong evidence that the mean increase for each breed is different. The average increase for Ayshires is significantly greater than the average increase for Jerseys.

$t \approx \dfrac{5}{0.945}$

$t \approx 5.29$

7. **Temperatures.**

Paired data assumption: The data are paired by city.
Randomization condition: These cities might not be representative of all European cities, so be cautious in generalizing the results.
10% condition: 12 cities are less than 10% of all European cities.
Normal population assumption: The histogram of differences between January and July mean temperature is roughly unimodal and symmetric.

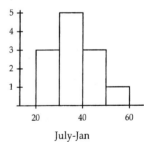
July-Jan

Since the conditions are satisfied, the sampling distribution of the difference can be modeled with a Student's *t*-model with 12 – 1 = 11 degrees of freedom. We will find a paired *t*-interval, with 90% confidence.

$\bar{d} \pm t^*_{n-1}\left(\dfrac{s_d}{\sqrt{n}}\right) = 36.8333 \pm t^*_{11}\left(\dfrac{8.66375}{\sqrt{12}}\right) \approx (32.3, 41.3)$

We are 90% confident that the average high temperature in European cities in July is an average of between 32.3° to 41.4° higher than in January.

8. **Marathons.**

Paired data assumption: The data are paired by year.
Randomization condition: Assume these years, at this marathon, are representative of all differences.
10% condition: 21 years is less than 10% of years for all marathons.
Normal population assumption: The histogram of differences between women's and men's times is roughly unimodal and symmetric.

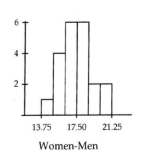
Women-Men

Since the conditions are satisfied, the sampling distribution of the difference can be modeled with a Student's *t*-model with 21 – 1 = 20 degrees of freedom. We will find a paired *t*-interval, with 90% confidence.

$$\bar{d} \pm t_{n-1}^* \left(\frac{s_d}{\sqrt{n}} \right) = 17.4619 \pm t_{20}^* \left(\frac{1.58729}{\sqrt{21}} \right) \approx (16.86, 18.06)$$

We are 90% confident women's winning marathon times are an average of between 16.86 and 18.06 minutes higher than men's winning times.

9. Push-ups.

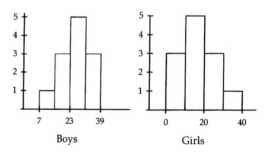

Boys   Girls

Independent groups assumption: The group of boys is independent of the group of girls.
Randomization condition: Assume that students are assigned to gym classes at random.
10% condition: 12 boys and 12 girls are less than 10% of all kids.
Nearly Normal condition: The histograms of the number of push-ups from each group are roughly unimodal and symmetric.

Since the conditions are satisfied, it is appropriate to model the sampling distribution of the difference in means with a Student's *t*-model, with 21 degrees of freedom (from the approximation formula). We will construct a two-sample *t*-interval, with 90% confidence.

$$(\bar{y}_B - \bar{y}_G) \pm t_{df}^* \sqrt{\frac{s_B^2}{n_B} + \frac{s_G^2}{n_G}} = (23.8333 - 16.5000) \pm t_{21}^* \sqrt{\frac{7.20900^2}{12} + \frac{8.93919^2}{12}} \approx (1.6, 13.0)$$

We are 90% confident that, at Gossett High, the mean number of push-ups that boys can do is between 1.6 and 13.0 more than the mean for the girls.

10. Exercise.

a) Paired data assumption: The data are paired by type of exercise machine.
Randomization condition: Assume that the men and women participating are representative of all men and women in terms of number of minutes of exercise required to burn 200 calories.
10% condition: The participants are less than 10% of all people.
Nearly Normal condition: The histogram of differences between women's and men's times is roughly unimodal and symmetric.

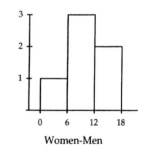

Women-Men

Since the conditions are satisfied, the sampling distribution of the difference can be modeled with a Student's *t*-model with 6 – 1 = 5 degrees of freedom. We will find a paired *t*-interval, with 95% confidence.

$$\bar{d} \pm t_{n-1}^* \left(\frac{s_d}{\sqrt{n}} \right) = 10 \pm t_5^* \left(\frac{4.97996}{\sqrt{6}} \right) \approx (4.77, 15.23)$$

We are 95% confident that women take an average of 4.8 to 15.2 minutes longer to burn 200 calories than men, when exercising at a light exertion rate.

b) **Nearly Normal condition:** There is no reason to think that this histogram does not represent differences drawn from a Normal population.

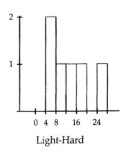

Light-Hard

Since the conditions are satisfied (some from part a), the sampling distribution of the difference can be modeled with a Student's t-model with 6 – 1 = 5 degrees of freedom. We will find a paired t-interval, with 95% confidence.

$$\bar{d} \pm t^*_{n-1}\left(\frac{s_d}{\sqrt{n}}\right) = 12.6667 \pm t^*_5\left(\frac{7.39369}{\sqrt{6}}\right) \approx (4.91, 20.42)$$

We are 95% confident that women exercising with light exertion take an average of 4.9 to 20.4 minutes longer to burn 200 calories than women exercising with hard exertion.

c) Since these data are averages, we expect the individual times to be more variable. Our standard error would be larger, resulting in a larger margin of error.

11. Job satisfaction.

a) Use a paired t-test.

Paired data assumption: The data are before and after job satisfaction rating for the same workers.
Randomization condition: The workers were randomly selected to participate.
10% condition: Assume that 10 workers are less than 10% of the workers at the company.
Nearly Normal conditon: The histogram of differences between before and after job satisfaction ratings is roughly unimodal and symmetric.

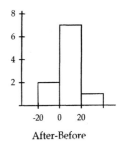

After-Before

b) H_0: The mean difference in before and after job satisfaction scores is zero, and the exercise program is not effective at improving job satisfaction. $(\mu_d = 0)$

H_A: The mean difference in before and after job satisfaction scores is greater than zero, and the exercise program is effective at improving job satisfaction. $(\mu_d > 0)$

Since the conditions are satisfied, the sampling distribution of the difference can be modeled with a Student's t-model with 10 – 1 = 9 degrees of freedom, $t_9\left(0, \frac{7.47217}{\sqrt{10}}\right)$.

We will use a paired t-test, with $\bar{d} = 8.5$.

Since the P-value = 0.0029 is low, we reject the null hypothesis. There is evidence that the mean job satisfaction rating has increased since the implementation of the exercise program.

$$t = \frac{\bar{d} - 0}{\frac{s_d}{\sqrt{n}}}$$

$$t = \frac{8.5 - 0}{\frac{7.47217}{\sqrt{10}}}$$

$$t \approx 3.60$$

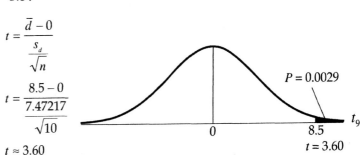

$P = 0.0029$

t_9

$t = 3.60$

12. Summer school.

a) H$_0$: The mean difference between August and June scores is zero, and the summer school program is not worthwhile. $(\mu_d = 0)$

   H$_A$: The mean difference between August and June scores is greater than zero, and the summer school program is worthwhile. $(\mu_d > 0)$

   Paired data assumption: The scores are paired by student.
   Randomization condition: Assume that these students are representative of all students who might attend this summer school in other years.
   10% condition: 6 students are less than 10% of all students.
   Normal population assumption: The histogram of differences between August and June scores shows a distribution that could have come from a Normal population.

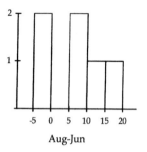

Aug-Jun

   Since the conditions are satisfied, the sampling distribution of the difference can be modeled with a Student's t-model with 6 – 1 = 5 degrees of freedom, $t_5\left(0, \dfrac{7.44759}{\sqrt{6}}\right)$.

   We will use a paired t-test, with $\bar{d} = 5.\overline{3}$.

   Since the P-value = 0.0699 is fairly high, we fail to reject the null hypothesis. There is not strong evidence that scores increased on average. The summer school program does not appear worthwhile, but the P-value is low enough that we should look at a larger sample to be more confident in our conclusion.

   $$t = \frac{\bar{d} - 0}{\frac{s_d}{\sqrt{n}}}$$

   $$t = \frac{5.\overline{3} - 0}{\frac{7.44759}{\sqrt{6}}}$$

   $$t \approx 1.75$$

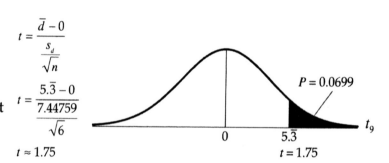

13. Sleep.

a) H$_0$: The mean number of additional hours of sleep is zero. $(\mu_d = 0)$

   H$_A$: The mean number of additional hours of sleep is greater than zero. $(\mu_d > 0)$

b) $P(t_9 > 3.680) = 0.0025$

c) If there is no gain of additional hours of sleep with the herb, the probability of observing a mean difference as large of larger than the one observed is 0.0025. (one-quarter percent)

d) Since the P-value = 0.0025 is low, we reject the null hypothesis. There is strong evidence that people taking the herb will get additional hours of sleep.

e) If we incorrectly reject the null hypothesis of no average gain in hours of sleep, we have committed a Type I error.

14. Gasoline.

a) H_0: The mean difference in mileage between premium and regular is zero. $\left(\mu_d = 0\right)$

H_A: The mean difference in mileage between premium and regular is greater than zero. $\left(\mu_d > 0\right)$

Paired data assumption: The mileage is paired by car.
Randomization condition: We randomized the order in which the different types of gasoline were used in each car.
10% condition: We are testing the mileage, not the car, so we don't need to check this condition.
Normal population assumption: The histogram of differences between premium and regular is roughly unimodal and symmetric.

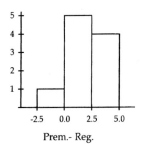

Prem.- Reg.

Since the conditions are satisfied, the sampling distribution of the difference can be modeled with a Student's *t*-model with 10 – 1 = 9 degrees of freedom, $t_9\left(0, \dfrac{1.41421}{\sqrt{10}}\right)$.

We will use a paired *t*-test, with $\bar{d} = 2$.

Since the *P*-value = 0.0008 is very low, we reject the null hypothesis. There is strong evidence of a mean increase in gas mileage between regular and premium.

$$t = \frac{\bar{d} - 0}{\dfrac{s_d}{\sqrt{n}}}$$

$$t = \frac{2 - 0}{\dfrac{1.41421}{\sqrt{10}}}$$

$$t \approx 4.47$$

$P = 0.0008$

t_9

0

2

$t = 4.47$

b) $\bar{d} \pm t^*_{n-1}\left(\dfrac{s_d}{\sqrt{n}}\right) = 2 \pm t^*_9\left(\dfrac{1.41421}{\sqrt{10}}\right) \approx (1.18,\ 2.82)$

We are 90% confident that the mean increase in gas mileage when using premium rather than regular gasoline is between 1.18 and 2.82 miles per gallon.

c) Premium gasoline costs more than regular gasoline. The increase in price might outweigh the increase in mileage.

d) With $t = 1.25$ and a *P*-value = 0.1144, we would have failed to reject the null hypothesis, and conclude that there was no evidence of a mean difference in mileage. The variation in performance of individual cars is greater than the variation related to the type of gasoline. This masked the true difference in mileage due to the gasoline. (Not to mention the fact that the two-sample test is not appropriate because we don't have independent samples!)

15. Yogurt.

H_0: The mean difference in calories between servings of strawberry and vanilla yogurt is zero. $(\mu_d = 0)$

H_A: The mean difference in calories between servings of strawberry and vanilla yogurt is different from zero. $(\mu_d \neq 0)$

Paired data assumption: The yogurt is paired by brand.
Randomization condition: Assume that these brands are representative of all brands.
10% condition: 12 brands of yogurt might not be less than 10% of all brands of yogurt. Proceed cautiously.
Normal population assumption: The histogram of differences in calorie content between strawberry and vanilla shows an outlier, Great Value. When the outlier is eliminated, the histogram of differences is roughly unimodal and symmetric.

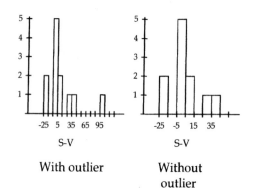

When Great Value yogurt is removed, the conditions are satisfied. The sampling distribution of the difference can be modeled with a Student's *t*-model with

$11 - 1 = 10$ degrees of freedom, $t_{10}\left(0, \dfrac{18.0907}{\sqrt{11}}\right)$.

We will use a paired *t*-test, with $\bar{d} \approx 4.54545$.

Since the *P*-value = 0.4241 is high, we fail to reject the null hypothesis. There is no evidence of a mean difference in calorie content between strawberry yogurt and vanilla yogurt.

$$t = \frac{\bar{d} - 0}{\frac{s_d}{\sqrt{n}}}$$

$$t = \frac{4.54545 - 0}{\frac{18.0907}{\sqrt{11}}}$$

$$t \approx 0.833$$

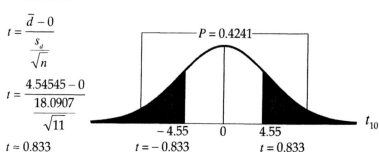

16. Caffeine.

a) H_0: The mean difference in memory score for caffeine drinkers is zero. $(\mu_d = 0)$

H_A: The mean difference in memory score for caffeine drinkers is not zero. $(\mu_d \neq 0)$

Paired data assumption: The subjects are matched with themselves using pretests and posttests.
Randomization condition: Volunteers were randomly assigned to the caffeine group or the no-caffeine group.
10% condition: We are testing the effects of caffeine, not the subjects, so this condition doesn't need to be checked.
Normal population assumption: Assumptions of Normality were deemed reasonable based on histograms of differences in scores. (Given)

Since the conditions are satisfied, the sampling distribution of the difference can be modeled with a Student's *t*-model with 15 – 1 = 14 degrees of freedom, $t_{14}\left(0, \frac{2.988}{\sqrt{15}}\right)$.

We will use a paired *t*-test, with $\bar{d} = -0.933$.

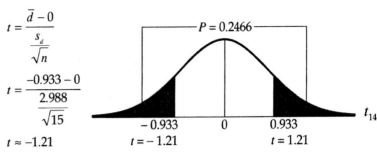

Since the *P*-value = 0.2466 is high, we fail to reject the null hypothesis. There is no evidence of a mean difference in memory score before and after drinking caffeine.

$$t = \frac{\bar{d} - 0}{\frac{s_d}{\sqrt{n}}}$$

$$t = \frac{-0.933 - 0}{\frac{2.988}{\sqrt{15}}}$$

$$t \approx -1.21$$

b) H$_0$: The mean difference in memory score for no-caffeine drinkers is zero. $(\mu_d = 0)$

H$_A$: The mean difference in memory score for no-caffeine drinkers is not zero. $(\mu_d \neq 0)$

Paired data assumption: The subjects are matched with themselves using pretests and posttests.
Randomization condition: Volunteers were randomly assigned to the caffeine group or the no-caffeine group.
10% condition: We are testing the effects of caffeine, not the subjects, so this condition doesn't need to be checked.
Normal population assumption: Assumptions of Normality were deemed reasonable based on histograms of differences in scores. (Given)

Since the conditions are satisfied, the sampling distribution of the difference can be modeled with a Student's *t*-model with 15 – 1 = 14 degrees of freedom, $t_{14}\left(0, \frac{2.441}{\sqrt{15}}\right)$.

We will use a paired *t*-test, with $\bar{d} = 1.429$.

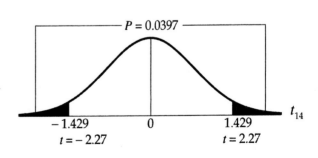

Since the *P*-value = 0.0397 is low, we reject the null hypothesis. There is evidence of a mean difference in memory score before and after drinking a drink with no caffeine. We expect people to score higher on the memory test after drinking the beverage with no caffeine.

$$t = \frac{\bar{d} - 0}{\frac{s_d}{\sqrt{n}}}$$

$$t = \frac{1.429 - 0}{\frac{2.441}{\sqrt{15}}}$$

$$t \approx 2.27$$

c) Answers may vary. This does not indicate that some mystery substance in noncaffeinated soda may aid memory. It is possible that the difference in mean memory score we have witnessed in the no-caffeine group is simply due to sampling variation. There may actually be no difference. If this is the case, we have committed a Type I error. Or, we could be witnessing the placebo effect. The subjects may have thought that there was a something in the drink that was supposed to help their memory.

17. Braking.

a) **Randomization Condition:** These stops are probably representative of all such stops for this type of car, but not for all cars.
10% Condition: 10 stops are less than 10% of all possible stops.
Nearly Normal Condition: A histogram of the stopping distances is roughly unimodal and symmetric.

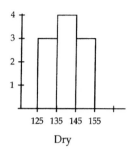

Dry

The stops in the sample had a mean stopping distance of 139.4 feet, and a standard deviation of 8.09938 feet. Since the conditions have been satisfied, construct a one-sample t-interval, with $10 - 1 = 9$ degrees of freedom, at 95% confidence.

$$\bar{y} \pm t^*_{n-1}\left(\frac{s}{\sqrt{n}}\right) = 139.4 \pm t^*_9\left(\frac{8.09938}{\sqrt{10}}\right) \approx (133.6, 145.2)$$

We are 95% confident that the mean dry pavement stopping distance for this type of car is between 133.6 and 145.2 feet.

b) **Independent Groups Assumption:** The wet pavement stops and dry pavement stops were made under different conditions and not paired in any way.
Randomization Condition: These stops are probably representative of all such stops for this type of car, but not for all cars.
10% Condition: 10 stops are less than 10% of all possible stops.
Nearly Normal Condition: The histogram of dry pavement stopping distances is roughly unimodal and symmetric (from part a), but the histogram of wet pavement stopping distances is a bit skewed. Since the Normal probability plot looks fairly straight, we will proceed.

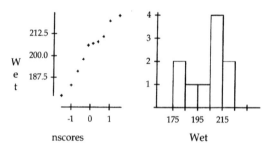

Since the conditions are satisfied, it is appropriate to model the sampling distribution of the difference in means with a Student's t-model, with 13.8 degrees of freedom (from the approximation formula). We will construct a two-sample t-interval, with 95% confidence.

$$(\bar{y}_W - \bar{y}_D) \pm t^*_{df}\sqrt{\frac{s^2_W}{n_W} + \frac{s^2_D}{n_D}} = (202.4 - 139.4) \pm t^*_{13.8}\sqrt{\frac{15.07168^2}{10} + \frac{8.09938^2}{10}} \approx (51.4, 74.6)$$

We are 95% confident that the mean stopping distance on wet pavement is between 51.4 and 74.6 feet longer than the mean stopping distance on dry pavement.

18. Brain waves.

a) H_0: The mean alpha-wave frequency for nonconfined inmates is the same as the mean alpha wave frequency for confined inmates. $\left(\mu_{NC} = \mu_C \text{ or } \mu_{NC} - \mu_C = 0\right)$

H_A: The mean alpha-wave frequency for nonconfined inmates is different from the mean alpha wave frequency for confined inmates. $\left(\mu_{NC} \neq \mu_C \text{ or } \mu_{NC} - \mu_C \neq 0\right)$

b) Independent Groups Assumption: The two groups of inmates were placed under different conditions, solitary confinement and not confined.
Randomization Condition: Inmates were randomly assigned to groups.
10% Condition: 10 confined and 10 nonconfined inmates are less than 10% of all inmates.
Nearly Normal Condition: The histograms of the alpha-wave frequencies are unimodal and symmetric.

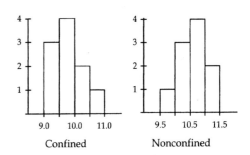

c) Since the conditions are satisfied, it is appropriate to model the sampling distribution of the difference in means with a Student's *t*-model, with 16.9 degrees of freedom (from the approximation formula). We will perform a two-sample *t*-test. We know:

$$\bar{y}_{NC} = 10.58 \qquad \bar{y}_{C} = 9.78$$
$$s_{NC} = 0.458984 \qquad s_{C} = 0.597774$$
$$n_{NC} = 10 \qquad n_{C} = 10$$

The sampling distribution model has mean 0, with standard error:

$$SE(\bar{y}_{NC} - \bar{y}_{C}) = \sqrt{\frac{0.458984^2}{10} + \frac{0.597774^2}{10}} \approx 0.2383.$$

The observed difference between the mean scores is 10.58 – 9.78 ≈ 0.80.

$$t = \frac{(\bar{y}_{NC} - \bar{y}_{C}) - (0)}{SE(\bar{y}_{NC} - \bar{y}_{C})}$$

$$t \approx \frac{0.80}{0.2382}$$

$$t \approx 3.357$$

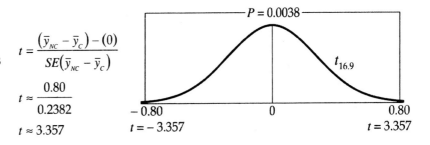

Since the *P*-value = 0.0038 is low, we reject the null hypothesis. There is evidence the mean alpha-wave frequency is different for nonconfined inmates and confined inmates.

d) The evidence suggests that the mean alpha-wave frequency for inmates subjected to confinement is lower than the mean alpha-wave frequency for inmates that are not confined.

19. Braking, test 2.

a) Randomization Condition: These cars are not a random sample, but are probably representative of all cars in terms of stopping distance.
10% Condition: These 10 cars are less than 10% of all cars.
Nearly Normal Condition: A histogram of the stopping distances is skewed to the right, but this may just be sampling variation from a Normal population. The "skew" is only a couple of stopping distances. We will proceed cautiously.

The cars in the sample had a mean stopping distance of 138.7 feet and a standard deviation of 9.66149 feet. Since the conditions have been satisfied, construct a one-sample *t*-interval, with 10 – 1 = 9 degrees of freedom, at 95% confidence.

$$\bar{y} \pm t^*_{n-1}\left(\frac{s}{\sqrt{n}}\right) = 138.7 \pm t^*_9\left(\frac{9.66149}{\sqrt{10}}\right) \approx (131.8,\ 145.6)$$

We are 95% confident that the mean dry pavement stopping distance for cars with this type of tires is between 131.8 and 145.6 feet. This estimate is based on an assumption that these cars are representative of all cars and that the population of stopping distances is Normal.

b) **Paired data assumption:** The data are paired by car.
Randomization condition: Assume that the cars are representative of all cars.
10% condition: The 10 cars tested are less than 10% of all cars.
Normal population assumption: The difference in stopping distance for car #4 is an outlier, at only 12 feet. After excluding this difference, the histogram of differences is unimodal and symmetric.

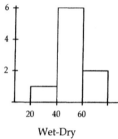
Wet-Dry

Since the conditions are satisfied, the sampling distribution of the difference can be modeled with a Student's *t*-model with 9 – 1 = 8 degrees of freedom. We will find a paired *t*-interval, with 95% confidence.

$$\bar{d} \pm t^*_{n-1}\left(\frac{s_d}{\sqrt{n}}\right) = 55 \pm t^*_8\left(\frac{10.2103}{\sqrt{9}}\right) \approx (47.2,\ 62.8)$$

With car #4 removed, we are 95% confident that the mean increase in stopping distance on wet pavement is between 47.2 and 62.8 feet. (If you leave the outlier in, the interval is 38.8 to 62.6 feet, but you should remove it! This procedure is sensitive to the presence of outliers!)

20. Tuition.

a) **Paired data assumption:** The data are paired by college.
Randomization condition: The colleges were selected randomly.
10% condition: The 19 colleges selected are less than 10% of all colleges.
Normal population assumption: The tuition difference for UC Irvine, at $7700, is an outlier. Once it has been removed, the histogram of the differences is roughly unimodal and symmetric.

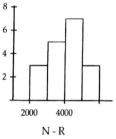
N - R

Since the conditions are satisfied, the sampling distribution of the difference can be modeled with a Student's *t*-model with 18 – 1 = 17 degrees of freedom. We will find a paired *t*-interval, with 90% confidence.

$$\bar{d} \pm t^*_{n-1}\left(\frac{s_d}{\sqrt{n}}\right) = 4127.78 \pm t^*_{17}\left(\frac{1134.39}{\sqrt{18}}\right) \approx (3662.65,\ 4592.91)$$

b) With UC Irvine removed, we are 90% confident that the mean increase in tuition for nonresidents versus residents is between about $3700 and $4600. (If you left UC Irvine in your data, the interval is about $3800 to $4900, but you should remove it! This procedure is sensitive to the presence of outliers!)

c) There is no evidence to suggest that the magazine made a false claim. An increase of $4000 for nonresidents is contained within our 90% confidence interval.

21. Strikes.

a) Since 60% of 50 pitches is 30 pitches, the Little Leaguers would have to throw an average of more than 30 strikes in order to give support to the claim made by the advertisements.

H₀: The mean number of strikes thrown by Little Leaguers who have completed the training is 30. $(\mu_A = 30)$

H_A: The mean number of strikes thrown by Little Leaguers who have completed the training is greater than 30. $(\mu_A > 30)$

Randomization Condition: Assume that these players are representative of all Little League pitchers.
10% Condition: 20 pitchers are less than 10% of all Little League pitchers.
Nearly Normal Condition: The histogram of the number of strikes thrown after the training is roughly unimodal and symmetric.

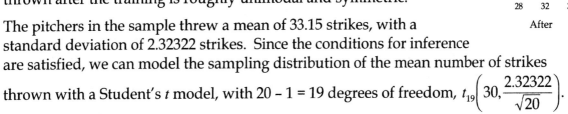

After

The pitchers in the sample threw a mean of 33.15 strikes, with a standard deviation of 2.32322 strikes. Since the conditions for inference are satisfied, we can model the sampling distribution of the mean number of strikes thrown with a Student's *t* model, with 20 – 1 = 19 degrees of freedom, $t_{19}\left(30, \dfrac{2.32322}{\sqrt{20}}\right)$.

We will perform a one-sample *t*-test.

$$t = \frac{\bar{y}_A - \mu_0}{\frac{s_A}{\sqrt{n_A}}}$$

$$t = \frac{33.15 - 30}{\frac{2.32322}{\sqrt{20}}}$$

$$t = 6.06$$

Since the *P*-value = 3.92×10^{-6} is very low, we reject the null hypothesis. There is strong evidence that the mean number of strikes that Little Leaguers can throw after the training is more than 30. (This test says nothing about the effectiveness of the training; just that Little Leaguers can throw more than 60% strikes on average after completing the training. This might not be an improvement.)

b) H₀: The mean difference in number of strikes thrown before and after the training is zero. $(\mu_d = 0)$

H_A: The mean difference in number of strikes thrown before and after the training is greater than zero. $(\mu_d > 0)$

Paired data assumption: The data are paired by pitcher.
Randomization condition: Assume that these players are representative of all Little League pitchers.
10% condition: 20 players are less than 10% of all players.
Normal population assumption: The histogram of differences is roughly unimodal and symmetric.

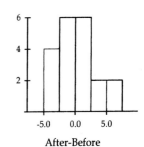

After-Before

Since the conditions are satisfied, the sampling distribution of the difference can be modeled with a Student's t-model with 20 – 1 = 19 degrees of freedom, $t_{19}\left(0, \dfrac{3.32297}{\sqrt{19}}\right)$.

We will use a paired t-test, with $\bar{d} = 0.1$.

Since the P-value = 0.4472 is high, we fail to reject the null hypothesis. There is no evidence of a mean difference in number of strikes thrown before and after the training. The training does not appear to be effective.

$$t = \frac{\bar{d} - 0}{\frac{s_d}{\sqrt{n}}}$$

$$t = \frac{0.1 - 0}{\frac{3.32297}{\sqrt{20}}}$$

$t \approx 0.135$

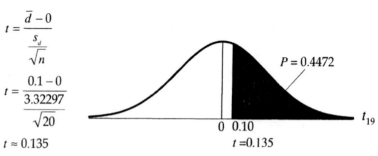

22. Uninsured.

H_0: The mean difference in percentage of Americans lacking health insurance in 1999-2000 and 2000-2001 is zero. $(\mu_d = 0)$

H_A: The mean difference in percentage of Americans lacking health insurance in 1999-2000 and 2000-2001 is greater than zero. $(\mu_d > 0)$

Paired data assumption: The data are paired by state.
Independence assumption: It is reasonable to think that the percentages of uninsured residents in each state are mutually independent.
Normal population assumption: The histogram of differences is roughly unimodal and symmetric, and the sample is large.

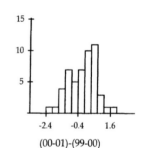

Since the conditions are satisfied, the sampling distribution of the difference can be modeled with a Student's t-model with 51 – 1 = 50 degrees of freedom, $t_{50}\left(0, \dfrac{0.846029}{\sqrt{51}}\right)$.

We will use a paired t-test, with $\bar{d} = -0.205882$.

Since the P-value = 0.956 is high, we fail to reject the null hypothesis. There is no evidence of a mean increase in the percentage of uninsured Americans from 1999-2000 to 2000-2001. In fact, the mean percentage of Americans who were uninsured was actually lower during the recession.

$$t = \frac{\bar{d} - 0}{\frac{s_d}{\sqrt{n}}}$$

$$t = \frac{-0.205882 - 0}{\frac{0.846029}{\sqrt{51}}}$$

$t \approx -1.74$

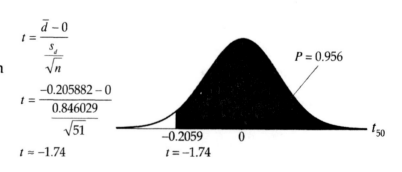

Review of Part VI – Learning About the World

1. Crawling.

a) H_0: The mean age at which babies begin to crawl is the same whether the babies were born in January or July. $\left(\mu_{Jan} = \mu_{July} \text{ or } \mu_{Jan} - \mu_{July} = 0\right)$

H_A: There is a difference in the mean age at which babies begin to crawl, depending on whether the babies were born in January or July. $\left(\mu_{Jan} \neq \mu_{July} \text{ or } \mu_{Jan} - \mu_{July} \neq 0\right)$

Independent groups assumption: The groups of January and July babies are independent.
Randomization condition: Although not specifically stated, we will assume that the babies are representative of all babies.
10% condition: 32 and 21 are less than 10% of all babies.
Nearly Normal condition: We don't have the actual data, so we can't check the distribution of the sample. However, the samples are fairly large. The Central Limit Theorem allows us to proceed.

Since the conditions are satisfied, it is appropriate to model the sampling distribution of the difference in means with a Student's t-model, with 43.68 degrees of freedom (from the approximation formula).

We will perform a two-sample t-test. The sampling distribution model has mean 0, with

standard error: $SE(\bar{y}_{Jan} - \bar{y}_{July}) = \sqrt{\dfrac{7.08^2}{32} + \dfrac{6.91^2}{21}} \approx 1.9596.$

The observed difference between the mean ages is 29.84 – 33.64 = – 3.8 weeks.

Since the *P*-value = 0.0590 is fairly low, we reject the null hypothesis. There is some evidence that mean age at which babies crawl is different for January and June babies. June babies appear to crawl a bit earlier than July babies, on average. Since the evidence is

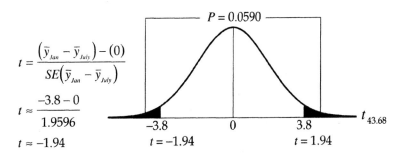

not strong, we might want to do some more research into this claim.

b) H_0: The mean age at which babies begin to crawl is the same whether the babies were born in April or October. $\left(\mu_{Apr} = \mu_{Oct} \text{ or } \mu_{Apr} - \mu_{Oct} = 0\right)$

H_A: There is a difference in the mean age at which babies begin to crawl, depending on whether the babies were born in April or October. $\left(\mu_{Apr} \neq \mu_{Oct} \text{ or } \mu_{Apr} - \mu_{Oct} \neq 0\right)$

The conditions (with minor variations) were checked in part a.

Since the conditions are satisfied, it is appropriate to model the sampling distribution of the difference in means with a Student's *t*-model, with 59.40 degrees of freedom (from the approximation formula).

We will perform a two-sample *t*-test. The sampling distribution model has mean 0, with standard error: $SE(\bar{y}_{Apr} - \bar{y}_{Oct}) = \sqrt{\dfrac{6.21^2}{26} + \dfrac{7.29^2}{44}} \approx 1.6404$.

The observed difference between the mean ages is $31.84 - 33.35 = -1.51$ weeks.

Since the *P*-value = 0.3610 is high, we fail to reject the null hypothesis. There is no evidence that mean age at which babies crawl is different for April and October babies.

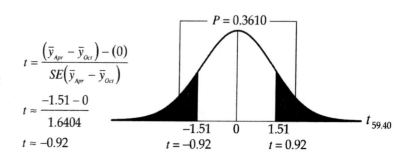

$$t = \dfrac{(\bar{y}_{Apr} - \bar{y}_{Oct}) - (0)}{SE(\bar{y}_{Apr} - \bar{y}_{Oct})}$$

$$t \approx \dfrac{-1.51 - 0}{1.6404}$$

$$t \approx -0.92$$

c) These results are not consistent with the researcher's claim. We have slight evidence in one test and no evidence in the other. The researcher will have to do better than this to convince us!

2. **Mazes and smells.**

H$_0$: The mean difference in maze times with and without the presence of a floral aroma is zero. $(\mu_d = 0)$

H$_A$: The mean difference in maze times with and without the presence of a floral aroma (unscented – scented) is greater than zero. $(\mu_d > 0)$

Paired data assumption: Each subject is paired with himself or herself.
Randomization condition: Subjects were randomized with respect to whether they did the scented trial first or second.
10% condition: We are testing the effects of the scent, not the subjects, so this condition doesn't need to be checked.
Nearly Normal condition: The histogram of differences between unscented and scented scores shows a distribution that could have come from a Normal population, and the sample size is fairly large.

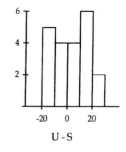

Since the conditions are satisfied, the sampling distribution of the difference can be modeled with a Student's *t*-model with $21 - 1 = 20$ degrees of freedom, $t_{20}\left(0, \dfrac{13.0087}{\sqrt{21}}\right)$.

We will use a paired *t*-test, (unscented – scented) with $\bar{d} = 3.85238$ seconds.

Since the *P*-value = 0.0949 is fairly high, we fail to reject the null hypothesis. There is little evidence that the mean difference in time required to complete the maze is greater than zero. The floral scent didn't appear to cause lower times.

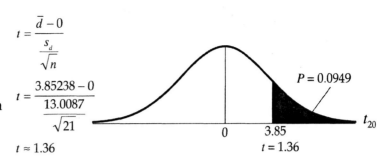

$$t = \frac{\bar{d} - 0}{\frac{s_d}{\sqrt{n}}}$$

$$t = \frac{3.85238 - 0}{\frac{13.0087}{\sqrt{21}}}$$

$$t \approx 1.36$$

3. Women.

H_0 : The percentage of businesses in the area owned by women is 26%. ($p = 0.26$)
H_A : The percentage of businesses in the area owned by women is not 26%. ($p \neq 0.26$)

Random condition: This is a random sample of 410 businesses in the Denver area.
10% condition: The sample of 410 businesses is less than 10% of all businesses.
Success/Failure condition: $np = (410)(0.26) = 106.6$ and $nq = (410)(0.74) = 303.4$ are both greater than 10, so the sample is large enough.

The conditions have been satisfied, so a Normal model can be used to model the sampling distribution of the proportion, with $\mu_{\hat{p}} = p = 0.26$ and $\sigma(\hat{p}) = \sqrt{\dfrac{pq}{n}} = \sqrt{\dfrac{(0.26)(0.74)}{410}} \approx 0.02166$.

We can perform a one-proportion *z*-test. The observed proportion of businesses owned by women is $\hat{p} = \dfrac{115}{410} \approx 0.2805$.

Since the *P*-value = 0.3443 is high, we fail to reject the null hypothesis. There is no evidence that the proportion of businesses in the Denver area owned by women is any different than the national figure of 26%.

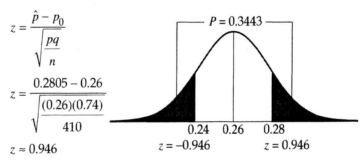

$$z = \frac{\hat{p} - p_0}{\sqrt{\frac{pq}{n}}}$$

$$z = \frac{0.2805 - 0.26}{\sqrt{\frac{(0.26)(0.74)}{410}}}$$

$$z \approx 0.946$$

4. Drugs.

a) **Paired data assumption:** The data are paired by drug.
   Randomization condition: These drugs DO NOT appear to be a random sample of all drugs. The names all begin with one of 5 letters! **10% condition:** These drugs are less than 10% of all drugs. **Nearly Normal condition:** The histogram of the differences in price is roughly unimodal and symmetric.

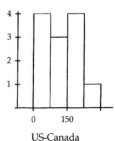

US-Canada

Since the conditions are satisfied, the sampling distribution of the difference can be modeled with a Student's *t*-model with 12 – 1 = 11 degrees of freedom. We will find a paired *t*-interval, with 95% confidence.

$$\bar{d} \pm t^*_{n-1}\left(\frac{s_d}{\sqrt{n}}\right) = 126 \pm t^*_{11}\left(\frac{76.2257}{\sqrt{12}}\right) \approx (77.57, 174.43)$$

We are 95% confident that the mean savings in drug cost when buying from the discount pharmacy is between about $77.60 and $174.40.

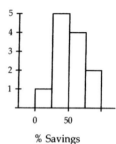

b) **Paired data assumption:** The data are paired by drug.
Randomization condition: These drugs DO NOT appear to be a random sample of all drugs. The names all begin with one of 5 letters!
10% condition: These drugs are less than 10% of all drugs.
Nearly Normal condition: The histogram of the percent savings is roughly unimodal and symmetric.

Since the conditions are satisfied, the sampling distribution of the difference can be modeled with a Student's *t*-model with 12 – 1 = 11 degrees of freedom. We will find a paired *t*-interval, with 95% confidence.

$$\bar{d} \pm t^*_{n-1}\left(\frac{s_d}{\sqrt{n}}\right) = 52.1667 \pm t^*_{11}\left(\frac{18.9681}{\sqrt{12}}\right) \approx (40.1\%, 64.2\%)$$

We are 95% confident the mean savings in drug cost when buying from the discount pharmacy is between 40.1% and 64.2%.

c) The analysis using the percents is more appropriate. It equalizes variability that is strictly due to the relative cost of the drugs. Some drugs are more expensive than others, and if just one of these drugs costs much less in Canada, it can pull the mean savings down.

d) These drugs are probably not a random sample of all drugs available. It is very unlikely that the 12 drugs chosen would have names beginning with one of a few letters. Also, the ad may have deliberately showcased those drugs that cost much less in Canada.

5. Pottery.

Independent groups assumption: The pottery samples are from two different sites.
Randomization condition: It is reasonable to think that the pottery samples are representative of all pottery at that site with respect to aluminum oxide content.
10% condition: The samples of 5 pieces are less than 10% of all pottery pieces.
Nearly Normal condition: The histograms of aluminum oxide content are roughly unimodal and symmetric.

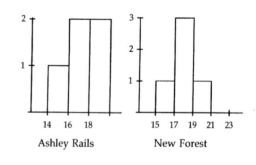

Ashley Rails   New Forest

Since the conditions are satisfied, it is appropriate to model the sampling distribution of the difference in means with a Student's *t*-model, with 7 degrees of freedom (from the approximation formula). We will construct a two-sample *t*-interval, with 95% confidence.

$$(\bar{y}_{AR} - \bar{y}_{NF}) \pm t^*_{df}\sqrt{\frac{s^2_{AR}}{n_{AR}} + \frac{s^2_{NF}}{n_{NF}}} = (17.32 - 18.18) \pm t^*_7\sqrt{\frac{1.65892^2}{5} + \frac{1.77539^2}{5}} \approx (-3.37, 1.65)$$

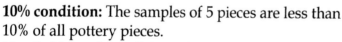

We are 95% confident that the difference in the mean percentage of aluminum oxide content of the pottery at the two sites is between –3.37% and 1.65%. Since 0 is in the interval, there is no evidence that the aluminum oxide content at the two sites is different. It would be reasonable for the archaeologists to think that the same ancient people inhabited the sites.

6. Streams.

Random condition: The researchers randomly selected 172 streams.
10% condition: 172 is less than 10% of all streams.
Success/Failure condition: $n\hat{p} = 69$ and $n\hat{q} = 103$ are both greater than 10, so the sample is large enough.

Since the conditions are met, we can use a one-proportion z-interval to estimate the percentage of Adirondack streams with a shale substrate.

$$\hat{p} \pm z^* \sqrt{\frac{\hat{p}\hat{q}}{n}} = \left(\frac{69}{172}\right) \pm 1.960 \sqrt{\frac{\left(\frac{69}{172}\right)\left(\frac{103}{172}\right)}{172}} = (32.8\%, 47.4\%)$$

We are 95% confident that between 32.8% and 47.4% of Adirondack streams have a shale substrate.

7. Gehrig.

H_0: The proportion of ALS patients who were athletes is the same as the proportion of patients with other disorders who were athletes. $\left(p_{ALS} = p_{Other} \text{ or } p_{ALS} - p_{Other} = 0\right)$

H_A: The proportion of ALS patients who were athletes is greater than the proportion of patients with other disorders who were athletes. $\left(p_{ALS} > p_{Other} \text{ or } p_{ALS} - p_{Other} > 0\right)$

Random condition: This is NOT a random sample. We must assume that these patients are representative of all patients with neurological disorders.
10% condition: 280 and 151 are both less than 10% of all patients with disorders.
Independent samples condition: The groups are independent.
Success/Failure condition: $n\hat{p}(\text{ALS}) = (280)(0.38) = 106$, $n\hat{q}(\text{ALS}) = (280)(0.72) = 174$, $n\hat{p}(\text{Other}) = (151)(0.26) = 39$, and $n\hat{q}(\text{Other}) = (151)(0.74) = 112$ are all greater than 10, so the samples are both large enough.

Since the conditions have been satisfied, we will model the sampling distribution of the difference in proportion with a Normal model with mean 0 and standard deviation estimated by

$$SE_{pooled}\left(\hat{p}_{ALS} - \hat{p}_{Other}\right) = \sqrt{\frac{\hat{p}_{pooled}\hat{q}_{pooled}}{n_{ALS}} + \frac{\hat{p}_{pooled}\hat{q}_{pooled}}{n_{Other}}} = \sqrt{\frac{(0.336)(0.664)}{280} + \frac{(0.336)(0.664)}{151}} \approx 0.0477.$$

The observed difference between the proportions is 0.38 – 0.26 = 0.12.

Since the *P*-value = 0.0060 is very low, we reject the null hypothesis. There is strong evidence that the proportion of ALS patients who are athletes is greater than the proportion of patients with other disorders who are athletes.

$$z = \frac{0.12 - 0}{0.0477}$$

$$z \approx 2.52$$

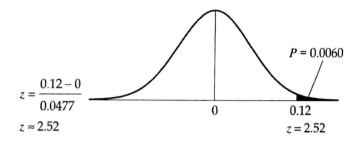

b) This was a retrospective observational study. In order to make the inference, we must assume that the patients studied are representative of all patients with neurological disorders.

8. **Teen drinking.**

Paired data assumption: The data are paired by country.
Randomization condition: We don't know the individuals in the different countries were chosen. We will assume that the individuals are representative of teens in each country.
10% condition: The teens sampled are less than 10% of all teens in the countries.
Normal population assumption: The histogram of differences is roughly unimodal and symmetric.

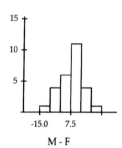

Since the conditions are satisfied, the sampling distribution of the difference can be modeled with a Student's *t*-model with 27 − 1 = 26 degrees of freedom. We will find a paired *t*-interval, with 95% confidence.

$$\bar{d} \pm t^*_{n-1}\left(\frac{s_d}{\sqrt{n}}\right) = 7.96296 \pm t^*_{26}\left(\frac{8.79459}{\sqrt{27}}\right) \approx (4.5\%, 11.4\%)$$

We are 95% confident that the mean percentage of 15-year-old boys in these countries who have been drunk at least twice is between 4.5% and 11.4% higher than the percentage of 15-year-old girls.

9. **Babies.**

H_0: The mean weight of newborns in the U.S. is 7.41 pounds, the same as the mean weight of Australian babies. $(\mu = 7.41)$

H_A: The mean weight of newborns in the U.S. is not the same as the mean weight of Australian babies. $(\mu \neq 7.41)$

Randomization condition: Assume that the babies at this Missouri hospital are representative of all U.S. newborns. (Given)
10% condition: 112 newborns are less than 10% of all newborns.
Nearly Normal condition: We don't have the actual data, so we cannot look at a graphical display, but since the sample is large, it is safe to proceed.

The babies in the sample had a mean weight of 7.68 pounds and a standard deviation in weight of 1.31 pounds. Since the conditions for inference are satisfied, we can model the sampling distribution of the mean weight of U.S. newborns with a Student's t model, with

$112 - 1 = 111$ degrees of freedom, $t_{111}\left(7.41, \dfrac{1.31}{\sqrt{112}}\right)$.

We will perform a one-sample t-test.

Since the P-value = 0.0313 is low, we reject the null hypothesis. If we believe that the babies at this Missouri hospital are representative of all U.S. babies, there is evidence to suggest that the mean weight of U.S. babies is different than the mean weight of Australian babies. U.S. babies appear to weigh more on average.

$t = \dfrac{\bar{y} - \mu_0}{SE(\bar{y})}$

$t = \dfrac{7.68 - 7.41}{\dfrac{1.31}{\sqrt{112}}}$

$t \approx 2.18$

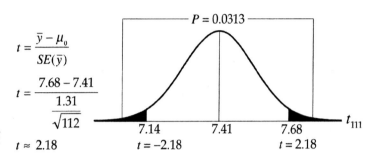

10. Petitions.

a) $\dfrac{1772}{2000} = 0.886 = 88.6\%$ of the sample signatures were valid.

b) $\dfrac{250,000}{304,266} \approx 0.822 \approx 82.2\%$ of the petition signatures must be valid in order to have the initiative certified by the Elections Committee.

c) If the Elections Committee commits a Type I error, a petition would be certified when there are not enough valid signatures.

d) If the Elections Committee commits a Type II error, a valid petition is not certified.

e) H_0: The percentage of valid signatures is 82.2% $(p = 0.822)$
H_A: The percentage of valid signatures is greater than 82.2% $(p > 0.822)$

Random Condition: This is a simple random sample of 2000 signatures.
10% condition: The sample of 2000 businesses is less than 10% of all signatures.
Success/Failure condition: $np = (2000)(0.822) = 1644$ and $nq = (2000)(0.178) = 356$ are both greater than 10, so the sample is large enough.

The conditions have been satisfied, so a Normal model can be used to model the sampling distribution of the proportion, with $\mu_{\hat{p}} = p = 0.822$ and

$\sigma(\hat{p}) = \sqrt{\dfrac{pq}{n}} = \sqrt{\dfrac{(0.822)(0.178)}{2000}} \approx 0.00855.$

We can perform a one-proportion z-test.

The observed proportion of valid signatures is $\hat{p} = \dfrac{1772}{2000} \approx 0.886.$

$z = \dfrac{\hat{p} - p_0}{\sigma(\hat{p})}$

$z \approx \dfrac{0.886 - 0.822}{0.008553}$

$z \approx 7.48$

Since the P-value = 3.64×10^{-14} is low, we reject the null hypothesis. There is strong evidence that the percentage of valid signatures is greater than 82.2%. The petition should be certified.

f) In order to increase the power of their test to detect valid petitions, the Elections Committee could sample more signatures.

11. Feeding fish.

a) If there is no difference in the average fish sizes, the chance of observing a difference this large, or larger, just by natural sampling variation is 0.1%.

b) There is evidence that largemouth bass that are fed a natural diet are larger. The researchers would advise people who raise largemouth bass to feed them a natural diet.

c) If the advice is incorrect, the researchers have committed a Type I error.

12. Risk.

These samples are independent, one sample of midsize cars and another of SUVs. The appropriate test is a two-sample t-test. There is no indication of random sampling. The back-to-back stemplot shows the distribution of the death rates for midsize cars is uniform, not unimodal and symmetric. The distribution of death rates for the SUVs has two outliers. With sample sizes this small, it is probably unwise to proceed with this inference.

Midsize		SUV
7	4	
4	5	5
4	6	02
6	7	6
8	8	
7	9	1
	10	9

$8\,|\,2 = 82$ deaths per million sales

13. Age.

a) **Independent groups assumption:** The group of patients with and without cardiac disease are not related in any way.
Randomization condition: Assume that these patients are representative of all people.
10% condition: 2397 patients without cardiac disease and 450 patients with cardiac disease are both less than 10% of all people.
Normal population assumption: We don't have the actual data, so we will assume that the population of ages of patients is Normal.

Since the conditions are satisfied, it is appropriate to model the sampling distribution of the difference in means with a Student's t-model, with 670 degrees of freedom (from the approximation formula). We will construct a two-sample t-interval, with 95% confidence.

$$(\bar{y}_{Card} - \bar{y}_{None}) \pm t^*_{df}\sqrt{\frac{s^2_{Card}}{n_{Card}} + \frac{s^2_{None}}{n_{None}}} = (74.0 - 69.8) \pm t^*_{670}\sqrt{\frac{7.9^2}{450} + \frac{8.7^2}{2397}} \approx (3.39, 5.01)$$

We are 95% confident that the mean age of patients with cardiac disease is between 3.39 and 5.01 years higher than the mean age of patients without cardiac disease.

b) Older patients are at greater risk for a variety of health problems. If an older patient does not survive a heart attack, the researchers will not know to what extent depression was involved, because there will be a variety of other possible variables influencing the death rate. Additionally, older patients may be more (or less) likely to be depressed than younger ones.

14. Smoking.

Randomization condition: Assume that these patients are representative of all people.

10% condition: 2397 patients without cardiac disease and 450 patients with cardiac disease are both less than 10% of all people.

Independent groups assumption: The group of patients with and without cardiac disease are not related in any way.

Success/Failure condition: $n\hat{p}$ (cardiac) = $(450)(0.32) = 144$, $n\hat{q}$ (cardiac) = $(450)(0.68) = 306$, $n\hat{p}$ (none) = $(2397)(0.237) = 568$, and $n\hat{q}$ (none) = $(2397)(0.763) = 1829$ are all greater than 10, so the samples are both large enough.

Since the conditions have been satisfied, we will find a two-proportion z-interval.

$$\left(\hat{p}_{Card} - \hat{p}_{None}\right) \pm z^* \sqrt{\frac{\hat{p}_{Card}\hat{q}_{Card}}{n_{Card}} + \frac{\hat{p}_{None}\hat{q}_{None}}{n_{None}}} = (0.32 - 0.237) \pm 1.960 \sqrt{\frac{(0.32)(0.68)}{450} + \frac{(0.237)(0.763)}{2397}}$$

$$= (0.0367, 0.1293)$$

We are 95% confident that the proportion of smokers is between 3.67% and 12.93% higher for patients with cardiac disease than for patients without cardiac disease.

b) Since the confidence interval does not contain 0, there is evidence that cardiac patients have a higher rate of smokers than the patients without cardiac disease. The two groups are different.

c) Smoking could be a confounding variable. Smokers have a higher risk of other health problems that may be associated with their ability to survive a heart attack.

15. Computer use.

a) It is unlikely that an equal number of boys and girls were contacted strictly by chance. It is likely that this was a stratified random sample, stratified by gender.

b) **Randomization condition:** The teens were selected at random.

10% condition: 620 boys and 620 girls are both less than 10% of all teens.

Independent groups assumption: The groups of boys and girls are not paired or otherwise related in any way.

Success/Failure condition: $n\hat{p}$ (boys) = $(620)(0.77) = 477$, $n\hat{q}$ (boys) = $(620)(0.23) = 143$, $n\hat{p}$ (girls) = $(620)(0.65) = 403$, and $n\hat{q}$ (girls) = $(620)(0.35) = 217$ are all greater than 10, so the samples are both large enough.

Since the conditions have been satisfied, we will find a two-proportion z-interval.

$$\left(\hat{p}_B - \hat{p}_G\right) \pm z^* \sqrt{\frac{\hat{p}_B\hat{q}_B}{n_B} + \frac{\hat{p}_G\hat{q}_G}{n_G}} = (0.77 - 0.65) \pm 1.960 \sqrt{\frac{(0.77)(0.33)}{620} + \frac{(0.65)(0.35)}{620}} = (0.070, 0.170)$$

We are 95% confident that the proportion of computer gamers is between 7.0% and 17.0% higher for boys than for girls.

c) Since the interval lies entirely above 0, there is evidence that a greater percentage of boys play computer games than girls.

16. Recruiting.

a) **Randomization condition:** Assume that these students are representative of all admitted graduate students.
10% condition: Two groups of 500 students are both less than 10% of all students.
Independent groups assumption: The groups are from two different years.
Success/Failure condition: $n\hat{p}$ (before) = (500)(0.52) = 260, $n\hat{q}$ (before) = (500)(0.48) = 240, $n\hat{p}$ (after) = (500)(0.54) = 270, and $n\hat{q}$ (after) = (500)(0.46) = 230 are all greater than 10, so the samples are both large enough.

Since the conditions have been satisfied, we will find a two-proportion z-interval.

$$\left(\hat{p}_A - \hat{p}_B\right) \pm z^* \sqrt{\frac{\hat{p}_A \hat{q}_A}{n_A} + \frac{\hat{p}_B \hat{q}_B}{n_B}} = (0.54 - 0.52) \pm 1.960 \sqrt{\frac{(0.54)(0.46)}{500} + \frac{(0.52)(0.48)}{500}} = (-0.042, 0.082)$$

We are 95% confident that the change in proportion of students who choose to enroll is between –4.2% and 8.2%.

b) Since 0 is contained in the interval, there is no evidence to suggest a change in the proportion of students who enroll. The program does not appear to be effective.

17. Hearing.

Paired data assumption: The data are paired by subject.
Randomization condition: The order of the tapes was randomized.
10% condition: We are testing the tapes, not the people, so this condition does not need to be checked.
Normal population assumption: The histogram of differences between List A and List B is roughly unimodal and symmetric.

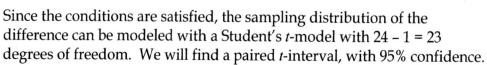

Since the conditions are satisfied, the sampling distribution of the difference can be modeled with a Student's t-model with 24 – 1 = 23 degrees of freedom. We will find a paired t-interval, with 95% confidence.

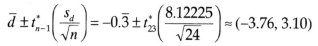

$$\bar{d} \pm t^*_{n-1}\left(\frac{s_d}{\sqrt{n}}\right) = -0.\overline{3} \pm t^*_{23}\left(\frac{8.12225}{\sqrt{24}}\right) \approx (-3.76, 3.10)$$

We are 95% confident that the mean difference in the number of words a person might misunderstand using these two lists is between – 3.76 and 3.10 words. Since 0 is contained in the interval, there is no evidence to suggest that that the two lists are different for the purposes of the hearing test when there is background noise. It is reasonable to think that the two lists are still equivalent.

18. Cesareans.

H_0: The proportion of births involving cesarean deliveries is the same in Vermont and New Hampshire. $\left(p_{VT} = p_{NH} \text{ or } p_{VT} - p_{NH} = 0\right)$

H_A: The proportion of births involving cesarean deliveries is different in Vermont and New Hampshire. $\left(p_{VT} \neq p_{NH} \text{ or } p_{VT} - p_{NH} \neq 0\right)$

Random condition: Hospitals were randomly selected.
10% condition: 223 and 186 are both less than 10% of all births in these states.
Independent samples condition: Vermont and New Hampshire are different states!
Success/Failure condition: $n\hat{p}(VT) = (223)(0.166) = 37$, $n\hat{q}(VT) = (223)(0.834) = 186$, $n\hat{p}(NH) = (186)(0.188) = 35$, and $n\hat{q}(NH) = (186)(0.812) = 151$ are all greater than 10, so the samples are both large enough.

Since the conditions have been satisfied, we will model the sampling distribution of the difference in proportion with a Normal model with mean 0 and standard deviation estimated by

$$SE_{pooled}\left(\hat{p}_{VT} - \hat{p}_{NH}\right) = \sqrt{\frac{\hat{p}_{pooled}\hat{q}_{pooled}}{n_{VT}} + \frac{\hat{p}_{pooled}\hat{q}_{pooled}}{n_{NH}}} = \sqrt{\frac{(0.176)(0.824)}{223} + \frac{(0.176)(0.824)}{186}} \approx 0.03782.$$

The observed difference between the proportions is $0.166 - 0.188 = -0.022$.

Since the *P*-value = 0.5563 is high, we fail to reject the null hypothesis. There is no evidence that the proportion of cesarean births in Vermont is different from the proportion of cesarean births in New Hampshire.

$$z = \frac{-0.22 - 0}{0.03782}$$

$$z \approx -0.59$$

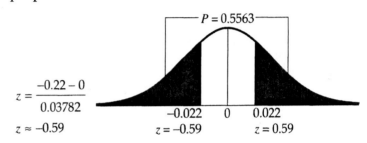

19. Newspapers.

a) An examination of a graphical display reveals Spain, Portugal, and Italy to be outliers. They are all Mediterranean countries, and all have a significantly higher percentage of men than women reading a newspaper daily.

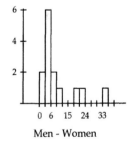

Men - Women

b) H_0: The mean difference in the percentage of men and women who read a daily newspaper in these countries is zero. $(\mu_d = 0)$

H_A: The mean difference in the percentage of men and women who read a daily newspaper in these countries is greater than zero. $(\mu_d > 0)$

Paired data assumption: The data are paired by country.
Randomization condition: Samples in each country were random.
10% condition: The 1000 respondents in each country are less than 10% of the people in these countries.
Nearly Normal condition: With three outliers removed, the distribution of differences is roughly unimodal and symmetric.

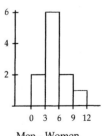

Men - Women

Since the conditions are satisfied, the sampling distribution of the difference can be modeled with a Student's *t*-model with $11 - 1 = 10$ degrees of freedom, $t_{10}\left(0, \frac{2.83668}{\sqrt{11}}\right)$.

We will use a paired *t*-test, (Men - Women) with $\bar{d} = 4.75455$.

Since the *P*-value = 0.0001 is very low, we reject the null hypothesis. There is strong evidence that the mean difference is greater than zero. The percentage of men in these countries who read the paper daily appears to be greater than the percentage of women who do so.

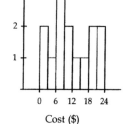

$$t = \frac{\bar{d} - 0}{\frac{s_d}{\sqrt{11}}}$$

$$t \approx \frac{4.75455 - 0}{\frac{2.83668}{\sqrt{11}}}$$

$$t \approx 5.56$$

20. Meals.

H_0: The college student's mean daily food expense is \$10. $(\mu = 10)$

H_A: The college student's mean daily food expense is greater than \$10. $(\mu > 10)$

Randomization condition: Assume that these days are representative of all days.
10% condition: 14 days are less than 10% of all days.
Nearly Normal condition: The histogram of daily expenses is fairly unimodal and symmetric. It is reasonable to think that this sample came from a Normal population.

The expenses in the sample had a mean of 11.4243 dollars and a standard deviation of 8.05794 dollars. Since the conditions for inference are satisfied, we can model the sampling distribution of the mean daily expense with a Student's *t* model, with 14 – 1 = 13 degrees of freedom, $t_{13}\left(10, \frac{8.05794}{\sqrt{14}}\right)$.

We will perform a one-sample *t*-test.

Since the *P*-value = 0.2600 is high, we fail to reject the null hypothesis. There is no evidence that the student's average spending is more than \$10 per day.

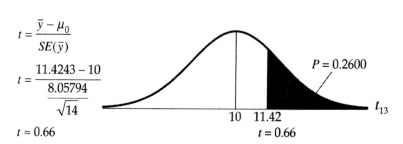

$$t = \frac{\bar{y} - \mu_0}{SE(\bar{y})}$$

$$t = \frac{11.4243 - 10}{\frac{8.05794}{\sqrt{14}}}$$

$$t \approx 0.66$$

21. Wall Street.

Random condition: 1002 American adults were randomly selected.
10% condition: 1002 is less than 10% of all American adults.
Success/Failure condition: $n\hat{p} = (1002)(0.60) = 601$ and $n\hat{q} = (1002)(0.40) = 401$ are both greater than 10, so the sample is large enough.

Since the conditions are met, we can use a one-proportion *z*-interval to estimate the percentage of American adults who agree with the statement.

$$\hat{p} \pm z^* \sqrt{\frac{\hat{p}\hat{q}}{n}} = (0.60) \pm 1.960 \sqrt{\frac{(0.60)(0.40)}{1002}} \approx (57.0\%, 63.0\%)$$

We are 95% confident that between 57.0% and 63.0% of American adults would agree with the statement about Wall Street.

22. Teach for America.

H_0: The mean score of students with certified teachers is the same as the mean score of students with uncertified teachers. $\left(\mu_C = \mu_U \ \text{or} \ \mu_C - \mu_U = 0\right)$

H_A: The mean score of students with certified teachers is greater than as the mean score of students with uncertified teachers. $\left(\mu_C > \mu_U \ \text{or} \ \mu_C - \mu_U > 0\right)$

Independent groups assumption: The certified and uncertified teachers are independent groups.
Randomization condition: Assume the students studied were representative of all students.
10% condition: Two samples of size 44 are both less than 10% of the population.
Nearly Normal condition: We don't have the actual data, so we can't look at the graphical displays, but the sample sizes are large, so we can proceed.

Since the conditions are satisfied, it is appropriate to model the sampling distribution of the difference in means with a Student's *t*-model, with 86 degrees of freedom (from the approximation formula).

We will perform a two-sample *t*-test. The sampling distribution model has mean 0, with standard error: $SE(\bar{y}_C - \bar{y}_U) = \sqrt{\dfrac{9.31^2}{44} + \dfrac{9.43^2}{44}} \approx 1.9977$.

The observed difference between the mean scores is 35.62 – 32.48 = 3.14.

Since the *P*-value = 0.0598 is fairly high, we fail to reject the null hypothesis. There is little evidence that students with certified teachers had mean scores higher than students with uncertified teachers. However, since the *P*-value is not extremely high, further investigation is recommended.

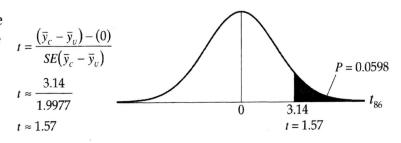

$$t = \frac{(\bar{y}_C - \bar{y}_U) - (0)}{SE(\bar{y}_C - \bar{y}_U)}$$

$$t \approx \frac{3.14}{1.9977}$$

$$t \approx 1.57$$

23. Legionnaires' disease.

Paired data assumption: The data are paired by room.
Randomization condition: We will assume that these rooms are representative of all rooms at the hotel.
10% condition: Assume that 8 rooms are less than 10% of the rooms. The hotel must have more than 80 rooms.
Nearly Normal condition: The histogram of differences between before and after measurements is roughly unimodal and symmetric.

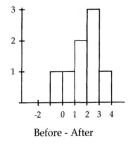

Since the conditions are satisfied, the sampling distribution of the difference can be modeled with a Student's *t*-model with 8 – 1 = 7 degrees of freedom. We will find a paired *t*-interval, with 95% confidence.

$$\bar{d} \pm t^*_{n-1}\left(\frac{s_d}{\sqrt{n}}\right) = 1.6125 \pm t^*_7\left(\frac{1.23801}{\sqrt{8}}\right) \approx (0.58,\ 2.65)$$

We are 95% confident that the mean difference in the bacteria counts is between 0.58 and 2.65 colonies per cubic foot of air. Since the entire interval is above 0, there is evidence that the new air-conditioning system was effective in reducing average bacteria counts.

24. Teach for America, Part II.

H_0: The mean score of students with certified teachers is the same as the mean score of students with uncertified teachers. $\left(\mu_C = \mu_U \text{ or } \mu_C - \mu_U = 0\right)$

H_A: The mean score of students with certified teachers is different than the mean score of students with uncertified teachers. $\left(\mu_C \neq \mu_U \text{ or } \mu_C - \mu_U \neq 0\right)$

Mathematics: Since the *P*-value = 0.0002 is low, we reject the null hypothesis. There is strong evidence that students with certified teachers have different mean math scores than students with uncertified teachers. Students with certified teachers do better.

Language: Since the *P*-value = 0.045 is fairly low, we reject the null hypothesis. There is evidence that students with certified teachers have different mean language scores than students with uncertified teachers. Students with certified teachers do better. However, since the *P*-value is not extremely low, further investigation is recommended.

25. Bipolar kids.

a) **Random condition:** Assume that the 89 children are representative of all children with bipolar disorder.
10% condition: 89 is less than 10% of all children with bipolar disorder.
Success/Failure condition: $n\hat{p} = 26$ and $n\hat{q} = 63$ are both greater than 10, so the sample is large enough.

Since the conditions are met, we can use a one-proportion *z*-interval to estimate the percentage of children with bipolar disorder who might be helped by this treatment.

$$\hat{p} \pm z^*\sqrt{\frac{\hat{p}\hat{q}}{n}} = \left(\frac{26}{89}\right) \pm 1.960\sqrt{\frac{\left(\frac{26}{89}\right)\left(\frac{63}{89}\right)}{89}} \approx (19.77\%, 38.66\%)$$

We are 95% confident that between 19.77% and 38.66% of children with bipolar disorder will be helped with medication and psychotherapy.

b)

$$ME = z^*\sqrt{\frac{\hat{p}\hat{q}}{n}}$$

$$0.06 = 1.960\sqrt{\frac{\left(\frac{26}{89}\right)\left(\frac{63}{89}\right)}{n}}$$

$$n = \frac{(1.960)^2\left(\frac{26}{89}\right)\left(\frac{63}{89}\right)}{(0.06)^2}$$

$$n \approx 221 \text{ children}$$

In order to estimate the proportion of children helped with medication and psychotherapy to within 6% with 95% confidence, we would need a sample of at least 221 children. All decimals in the final answer must be rounded up, to the next person.
(For a more cautious answer, let $\hat{p} = \hat{q} = 0.5$. This method results in a required sample of 267 children.)

26. Online testing.

a) H_0: The mean difference in the scores between Test A and Test B is zero. $\left(\mu_d = 0\right)$

H_A: The mean difference in the scores between Test A and Test B is not zero. $\left(\mu_d \neq 0\right)$

Paired data assumption: The data are paired by student.
Randomization condition: The volunteers were randomized with respect to the order in which they took the tests and which form they took in each environment.
10% condition: We are testing the difficulty of the tests, not the people, so we don't need to check this condition.
Nearly Normal condition: The distribution of differences is roughly unimodal and symmetric.

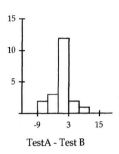

TestA - Test B

Since the conditions are satisfied, the sampling distribution of the difference can be modeled with a Student's t-model with 20 – 1 = 19 degrees of freedom, $t_{19}\left(0, \dfrac{3.52584}{\sqrt{20}}\right)$.

We will use a paired t-test, (Test A – Test B) with $\bar{d} = 0.3$.

Since the *P*-value = 0.7078 is very high, we fail to reject the null hypothesis. There is no evidence of that the mean difference in score is different from zero. It is reasonable to think that Test A and Test B are equivalent in terms of difficulty.

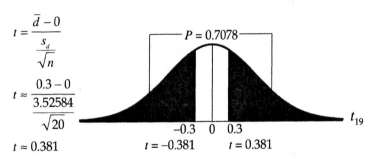

$$t = \frac{\bar{d} - 0}{\frac{s_d}{\sqrt{n}}}$$

$$t \approx \frac{0.3 - 0}{\frac{3.52584}{\sqrt{20}}}$$

$$t \approx 0.381$$

$P = 0.7078$

−0.3 0 0.3
$t = -0.381$ $t = 0.381$

t_{19}

b) H_0: The mean difference between paper scores and online scores is zero. $\left(\mu_d = 0\right)$

H_A: The mean difference between paper scores and online scores is not zero. $\left(\mu_d \neq 0\right)$

Paired data assumption: The data are paired by student.
Randomization condition: The volunteers were randomized with respect to the order in which they took the tests and which form they took in each environment.
10% condition: We are testing the formats of the test (paper/online), not the people, so we don't need to check this condition.

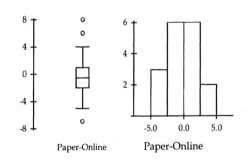

Paper-Online   Paper-Online

Nearly Normal condition: The boxplot of the distribution of differences is shows three outliers: students 3, 10, and 17. With these outliers removed, the histogram is roughly unimodal and symmetric.

Since the conditions are satisfied, the sampling distribution of the difference can be modeled with a Student's t-model with 17 – 1 = 16 degrees of freedom, $t_{16}\left(0, \dfrac{2.26222}{\sqrt{17}}\right)$.

We will use a paired t-test, (Paper – Online) with $\bar{d} = -0.647059$.

Since the *P*-value = 0.2555 is high, we fail to reject the null hypothesis. There is no evidence that the mean difference in score is different from zero. It is reasonable to think that paper and online tests are equivalent in terms of difficulty.

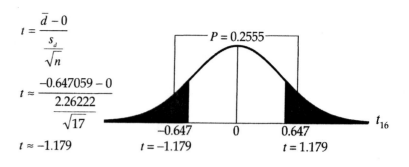

27. Bread.

a) Since the histogram shows that the distribution of the number of loaves sold per day is skewed strongly to the right, we can't use the Normal model to estimate the number of loaves sold on the busiest 10% of days.

b) **Randomization condition:** Assume that these days are representative of all days.
   10% condition: 100 days are less than 10% of all days.
   Nearly Normal condition: The histogram is skewed strongly to the right. However, since the sample size is large, the Central Limit Theorem guarantees that the distribution of averages will be approximately Normal.

   The days in the sample had a mean of 103 loaves sold and a standard deviation of 9 loaves sold. Since the conditions are satisfied, the sampling distribution of the mean can be modeled by a Student's *t*- model, with 103 – 1 = 103 degrees of freedom. We will use a one-sample *t*-interval with 95% confidence for the mean number of loaves sold. (By hand, use $t^*_{50} \approx 2.403$ from the table.)

c) $$\bar{y} \pm t^*_{n-1}\left(\frac{s}{\sqrt{n}}\right) = 103 \pm t^*_{102}\left(\frac{9}{\sqrt{103}}\right) \approx (101.2,\ 104.8)$$

   We are 95% confident that the mean number of loaves sold per day at the Clarksburg Bakery is between 101.2 and 104.8.

d) We know that in order to cut the margin of error in half, we need to a sample four times as large. If we allow a margin of error that is twice as wide, that would require a sample only one-fourth the size. In this case, our original sample is 100 loaves; so 25 loaves would be a sufficient number to estimate the mean with a margin of error twice as wide.

e) Since the interval is completely above 100 loaves, there is strong evidence that the estimate was incorrect. The evidence suggests that the mean number of loaves sold per day is greater than 100. This difference is statistically significant, but may not be practically significant. It seems like the owners made a pretty good estimate!

28. Irises.

a) Parallel boxplots of the distributions of petal lengths for the two species of flower are at the right. No units are specified, but millimeters seems like a reasonable guess.

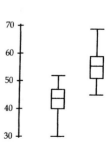

b) The petals of *versicolor* are generally longer than the petals of *virginica*. Both distributions have about the same range, and both distributions are fairly symmetric.

c) **Independent groups assumption:** The two species of flowers are independent.
Randomization condition: It is reasonable to assume that these flowers are representative of their species.
10% condition: Two samples of 50 flowers each are less than 10% of all flowers.
Nearly Normal condition: The boxplots show distributions of petal lengths that are reasonably symmetric with no outliers. Additionally, the samples are large.

Since the conditions are satisfied, it is appropriate to model the sampling distribution of the difference in means with a Student's *t*-model, with 97.92 degrees of freedom (from the approximation formula). We will construct a two-sample *t*-interval, with 95% confidence.

$$(\bar{y}_{Ver} - \bar{y}_{Vir}) \pm t^*_{df}\sqrt{\frac{s^2_{Ver}}{n_{Ver}} + \frac{s^2_{Vir}}{n_{Vir}}} = (55.52 - 43.22) \pm t^*_{97.92}\sqrt{\frac{5.51895^2}{50} + \frac{5.36158^2}{50}} \approx (10.14, 14.46)$$

d) We are 95% confident the mean petal length of *versicolor* irises is between 10.14 and 14.46 millimeters longer than the mean petal length of *virginica* irises.

e) Since the interval is completely above 0, there is strong evidence that the mean petal length of *versicolor* irises is greater than the mean petal length of *virginica* irises.

29. Insulin and diet.

a) H_0: People with high dairy consumption have IRS at the same rate as those with low dairy consumption. $\left(p_{High} = p_{Low} \text{ or } p_{High} - p_{Low} = 0\right)$

H_A: People with high dairy consumption have IRS at a different rate than those with low dairy consumption. $\left(p_{High} \neq p_{Low} \text{ or } p_{High} - p_{Low} \neq 0\right)$

Random condition: Assume that the people studied are representative of all people.
10% condition: 102 and 190 are both less than 10% of all people.
Independent samples condition: The two groups are not related.
Success/Failure condition: $n\hat{p}$(high) = 24, $n\hat{q}$(high) = 78, $n\hat{p}$(low) = 85, and $n\hat{q}$(low) = 105 are all greater than 10, so the samples are both large enough.

Since the conditions have been satisfied, we will model the sampling distribution of the difference in proportion with a Normal model with mean 0 and standard deviation estimated by

$$SE_{pooled}\left(\hat{p}_{High} - \hat{p}_{Low}\right) = \sqrt{\frac{\hat{p}_{pooled}\hat{q}_{pooled}}{n_{High}} + \frac{\hat{p}_{pooled}\hat{q}_{pooled}}{n_{Low}}} = \sqrt{\frac{(0.373)(0.627)}{102} + \frac{(0.373)(0.627)}{190}} \approx 0.05936.$$

The observed difference between the proportions is 0.2352 − 0.4474 = − 0.2122.

Since the *P*-value = 0.0004 is very low, we reject the null hypothesis. There is strong evidence that the proportion of people with IRS is different for those who with high dairy consumption compared to those with low dairy consumption. People who

$$z = \frac{-0.2122 - 0}{0.05936}$$

$$z \approx -3.57$$

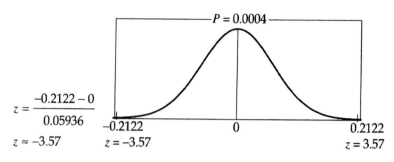

consume dairy products more than 35 times per week appear less likely to have IRS than those who consume dairy products fewer than 10 times per week.

b) There is evidence of an association between the low consumption of dairy products and IRS, but that does not prove that dairy consumption influences the development of IRS. This is an observational study, and a controlled experiment is required to prove cause and effect.

30. Speeding.

a) H_0 : The percentage of speeding tickets issued to black drivers is 16%, the same as the percentage of registered drivers who are black. ($p = 0.16$)

H_A : The percentage of speeding tickets issued to black drivers is greater than 16%, the percentage of registered drivers who are black. ($p > 0.16$)

Random condition: Assume that this month is representative of all months with respect to the percentage of tickets issued to black drivers.
10% condition: 324 speeding tickets are less than 10% of all tickets.
Success/Failure condition: $np = (324)(0.16) = 52$ and $nq = (324)(0.84) = 272$ are both greater than 10, so the sample is large enough.

The conditions have been satisfied, so a Normal model can be used to model the sampling distribution of the proportion, with $\mu_{\hat{p}} = p = 0.16$ and

$$\sigma(\hat{p}) = \sqrt{\frac{pq}{n}} = \sqrt{\frac{(0.16)(0.84)}{324}} \approx 0.02037.$$

We can perform a one-proportion *z*-test.
The observed proportion of tickets issued to black drivers is $\hat{p} = 0.25$.

$$z = \frac{\hat{p} - p_0}{\sigma(\hat{p})}$$

Since the *P*-value = 4.96×10^{-6} is very low, we reject the null hypothesis. There is strong evidence that the percentage of speeding tickets issued to black drivers is greater than 16%.

$$z \approx \frac{0.25 - 0.16}{0.02037}$$

$$z \approx 4.42$$

b) There is strong evidence of an association between the receipt of a speeding ticket and race. Black drivers appear to be issued tickets at a higher rate than expected. However, this does not prove that racial profiling exists. There may be other factors present.

c) Answers may vary. The primary statistic of interest is the percentage of black motorists on this section of the New Jersey Turnpike. For example, if 80% of drivers on this section are black, then 25% of the speeding tickets being issued to black motorists is not an usually high percentage. In fact, it is probably unusually low. On the other hand, if only 3% of the motorists on this section of the turnpike are black, then there is even more evidence that racial profiling may be occurring.

31. Rainmakers?

Independent groups assumption: The two groups of clouds are independent.
Randomization condition: Researchers randomly assigned clouds to be seeded with silver iodide or not seeded.
10% condition: We are testing cloud seeding, not clouds themselves, so this condition doesn't need to be checked.
Nearly Normal condition: We don't have the actual data, so we can't look at the distributions, but the means of group are significantly higher than the medians. This is an indication that the distributions are skewed to the right, with possible outliers. The samples sizes of 26 each are fairly large, so it should be safe to proceed, but we should be careful making conclusions, since there may be outliers.

Since the conditions are satisfied, it is appropriate to model the sampling distribution of the difference in means with a Student's *t*-model, with 33.86 degrees of freedom (from the approximation formula). We will construct a two-sample *t*-interval, with 95% confidence.

$$(\bar{y}_S - \bar{y}_U) \pm t^*_{df}\sqrt{\frac{s_S^2}{n_S} + \frac{s_U^2}{n_U}} = (441.985 - 164.588) \pm t^*_{33.86}\sqrt{\frac{650.787^2}{26} + \frac{278.426^2}{26}} \approx (-4.76, 559.56)$$

We are 95% confident the mean amount of rainfall produced by seeded clouds is between 4.76 acre-feet less than and 559.56 acre-feet more than the mean amount of rainfall produced by unseeded clouds.

Since the interval contains 0, there is little evidence that the mean rainfall produced by seeded clouds is any different from the mean rainfall produced by unseeded clouds. However, we shouldn't place too much faith in this conclusion. It is based on a procedure that is sensitive to outliers, and there may have been outliers present.

32 Fritos.

a) H_0: The mean weight of bags of Fritos is 35.4 grams. $(\mu = 35.4)$

H_A: The mean weight of bags of Fritos is less than 35.4 grams. $(\mu < 35.4)$

b) **Randomization condition:** It is reasonable to think that the 6 bags are representative of all bags of Fritos.
10% condition: 6 bags are less than 10% of all bags.
Nearly Normal condition: The histogram of bags weights shows one unusually heavy bag. Although not technically an outlier, it probably should be excluded for the purposes of the test. (We will leave it in for the preliminary test, then remove it and test again.)

c) The bags in the sample had a mean weight of 35.5333 grams and a standard deviation in weight of 0.450185 grams. Since the conditions for inference are satisfied, we can model the sampling distribution of the mean weight of bags of Fritos with a Student's t model, with $6 - 1 = 5$ degrees of freedom, $t_5\left(35.4, \dfrac{0.450185}{\sqrt{6}}\right)$.

We will perform a one-sample t-test.

Since the P-value = 0.7497 is high, we fail to reject the null hypothesis. There is no evidence to suggest that the mean weight of bags of Fritos is less than 35.4 grams.

$$t = \dfrac{\bar{y} - \mu_0}{SE(\bar{y})}$$

$$t = \dfrac{35.5333 - 35.4}{\dfrac{0.450185}{\sqrt{6}}}$$

$$t = 0.726$$

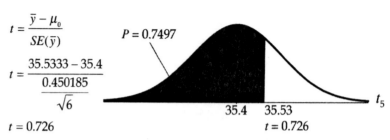

$P = 0.7497$

35.4   35.53

$t = 0.726$

t_5

d) With the one unusually high value removed, the mean weight of the 5 remaining bags is 35.36 grams, with a standard deviation in weight of 0.167332 grams. Since the conditions for inference are satisfied, we can model the sampling distribution of the mean weight of bags of Fritos with a Student's t model, with $5 - 1 = 4$ degrees of freedom, $t_4\left(35.4, \dfrac{0.167332}{\sqrt{5}}\right)$.

We will perform a one-sample t-test.

Since the P-value = 0.3107 is high, we fail to reject the null hypothesis. There is no evidence to suggest that the mean weight of bags of Fritos is less than 35.4 grams.

$$t = \dfrac{\bar{y} - \mu_0}{SE(\bar{y})}$$

$$t = \dfrac{35.36 - 35.4}{\dfrac{0.167332}{\sqrt{5}}}$$

$$t = -0.53$$

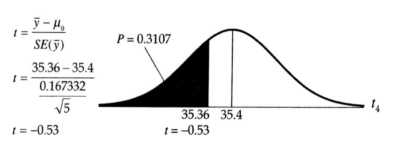

$P = 0.3107$

35.36   35.4

$t = -0.53$

t_4

e) Neither test provides evidence that the mean weight of bags of Fritos is less than 35.4 grams. It is reasonable to believe that the mean weight of the bags is the same as the stated weight. However, the sample sizes are very small, and the tests have very little power to detect lower mean weights. It would be a good idea to weigh more bags.

33. Color or text?

a) By randomizing the order of the cards (shuffling), the researchers are attempting to avoid bias that may result from volunteers remembering the order of the cards from the first task. Although mentioned, hopefully the researchers are randomizing the order in which the tasks are performed. For example, if each volunteer performs the color task first, they may all do better on the text task, simply because they have had some practice in memorizing cards. By randomizing the order, that bias is controlled.

b) H_0: The mean difference between color and word scores is zero. $(\mu_d = 0)$

H_A: The mean difference between color and word scores is not zero. $(\mu_d \neq 0)$

c) **Paired data assumption:** The data are paired by volunteer.
Randomization condition: Hopefully, the volunteers were randomized with respect to the order in which they performed the tasks, and the cards were shuffled between tasks.
10% condition: We are testing the format of the task (color/word), not the people, so we don't need to check this condition.
Nearly Normal condition: The histogram of the differences is roughly unimodal and symmetric.

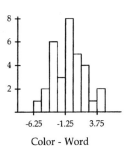

Color - Word

d) Since the conditions are satisfied, the sampling distribution of the difference can be modeled with a Student's t-model with $32 - 1 = 31$ degrees of freedom, $t_{31}\left(0, \dfrac{2.50161}{\sqrt{32}}\right)$.

We will use a paired t-test, (color – word) with $\bar{d} = -0.75$.

e) Since the P-value $= 0.0999$ is high, we fail to reject the null hypothesis. There is no evidence that the mean difference in score is different from zero. It is reasonable to think that neither color nor word dominates perception. However, if the order in which volunteers took

$$t = \frac{\bar{d} - 0}{\frac{s_d}{\sqrt{n}}}$$

$$t \approx \frac{-0.75 - 0}{\frac{2.50161}{\sqrt{32}}}$$

$$t \approx -1.70$$

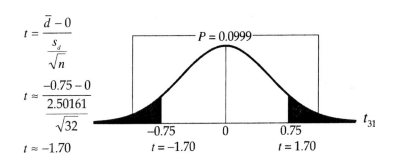

the test was not randomized, making conclusions from this test may be risky.

34. And it means?

a) The margin of error is $\dfrac{(2391 - 1644)}{2} = \$373.50$.

b) The insurance agent is 95% confident that the mean loss claimed by clients after home burglaries is between $1644 and $2391.

c) 95% of all random samples of this size will produce intervals that contain the true mean loss claimed.

35. Batteries.

a) Different samples have different means. Since this is a fairly small sample, the difference may be due to natural sampling variation. Also, we have no idea how to quantify "a lot less" with out considering the variation as measured by the standard deviation.

b) H_0: The mean life of a battery is 100 hours. $(\mu = 100)$

H_A: The mean life of a battery is less than 100 hours. $(\mu < 100)$

c) **Randomization condition:** It is reasonable to think that these 16 batteries are representative of all batteries of this type
10% condition: 16 batteries are less than 10% of all batteries.
Normal population assumption: Since we don't have the actual data, we can't check a graphical display, and the sample is not large. Assume that the population of battery lifetimes is Normal.

d) The batteries in the sample had a mean life of 97 hours and a standard deviation of 12 hours. Since the conditions for inference are satisfied, we can model the sampling distribution of the mean battery life with a Student's *t* model, with 16 – 1 = 15 degrees of freedom, $t_{15}\left(100, \dfrac{12}{\sqrt{16}}\right)$, or $t_{15}(100,3)$

We will perform a one-sample *t*-test.

Since the *P*-value = 0.1666 is greater than $\alpha = 0.05$, we fail to reject the null hypothesis. There is no evidence to suggest that the mean battery life is less than 100 hours.

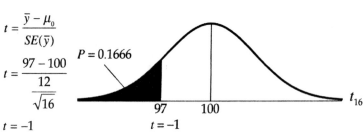

$$t = \frac{\bar{y} - \mu_0}{SE(\bar{y})}$$

$$t = \frac{97 - 100}{\dfrac{12}{\sqrt{16}}}$$

$$t = -1$$

P = 0.1666

$t = -1$

e) If the mean life of the company's batteries is only 98 hours, then the mean life is less than 100, and the null hypothesis is false. We failed to reject a false null hypothesis, making a Type II error.

36. Hamsters.

a) **Randomization condition:** Assume that these litters are representative of all litters.
10% condition: 47 litters are less than 10% of all litters.
Nearly Normal condition: We don't have the actual data, so we can't look at a graphical display. However, since the sample size is large, the Central Limit Theorem guarantees that the distribution of averages will be approximately Normal, as long as there are no outliers.

The litters in the sample had a mean size of 7.72 baby hamsters and a standard deviation of 2.5 baby hamsters. Since the conditions are satisfied, the sampling distribution of the mean can be modeled by a Student's *t*- model, with 47 – 1 = 46 degrees of freedom. We will use a one-sample *t*-interval with 90% confidence for the mean number of baby hamsters per litter.

$$\bar{y} \pm t^*_{n-1}\left(\frac{s}{\sqrt{n}}\right) = 7.72 \pm t^*_{46}\left(\frac{2.5}{\sqrt{47}}\right) \approx (7.11, 8.33)$$

We are 90% confident that the mean number of baby hamsters per litter is between 7.11 and 8.33.

b) A 98% confidence interval would have a larger margin of error. Higher levels of confidence come at the price of less precision in the estimate.

c) A quick estimate using *z* gives us a sample size of about 25 litters. Using this estimate, $t^*_{24} = 2.064$ at 95% confidence. We need a sample of about 27 litters in order to estimate the number of baby hamsters per litter to within 1 baby hamster.

$$ME = t^*_{24}\left(\frac{s}{\sqrt{n}}\right)$$

$$1 = 2.064\left(\frac{2.5}{\sqrt{n}}\right)$$

$$n = \frac{(2.064)^2(2.5)^2}{(1)^2}$$

$$n \approx 27$$

Chapter 26 – Comparing Counts

1. **Dice.**

   a) If the die were fair, you'd expect each face to show 10 times.

   b) Use a chi-square test for goodness-of-fit. We are comparing the distribution of a single variable (outcome of a die roll) to an expected distribution.

   c) H_0: The die is fair. (All faces have the same probability of coming up.)

   H_A: The die is not fair. (Some faces are more or less likely to come up than others.)

   d) **Counted data condition:** We are counting the number of times each face comes up.
   Randomization condition: Die rolls are random and independent of each other.
   Expected cell frequency condition: We expect each face to come up 10 times, and 10 is greater than 5.

   e) Under these conditions, the sampling distribution of the test statistic is χ^2 on $6 - 1 = 5$ degrees of freedom. We will use a chi-square goodness-of-fit test.

Face	Observed	Expected	Residual = $(Obs - Exp)$	$(Obs - Exp)^2$	Component = $\dfrac{(Obs - Exp)^2}{Exp}$
1	11	10	1	1	0.1
2	7	10	– 3	9	0.9
3	9	10	– 1	1	0.1
4	15	10	5	25	2.5
5	12	10	2	4	0.4
6	6	10	– 4	16	1.6

$$\sum = 5.6$$

Since the *P*-value = 0.3471 is high, we fail to reject the null hypothesis. There is no evidence that the die is unfair.

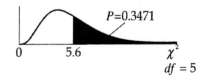

2. **M&M's**

   a) There are $29 + 23 + 12 + 14 + 8 + 20 = 106$ M&M's in the bag. The expected number of M&M's of each color is: $106(0.20) = 21.2$ red, $106(0.20) = 21.2$ yellow, $106(0.10) = 10.6$ orange, $106(0.10) = 10.6$ blue, $106(0.10) = 10.6$ green, and $106(0.30) = 31.8$ brown.

   b) Use a chi-square test for goodness-of-fit. We are comparing the distribution of a single variable (color) to an expected distribution.

   c) H_0: The distribution of colors of M&M's is as specified by the company.

   H_A: The distribution of colors of M&M's is different than specified by the company.

d) Counted data condition: The author counted the M&M's in the bag.
Randomization condition: These M&M's are mixed thoroughly at the factory.
Expected cell frequency condition: The expected counts (calculated in part b) are all greater than 5.

e) Since there are 6 different colors, there are 6 – 1 = 5 degrees of freedom.

f) Under these conditions, the sampling distribution of the test statistic is χ^2 on 6 – 1 = 5 degrees of freedom. We will use a chi-square goodness-of-fit test.

Color	Observed	Expected	Residual = $(Obs - Exp)$	$(Obs - Exp)^2$	Component = $\dfrac{(Obs - Exp)^2}{Exp}$
yellow	29	21.2	7.8	60.84	2.8698
red	23	21.2	1.8	3.24	0.1528
orange	12	10.6	1.4	1.96	0.1849
blue	14	10.6	3.4	11.56	1.0906
green	8	10.6	– 2.6	6.76	0.6377
brown	20	31.8	– 11.8	139.24	4.3786

$$\sum \approx 9.314$$

$\chi^2 = 9.314$. Since the *P*-value = 0.0972 is high, we fail to reject the null hypothesis.

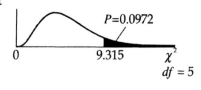

g) There is no evidence that the distribution of colors is anything other than the distribution specified by the company.

3. Nuts.

a) The weights of the nuts are quantitative. Chi-square goodness-of-fit requires counts.

b) In order to use a chi-square test, you could count the number of each type of nut. However, it's not clear whether the company's claim was a percentage by number or a percentage by weight.

4. Mileage.

The average number of miles traveled is quantitative date, not categorical. Chi-square is for comparing counts.

5. NYPD and race.

H_0: The distribution of ethnicities in the police department represents the distribution of ethnicities of the youth of New York City.

H_A: The distribution of ethnicities in the police department does not represent the distribution of ethnicities of the youth of New York City.

Counted data condition: The percentages reported must be converted to counts.
Randomization condition: Assume that the current NYPD is representative of recent departments with respect to ethnicity.
Expected cell frequency condition: The expected counts are all much greater than 5.

Ethnicity	Observed	Expected
White	16965	7644.852
Black	3796	7383.042
Latino	5001	8247.015
Asian	367	2382.471
Other	52	523.620

(Note: The observed counts should be whole numbers. They are actual policemen. The expected counts may be decimals, since they behave like averages.)

Under these conditions, the sampling distribution of the test statistic is χ^2 on $5 - 1 = 4$ degrees of freedom. We will use a chi-square goodness-of-fit test.

$$\chi^2 = \sum_{all\,cells} \frac{(Obs - Exp)^2}{Exp} = \frac{(16965 - 7644.852)^2}{7644.852} + \frac{(3796 - 7383.042)^2}{7383.042} + \frac{(5001 - 8247.015)^2}{8247.015}$$

$$+ \frac{(367 - 2382.471)^2}{2382.471} + \frac{(52 - 523.620)^2}{523.620} \approx 16,500$$

With χ^2 of over 16,500, on 4 degrees of freedom, the *P*-value is essentially 0, so we reject the null hypothesis. There is strong evidence that the distribution of ethnicities of NYPD officers does not represent the distribution of ethnicities of the youth of New York City. Specifically, the proportion of white officers is much higher than the proportion of white youth in the community. As one might expect, there are also lower proportions of officers who are black, Latino, Asian, and other ethnicities than we see in the youth in the community.

6. Violence against women.

H_0: The weapon use rates in murders of women have the same distribution as the weapon use rates of all murders.

H_A: The weapon use rates in murders of women have a different distribution than the weapon use rates of all murders.

Counted data condition: The percentages reported must be converted to counts.
Randomization condition: Assume that the weapon use rates from 1996 are representative of the weapon use rates for all recent years.
Expected cell frequency condition: The expected counts are all much greater than 5.

weapon	Observed	Expected
guns	1139	1276.242
knives	372	263.703
other	158	338.184
personal	344	134.871

(Note: The observed counts should be whole numbers. They are actual murders. The expected counts may be decimals, since they behave like averages.)

Under these conditions, the sampling distribution of the test statistic is χ^2 on $4 - 1 = 3$ degrees of freedom. We will use a chi-square goodness-of-fit test.

$$\chi^2 = \sum_{all\ cells} \frac{(Obs - Exp)^2}{Exp} = \frac{(1139 - 1276.242)^2}{1276.242} + \frac{(372 - 263.703)^2}{263.703} + \frac{(158 - 338.184)^2}{338.184} + \frac{(344 - 134.871)^2}{134.871} \approx 479.508$$

With $\chi^2 \approx 479.508$, on 3 degrees of freedom, the *P*-value is essentially 0, so we reject the null hypothesis. There is strong evidence that the distribution of weapon use rates is different for murders of women than for all murders. Women are much more likely to be killed by personal attacks and less likely to be killed by other weapons.

7. **Fruit flies.**

   a) H_0: The ratio of traits in this type of fruit fly is 9:3:3:1, as genetic theory predicts.

   H_A: The ratio of traits in this type of fruit fly is not 9:3:3:1.

   Counted data condition: The data are counts.
   Randomization condition: Assume that these flies are representative of all fruit flies of this type.
   Expected cell frequency condition: The expected counts are all greater than 5.

trait	Observed	Expected
YN	59	56.25
YS	20	18.75
EN	11	18.75
ES	10	6.25

   Under these conditions, the sampling distribution of the test statistic is χ^2 on $4 - 1 = 3$ degrees of freedom. We will use a chi-square goodness-of-fit test.

   $$\chi^2 = \sum_{all\ cells} \frac{(Obs - Exp)^2}{Exp} = \frac{(59 - 56.25)^2}{56.25} + \frac{(20 - 18.75)^2}{18.75} + \frac{(11 - 18.75)^2}{18.75} + \frac{(10 - 6.25)^2}{6.25} \approx 5.671$$

   With $\chi^2 \approx 5.671$, on 3 degrees of freedom, the *P*-value = 0.1288 is high, so we fail to reject the null hypothesis. There is no evidence that the ratio of traits is different than the theoretical ratio predicted by the genetic model. The observed results are consistent with the genetic model.

   b) With $\chi^2 \approx 11.342$, on 3 degrees of freedom, the *P*-value = 0.0100 is low, so we reject the null hypothesis. There is strong evidence that the ratio of traits is different than the theoretical ratio predicted by the genetic model. Specifically, there is evidence that the normal wing length occurs less frequently than expected and the short wing length occurs more frequently than expected.

trait	Observed	Expected
YN	118	112.5
YS	40	37.5
EN	22	37.5
ES	20	12.5

   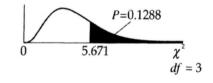

   c) At first, this seems like a contradiction. We have two samples with exactly the same ratio of traits. The smaller of the two provides no evidence of a difference, yet the larger one provides strong evidence of a difference. This is explained by the sample size. In general, large samples decrease the proportion of variation from the true ratio. Because of the relatively small sample in the first test, we are unwilling to say that there is a difference. There just isn't enough evidence. But the larger sample allows to be more certain about the difference.

8. Pi.

H_0: Digits of π are uniformly distributed (all occur with frequency 1/10).

H_A: Digits of π are not uniformly distributed.

Counted data condition: The data are counts.
Randomization condition: Assume that the first million digits of π are representative of all digits.
Expected cell frequency condition: The expected counts are all greater than 5.

Digit	Observed	Expected
0	99959	100000
1	99758	100000
2	100026	100000
3	100229	100000
4	100230	100000
5	100359	100000
6	99548	100000
7	99800	100000
8	99985	100000
9	100106	100000

Under these conditions, the sampling distribution of the test statistic is χ^2 on $10 - 1 = 9$ degrees of freedom. We will use a chi-square goodness-of-fit test.

With $\chi^2 \approx 5.509$, on 9 degrees of freedom, the P-value = 0.7879 is high, so we fail to reject the null hypothesis. There is no evidence that the digits of π are not uniformly distributed. These data are consistent with the null hypothesis.

9. Titanic.

a) $P(\text{crew}) = \dfrac{885}{2201} \approx 0.402$

b) $P(\text{third and alive}) = \dfrac{178}{2201} \approx 0.081$

c) $P(\text{alive} \mid \text{first}) = \dfrac{P(\text{alive and first})}{P(\text{first})} = \dfrac{^{202}\!/_{2201}}{^{325}\!/_{2201}} = \dfrac{202}{325} \approx 0.622$

d) The overall chance of survival is $\dfrac{710}{2201} \approx 0.323$, so we would expect about 32.3% of the crew, or about 285.48 members of the crew, to survive.

e) H_0: Survival was independent of status on the ship.

H_A: Survival depended on status on the ship.

f) The table has 2 rows and 4 columns, so there are $(2-1) \times (4-1) = 3$ degrees of freedom.

g) With $\chi^2 \approx 187.8$, on 3 degrees of freedom, the P-value is essentially 0, so we reject the null hypothesis. There is strong evidence survival depended on status. First-class passengers were more likely to survive than any other class or crew.

10. NYPD and gender.

a) $P(\text{female}) = \dfrac{5613}{37,379} \approx 0.150$

b) $P(\text{detective}) = \dfrac{4864}{37,379} \approx 0.130$

c) The overall percentage of females is 15%, so we would expect about 15% of the detectives, or about 729.6 detectives, to be female.

d) We have one group, categorized according to two variables, rank and gender, so we will perform a chi-square test for independence.

e) H_0: Rank is independent of gender in the NYPD.

H_A: Rank is associated with gender in the NYPD.

f) **Counted data condition:** The data are counts. **Randomization condition:** These data are not a random sample, but all NYPD officers. Assume that these officers are representative with respect to the recent distribution of gender and rank in the NYPD. **Expected cell frequency condition:** The expected counts are all greater than 5.

Expected counts

Rank	Male	Female
Officer	22249.5	3931.5
Detective	4133.6	730.4
Sergeant	3665.3	647.7
Lieutenant	1208.5	213.5
Captain	315.3	55.7
Higher ranks	193.8	34.2

g) The table has 6 rows and 2 columns, so there are $(6-1) \times (2-1) = 5$ degrees of freedom.

h) $\chi^2 = \sum_{all\,cells} \frac{(Obs - Exp)^2}{Exp} \approx 290.131$, and the *P*-value is essentially 0.

i) Since the *P*-value is so low, we reject the null hypothesis. There is strong evidence of an association between gender and rank in the NYPD.

j) A table of the standardized residuals is at the right, calculated by using $c = \frac{Obs - Exp}{\sqrt{Exp}}$.

There is evidence that women are over-represented in the lower ranks and under-represented in every rank from sergeant on up.

Rank	Male	Female
Officer	-2.3434	5.5747
Detective	-1.1759	2.7973
Sergeant	3.8429	-9.1421
Lieutenant	3.5824	-8.5223
Captain	2.4617	-5.8563
Higher ranks	1.7412	-4.1423

11. Cranberry juice.

a) This is an experiment. Volunteers were assigned to treatment groups, each of which drank a different beverage.

b) We are concerned with the proportion of urinary tract infections among three different groups. We will use a chi-square test for homogeneity.

c) H_0: The proportion of urinary tract infection is the same for each group.

H_A: The proportion of urinary tract infection is different among the groups.

d) **Counted data condition:** The data are counts. **Randomization condition:** Although not specifically stated, we will assume that the women were randomly assigned to treatments. **Expected cell frequency condition:** The expected counts are all greater than 5.

	Cranberry (Obs/Exp)	Lactobacillus (Obs/Exp)	Control (Obs/Exp)
Infection	8 / 15.333	20 / 15.333	18 / 15.333
No infection	42 / 34.667	30 / 34.667	32 / 34.667

e) The table has 2 rows and 3 columns, so there are $(2-1) \times (3-1) = 2$ degrees of freedom.

f) $\chi^2 = \sum\limits_{all\ cells} \dfrac{(Obs - Exp)^2}{Exp} \approx 7.776$

P-value ≈ 0.020.

g) Since the P-value is low, we reject the null hypothesis. There is strong evidence of difference in the proportion of urinary tract infections for cranberry juice drinkers, lactobacillus drinkers, and women that drink neither of the two beverages.

h) A table of the standardized residuals is below, calculated by using $c = \dfrac{Obs - Exp}{\sqrt{Exp}}$.

	Cranberry	Lactobacillus	Control
Infection	–1.87276	1.191759	0.681005
No infection	1.245505	–0.79259	–0.45291

There is evidence that women who drink cranberry juice are less likely to develop urinary tract infections, and women who drank lactobacillus are more likely to develop urinary tract infections.

12. Cars.

a) We have two groups, staff and students (selected from different lots), and we are concerned with the distribution of one variable, origin of car. We will perform a chi-square test for homogeneity.

b) H₀: The distribution of car origin is the same for students and staff.

Hₐ: The distribution of car origin is different for students and staff.

c) **Counted data condition:** The data are counts.
Randomization condition: Cars were surveyed randomly.
Expected cell frequency condition: The expected counts are all greater than 5.

	Driver	
Origin	**Student** (Obs / Exp)	**Staff** (Obs / Exp)
American	107 / 115.15	105 / 96.847
European	33 / 24.443	12 / 20.557
Asian	55 / 55.404	47 / 46.596

d) Under these conditions, the sampling distribution of the test statistic is χ^2 on 2 degrees of freedom. We will use a chi-square test for homogeneity.

With $\chi^2 = \sum\limits_{all\ cells} \dfrac{(Obs - Exp)^2}{Exp} \approx 7.828$, on 2 degrees of freedom, the P-value ≈ 0.020.

e) Since *P*-value = 0.020 is low, we reject the null hypothesis. There is strong evidence that the distribution of car origins at this university differs between students and staff. Students are more likely to drive European cars than staff and less likely than staff to drive American cars.

13. Montana.

a) We have one group, categorized according to two variables, political party and gender, so we will perform a chi-square test for independence.

b) H_0: Political party is independent of gender in Montana.

H_A: There is an association between political party and gender in Montana.

c) **Counted data condition:** The data are counts.
Randomization condition: Although not specifically stated, we will assume that the poll was conducted randomly.
Expected cell frequency condition: The expected counts are all greater than 5.

	Democrat (Obs/Exp)	Republican (Obs/Exp)	Independent (Obs/Exp)
Male	36 / 43.663	45 / 40.545	24 / 20.792
Female	48 / 40.337	33 / 37.455	16 / 19.208

d) Under these conditions, the sampling distribution of the test statistic is χ^2 on 2 degrees of freedom. We will use a chi-square test for independence.

$$\chi^2 = \sum_{all\ cells} \frac{(Obs - Exp)^2}{Exp} \approx 4.851$$

The *P*-value ≈ 0.0884

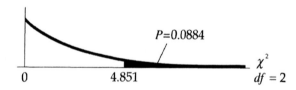

e) Since the *P*-value ≈ 0.0884 is fairly high, we fail to reject the null hypothesis. There is little evidence of an association between gender and political party in Montana.

14. Fish diet.

a) This is an observational prospective study. Swedish men were selected, and then followed for 30 years.

b) We have one group, categorized according to two variables, fish consumption and incidence of prostate cancer, so we will perform a chi-square test for independence.

c) H_0: Prostate cancer and fish consumption are independent.

H_A: There is an association between prostate cancer and fish consumption.

Counted data condition: The data are counts.
Randomization condition: Assume that these men are representative of all men.
Expected cell frequency condition: The expected counts are all greater than 5.

Fish Consumption	Prostate Cancer (Obs / Exp)	No Prostate Cancer (Obs / Exp)
Never/Seldom	14 / 9.21	110 / 114.79
Small part	201 /194.74	2420 / 2426.3
Moderate part	209 / 221.26	2769 / 2756.7
Large part	42 / 40.79	507 / 508.21

Under these conditions, the sampling distribution of the test statistic is χ^2 on 3 degrees of freedom. We will use a chi-square test for independence.

$$\chi^2 = \sum_{all\,cells} \frac{(Obs - Exp)^2}{Exp} \approx 3.677,\text{ and the }P\text{-value} \approx 0.2985.$$

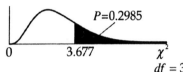

Since the *P*-value ≈ 0.2985 is high, we fail to reject the null hypothesis. There is no evidence of an association between prostate cancer and fish consumption.

d) This does not prove that eating fish does not prevent prostate cancer. There is merely a lack of evidence of a relationship. Furthermore, association (or lack thereof) does not prove a cause-and-effect relationship. We would need to conduct a controlled experiment before anything could be proven.

15. Montana revisited.

H₀: Political party is independent of region in Montana.

H_A: There is an association between political party and region in Montana.

Counted data condition: The data are counts.
Randomization condition: Although not specifically stated, we will assume that the poll was conducted randomly.
Expected cell frequency condition: The expected counts are all greater than 5.

	Democrat (Obs / Exp)	Republican (Obs / Exp)	Independent (Obs / Exp)
West	39 / 28.277	17 / 26.257	12 / 13.465
Northeast	15 / 23.703	30 / 22.01	12 / 11.287
Southeast	30 /32.02	31 / 29.733	16 / 15.248

Under these conditions, the sampling distribution of the test statistic is χ^2 on 4 degrees of freedom. We will use a chi-square test for independence.

$$\chi^2 = \sum_{all\ cells} \frac{(Obs - Exp)^2}{Exp} \approx 13.849,\ \text{and the } P\text{-value} \approx 0.0078$$

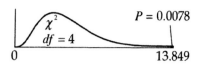

Since the P-value ≈ 0.0078 is low, reject the null hypothesis.
There is strong evidence of an association between region and political party in Montana. Residents in the West are more likely to be Democrats than Republicans, and residents in the Northeast are more likely to be Republicans than Democrats.

16. Working parents.

a) This is a survey of adults. The Gallup Poll simply asked two groups of randomly selected adults a question. Since the same adults were not asked in 1991 and 2001, this is NOT a prospective study of opinion.

b) We have two groups, 1991 and 2001, and we are concerned with the distribution of one variable, attitude about the ideal family. We will perform a chi-square test for homogeneity.

c) H_0: The distribution of attitudes about the ideal family was the same in 1991 and 2001.

H_A: The distribution of attitudes about the ideal family was not the same in 1991 and 2001.

Counted data condition: The data are counts.
Randomization condition: Adults were surveyed randomly.
Expected cell frequency condition: The expected counts are all greater than 5.

	1991 (Obs / Exp)	2001 (Obs / Exp)
Both work full time	142 / 136.5	131/ 136.5
One works full time, other part time	274/ 259	244 / 259
One works, other works at home	152 / 162.5	173/ 162.5
One works, other stays home for kids	396 / 406	416 / 406
No opinion	51 / 51	51 / 51

Under these conditions, the sampling distribution of the test statistic is χ^2 on 4 degrees of freedom. We will use a chi-square test for homogeneity.

With $\chi^2 = \sum_{all\ cells} \frac{(Obs - Exp)^2}{Exp} \approx 4.030$, the P-value ≈ 0.4019.

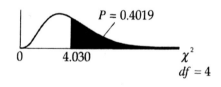

Since P-value = 0.4019 is high, we fail to reject the null hypothesis. There is no evidence of a change in the distribution of attitudes about the ideal family between 1991 and 2001.

17. Grades.

a) We have two groups, students of Professor Alpha and students of Professor Beta, and we are concerned with the distribution of one variable, grade. We will perform a chi-square test for homogeneity.

b) H_0: The distribution of grades is the same for the two professors.

H_A: The distribution of grades is different for the two professors.

The expected counts are organized in the table below:

	Prof. Alpha	Prof. Beta
A	6.667	5.333
B	12.778	10.222
C	12.222	9.778
D	6.111	4.889
F	2.222	1.778

Since three cells have expected counts less than 5, the chi-square procedures are not appropriate. Cells would have to be combined in order to proceed. (We will do this in Exercise 19.)

18. Full moon.

a) We have two groups, weeks of six full moons and six other weeks, and we are concerned with the distribution of one variable, type of offense. We will perform a chi-square test for homogeneity.

b) H_0: The distribution of type of offense is the same for full moon weeks as it is for weeks in which there is not a full moon.

H_A: The distribution of type of offense is different for full moon weeks than it is for weeks in which there is not a full moon.

The expected counts are organized in the table below:

Offense	Full Moon	Not Full
Violent	2.558	2.442
Property	19.442	18.558
Drugs / Alcohol	23.535	22.465
Domestic Abuse	12.791	12.209
Other offenses	7.674	7.326

Since two cells have expected counts less than 5, the chi-square procedures are not appropriate. Cells would have to be combined in order to proceed. (We will do this in Exercise 20.)

19. Grades again.

a) **Counted data condition:** The data are counts. **Randomization condition:** Assume that these students are representative of all students that have ever taken courses from the professors. **Expected cell frequency condition:** The expected counts are all greater than 5.

	Prof. Alpha (Obs / Exp)	Prof. Beta (Obs / Exp)
A	3 / 6.667	9 / 5.333
B	11 / 12.778	12 / 10.222
C	14 / 12.222	8 / 9.778
Below C	12 / 8.333	3 / 6.667

b) Under these conditions, the sampling distribution of the test statistic is χ^2 on 3 degrees of freedom, instead of 4 degrees of freedom before the change in the table. We will use a chi-square test for homogeneity.

c) With $\chi^2 = \sum\limits_{all\,cells} \dfrac{(Obs - Exp)^2}{Exp} \approx 9.306$, the *P*-value ≈ 0.0255.

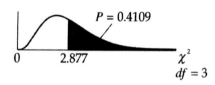

Since *P*-value = 0.0255 is low, we reject the null hypothesis.
There is evidence that the grade distributions for the two professors are different.
Professor Alpha gives fewer As and more grades below C than Professor Beta.

20. Full moon, next phase.

a) **Counted data condition:** The data are counts.
Randomization condition: Assume that these
weeks are representative of all weeks.
Expected cell frequency condition:
It seems reasonable to combine the violent
offenses and domestic abuse, since both
involve some sort of violence. Combining
violent crimes with the "other offenses" is

Offense	Full Moon (Obs / Exp)	Not Full (Obs / Exp)
Violent / Domestic Abuse	13 / 15.349	17 / 14.651
Property	17 / 19.442	21 / 18.558
Drugs / Alcohol	27 / 23.535	19 / 22.465
Other offenses	9 / 7.674	6 / 7.326

okay, but that may put very minor offenses in with violent offenses, which doesn't seem
right. Once the cells are combined, all expected counts are greater than 5.

b) Under these conditions, the sampling distribution of the test statistic is χ^2 on 3 degrees of
freedom, instead of 4 degrees of freedom before the change in the table. We will use a chi-
square test for homogeneity.

c) With $\chi^2 = \sum\limits_{all\,cells} \dfrac{(Obs - Exp)^2}{Exp} \approx 2.877$, the *P*-value ≈ 0.4109.

Since *P*-value = 0.4109 is high, we fail to reject the null hypothesis. There is no evidence
that the distribution of offenses is different during the full moon than during other phases.

21. Racial steering.

H_0: There is no association between race and the section of the apartment complex in which
people live.

H_A: There is an association between race and the section of the apartment complex in
which people live.

Counted data condition: The data are counts.
Randomization condition: Assume that the
recently rented apartments are representative of
all apartments in the complex.
Expected cell frequency condition: The expected
counts are all greater than 5.

	White (Obs / Exp)	Black (Obs / Exp)
Section A	87 / 76.179	8 / 18.821
Section B	83 / 93.821	34 / 23.179

Under these conditions, the sampling distribution of the test statistic is χ^2 on 1 degree of
freedom. We will use a chi-square test for independence.

$$\chi^2 = \sum_{all\,cells} \frac{(Obs - Exp)^2}{Exp} \approx \frac{(87 - 76.179)^2}{76.179} + \frac{(8 - 18.821)^2}{18.821} + \frac{(83 - 93.821)^2}{93.821} + \frac{(34 - 23.179)^2}{23.179}$$

$$\approx 1.5371 + 6.2215 + 1.2481 + 5.0517$$

$$\approx 14.058$$

With $\chi^2 \approx 14.058$, on 1 degree of freedom, the *P*-value ≈ 0.0002.

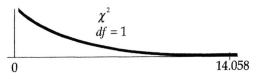

Since the *P*-value ≈ 0.0002 is low, we reject the null hypothesis. There is strong evidence of an association between race and the section of the apartment complex in which people live. An examination of the components shows us that whites are much more likely to rent in Section A (component = 6.2215), and blacks are much more likely to rent in Section B (component = 5.0517).

22. Titanic, redux.

H_0: Survival was independent of gender on the Titanic.

H_A: There is an association between survival and gender on the Titanic.

Counted data condition: The data are counts.
Randomization condition: We have the entire population of the Titanic.
Expected cell frequency condition: The expected counts are all greater than 5.

	Female (Obs / Exp)	Male (Obs / Exp)
Alive	343 / 151.613	367 / 558.387
Dead	127 / 318.387	1364 / 1172.613

Under these conditions, the sampling distribution of the test statistic is χ^2 on 1 degree of freedom. We will use a chi-square test for independence.

$$\chi^2 = \sum_{all\,cells} \frac{(Obs - Exp)^2}{Exp} \approx \frac{(343 - 151.613)^2}{151.613} + \frac{(367 - 558.387)^2}{558.387} + \frac{(127 - 318.387)^2}{318.387} + \frac{(1364 - 1172.613)^2}{1172.613}$$

$$\approx 241.5953 + 65.5978 + 115.0455 + 31.2371$$

$$\approx 453.476$$

With $\chi^2 \approx 453.476$, on 1 degree of freedom, the *P*-value is essentially 0.

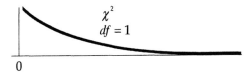

Since the *P*-value is so low, we reject the null hypothesis. There is strong evidence of an association between survival and gender on the Titanic. Females were much more likely to survive than males.

23. Steering revisited.

a) H_0: The proportion of whites who live in Section A is the same as the proportion of blacks who live in Section A. $\left(p_W = p_B \text{ or } p_W - p_B = 0\right)$

H_A: The proportion of whites who live in Section A is different than the proportion of blacks who live in Section A. $\left(p_W \neq p_B \text{ or } p_W - p_B \neq 0\right)$

Independence assumption: Assume that people rent apartments independent of the section.
Independent samples condition: The groups are not associated.
Success/Failure condition: $n\hat{p}$ (white) = 87, $n\hat{q}$ (white) = 83, $n\hat{p}$ (black) = 8, and $n\hat{q}$ (black) = 34. These are not all greater than 10, since the number of black renters in Section A is only 8, but it is close to 10, and the others are large. It should be safe to proceed.

Since the conditions have been satisfied, we will model the sampling distribution of the difference in proportion with a Normal model with mean 0 and standard deviation

estimated by $SE_{pooled}(\hat{p}_W - \hat{p}_B) = \sqrt{\dfrac{\hat{p}_{pooled}\hat{q}_{pooled}}{n_W} + \dfrac{\hat{p}_{pooled}\hat{q}_{pooled}}{n_B}} = \sqrt{\dfrac{\left(\frac{95}{212}\right)\left(\frac{117}{212}\right)}{170} + \dfrac{\left(\frac{95}{212}\right)\left(\frac{117}{212}\right)}{42}} \approx 0.0856915.$

The observed difference between the proportions is:
0.5117647 – 0.1904762 = 0.3212885.

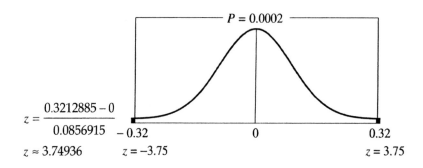

$z = \dfrac{0.3212885 - 0}{0.0856915}$

$z \approx 3.74936$

(You have to use a *ridiculous* number of decimal places to get this to come out "right". This is to done to illustrate the point of the question. DO NOT DO THIS! Use technology.)

Since the *P*-value = 0.0002 is low, we reject the null hypothesis. There is strong evidence of a difference in the proportion of whites and blacks living in Section A. The evidence suggests that the proportion of all whites living in Section A is much higher than the proportion of all black residents living in Section A.

The value of z for this test was approximately 3.74936. $z^2 \approx (3.74936)^2 \approx 14.058$, the same as the value for χ^2 in Exercise 21.

b) The resulting *P*-values were both approximately 0.0002. The two tests are equivalent.

24. Survival and gender, one more time.

a) H_0: The proportion of females who survived is the same as the proportion of males who survived. $\left(p_F = p_M \text{ or } p_F - p_M = 0\right)$

H_A: The proportion of females who survived is different than the proportion of males who survived. $\left(p_F \neq p_M \text{ or } p_F - p_M \neq 0\right)$

Independence assumption: Assume that survival and gender are independent.
Independent samples condition: The groups are not associated.
Success/Failure condition: $n\hat{p}$ (female) = 343, $n\hat{q}$ (female) = 127, $n\hat{p}$ (male) = 367, and $n\hat{q}$ (male) = 1364. All are greater than 10.

Since the conditions have been satisfied, we will model the sampling distribution of the difference in proportion with a Normal model with mean 0 and standard deviation estimated by

$$SE_{pooled}(\hat{p}_F - \hat{p}_M) = \sqrt{\frac{\hat{p}_{pooled}\hat{q}_{pooled}}{n_F} + \frac{\hat{p}_{pooled}\hat{q}_{pooled}}{n_M}} = \sqrt{\frac{\left(\frac{710}{2201}\right)\left(\frac{1491}{2201}\right)}{470} + \frac{\left(\frac{710}{2201}\right)\left(\frac{1491}{2201}\right)}{1731}} \approx 0.0243142295.$$

The observed difference between the proportions is: $0.729787234 - 0.212016176 = 0.517771058.$

$$z = \frac{0.517771058 - 0}{0.0243142295}$$

$$z \approx 21.29498$$

(You have to use a *ridiculous* number of decimal places to get this to come out "right". This is to done to illustrate the point of the question. DO NOT DO THIS! Use technology.)

Since the *P*-value is essentially 0, we reject the null hypothesis. There is strong evidence of a difference between the proportions of women and men who survived on the Titanic. Women survived at a much higher rate.

b) The value of *z* for this test was approximately 21.29498. $z^2 \approx (21.29498)^2 \approx 453.476$, the same as the value for χ^2 in Exercise 22.

c) The resulting *P*-values were both essentially 0. The two tests are equivalent.

25. Race and education.

H_0: Race is independent of education level.

H_A: There is an association between race and education level.

Counted data condition: The data are counts.
Randomization condition: Assume that the sample was taken randomly.
Expected cell frequency condition: The expected counts are all greater than 5.

	Not HS Grad (Obs/Exp)	HS Diploma (Obs/Exp)	College Grad (Obs/Exp)	Adv. Degree (Obs/Exp)
White	810 / 1469	6429 / 6523.8	4725 / 4055.7	1127 / 1042.5
Black	263 / 283.57	1598 / 1259.3	549 / 782.88	117 / 201.24
Hispanic	1031 / 315.43	1269 / 1400.8	412 / 870.86	99 / 223.86
Other	66 / 102	341 / 453	305 / 281.61	197 / 72.389

Under these conditions, the sampling distribution of the test statistic is χ^2 on 9 degrees of freedom. We will use a chi-square test for independence.

With $\chi^2 = \sum_{all\,cells} \frac{(Obs - Exp)^2}{Exp} \approx 2815.968$, on 9 degrees of freedom, the *P*-value is essentially 0.

Since the *P*-value is so low, we reject the null hypothesis. There is strong evidence of an association between race and education level. Hispanics are more likely to not have a high school diploma, and fewer whites fail to graduate from high school than expected.

26. Pregancies.

H_0: Pregnancy outcome is independent of age.

H_A: There is an association between pregnancy outcome and age.

Counted data condition: The data are counts.
Randomization condition: Assume that these women are representative of all pregnant women.
Expected cell frequency condition: The expected counts are all greater than 5.

	Live Births (Obs/Exp)	Abortions (Obs/Exp)	Fetal losses (Obs/Exp)
Under 20	49 / 55.286	26 / 19	13 / 13.714
20 – 29	201 / 199.15	75 / 68.443	41 / 49.403
30 – 34	88 / 79.787	18 / 27.42	21 / 19.792
35 and over	49 / 52.773	14 / 18.136	21 / 13.091

Under these conditions, the sampling distribution of the test statistic is χ^2 on 6 degrees of freedom. We will use a chi-square test for independence.

With $\chi^2 = \displaystyle\sum_{all\ cells} \frac{(Obs - Exp)^2}{Exp} \approx 15.552$, on 6 degrees of freedom, the P-value ≈ 0.0164.

Since the P-value is so low, we reject the null hypothesis. There is strong evidence of an association between pregnancy outcome and age. Women over 35 have significantly more fetal losses than expected, and women in the 30 – 34 age group have significantly fewer abortions than expected.

27. Race and education, part 2.

H_0: Race is independent of education level.

H_A: There is an association between race and education level.

Counted data condition: The data are counts.
Randomization condition: Assume that the sample was taken randomly.
Expected cell frequency condition: The expected counts are all greater than 5.

	HS Diploma (Obs/Exp)	College Grad (Obs/Exp)	Adv. Degree (Obs/Exp)
Black	1598 / 1605.1	549 / 538.01	117 / 120.93
Hispanic	1269 / 1261.9	412 / 422.99	99 / 95.074

Under these conditions, the sampling distribution of the test statistic is χ^2 on 2 degrees of freedom. We will use a chi-square test for independence.

With $\chi^2 = \displaystyle\sum_{all\,cells} \dfrac{(Obs - Exp)^2}{Exp} \approx 0.870$, on 2 degrees of freedom, the *P*-value ≈ 0.6471.

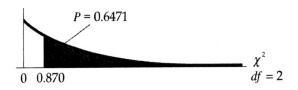

Since the *P*-value is high, we fail to reject the null hypothesis. There is no evidence of an association between race and educational performance. Blacks and Hispanics appear to have similar distributions of educational opportunity.

28. Education by age.

H_0: The distribution of education level attained is the same for different age groups.

H_A: The distribution of education level attained is different for different age groups.

Counted data condition: The data are counts.
Randomization condition: Assume that the sample was taken randomly.
Expected cell frequency condition: The expected counts are all greater than 5.

	25 – 34 (Obs / Exp)	35 – 44 (Obs / Exp)	45 – 54 (Obs / Exp)	55 – 64 (Obs / Exp)	65 and older (Obs / Exp)
Not HS Grad	27 / 60.2	50 / 60.2	52 / 60.2	71 / 60.2	101 / 60.2
HS	82 / 66.2	19 / 66.2	88 / 66.2	83 / 66.2	59 / 66.2
1 – 3 years college	43 / 33	56 / 33	26 / 33	20 / 33	20 / 33
4+ years college	48 / 40.6	75 / 40.6	34 / 40.6	26 / 40.6	20 / 40.6

Under these conditions, the sampling distribution of the test statistic is χ^2 on 12 degrees of freedom. We will use a chi-square test for homogeneity. (There are 200 people in each age group, an indication that we are examining 5 age groups, with respect to one variable, education level attained.)

With $\chi^2 = \displaystyle\sum_{all\,cells} \dfrac{(Obs - Exp)^2}{Exp} \approx 178.453$, on 12 degrees of freedom, the *P*-value is essentially 0.

Since the *P*-value is so low, we reject the null hypothesis. There is strong evidence that the distribution of education level attained is different between the groups. Generally, younger people are more likely to have higher levels of education than older people, who are themselves over represented at the lower education levels. Specifically, people in the 35 – 44 age group were less likely to have only a high school diploma, and more likely to have at least four years of college.

Chapter 27 – Inferences for Regression

1. **Ms. President?**

   a) The equation of the line of best fit for these data points is $\%\hat{Y}es = -5.583 + 0.999(Year)$, where *Year* is measured in years since 1900. According to the linear model, the percentage of people willing to vote for a female candidate for president is increasing by about 1% each year.

   b) H_0: There has been no change in the percentage of people willing to vote for a female candidate for president. $(\beta_1 = 0)$

   H_A: There has been a change in the percentage of people willing to vote for a female candidate for president. $(\beta_1 \neq 0)$

   c) Assuming the conditions have been met, the sampling distribution of the regression slope can be modeled by a Student's *t*-model with (16 – 2) = 14 degrees of freedom. We will use a regression slope *t*-test.

   The value of $t = 15.1$. The *P*-value ≤ 0.0001 means that the association we see in the data is unlikely to occur by chance. We reject the null hypothesis, and conclude that there is strong evidence that the percentage of people willing to vote for a female presidential candidate is increasing.

   d) 94.2% of the variation in the percentage of people willing to vote for a female president is explained by the year.

2. **Drug use.**

   a) The equation of the line of best fit for these data points is $\%Oth\hat{e}rDrugs = -3.068 + 0.615(\%Marijuana)$. According to the linear model, the percentage of ninth graders in these countries who use other drugs increases by about 0.615% for each additional 1% of ninth graders who use marijuana.

   b) H_0: There is no linear relationship between marijuana use and use of other drugs. $(\beta_1 = 0)$

   H_A: There is a linear relationship between marijuana use and use of other drugs. $(\beta_1 \neq 0)$

   c) Assuming the conditions have been met, the sampling distribution of the regression slope can be modeled by a Student's *t*-model with (11 – 2) = 9 degrees of freedom. We will use a regression slope *t*-test.

   The value of $t = 7.85$. The *P*-value of 0.0001 means that the association we see in the data is unlikely to occur by chance. We reject the null hypothesis, and conclude that there is strong evidence that the percentage of ninth graders who use other drugs is related to the percentage of ninth graders who use marijuana. Countries with a high percentage of ninth graders using marijuana tend to have a high percentage of ninth graders using other drugs.

   d) 87.3% of the variation in the percentage of ninth graders using other drugs can be accounted for by the percentage of ninth graders using marijuana.

e) The use of other drugs is associated with marijuana use, but there is no proof of a cause-and-effect relationship between the two variables. There may be lurking variables present.

3. No opinion.

a) H_0: There has been no change in the percentage of people expressing no opinion about their willingness to vote for a female candidate for president. $(\beta_1 = 0)$

H_A: There has been a change in the percentage of people expressing no opinion about their willingness to vote for a female candidate for president. $(\beta_1 \neq 0)$

b) Assuming the conditions have been met, the sampling distribution of the regression slope can be modeled by a Student's *t*-model with (16 – 2) = 14 degrees of freedom. We will use a regression slope *t*-test.

The equation of the line of best fit for these data points is:
$\%No\hat{O}pinion = 7.693 - 0.0427(Year)$, where
Year is measured in years since 1900.

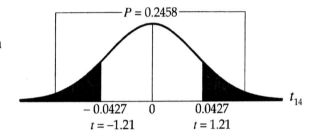

The value of *t* = –1.21. The *P*-value of 0.2458 means that the association we see in the data is likely to occur by chance alone. We fail to reject the null hypothesis, and conclude that there is no evidence of a change in the percentage of people expressing no opinion about their willingness to vote for a female presidential candidate.

c) The plot indicates no real trend. The relatively high value in 1945 and the relatively low value around 2000 are influential points. The true slope does not appear to be negative after discounting influential points at the ends. However, this does not change the conclusion of the hypotheses test in part b. We failed to find evidence of a relationship anyway.

4. Cholesterol.

a) H_0: There is no linear relationship between age and cholesterol. $(\beta_1 = 0)$

H_A: Cholesterol levels tend to increase with age. $(\beta_1 > 0)$

b) Assuming the conditions have been met, the sampling distribution of the regression slope can be modeled by a Student's *t*-model with (294 – 2) = 292 degrees of freedom. We will use a regression slope *t*-test. The equation of the line of best fit for these data points is:
$Chol\hat{e}sterol = 196.619 + 0.746(Age)$

The value of *t* = 1.23. The *P*-value of 0.1103 (remember that regression output usually gives two-sided *P*-values, so the one-sided *P*-value = 0.2206/2) means that the association we see in the data is fairly likely to occur by chance alone. We fail to reject the null

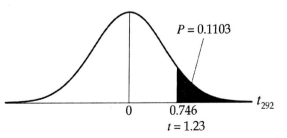

hypothesis, and conclude that there is no evidence of a positive linear relationship between age and cholesterol. The plot should be examined to see if a non-linear relationship is present.

5. Marriage age.

a) H_0: The difference in age between men and women at first marriage has not been decreasing since 1975. $(\beta_1 = 0)$

H_A: The difference in age between men and women at first marriage has been decreasing since 1975. $(\beta_1 < 0)$

b) **Straight enough condition:** The scatterplot is not provided, but the residuals plot looks unpatterned. The scatterplot is likely to be straight enough.
Independence assumption: We are examining a relationship over time, so there is reason to be cautious, but the residuals plot shows no evidence of dependence.
Does the plot thicken? condition: The residuals plot shows no obvious trends in the spread.
Nearly Normal condition: The histogram is not particularly unimodal and symmetric, but shows no obvious skewness of outliers.

Since conditions have been satisfied, the sampling distribution of the regression slope can be modeled by a Student's t-model with $(24 - 2) = 22$ degrees of freedom. We will use a regression slope t-test. The equation of the line of best fit for these data points is:
$(Men - \hat{Women}) = 49.9 - 0.024(Year)$

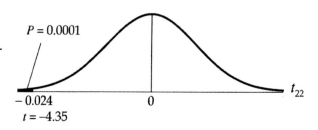

The value of $t = -4.35$. The P-value of about 0.0001 (remember that regression output usually gives two-sided P-values, so the one-sided P-value $= 0.0003/2$) means that the association we see in the data is unlikely to occur by chance. We reject the null hypothesis, and conclude that there is strong evidence of a negative linear relationship between difference in age at first marriage and year. The difference in marriage age between men and women appears to be decreasing over time.

d) $b_1 \pm t_{n-2}^* \times SE(b_1) = -0.024 \pm (2.074) \times 0.0055 \approx (-0.035, -0.013)$

We are 95% confident that the mean difference in age between men and women at first marriage decreases by between 0.013 and 0.035 years in age for each year that passes.

6. Used cars.

a) A scatterplot of the used cars data is at the right.

b) A linear model is probably appropriate. The plot appears to be linear.

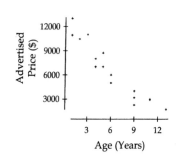

c)

```
Dependent variable is:   Price
No Selector
R squared = 89.4%    R squared (adjusted) = 88.7%
s =  1221  with  17 - 2 = 15  degrees of freedom
```

Source	Sum of Squares	df	Mean Square	F-ratio
Regression	187830720	1	187830720	126
Residual	22346074	15	1489738	

Variable	Coefficient	s.e. of Coeff	t-ratio	prob
Constant	12319.6	575.7	21.4	≤ 0.0001
Age (years)	-924.000	82.29	-11.2	≤ 0.0001

The equation of the regression line is:
$\hat{Price} = 12319.6 - 924(Age)$.
According to the model, the average asking price for a used Toyota Corolla decreases by about $924 dollars for each additional year in age. Let's take a closer look.

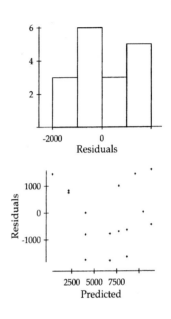

d) **Straight enough condition:** The scatterplot is straight enough to try a linear model.
Independence assumption: Prices of Toyota Corollas of different ages might be related, but the residuals plot looks fairly scattered. (The fact that there are several prices for some years draws our eyes to some patterns that may not exist.)
Does the plot thicken? condition: The residuals plot shows no obvious patterns in the spread.
Nearly Normal condition: The histogram is reasonably unimodal and symmetric, and shows no obvious skewness or outliers.

e) Since conditions have been satisfied, the sampling distribution of the regression slope can be modeled by a Student's *t*-model with (17 – 2) = 15 degrees of freedom. We will use a regression slope *t*-interval, with 95% confidence.

$$b_1 \pm t^*_{n-2} \times SE(b_1) = -924 \pm (2.131) \times 82.29 \approx (-1099.4, -748.6)$$

f) We are 95% confident that the advertised price of a used Toyota Corolla is decreasing by an average of between $748.60 and $1099.40 for each additional year in age.

7. Fuel economy.

a) H₀: There is no linear relationship between the weight of a car and its mileage. $(\beta_1 = 0)$

Hₐ: There is a linear relationship between the weight of a car and its mileage. $(\beta_1 \neq 0)$

b) **Straight enough condition:** The scatterplot is straight enough to try a linear model.
Independence assumption: The residuals plot is scattered.
Does the plot thicken? condition: The residuals plot indicates some possible "thickening" as the predicted values increases, but it's probably not enough to worry about.
Nearly Normal condition: The histogram of residuals is unimodal and symmetric, with one possible outlier. With the large sample size, it is okay to proceed.

Since conditions have been satisfied, the sampling distribution of the regression slope can be modeled by a Student's *t*-model with (50 – 2) = 48 degrees of freedom. We will use a regression slope *t*-test. The equation of the line of best fit for these data points is:
$\hat{MPG} = 48.7393 - 8.21362(Weight)$, where *Weight* is measured in thousands of pounds.

The value of $t = -12.2$. The *P*-value of less than 0.0001 means that the association we see in the data is unlikely to occur by chance. We reject the null hypothesis, and conclude that there is strong evidence of a linear relationship between weight of a car and its mileage. Cars that weigh more tend to have lower gas mileage.

8. **SAT scores.**

   a) H_0: There is no linear relationship between SAT Verbal and Math scores. $(\beta_1 = 0)$

   H_A: There is a linear relationship between SAT Verbal and Math scores. $(\beta_1 \neq 0)$

   b) **Straight enough condition:** The scatterplot is straight enough to try a linear model.
   Independence assumption: The residuals plot is scattered.
   Does the plot thicken? condition: The spread of the residuals is consistent.
   Nearly Normal condition: The histogram of residuals is unimodal and symmetric, with one possible outlier. With the large sample size, it is okay to proceed.

   Since conditions have been satisfied, the sampling distribution of the regression slope can be modeled by a Student's *t*-model with $(162 - 2) = 160$ degrees of freedom. We will use a regression slope *t*-test. The equation of the line of best fit for these data points is:
   $\widehat{Math} = 209.554 + 0.675075(Verbal)$.

   The value of $t = 11.9$. The *P*-value of less than 0.0001 means that the association we see in the data is unlikely to occur by chance. We reject the null hypothesis, and conclude that there is strong evidence of a linear relationship between SAT Verbal and Math scores. Students with higher SAT-Verbal scores tend to have higher SAT-Math scores.

9. **Fuel economy, part II.**

   a) Since conditions have been satisfied in Exercise 7, the sampling distribution of the regression slope can be modeled by a Student's *t*-model with $(50 - 2) = 48$ degrees of freedom. (Use $t^*_{45} = 2.014$ from the table.) We will use a regression slope *t*-interval, with 95% confidence.

   $b_1 \pm t^*_{n-2} \times SE(b_1) = -8.21362 \pm (2.014) \times 0.6738 \approx (-9.57, -6.86)$

   b) We are 95% confident that the mean mileage of cars decreases by between 6.86 and 9.57 miles per gallon for each additional 1000 pounds of weight.

10. **SAT, part II.**

    a) Since conditions have been satisfied in Exercise 8, the sampling distribution of the regression slope can be modeled by a Student's *t*-model with $(162 - 2) = 160$ degrees of freedom. (Use $t^*_{140} = 1.656$ from the table.) We will use a regression slope *t*-interval, with 90% confidence.

    $b_1 \pm t^*_{n-2} \times SE(b_1) = 0.675075 \pm (1.656) \times 0.0568 \approx (0.581, 0.769)$

    b) We are 90% confident that the mean Math SAT scores increase by between 0.581 and 0.769 point for each additional point scored on the Verbal test.

11. MPG revisited.

a) The regression equation predicts that cars that weigh 2500 pounds will have a mean fuel efficiency of $48.7393 - 8.21362(2.5) = 28.20525$ miles per gallon.

$$\hat{y}_v \pm t^*_{n-2}\sqrt{SE^2(b_1) \cdot (x_v - \bar{x})^2 + \frac{s_e^2}{n}}$$

$$= 28.20525 \pm (2.014)\sqrt{0.6738^2 \cdot (2.5 - 2.8878)^2 + \frac{2.413^2}{50}}$$

$$\approx (27.34, 29.07)$$

We are 95% confident that cars weighing 2500 pounds will have mean fuel efficiency between 27.34 and 29.07 miles per gallon.

b) The regression equation predicts that cars that weigh 3450 pounds will have a mean fuel efficiency of $48.7393 - 8.21362(3.45) = 20.402311$ miles per gallon.

$$\hat{y}_v \pm t^*_{n-2}\sqrt{SE^2(b_1) \cdot (x_v - \bar{x})^2 + \frac{s_e^2}{n} + s_e^2}$$

$$= 20.402311 \pm (2.014)\sqrt{0.6738^2 \cdot (3.45 - 2.8878)^2 + \frac{2.413^2}{50} + 2.413^2}$$

$$\approx (15.44, 25.37)$$

We are 95% confident that a car weighing 3450 pounds will have fuel efficiency between 15.44 and 25.37 miles per gallon.

12. SATs again.

a) The regression equation predicts that students with an SAT-Verbal score of 500 will have a mean SAT-Math score of $209.554 + 0.675075(500) = 547.0915$.

$$\hat{y}_v \pm t^*_{n-2}\sqrt{SE^2(b_1) \cdot (x_v - \bar{x})^2 + \frac{s_e^2}{n}}$$

$$= 547.0915 \pm (1.656)\sqrt{0.0568^2 \cdot (500 - 596.292)^2 + \frac{71.75^2}{162}}$$

$$\approx (534.09, 560.10)$$

We are 90% confident that students with scores of 500 on the SAT-Verbal will have a mean SAT-Math score between 534.09 and 560.10.

b) The regression equation predicts that students with an SAT-Verbal score of 710 will have a mean SAT-Math score of $209.554 + 0.675075(710) = 688.85725$.

$$\hat{y}_v \pm t^*_{n-2}\sqrt{SE^2(b_1) \cdot (x_v - \bar{x})^2 + \frac{s_e^2}{n} + s_e^2}$$

$$= 688.85725 \pm (1.656)\sqrt{0.0568^2 \cdot (710 - 596.296)^2 + \frac{71.75^2}{162} + 71.75^2}$$

$$\approx (569.19, 808.52)$$

We are 90%confident that a student scoring 710 on the SAT-Verbal would have an SAT-Math score of between 569.19 and 808.52. Since we are talking about individual scores, and not means, it is reasonable to restrict ourselves to possible scores, so we are 90% confident that the class president scored between 570 and 800 on the SAT-Math test.

13. Cereal.

a) H_0: There is no linear relationship between the number of calories and the sodium content of cereals. $(\beta_1 = 0)$

H_A: There is a linear relationship between the number of calories and the sodium content of cereals. $(\beta_1 \neq 0)$

Since these data were judged acceptable for inference, the sampling distribution of the regression slope can be modeled by a Student's t-model with (77 – 2) = 75 degrees of freedom. We will use a regression slope t-test. The equation of the line of best fit for these data points is: $\widehat{Sodium} = 21.4143 + 1.29357(Calories)$.

The value of t = 2.73. The P-value of 0.0079 means that the association we see in the data is unlikely to occur by chance. We reject the null hypothesis, and conclude that there is strong evidence of a linear relationship between the number of calories and sodium content of cereals. Cereals with higher numbers of calories tend to have higher sodium contents.

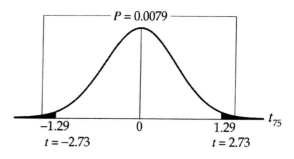

b) Only 9% of the variability in sodium content can be explained by the number of calories. The residual standard deviation is 80.49 mg, which is pretty large when you consider that the range of sodium content is only 320 mg. Although there is strong evidence of a linear association, it is too weak to be of much use. Predictions would tend to be very imprecise.

14. Brain size.

a) H_0: There is no linear relationship between brain size and IQ. $(\beta_1 = 0)$

H_A: There is a linear relationship between brain size and IQ. $(\beta_1 \neq 0)$

Since these data were judged acceptable for inference, the sampling distribution of the regression slope can be modeled by a Student's t-model with (21 – 2) = 19 degrees of freedom. (There are 21 dots on the scatterplot. I counted!) We will use a regression slope t-test. The equation of the line of best fit for these data points is:
$\widehat{IQ\_Verbal} = 24.1835 + 0.098842(Size)$.

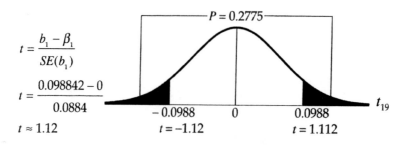

The value of $t \approx 1.12$. The P-value of 0.2775 means that the association we see in the data is likely to occur by chance. We fail to reject the null hypothesis, and conclude that there is no evidence of a linear relationship between brain size and verbal IQ score.

b) Since $R^2 = 6.5\%$, only 6.5% of the variability in verbal IQ can be explained by brain size. This association is very weak. There are three students with large brains who scored high on the IQ test. Without them, there appears to be no association at all.

15. Another bowl.

Straight enough condition: The scatterplot is not straight.
Independence assumption: The residuals plot shows a curved pattern.
Does the plot thicken? condition: The spread of the residuals is not consistent. The residuals plot "thickens" as the predicted values increase.
Nearly Normal condition: The histogram of residuals is skewed to the right, with an outlier.

These data are not appropriate for inference.

16. Winter.

Straight enough condition: The scatterplot is not straight.
Independence assumption: The residuals plot shows a curved pattern.
Does the plot thicken? condition: The spread of the residuals is not consistent. The residuals plot shows decreasing variability as the predicted values increase.
Nearly Normal condition: The histogram of residuals is skewed to the right, with an outlier.

These data are not appropriate for inference.

17. Acid rain.

a) H_0: There is no linear relationship between BCI and pH. $\left(\beta_1 = 0\right)$

H_A: There is a linear relationship between BCI and pH. $\left(\beta_1 \neq 0\right)$

Assuming the conditions for inference are satisfied, the sampling distribution of the regression slope can be modeled by a Student's t-model with $(163 - 2) = 161$ degrees of freedom. We will use a regression slope t-test. The equation of the line of best fit for these data points is: $\hat{BCI} = 2733.37 - 197.694(pH)$.

$t = \dfrac{b_1 - \beta_1}{SE(b_1)}$

$t = \dfrac{-197.694 - 0}{25.57}$

$t \approx -7.73$

The value of $t \approx -7.73$. The P-value (two-sided!) of essentially 0 means that the association we see in the data is unlikely to occur by chance. We reject the null hypothesis, and conclude that there is strong evidence of a linear relationship between BCI and pH. Streams with higher pH tend to have lower BCI.

18. El Niño.

a) The regression equation is $\hat{Temp} = 15.3066 + 0.004(CO_2)$, with CO_2 concentration measured in parts per million from the top of Mauna Loa in Hawaii, and temperature in degrees Celsius.

b) H_0: There is no linear relationship between temperature and CO_2 concentration. $(\beta_1 = 0)$

H_A: There is a linear relationship between temperature and CO_2 concentration. $(\beta_1 \neq 0)$

Since the scatterplots and residuals plots showed that the data were appropriate for inference, the sampling distribution of the regression slope can be modeled by a Student's t-model with $(37 - 2) = 35$ degrees of freedom. We will use a regression slope t-test.

$$t = \frac{b_1 - \beta_1}{SE(b_1)}$$

$$t = \frac{0.004 - 0}{0.0009}$$

$$t \approx 4.44$$

The value of $t \approx 4.44$. The P-value (two-sided!) of about 0.00008 means that the association we see in the data is unlikely to occur by chance. We reject the null hypothesis, and conclude that there is strong evidence of a linear relationship between CO_2 concentration and temperature. Years with higher CO_2 concentration tend to be warmer, on average.

c) Since $R^2 = 33.4\%$, only 33.4% of the variability in temperature can be explained by the CO_2 concentration. Although there is strong evidence of a linear association, it is too weak to be of much use. Predictions would tend to be very imprecise.

19. Ozone.

a) H_0: There is no linear relationship between population and ozone level. $(\beta_1 = 0)$

H_A: There is a positive linear relationship between population and ozone level. $(\beta_1 > 0)$

Assuming the conditions for inference are satisfied, the sampling distribution of the regression slope can be modeled by a Student's t-model with $(16 - 2) = 14$ degrees of freedom. We will use a regression slope t-test. The equation of the line of best fit for these data points is: $Oz\hat{o}ne = 18.892 + 6.650(Population)$, where ozone level is measured in parts per million and population is measured in millions.

The value of $t \approx 3.48$. The P-value of 0.0018 means that the association we see in the data is unlikely to occur by chance. We reject the null hypothesis, and conclude that there is strong evidence of a

$$t = \frac{b_1 - \beta_1}{SE(b_1)}$$

$$t = \frac{6.650 - 0}{1.910}$$

$$t \approx 3.48$$

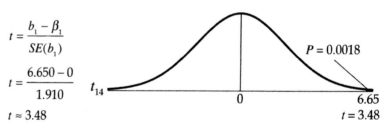

positive linear relationship between ozone level and population. Cities with larger populations tend to have higher ozone levels.

b) $b_1 \pm t^*_{n-2} \times SE(b_1) = 6.65 \pm (1.761) \times 1.910 \approx (3.29, 10.01)$

We are 90% confident that each additional million people will increase mean ozone levels by between 3.29 and 10.01 parts per million.

c) The regression equation predicts that cities with a population of 600,000 people will have ozone levels of $18.892 + 6.650(0.6) = 22.882$ parts per million.

$$\hat{y}_v \pm t^*_{n-2}\sqrt{SE^2(b_1)\cdot(x_v-\bar{x})^2+\frac{s_e^2}{n}}$$

$$= 22.882 \pm (1.761)\sqrt{1.91^2\cdot(0.6-1.7)^2+\frac{5.454^2}{16}}$$

$$\approx (18.47,\ 27.29)$$

We are 90% confident that the mean ozone level for cities with populations of 600,000 will be between 18.47 and 27.29 parts per million.

20. Sales and profits.

a) H_0: There is no linear relationship between sales and profit. $(\beta_1 = 0)$

H_A: There is a linear relationship between sales and profit. $(\beta_1 \neq 0)$

Assuming the conditions for inference are satisfied, the sampling distribution of the regression slope can be modeled by a Student's t-model with $(79 - 2) = 77$ degrees of freedom. We will use a regression slope t-test. The equation of the line of best fit for these data points is: $\widehat{Profits} = -176.644 + 0.092498(Sales)$, with both profits and sales measured in millions of dollars.

The value of $t \approx 12.33$. The P-value of essentially 0 means that the association we see in the data is unlikely to occur by chance. We reject the null hypothesis, and conclude that there is strong evidence of a linear relationship between sales and profits. Companies with higher sales tend to have higher profits.

$$t = \frac{b_1 - \beta_1}{SE(b_1)}$$

$$t = \frac{0.092498 - 0}{0.0075}$$

$$t \approx 12.33$$

b) $R^2 = 66.2\%$, so 66.2% of the variability in profits can be explained by sales.

c) There are 77 degrees of freedom, so use $t^*_{75} = 1.992$ as a conservative estimate from the table.

$$b_1 \pm t^*_{n-2} \times SE(b_1) = 0.092498 \pm (1.992) \times 0.0075 \approx (0.078,\ 0.107)$$

We are 95% confident that each additional million dollars in sales will increase mean profits by between \$78,000 and \$107,000.

d) The regression equation predicts that corporations with sales of \$9,000 million dollars will have profits of $-176.644 + 0.092498(9000) = 655.838$ million dollars.

$$\hat{y}_v \pm t^*_{n-2}\sqrt{SE^2(b_1)\cdot(x_v-\bar{x})^2+\frac{s_e^2}{n}+s_e^2}$$

$$= 655.838 \pm (1.992)\sqrt{0.0075^2\cdot(9000-4178.29)^2+\frac{466.2^2}{79}+466.2^2}$$

$$\approx (-281.46,\ 1593.14)$$

We are 95% confident that the Eli Lilly's profits will be between –\$281,460,000 and \$1,593,140,000. This interval is too wide to be of any use.

(If you use $t^*_{77} = 1.991297123$, your interval will be $(-281.1,\ 1592.8)$)

21. Start the car!

a) Since the number of degrees of freedom is 33 – 2 = 31, there were 33 batteries tested.

b) **Straight enough condition:** The scatterplot is roughly straight, but very scattered.
Independence assumption: The residuals plot shows no pattern.
Does the plot thicken? condition: The spread of the residuals is consistent.
Nearly Normal condition: The Normal probability plot of residuals is reasonably straight.

c) H_0: There is no linear relationship between cost and power. $\left(\beta_1 = 0\right)$

H_A: There is a positive linear relationship between cost and power. $\left(\beta_1 > 0\right)$

Since the conditions for inference are satisfied, the sampling distribution of the regression slope can be modeled by a Student's t-model with (33 – 2) = 31 degrees of freedom. We will use a regression slope t-test. The equation of the line of best fit for these data points is: $\hat{Power} = 384.594 + 4.14649(Cost)$, with power measured in cold cranking amps, and cost measured in dollars.

The value of $t \approx 3.23$. The P-value of 0.0015 means that the association we see in the data is unlikely to occur by chance. We reject the null hypothesis, and conclude that there is strong evidence of a positive linear relationship between cost and power. Batteries that cost more tend to have more power.

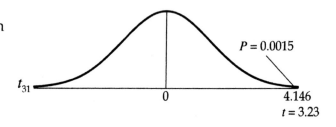

d) Since $R^2 = 25.2\%$, only 25.2% of the variability in power can be explained by cost. The residual standard deviation is 116 amps. That's pretty large, considering the range battery power is only about 400 amps. Although there is strong evidence of a linear association, it is too weak to be of much use. Predictions would tend to be very imprecise.

e) The equation of the line of best fit for these data points is: $\hat{Power} = 384.594 + 4.14649(Cost)$, with power measured in cold cranking amps, and cost measured in dollars.

f) There are 31 degrees of freedom, so use $t^*_{30} = 1.697$ as a conservative estimate from the table.

$$b_1 \pm t^*_{n-2} \times SE(b_1) = 4.14649 \pm (1.697) \times 1.282 \approx (1.97, 6.32)$$

g) We are 95% confident that the mean power increases by between 1.97 and 6.32 cold cranking amps for each additional dollar in cost.

22. Crawling.

a) If the data had been plotted for individual babies, the association would appear to be weaker, since individuals are more variable than averages.

b) H_0: There is no linear relationship between 6-month temperature and crawling age. $\left(\beta_1 = 0\right)$

H_A: There is a linear relationship between 6-month temperature and crawling age. $\left(\beta_1 \neq 0\right)$

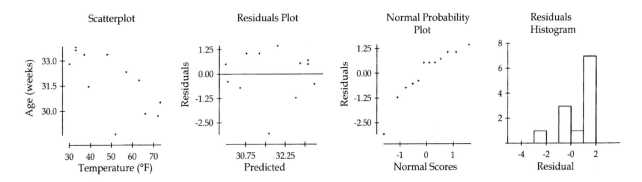

Straight enough condition: The scatterplot is straight enough to try linear regression.

Independence assumption: The residuals plot shows no pattern, but there may be an outlier. If the month of May were just one data point, it would be removed. However, since it represents the average crawling age of several babies, there is no justification for its removal.

Does the plot thicken? condition: The spread of the residuals is consistent

Nearly Normal condition: The Normal probability plot of residuals isn't very straight, largely because of the data point for May. The histogram of residuals also shows this outlier.

Since we had difficulty with the conditions for inference, we will proceed cautiously. These data may not be appropriate for inference. The sampling distribution of the regression slope can be modeled by a Student's t-model with $(12 - 2) = 10$ degrees of freedom. We will use a regression slope t-test.

Dependent variable is: **Age**
No Selector
R squared = 49.0%   R squared (adjusted) = 43.9%
s = 1.319 with 12 - 2 = 10 degrees of freedom

The equation of the line of best fit for these data points is: $A\hat{g}e = 35.6781 - 0.077739(Temp)$, with average crawling age measured in weeks and average temperature in °F.

Source	Sum of Squares	df	Mean Square	F-ratio
Regression	16.6933	1	16.6933	9.59
Residual	17.4028	10	1.74028	

Variable	Coefficient	s.e. of Coeff	t-ratio	prob
Constant	35.6781	1.318	27.1	≤ 0.0001
Temp	-0.077739	0.0251	-3.10	0.0113

The value of $t \approx -3.10$. The P-value of 0.0113 means that the association we see in the data is unlikely to occur by chance. We reject the null hypothesis, and conclude that there is strong evidence of a linear relationship between average temperature and average crawling age. Babies who reach six months of age in warmer temperatures tend to crawl at earlier ages than babies who reach six months of age in colder temperatures.

$P = 0.0113$

t_{10}

-0.078 0   0.078
$t = -3.10$ $t = 3.10$

c) $b_1 \pm t^*_{n-2} \times SE(b_1) = -0.077739 \pm (2.228) \times 0.0251 \approx (-0.134, -0.022)$

We are 95% that the average crawling age decreases by between 0.022 weeks and 1.34 weeks when the average temperature increases by 10°F.

23. Printers.

H_0: There is no linear relationship between speed and cost of printers. $(\beta_1 = 0)$

H_A: There is a linear relationship between speed and cost of printers. $(\beta_1 \neq 0)$

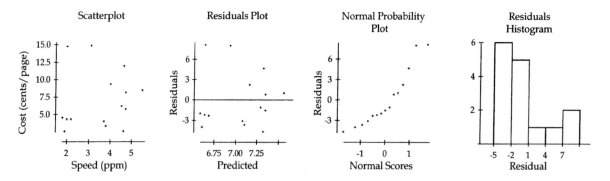

Straight enough condition: The scatterplot is straight enough to try linear regression, although it looks very scattered, and there doesn't appear to be any association.
Independence assumption: The residuals plot shows no pattern.
Does the plot thicken? condition: The spread of the residuals is consistent
Nearly Normal condition: The Normal probability plot of residuals isn't very straight, and the histogram of the residuals is strongly skewed to the right.

Since we had difficulty with the conditions, inference is not appropriate.
(If you had proceeded with the test, the *P*-value of 0.7756 would have failed to provide evidence of a linear relationship.)

24. Strike two.

H_0: The effectiveness of the video is independent of the player's initial ability. $(\beta_1 = 0)$

H_A: The effectiveness of the video depends on the player's initial ability. $(\beta_1 \neq 0)$

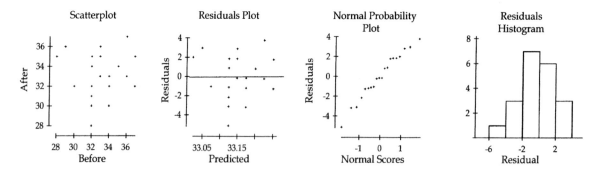

Straight enough condition: The scatterplot is straight enough to try linear regression, although it looks very scattered, and there doesn't appear to be any association.
Independence assumption: The residuals plot shows no pattern.
Does the plot thicken? condition: The spread of the residuals is consistent.
Nearly Normal condition: The Normal probability plot of residuals is very straight, and the histogram of the residuals is unimodal and symmetric with no outliers.

Since the conditions for inference are inference are satisfied, the sampling distribution of the regression slope can be modeled by a Student's *t*-model with (20 – 2) = 18 degrees of freedom. We will use a regression slope *t*-test.

Dependent variable is: **After**
No Selector
R squared = 0.1%   R squared (adjusted) = -5.5%
s = 2.386 with 20 - 2 = 18 degrees of freedom

Source	Sum of Squares	df	Mean Square	F-ratio
Regression	0.071912	1	0.071912	0.013
Residual	102.478	18	5.69323	

Variable	Coefficient	s.e. of Coeff	t-ratio	prob
Constant	32.3161	7.439	4.34	0.0004
Before	0.025232	0.2245	0.112	0.9118

The equation of the line of best fit for these data points is: $After = 32.3161 + 0.025232(Before)$, where we are counting the number of strikes thrown before and after the training program.

The value of $t \approx 0.112$. The *P*-value of 0.9118 means that the association we see in the data is quite likely to occur by chance. We fail to reject the null hypothesis, and conclude that there is no evidence of a linear relationship between the number of strikes thrown before the training program and the number of strikes thrown after the program. The effectiveness of the program does not appear to depend on the initial ability of the player.

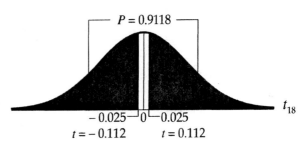

25. Body fat.

a) H_0: There is no linear relationship between waist size and percent body fat. $(\beta_1 = 0)$

H_A: Percent body fat increases with waist size. $(\beta_1 > 0)$

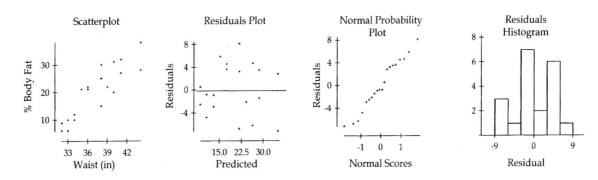

Straight enough condition: The scatterplot is straight enough to try linear regression.
Independence assumption: The residuals plot shows no pattern.
Does the plot thicken? condition: The spread of the residuals is consistent.
Nearly Normal condition: The Normal probability plot of residuals is straight, and the histogram of the residuals is unimodal and symmetric with no outliers.

Since the conditions for inference are inference are satisfied, the sampling distribution of the regression slope can be modeled by a Student's *t*-model with (20 – 2) = 18 degrees of freedom. We will use a regression slope *t*-test.

Dependent variable is: **Body Fat %**
No Selector
R squared = 78.7%   R squared (adjusted) = 77.5%
s = 4.540 with 20 - 2 = 18 degrees of freedom

Source	Sum of Squares	df	Mean Square	F-ratio
Regression	1366.79	1	1366.79	66.3
Residual	370.960	18	20.6089	

The equation of the line of best fit for these data points is: $\%BodyFat = -62.5573 + 2.22152(Waist)$.

Variable	Coefficient	s.e. of Coeff	t-ratio	prob
Constant	-62.5573	10.16	-6.16	≤ 0.0001
Waist (in)	2.22152	0.2728	8.14	≤ 0.0001

The value of $t \approx 8.14$. The *P*-value of essentially 0 means that the association we see in the data is unlikely to occur by chance. We reject the null hypothesis, and conclude that there is strong evidence of a linear relationship between waist size and percent body fat. People with larger waists tend to have a higher percentage of body fat.

b) The regression equation predicts that people with 40-inch waists will have $-62.5573 + 2.22152(40) = 20.3035\%$ body fat. The average waist size of the people sampled was approximately 37.05 inches.

$$\hat{y}_v \pm t^*_{n-2}\sqrt{SE^2(b_1)\cdot(x_v-\bar{x})^2 + \frac{s_e^2}{n}}$$

$$= 20.3035 \pm (2.101)\sqrt{0.2728^2\cdot(40-37.05)^2 + \frac{4.54^2}{20}}$$

$$\approx (17.58, 23.03)$$

We are 95% confident that the mean percent body fat for people with 40-inch waists is between 17.58% and 23.03%.

26. Body fat, again.

a)

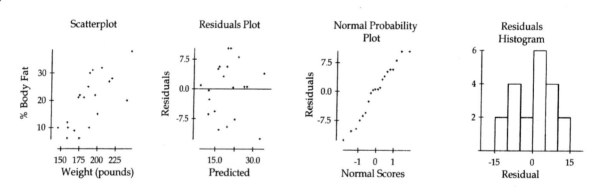

Straight enough condition: The scatterplot is straight enough to try linear regression.
Independence assumption: The residuals plot shows no pattern.
Does the plot thicken? condition: The spread of the residuals is consistent.
Nearly Normal condition: The Normal probability plot of residuals is straight, and the histogram of the residuals is unimodal and symmetric with no outliers.

Since the conditions for inference are inference are satisfied, the sampling distribution of the regression slope can be modeled by a Student's *t*-model with (20 – 2) = 18 degrees of freedom. We will use a regression slope *t*-interval.

```
Dependent variable is:    Body Fat %
No Selector
R squared = 48.5%     R squared (adjusted) = 45.7%
s = 7.049  with  20 - 2 = 18  degrees of freedom

Source        Sum of Squares   df   Mean Square   F-ratio
Regression    843.325          1    843.325       17.0
Residual      894.425          18   49.6903

Variable    Coefficient   s.e. of Coeff   t-ratio   prob
Constant    -27.3763      11.55           -2.37     0.0291
Weight (lb) 0.249874      0.0607          4.12      0.0006
```

The equation of the line of best fit for these data points is: $\%\hat{BodyFat} = -27.3763 + 0.249874(Weight)$.

$$b_1 \pm t^*_{n-2} \times SE(b_1) = 0.249874 \pm (1.734) \times 0.0607 \approx (0.145, 0.355)$$

b) We are 90% confident that the mean percent body fat increases between 1.45% and 3.55% for an additional 10 pounds in weight.

c) The regression equation predicts that a person weighing 165 pounds would have $-27.3763 + 0.249874(165) = 13.85291\%$ body fat. The average weight of the people sampled was 188.6 pounds.

$$\hat{y}_v \pm t^*_{n-2} \sqrt{SE^2(b_1) \cdot (x_v - \bar{x})^2 + \frac{s_e^2}{n} + s_e^2}$$

$$= 13.85291 \pm (2.101) \sqrt{0.0607^2 \cdot (165 - 188.6)^2 + \frac{7.049^2}{20} + 7.049^2}$$

$$\approx (-1.61, 29.32)$$

We are 95% confident that a person weighing 165 pounds would have between 0% (–1.61%) and 29.32% body fat.

27. Education and mortality.

a) **Straight enough condition:** The scatterplot is straight enough to try linear regression.
Independence assumption: The residuals plot shows no pattern. If these cities are representative of other cities, we can generalize our results.
Does the plot thicken? condition: The spread of the residuals is consistent.
Nearly Normal condition: The histogram of the residuals is unimodal and symmetric with no outliers.

b) H$_0$: There is no linear relationship between education and mortality. $(\beta_1 = 0)$

H$_A$: Cities with higher average education level have lower mortality rates. $(\beta_1 < 0)$

Since the conditions for inference are inference are satisfied, the sampling distribution of the regression slope can be modeled by a Student's t-model with (58 – 2) = 56 degrees of freedom. We will use a regression slope t-test. The equation of the line of best fit for these data points is: $\hat{Mortality} = 1493.26 - 49.9202(Education)$.

The value of $t \approx -6.24$. The P-value of essentially 0 means that the association we see in the data is unlikely to occur by chance. We reject the null hypothesis, and conclude that there is strong evidence of a linear relationship between the level of education in a city and its mortality rate. Cities with lower education levels tend to have higher mortality rates.

$$t = \frac{b_1 - \beta_1}{SE(b_1)}$$

$$t = \frac{-49.9202 - 0}{8.000}$$

$$t \approx -6.24$$

c) For 95% confidence, $t^*_{56} \approx 2.00327$.

$$b_1 \pm t^*_{n-2} \times SE(b_1) = -49.9202 \pm (2.003) \times 8.000 \approx (-65.95, -33.89)$$

d) We are 95% confident that the mean number of deaths per 100,000 people decreases by between 33.89 and 65.95 deaths for an increase of one year in average education level.

e) The regression equation predicts that cities with an adult population with an average of 12 years of school will have a mortality rate of $1493.26 - 49.9202(12) = 894.2176$ deaths per 100,000. The average education level was 11.0328 years.

$$\hat{y}_v \pm t^*_{n-2} \sqrt{SE^2(b_1) \cdot (x_v - \bar{x})^2 + \frac{s_e^2}{n}}$$

$$= 894.2176 \pm (2.003) \sqrt{8.00^2 \cdot (12 - 11.0328)^2 + \frac{47.92^2}{58}}$$

$$\approx (874.239, 914.196)$$

We are 95% confident that the mean mortality rate for cities with an average of 12 years of schooling is between 874.239 and 914.196 deaths per 100,000 residents.

28. Property assessments.

a) **Straight enough condition:** The scatterplot is straight enough to try linear regression.
Independence assumption: The residuals plot shows no pattern. If these cities are representative of other cities, we can generalize our results.
Does the plot thicken? condition: The spread of the residuals is consistent
Nearly Normal condition: The Normal probability plot is fairly straight.

b) H_0: There is no linear relationship between size and assessed valuation. $(\beta_1 = 0)$

H_A: Larger houses have higher assessed values. $(\beta_1 > 0)$

Since the conditions for inference are inference are satisfied, the sampling distribution of the regression slope can be modeled by a Student's t-model with $(18 - 2) = 16$ degrees of freedom. We will use a regression slope t-test. The equation of the line of best fit for these data points is: $\hat{Asse\$\$} = 37,108.8 + 11.8987(SqFt)$.

The value of $t \approx 2.77$.
The P-value of 0.0068 means that the association we see in the data is unlikely to occur by chance. We reject the null hypothesis, and conclude that there is strong evidence of a linear relationship between the size of a home and its assessed value. Larger homes tend to have higher assessed values.

$$t = \frac{b_1 - \beta_1}{SE(b_1)}$$

$$t = \frac{11.8987 - 0}{4.290} \quad t_{16}$$

$$t \approx 2.77$$

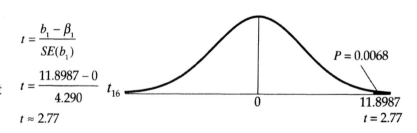

$P = 0.0068$

11.8987
$t = 2.77$

c) $R^2 = 32.5\%$. This model explains 32.5% of the variability in assessments.

d) For 90% confidence, $t_{16}^* \approx 1.746$.

$$b_1 \pm t_{n-2}^* \times SE(b_1) = 11.8987 \pm (1.746) \times 4.290 \approx (4.41, 19.39)$$

d) We are 90% confident that the mean assessed value increases by between \$441 and \$1939 for each additional 100 square feet in size.

e) The regression equation predicts that houses measuring 2100 square feet will have an assessed value of $37108.8 + 11.8987(2100) = \$62,096.07$. The average size of the houses sampled is 2003.39 square feet.

$$\hat{y}_v \pm t_{n-2}^* \sqrt{SE^2(b_1) \cdot (x_v - \bar{x})^2 + \frac{s_e^2}{n} + s_e^2}$$

$$= 62096.07 \pm (2.120)\sqrt{4.290^2 \cdot (2100 - 2003.39)^2 + \frac{4682^2}{18} + 4682^2}$$

$$\approx (51860, 72332)$$

We are 95% confident that the assessed value of a home measuring 2100 square feet will have an assessed value between \$51,860 and \$72,332. There is no evidence that this home has an assessment that is too high. The assessed value of \$70,200 falls within the prediction interval.

The homeowner might counter with an argument based on the mean assessed value of all homes such as this one.

$$\hat{y}_v \pm t_{n-2}^* \sqrt{SE^2(b_1) \cdot (x_v - \bar{x})^2 + \frac{s_e^2}{n}}$$

$$= 62096.07 \pm (2.120)\sqrt{4.290^2 \cdot (2100 - 2003.39)^2 + \frac{4682^2}{18}}$$

$$\approx (\$59,597, \$64,595)$$

The homeowner might ask the city assessor to explain why his home is assessed at \$70,200, if a typical 2100-square-foot home is assessed at between \$59,597 and \$64,595.

Review of Part VII – Inference When Variables Are Related

1. Genetics.

H_0: The proportions of traits are as specified by the ratio 1:3:3:9.
H_A: The proportions of traits are not as specified.

Counted data condition: The data are counts.
Randomization condition: Assume that these students are representative of all people.
Expected cell frequency condition: The expected counts (shown in the table) are all greater than 5.

Under these conditions, the sampling distribution of the test statistic is χ^2 on $4 - 1 = 3$ degrees of freedom. We will use a chi-square goodness-of-fit test.

Trait	Observed	Expected	Residual = $(Obs - Exp)$	$(Obs - Exp)^2$	Component = $\dfrac{(Obs - Exp)^2}{Exp}$
Attached, noncurling	10	7.625	2.375	5.6406	0.73975
Attached, curling	22	22.875	- 0.875	0.7656	0.03347
Free, noncurling	31	22.875	8.125	66.0156	2.8859
Free, curling	59	68.625	- 9.625	92.6406	1.35

$$\sum \approx 5.01$$

$\chi^2 = 5.01$. Since the P-value = 0.1711 is high, we fail to reject the null hypothesis.

There is no evidence that the proportions of traits are anything other than 1:3:3:9.

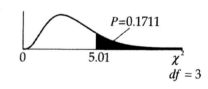

2. Tableware.

a) Since there are 57 degrees of freedom, there were 59 different products in the analysis.

b) 84.5% of the variation in retail price is explained by the polishing time.

c) Assuming the conditions have been met, the sampling distribution of the regression slope can be modeled by a Student's t-model with $(59 - 2) = 57$ degrees of freedom. We will use a regression slope t-interval. For 95% confidence, use $t^*_{57} \approx 2.0025$, or estimate from the table $t^*_{50} \approx 2.009$.

$$b_1 \pm t^*_{n-2} \times SE(b_1) = 2.49244 \pm (2.0025) \times 0.1416 \approx (2.21, 2.78)$$

d) We are 95% confident that the average price increases between \$2.21 and \$2.78 for each additional minute of polishing time.

3. Hard water.

a) H_0: There is no linear relationship between calcium concentration in water and mortality rates for males. $(\beta_1 = 0)$

H_A: There is a linear relationship between calcium concentration in water and mortality rates for males. $(\beta_1 \neq 0)$

b) Assuming the conditions have been satisfied, the sampling distribution of the regression slope can be modeled by a Student's *t*-model with (61 – 2) = 59 degrees of freedom. We will use a regression slope *t*-test. The equation of the line of best fit for these data points is: $\widehat{Mortality} = 1676 - 3.23(Calcium)$, where mortality is measured in deaths per 100,000, and calcium concentration is measured in parts per million.

$$t = \frac{b_1 - \beta_1}{SE(b_1)}$$

$$t = \frac{-3.23 - 0}{0.48}$$

$$t \approx -6.73$$

The value of *t* = – 6.73. The *P*-value of less than 0.0001 means that the association we see in the data is unlikely to occur by chance. We reject the null hypothesis, and conclude that there is strong evidence of a linear relationship between calcium concentration and mortality. Towns with higher calcium concentrations tend to have lower mortality rates.

c) For 95% confidence, use $t_{59}^* \approx 2.001$, or estimate from the table $t_{50}^* \approx 2.009$.

$$b_1 \pm t_{n-2}^* \times SE(b_1) = -3.23 \pm (2.001) \times 0.48 \approx (-4.19, -2.27)$$

d) We are 95% confident that the average mortality rate decreases by between 2.27 and 4.19 deaths per 100,000 for each additional part per million of calcium in drinking water.

4. Mutual funds.

a) **Paired data assumption:** These data are paired by mutual fund.
Randomization condition: Assume that these funds are representative of all large cap mutual funds.
10% condition: 15 mutual funds are less than 10% of all large cap mutual funds.
Nearly Normal condition: The histogram of differences is unimodal and symmetric.

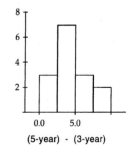

(5-year) - (3-year)

Since the conditions are satisfied, the sampling distribution of the difference can be modeled with a Student's *t*-model with 15 – 1 = 14 degrees of freedom. We will find a paired *t*-interval, with 95% confidence.

$$\bar{d} \pm t_{n-1}^* \left(\frac{s_d}{\sqrt{n}} \right) = 4.54 \pm t_{14}^* \left(\frac{2.50508}{\sqrt{15}} \right) \approx (3.15, 5.93)$$

Provided that these mutual funds are representative of all large cap mutual funds, we are 95% confident that, on average, 5-year yields are between 3.15% and 5.93% higher than 3-year yields.

b) H_0: There is no linear relationship between 3-year and 5-year rates of return. $(\beta_1 = 0)$

H_A: There is a linear relationship between 3-year and 5-year rates of return. $(\beta_1 \neq 0)$

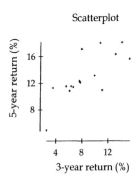

Scatterplot

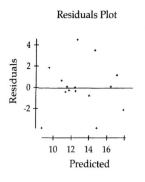

Residuals Plot

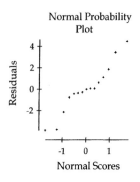

Normal Probability Plot

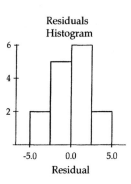

Residuals Histogram

Straight enough condition: The scatterplot is straight enough to try linear regression.
Independence assumption: The residuals plot shows no pattern.
Does the plot thicken? condition: The spread of the residuals is consistent.
Nearly Normal condition: The Normal probability plot of residuals isn't very straight. However, the histogram of residuals is unimodal and symmetric. With a sample size of 15, it is probably okay to proceed.

Since the conditions for inference are satisfied, the sampling distribution of the regression slope can be modeled by a Student's t-model with $(15 - 2) = 13$ degrees of freedom. We will use a regression slope t-test.

Dependent variable is: 5-year
No Selector
R squared = 58.4%   R squared (adjusted) = 55.2%
s = 2.360 with 15 - 2 = 13 degrees of freedom

Source	Sum of Squares	df	Mean Square	F-ratio
Regression	101.477	1	101.477	18.2
Residual	72.3804	13	5.56773	

Variable	Coefficient	s.e. of Coeff	t-ratio	prob
Constant	6.92904	1.557	4.45	0.0007
3-year	0.719157	0.1685	4.27	0.0009

The equation of the line of best fit for these data points is: $(5\hat{y}ear) = 6.92904 + 0.719157(3year)$.

The value of $t \approx 4.27$. The P-value of 0.0009 means that the association we see in the data is unlikely to occur by chance. We reject the null hypothesis, and conclude that there is strong evidence of a linear relationship between the rates of return for 3-year and 5-year periods. Provided that

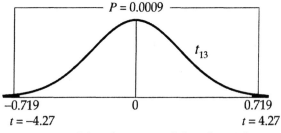

$P = 0.0009$

t_{13}

-0.719 0 0.719
$t = -4.27$ $t = 4.27$

these mutual funds are representative of all large cap mutual funds, mutual funds with higher 3-year returns tend to have higher 5-year returns.

5. Resume fraud.

In order to estimate the true percentage of people have misrepresented their backgrounds to within ± 5%, the company would have to perform about 406 random checks.

$$ME = z^* \sqrt{\frac{\hat{p}\hat{q}}{n}}$$

$$0.05 = 2.326\sqrt{\frac{(0.25)(0.75)}{n}}$$

$$n = \frac{(2.326)^2(0.25)(0.75)}{(0.05)^2}$$

$$n \approx 406 \text{ random checks}$$

6. **Paper airplanes.**

   a) It is reasonable to think that the flight distances are independent of one another. The histogram of flight distances (given) is unimodal and symmetric. Since the conditions are satisfied, the sampling distribution of the mean can be modeled by a Student's t model, with 11 – 1 = 10 degrees of freedom. We will use a one-sample t-interval with 95% confidence for the mean flight distance.

   $$\bar{y} \pm t^*_{n-1}\left(\frac{s}{\sqrt{n}}\right) = 48.3636 \pm t^*_{10}\left(\frac{18.0846}{\sqrt{11}}\right) \approx (36.21, 60.51)$$

   We are 95% confident that the mean distance the airplane may fly is between 36.21 and 60.51 feet.

   b) Since 40 feet is contained within our 95% confidence interval, it is plausible that the mean distance is 40 feet.

   c) A 99% confidence interval would be wider. Intervals with greater confidence are less precise.

   d) In order to cut the margin of error in half, she would need a sample size four times as large, or 44 flights.

7. **Back to Montana.**

   H_0: Political party is independent of income level in Montana.

   H_A: There is an association between political party and income level in Montana.

   Counted data condition: The data are counts.
   Randomization condition: Although not specifically stated, we will assume that the poll was conducted randomly.
   Expected cell frequency condition: The expected counts are all greater than 5.

	Democrat (Obs / Exp)	Republican (Obs / Exp)	Independent (Obs / Exp)
Low	30 / 24.119	16 / 22.396	12 / 14.485
Middle	28 / 30.772	24 / 28.574	22 / 14.653
High	26 / 29.109	38 / 27.03	6 / 13.861

   Under these conditions, the sampling distribution of the test statistic is χ^2 on 4 degrees of freedom. We will use a chi-square test for independence.

   $$\chi^2 = \sum_{all\,cells} \frac{(Obs - Exp)^2}{Exp} \approx 17.19$$

   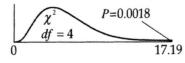

   The P-value ≈ 0.0018

   Since the P-value ≈ 0.0018 is low, we reject the null hypothesis. There is strong evidence of an association between income level and political party in Montana. An examination of the components shows that Democrats are more likely to have low incomes, Independents are more likely to have middle incomes, and Republicans are more likely to have high incomes.

8. **Wild horses.**

   a) Since there are 36 degrees of freedom, 38 herds of wild horses were studied.

   b) **Straight enough condition:** The scatterplot is straight enough to try linear regression.
   Independence assumption: The residuals plot shows no pattern.
   Does the plot thicken? condition: The spread of the residuals is consistent.
   Nearly Normal condition: The histogram of residuals is unimodal and symmetric.

   c) Since the conditions for inference are satisfied, the sampling distribution of the regression slope can be modeled by a Student's t-model with $(38 - 2) = 36$ degrees of freedom. We will use a regression slope t-interval, with 95% confidence. Use $t^*_{35} \approx 2.030$ as an estimate.

   $$b_1 \pm t^*_{n-2} \times SE(b_1) = 0.153969 \pm (2.030) \times 0.0114 \approx (0.131, 0.177)$$

   d) We are 95% confident that the mean number of foals in a herd increases by between 0.131 and 0.177 foals for each additional adult horse.

   e) The regression equation predicts that herds with 80 adults will have $-1.57835 + 0.153969(80) = 10.73917$ foals. The average size of the herds sampled is 110.237 adult horses. Use $t^*_{36} \approx 1.6883$, or use an estimate of $t^*_{35} \approx 1.690$, from the table.

   $$\hat{y}_v \pm t^*_{n-2} \sqrt{SE^2(b_1) \cdot (x_v - \bar{x})^2 + \frac{s_e^2}{n} + s_e^2}$$

   $$= 10.73917 \pm (1.6883) \sqrt{0.0114^2 \cdot (80 - 110.237)^2 + \frac{4.941^2}{38} + 4.941^2}$$

   $$\approx (2.26, 19.21)$$

   We are 95% confident that number of foals in a herd of 80 adult horses will be between 2.26 and 19.21. This prediction interval is too wide to be of much use.

9. **Lefties and music.**

   H_0: The proportion of right-handed people who can match the tone is the same as the proportion of left-handed people who can match the tone. $\left(p_L = p_R \text{ or } p_L - p_R = 0\right)$

   H_A: The proportion of right-handed people who can match the tone is different from the proportion of left-handed people who can match the tone. $\left(p_L \neq p_R \text{ or } p_L - p_R \neq 0\right)$

   Random condition: Assume that the people tested are representative of all people.
   10% condition: 76 and 53 are both less than 10% of all people.
   Independent samples condition: The groups are not associated.
   Success/Failure condition: $n\hat{p}$ (right) = 38, $n\hat{q}$ (right) = 38, $n\hat{p}$ (left) = 33, and $n\hat{q}$ (left) = 20 are all greater than 10, so the samples are both large enough.

   Since the conditions have been satisfied, we will model the sampling distribution of the difference in proportion with a Normal model with mean 0 and standard deviation

   $$\text{estimated by } SE_{pooled}(\hat{p}_L - \hat{p}_R) = \sqrt{\frac{\hat{p}_{pooled}\hat{q}_{pooled}}{n_L} + \frac{\hat{p}_{pooled}\hat{q}_{pooled}}{n_R}} = \sqrt{\frac{\left(\frac{71}{129}\right)\left(\frac{58}{129}\right)}{53} + \frac{\left(\frac{71}{129}\right)\left(\frac{58}{129}\right)}{76}} \approx 0.089.$$

The observed difference between the proportions is:
0.6226 – 0.5 = 0.1226.

Since the *P*-value = 0.1683 is high, we
fail to reject the null hypothesis. There
is no evidence that the proportion of
people able to match the tone differs
between right-handed and left-handed
people.

$$z = \frac{0.1226 - 0}{0.089}$$
$$z \approx 1.38$$

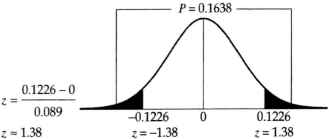

10. AP Statistics scores.

a) H_0: The distribution of AP Statistics scores at Ithaca High School is the same as it is
nationally.
H_A: The distribution of AP Statistics scores at Ithaca High School is different than it is
nationally.

Counted data condition: The data are counts.
Randomization condition: Assume that this group of students is representative of all years
at Ithaca High School.
Expected cell frequency condition: The expected counts (shown in the table) are all greater
than 5.

Under these conditions, the sampling distribution of the test statistic is χ^2 on 5 – 1 = 4
degrees of freedom. We will use a chi-square goodness-of-fit test.

Score	Observed	Expected	Residual = (Obs – Exp)	Standardized Residual = $\frac{(Obs - Exp)}{\sqrt{Exp}}$	Component = $\frac{(Obs - Exp)^2}{Exp}$
5	26	11.155	14.845	4.445	19.756
4	36	22.698	13.302	2.792	7.7955
3	19	24.153	– 5.153	– 1.049	1.0994
2	10	18.527	– 8.527	– 1.981	3.9245
1	6	20.467	– 14.47	– 3.198	10.226

$$\sum \approx 42.801$$

$\chi^2 \approx 42.801$. Since the *P*-value is essentially 0, we reject the null hypothesis.

There is strong evidence that the distribution of scores at Ithaca High School is different
than the national distribution. Students at IHS get fewer scores of 2 and 1 than expected,
and more scores of 4 and 5 than expected.

b) H_0: Gender and AP Statistics score are independent at Ithaca High School.

H_A: There is an association between gender and AP Statistics score at Ithaca High School.

Counted data condition: The data are counts.
Randomization condition: Assume this year's students are representative of all years.
Expected cell frequency condition: After combining the cells for scores of 2 and 1, the
expected counts are all greater than 5.

	Boys (Obs/Exp)	Girls (Obs/Exp)
5	13 / 13.67	13 / 12.33
4	21 / 18.928	15 / 17.072
3	6 / 9.9897	13 / 9.0103
2 or 1	11 / 8.4124	5 / 7.5876

Under these conditions, the sampling distribution of the test statistic is χ^2 on 3 degrees of freedom. We will use a chi-square test for independence. (This is a test for independence, since we have one group that has been classified according to two variables, gender and score. However, if you said it was a test for homogeneity, since you were comparing two groups, no one would get terribly upset!)

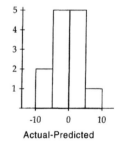

With $\chi^2 = \sum\limits_{all\ cells} \dfrac{(Obs - Exp)^2}{Exp} \approx 5.59$, the P-value ≈ 0.1336.

Since P-value ≈ 0.1336 is high, we fail to reject the null hypothesis. There is no evidence of an association between gender and score at Ithaca High School. The boys seem to do just as well as the girls.

11. Polling.

a) H_0: The mean difference in the number of predicted Democrats and the number of actual Democrats is zero. $(\mu_d = 0)$

H_A: The mean difference in the number of predicted Democrats and the number of actual Democrats is different than zero. $(\mu_d \neq 0)$

Paired data assumption: The data are paired by year.
Randomization condition: Assume these predictions are representative of other predictions.
10% condition: We are testing the predictions, not the years.
Nearly Normal condition: The histogram of differences between the predicted number of Democrats and the actual number of Democrats is roughly unimodal and symmetric. The year 1958 is an outlier, and was removed.

Since the conditions are satisfied, the sampling distribution of the difference can be modeled with a Student's t-model with 13 – 1 = 12 degrees of freedom, $t_{12}\left(0, \dfrac{3.57878}{\sqrt{13}}\right)$.

We will use a paired t-test, with $\bar{d} = -0.846$.

Since the P-value = 0.4106 is high, we fail to reject the null hypothesis. There is no evidence that the mean difference between the actual and predicted number of Democrats was anything other than 0.

$$t = \frac{\bar{d} - 0}{\dfrac{s_d}{\sqrt{n}}}$$

$$t = \frac{-0.846 - 0}{\dfrac{3.57878}{\sqrt{13}}}$$

$$t \approx -0.85$$

b) H_0: There is no linear relationship between Gallup's predictions and the actual number of Democrats. $(\beta_1 = 0)$

H_A: There is a linear relationship between Gallup's predictions and the actual number of Democrats. $(\beta_1 \neq 0)$

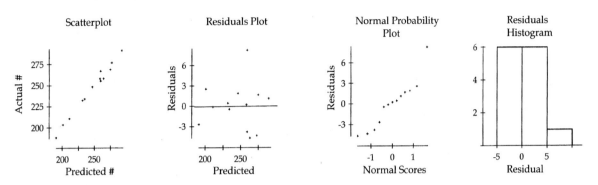

Straight enough condition: The scatterplot is straight enough to try linear regression.
Independence assumption: The residuals plot shows no pattern.
Does the plot thicken? condition: The spread of the residuals is consistent.
Nearly Normal condition: After an outlier in 1958 is removed, the Normal probability plot of residuals still isn't very straight. However, the histogram of residuals is roughly unimodal and symmetric. With a sample size of 13, it is probably okay to proceed.

Since the conditions for inference are satisfied, the sampling distribution of the regression slope can be modeled by a Student's t-model with (13 – 2) = 11 degrees of freedom. We will use a regression slope t-test.

Dependent variable is: **Actual**
No Selector
R squared = 98.7%   R squared (adjusted) = 98.6%
s = 3.628 with 13 - 2 = 11 degrees of freedom

The equation of the line of best fit for these data points is: $\hat{Actual} = 6.00180 + 0.972206(Predicted)$.

Source	Sum of Squares	df	Mean Square	F-ratio
Regression	10874.4	1	10874.4	826
Residual	144.805	11	13.1641	

Variable	Coefficient	s.e. of Coeff	t-ratio	prob
Constant	6.00180	8.395	0.715	0.4895
Predicted	0.972206	0.0338	28.7	≤ 0.0001

The value of $t \approx 28.7$. The P-value of essentially 0 means that the association we see in the data is unlikely to occur by chance. We reject the null hypothesis, and conclude that there is strong evidence of a linear relationship between the number of Democrats predicted by Gallup and the number of Democrats actually in the House of Representatives. Years in which the predicted number was high tend to have high actual numbers also. The high value of $R^2 = 98.7\%$ indicates a very strong model. Gallup polls seem very accurate.

12. Twins.

H_0: There is no association between duration of pregnancy and level of prenatal care.

H_A: There is an association between duration of pregnancy and level of prenatal care.

Counted data condition: The data are counts.
Randomization condition: Assume that these pregnancies are representative of all twin births.
Expected cell frequency condition: The expected counts are all greater than 5.

	Preterm (induced or Cesarean) (Obs/Exp)	Preterm (without procedures) (Obs/Exp)	Term or postterm (Obs/Exp)
Intensive	18 /16.676	15 / 15.579	28 / 28.745
Adequate	46 / 42.101	43 / 39.331	65 / 72.568
Inadequate	12 / 17.223	13 / 16.090	38 / 29.687

Under these conditions, the sampling distribution of the test statistic is χ^2 on 4 degrees of freedom. We will use a chi-square test for independence.

$$\chi^2 = \sum_{all\ cells} \frac{(Obs - Exp)^2}{Exp} \approx 6.14,\text{ and the }P\text{-value} \approx 0.1887.$$

Since the P-value ≈ 0.1887 is high, we fail to reject the null hypothesis. There is no evidence of an association between duration of pregnancy and level of prenatal care in twin births.

13. Twins, again.

H_0: The distributions of pregnancy durations are the same for the three years.

H_A: The distributions of pregnancy durations are different for the three years.

Counted data condition: The data are counts.
Independence assumption: Assume that the durations of the pregnancies are mutually indpendent.
Expected cell frequency condition: The expected counts are all greater than 5.

	1990 (Obs/Exp)	1995 (Obs/Exp)	2000 (Obs/Exp)
Preterm (induced or Cesarean)	11 /12.676	13 / 13.173	19 / 17.150
Preterm (without procedures)	13 / 13.266	14 / 13.786	18 / 17.948
Term or postterm	27 / 25.058	26 / 26.04	32 / 33.902

Under these conditions, the sampling distribution of the test statistic is χ^2 on 4 degrees of freedom. We will use a chi-square test for homogeneity.

$$\chi^2 = \sum_{all\ cells} \frac{(Obs - Exp)^2}{Exp} \approx 0.69,\text{ and the }P\text{-value} \approx 0.9526.$$

Since the P-value ≈ 0.9526 is high, we fail to reject the null hypothesis. There is no evidence that the distributions of the durations of pregnancies are different for the three years. It does not appear that they way the hospital deals with twin pregnancies has changed.

14. Preemies.

H_0: The proportion of "preemies" who are of "subnormal height" as adults is the same as the proportion of normal birth weight babies who are. $(p_P = p_N \text{ or } p_P - p_N = 0)$

H_A: The proportion of "preemies" who are of "subnormal height" as adults is greater than the proportion of normal birth weight babies who are. $(p_P > p_N \text{ or } p_P - p_N > 0)$

Random condition: Assume that these children are representative of all children.
10% condition: 242 and 233 are both less than 10% of all children.
Independent samples condition: The groups are not associated.
Success/Failure condition: $n\hat{p}$ (preemies) = 24, $n\hat{q}$ (preemies) = 218, $n\hat{p}$ (normal) = 12, and $n\hat{q}$ (normal) = 221 are all greater than 10, so the samples are both large enough.

Since the conditions have been satisfied, we will model the sampling distribution of the difference in proportion with a Normal model with mean 0 and standard deviation estimated by $SE_{pooled}(\hat{p}_P - \hat{p}_N) = \sqrt{\dfrac{\hat{p}_{pooled}\hat{q}_{pooled}}{n_P} + \dfrac{\hat{p}_{pooled}\hat{q}_{pooled}}{n_N}} = \sqrt{\dfrac{\left(\frac{36}{475}\right)\left(\frac{439}{475}\right)}{242} + \dfrac{\left(\frac{36}{475}\right)\left(\frac{439}{475}\right)}{233}} \approx 0.02429.$

The observed difference between the proportions is:
$0.09917 - 0.05150 = 0.04767$.

Since the *P*-value = 0.0249 is low, we reject the null hypothesis. There is moderate evidence that "preemies" are more likely to be of "subnormal height" as adults than children of normal birth weight.

$z = \dfrac{0.04767 - 0}{0.02429}$

$z \approx 1.96$

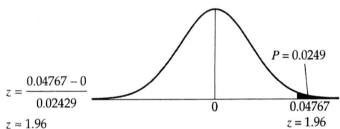

$P = 0.0249$

$0 \qquad 0.04767$
$z = 1.96$

15. LA rainfall.

a) **Independence assumption:** Annual rainfall is independent from year to year.
Nearly Normal condition: The histogram of the rainfall totals is skewed to the right, but the sample is fairly large, so it is safe to proceed.

The mean annual rainfall is 14.5165 inches, with a standard deviation 7.82044 inches. Since the conditions have been satisfied, construct a one-sample *t*-interval, with 22 – 1 = 21 degrees of freedom, at 90% confidence.

$$\bar{y} \pm t^*_{n-1}\left(\frac{s}{\sqrt{n}}\right) = 14.5164 \pm t^*_{21}\left(\frac{7.82044}{\sqrt{22}}\right) \approx (11.65, 17.39)$$

We are 90% confident that the mean annual rainfall in LA is between 11.65 and 17.39 inches.

b) Start by making an estimate, either using $z^* = 1.645$ or $t^*_{21} = 1.721$ from above. Either way, your estimate is around 40 people. Make a better estimate using $t^*_{40} = 1.684$. You would need about 44 years' data to estimate the annual rainfall in LA to within 2 inches.

$ME = t^*_{40}\left(\dfrac{s}{\sqrt{n}}\right)$

$2 = 1.684\left(\dfrac{7.82044}{\sqrt{n}}\right)$

$n = \dfrac{(2.064)^2(7.82044)^2}{(2)^2}$

$n \approx 44$ years

c) H_0: There is no linear relationship between year and annual LA rainfall. $(\beta_1 = 0)$

   H_A: There is a linear relationship between year and annual LA rainfall. $(\beta_1 \neq 0)$

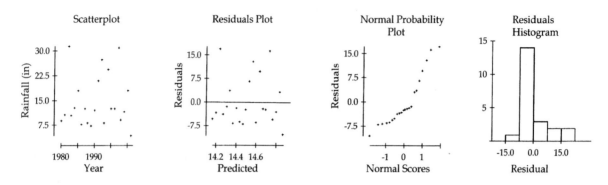

Straight enough condition: The scatterplot is straight enough to try linear regression, although there is no apparent pattern.

Independence assumption: The residuals plot shows no pattern.

Does the plot thicken? condition: The spread of the residuals is consistent.

Nearly Normal condition: The Normal probability plot of residuals is not straight, and the histogram of the residuals is skewed to the right, but a sample of 22 years is large enough to proceed.

Since the conditions for inference are satisfied, the sampling distribution of the regression slope can be modeled by a Student's t-model with (22 – 2) = 20 degrees of freedom. We will use a regression slope t-test.

```
Dependent variable is:   Rain  (in.)
No Selector
R squared = 0.1%    R squared (adjusted) = -4.9%
s = 8.011  with  22 - 2 = 20  degrees of freedom
```

Source	Sum of Squares	df	Mean Square	F-ratio
Regression	0.979449	1	0.979449	0.015
Residual	1283.37	20	64.1684	

Variable	Coefficient	s.e. of Coeff	t-ratio	prob
Constant	-51.6838	535.8	-0.096	0.9241
Year	0.033258	0.2692	0.124	0.9029

The equation of the line of best fit for these data points is: $\hat{Rain} = -51.6838 + 0.033258(Year)$.

The value of $t \approx 0.124$. The P-value of 0.92029 means that the association we see in the data is quite likely to occur by chance. We fail to reject the null hypothesis, and conclude that there is no evidence of a linear relationship between the annual rainfall in LA and the year.

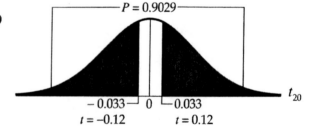

16. Age and party.

a) There is one sample, classified according to two different variables, so we will perform a chi-square test for independence.

b) H_0: There is no association between age and political party.

   H_A: There is an association between age and political party.

Counted data condition: The data are counts.
Randomization condition: These data are from a representative phone survey.
Expected cell frequency condition: The expected counts are all greater than 5.

	Republican (Obs/Exp)	Democrat (Obs/Exp)	Independent (Obs/Exp)
18 – 29	241 / 275.39	351 /351.18	409 / 374.44
30 – 49	299 / 274.84	330 / 350.47	370 / 373.69
50 – 64	282 / 274.56	341 / 350.12	375 / 373.31
65 +	279 / 276.21	382 / 352.23	343 / 375.56

Under these conditions, the sampling distribution of the test statistic is χ^2 on 6 degrees of freedom. We will use a chi-square test for independence.

$$\chi^2 = \sum_{all\,cells} \frac{(Obs - Exp)^2}{Exp} \approx 16.66, \text{ and the } P\text{-value} \approx 0.0106.$$

c) Since the P-value ≈ 0.0106 is low, we reject the null hypothesis. There is strong evidence of an association between age and political party.

The table of standardized residuals is useful for the analysis of the differences. Looking for the largest standardized residuals, we can see there are fewer Republicans and more Independents among those 18 – 29 than we expect, and fewer Independents than we expect among those 65 and older.

	Standardized Residuals		
	Republican	Democrat	Independent
18 – 29	–2.07	–0.01	1.79
30 – 49	1.46	–1.09	–0.19
50 – 64	0.45	–0.49	0.10
65 +	0.17	1.59	–1.68

17. Eye and hair color.

a) This is an attempt at linear regression. Regression inference is meaningless here, since eye and hair color are categorical variables.

b) This is an analysis based upon a chi-square test for independence.

H_0: Eye color and hair color are independent.

H_A: There is an association between eye color and hair color.

Since we have two categorical variables, this analysis seems appropriate. However, if you check the expected counts, you will find that 4 of them are less than 5. We would have to combine several cells in order to perform the analysis. (Always check the conditions!)

Since the value of chi-square is so high, it is likely that we would find an association between eye and hair color, even after the cells were combined. There are many cells of interest, but some of the most striking differences that would not be affected by cell combination involve people with fair hair. Blonds are likely to have blue eyes, and not likely to have brown eyes. Those with red hair are not likely to have brown eyes. Additionally, those with black hair are much more likely to have brown eyes than blue.

18. Depression and the Internet.

a) H_0: There is no linear relationship between depression and Internet usage. $\left(\beta_1 = 0\right)$

H_A: There is a linear relationship between depression and Internet usage. $\left(\beta_1 \neq 0\right)$

Since the conditions for inference are satisfied (given), the sampling distribution of the regression slope can be modeled by a Student's t-model with $(162 - 2) = 160$ degrees of freedom. We will use a regression slope t-test. The equation of the line of best fit for these data points is: $DepressionAfter = 0.565485 + 0.019948(InternetUsage)$.

The value of $t \approx 2.76$. The P-value of 0.0064 means that the association we see in the data is unlikely to occur by chance. We reject the null hypothesis, and conclude that there is strong evidence of a linear relationship between depression and Internet usage. Those with high levels of Internet usage tend to have high levels of depression. It should be noted, however, that although the

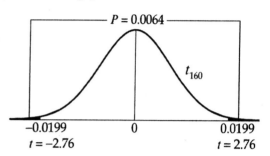

evidence is strong, the association is quite weak, with $R^2 = 4.6\%$. The regression analysis only explains 4.6% of the variation in depression level.

b) The study says nothing about causality, merely association. Furthermore, there are almost certainly other factors involved. In fact, if 4.6% of the variation in depression level is related to Internet usage, the other 95.4% of the variation must be related to something else!

c) H_0: The mean difference in depression before and after the experiment is zero. $\left(\mu_d = 0\right)$

H_A: The mean difference in depression before and after the experiment is greater than zero. $\left(\mu_d > 0\right)$

Since the conditions are satisfied (given), the sampling distribution of the difference can be modeled with a Student's t-model with $162 - 1 = 161$ degrees of freedom, $t_{161}\left(0, \dfrac{0.552417}{\sqrt{162}}\right)$.

We will use a paired t-test, with $\bar{d} = -0.118457$.

Since the P-value = 0.9965 is very high, we fail to reject the null hypothesis. There is no evidence that the mean depression level increased over the course of the experiment. In fact, these data suggest that depression levels actually decreased.

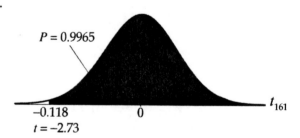

19. Pregnancy.

a) H_0: The proportion of live births is the same for women under the age of 38 as it is for women over the age of 38. $\left(p_{<38} = p_{\geq 38} \text{ or } p_{<38} - p_{\geq 38} = 0\right)$

H_A : The proportion of live births is different for women under the age of 38 than for women over the age of 38. $\left(p_{<38} \neq p_{\geq 38} \text{ or } p_{<38} - p_{\geq 38} \neq 0\right)$

Random condition: Assume that the women studied are representative of all women.
10% condition: 157 and 89 are both less than 10% of all women.
Independent samples condition: The groups are not associated.
Success/Failure condition: $n\hat{p}$ (under 38) = 42, $n\hat{q}$ (under 38) = 115, $n\hat{p}$ (38 and over) = 7, and $n\hat{q}$ (38 and over) = 82 are not all greater than 10, since the observed number of live births is only 7. However, if we check the pooled value, $n\hat{p}_{pooled}$ (38 and over) = (89)(0.191) = 17. All of the samples are large enough.

Since the conditions have been satisfied, we will model the sampling distribution of the difference in proportion with a Normal model with mean 0 and standard deviation

estimated by $SE_{pooled}\left(\hat{p}_{<38} - \hat{p}_{\geq 38}\right) = \sqrt{\dfrac{\hat{p}_{pooled}\hat{q}_{pooled}}{n_{<38}} + \dfrac{\hat{p}_{pooled}\hat{q}_{pooled}}{n_{\geq 38}}} = \sqrt{\dfrac{\left(\frac{49}{246}\right)\left(\frac{197}{246}\right)}{157} + \dfrac{\left(\frac{49}{246}\right)\left(\frac{197}{246}\right)}{89}} \approx 0.0530.$

The observed difference between the proportions is:
0.2675 − 0.0787 = 0.1888.

Since the *P*-value = 0.0004 is low, we reject the null hypothesis. There is strong evidence to suggest a difference in the proportion of live births for women under 38 and women 38 and over at this clinic. In fact, the evidence suggests that women under 38 have a higher proportion of live births.

$z = \dfrac{0.1888 - 0}{0.0530}$

$z \approx 3.56$

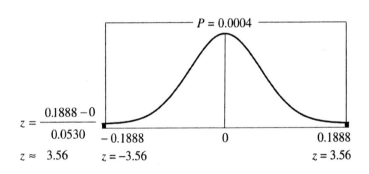

b) H_0: Age and birth rate are independent.

H_A: There is an association between age and birth rate

Counted data condition: The data are counts.
Randomization condition: Assume that these women are representative of all women.
Expected cell frequency condition: The expected counts are all greater than 5.

	Live birth (Obs / Exp)	No live birth (Obs / Exp)
Under 38	42 / 31.272	115 / 125.73
38 and over	7 / 17.728	82 / 71.27

Under these conditions, the sampling distribution of the test statistic is χ^2 on 1 degree of freedom. We will use a chi-square test for independence.

$\chi^2 = \displaystyle\sum_{all\ cells} \dfrac{(Obs - Exp)^2}{Exp} \approx 12.70$, and the *P*-value ≈ 0.0004.

Since the *P*-value ≈ 0.0004 is low, we reject the null hypothesis. There is strong evidence of an association between age and birth rate. Younger mothers tend to have higher birth rates.

c) A two-proportion z-test and a chi-square test for independence with 1 degree of freedom are equivalent. $z^2 = (3.563944)^2 = 12.70 = \chi^2$. The P-values are both the same.

20. Family planning.

H_0: Unplanned pregnancies and education level are independent.

H_A: There is an association between unplanned pregnancies and education level.

Counted data condition: The percentages must be converted to counts.
Randomization condition: Assume that these women are representative of all women.
Expected cell frequency condition: The expected counts are all greater than 5.

	<3 years HS (Obs/Exp)	3+ years HS (Obs/Exp)	Some college (Obs/Exp)
Planned	200 / 249.88	271 / 257.07	137 / 101.05
Unplanned	391 / 341.12	337 / 350.93	102 / 137.95

Under these conditions, the sampling distribution of the test statistic is χ^2 on 1 degree of freedom. We will use a chi-square test for independence.

$$\chi^2 = \sum_{all\ cells} \frac{(Obs - Exp)^2}{Exp} \approx 40.71,$$

and the P-value is essentially 0.

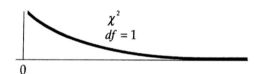

Since the P-value is essentially 0, we reject the null hypothesis. There is strong evidence of an association between unplanned pregnancies and education level. More educated women tend to have fewer unplanned pregnancies.

21. Old Faithful.

a) There is a moderate, linear, positive association between duration of the previous eruption and interval between eruptions for Old Faithful. Relatively long eruptions appear to be associated with relatively long intervals until the next eruption.

b) H_0: There is no linear relationship between duration of the eruption and interval until the next eruption. $\left(\beta_1 = 0\right)$

H_A: There is a linear relationship between duration of the eruption and interval until the next eruption. $\left(\beta_1 \neq 0\right)$

c) **Straight enough condition:** The scatterplot is straight enough to try linear regression.
Independence assumption: The residuals plot shows no pattern.
Does the plot thicken? condition: The spread of the residuals is consistent.
Nearly Normal condition: The histogram of residuals is unimodal and symmetric.

Since the conditions for inference are satisfied, the sampling distribution of the regression slope can be modeled by a Student's t-model with (222 – 2) = 220 degrees of freedom. We will use a regression slope t-test. The equation of the line of best fit for these data points is: $\widehat{Interval} = 33.9668 + 10.3582(Duration)$.

d) The value of $t \approx 27.1$. The *P*-value of essentially 0 means that the association we see in the data is unlikely to occur by chance. We reject the null hypothesis, and conclude that there is strong evidence of a linear relationship between duration and interval. Relatively long eruptions tend to be associated with relatively long intervals until the next eruption.

e) The regression equation predicts that an eruption with duration of 2 minutes will have an interval until the next eruption of $33.9668 + 10.3582(2) = 54.6832$ minutes. $(t_{220}^* \approx 1.9708)$

$$\hat{y}_v \pm t_{n-2}^* \sqrt{SE^2(b_1) \cdot (x_v - \bar{x})^2 + \frac{s_e^2}{n}}$$

$$= 54.6832 \pm (1.9708)\sqrt{0.3822^2 \cdot (2 - 3.57613)^2 + \frac{6.159^2}{222}}$$

$$\approx (53.24, 56.12)$$

We are 95% confident that, after a 2-minute eruption, the mean length of time until the next eruption will be between 53.24 and 56.12 minutes.

f) The regression equation predicts that an eruption with duration of 4 minutes will have an interval until the next eruption of $33.9668 + 10.3582(4) = 75.3996$ minutes. $(t_{220}^* \approx 1.9708)$

$$\hat{y}_v \pm t_{n-2}^* \sqrt{SE^2(b_1) \cdot (x_v - \bar{x})^2 + \frac{s_e^2}{n} + s_e^2}$$

$$= 75.3996 \pm (1.9708)\sqrt{0.3822^2 \cdot (4 - 3.57613)^2 + \frac{6.159^2}{222} + 6.159^2}$$

$$\approx (63.23, 87.57)$$

We are 95% confident that the length of time until the next eruption will be between 63.23 and 87.57 minutes, following a 4-minute eruption.

22. Togetherness.

a) H_0: There is no linear relationship number of meals eaten as a family and grades. $(\beta_1 = 0)$
H_A: There is a linear relationship number of meals eaten as a family and grades. $(\beta_1 \neq 0)$

Since the conditions for inference are satisfied (given), the sampling distribution of the regression slope can be modeled by a Student's *t*-model with (142 – 2) = 140 degrees of freedom. We will use a regression slope *t*-test. The equation of the line of best fit for these data points is: $\hat{GPA} = 2.7288 + 0.1093(Meals / Week)$.

$$t = \frac{b_1 - \beta_1}{SE(b_1)}$$

$$t = \frac{0.1093 - 0}{0.0263}$$

$$t \approx 4.16$$

The value of $t \approx 4.16$. The *P*-value of less than 0.0001 means that the association we see in the data is unlikely to occur by chance. We reject the null hypothesis, and conclude that there is strong evidence of a linear relationship between grades and the number of meals eaten as a family. Students whose families eat together relatively frequently tend to have higher grades than those whose families don't eat together as frequently.

b) This relationship would not be particularly useful for predicting a student's grade point average. $R^2 = 11.0\%$, which means that only 11% of the variation in GPA can be explained by the number of meals eaten together per week.

c) These conclusions are not contradictory. There is strong evidence that the slope is not zero, and that means strong evidence of a linear relationship. This does not mean that the relationship itself is strong, or useful for predictions.

23. Learning math.

a) H₀: The mean score of Accelerated Math students is the same as the mean score of traditional students. $\left(\mu_A = \mu_T \text{ or } \mu_A - \mu_T = 0\right)$

H_A: The mean score of Accelerated Math students is different from the mean score of traditional students. $\left(\mu_A \neq \mu_T \text{ or } \mu_A - \mu_T \neq 0\right)$

Independent groups assumption: Scores of students from different classes should be independent.
Randomization condition: Although not specifically stated, classes in this experiment were probably randomly assigned to learn either Accelerated Math or traditional curricula.
10% condition: 231 and 245 are less than 10% of all students.
Nearly Normal condition: We don't have the actual data, so we can't check the distribution of the sample. However, the samples are large. The Central Limit Theorem allows us to proceed.

Since the conditions are satisfied, it is appropriate to model the sampling distribution of the difference in means with a Student's t-model, with 459.24 degrees of freedom (from the approximation formula).

We will perform a two-sample t-test. The sampling distribution model has mean 0, with

standard error: $SE(\bar{y}_A - \bar{y}_T) = \sqrt{\dfrac{84.29^2}{231} + \dfrac{74.68^2}{245}} \approx 7.3158.$

The observed difference between the mean scores is 560.01 − 549.65 = 10.36

Since the P-value = 0.1574, we fail to reject the null hypothesis. There is no evidence that the Accelerated Math students have a different mean score on the pretest than the traditional students.

$$t = \frac{(\bar{y}_A - \bar{y}_T) - (0)}{SE(\bar{y}_A - \bar{y}_T)}$$

$$t \approx \frac{10.36}{7.3158}$$

$$t \approx 1.42$$

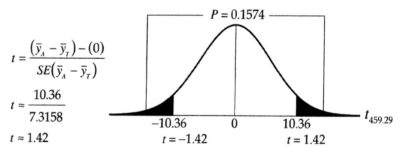

b) H₀: Accelerated Math students do not show significant improvement in test scores. The mean individual gain for Accelerated Math is zero. $\left(\mu_d = 0\right)$

H_A: Accelerated Math students show significant improvement in test scores. The mean individual gain for Accelerated Math is greater than zero. $\left(\mu_d > 0\right)$

Paired data assumption: The data are paired by student.
Randomization condition: Although not specifically stated, classes in this experiment were probably randomly assigned to learn either Accelerated Math or traditional curricula.
10% condition: We are testing the Accelerated Math program, not the students.

Nearly Normal condition: We don't have the actual data, so we cannot look at a graphical display, but since the sample is large, it is safe to proceed.

The Accelerated Math students had a mean individual gain of $\bar{d} = 77.53$ points and a standard deviation of 78.01 points. Since the conditions for inference are satisfied, we can model the sampling distribution of the mean individual gain with a Student's t model, with $231 - 1 = 230$ degrees of freedom, $t_{230}\left(0, \frac{78.01}{\sqrt{231}}\right)$. We will perform a paired t-test.

$$t = \frac{\bar{d} - 0}{\frac{s_d}{\sqrt{n}}}$$

Since the P-value is essentially 0, we reject the null hypothesis. There is strong evidence that the mean individual gain is greater than zero. The Accelerated Math students showed significant improvement.

$$t = \frac{77.53 - 0}{\frac{78.01}{\sqrt{231}}}$$

$t \approx 15.11$

c) H_0: Students taught using traditional methods do not show significant improvement in test scores. The mean individual gain for traditional methods is zero. $(\mu_d = 0)$

H_A: Students taught using traditional methods show significant improvement in test scores. The mean individual gain for traditional methods is greater than zero. $(\mu_d > 0)$

Paired data assumption: The data are paired by student.
Randomization condition: Although not specifically stated, classes in this experiment were probably randomly assigned to learn either Accelerated Math or traditional curricula.
10% condition: We are testing the program, not the students.
Nearly Normal condition: We don't have the actual data, so we cannot look at a graphical display, but since the sample is large, it is safe to proceed.

The students taught using traditional methods had a mean individual gain of $\bar{d} = 39.11$ points and a standard deviation of 66.25 points. Since the conditions for inference are satisfied, we can model the sampling distribution of the mean individual gain with a Student's t model, with $245 - 1 = 244$ degrees of freedom, $t_{244}\left(0, \frac{66.25}{\sqrt{245}}\right)$.

We will perform a paired t-test.

$$t = \frac{\bar{d} - 0}{\frac{s_d}{\sqrt{n}}}$$

Since the P-value is essentially 0, we reject the null hypothesis. There is strong evidence that the mean individual gain is greater than zero. The students taught using traditional methods showed significant improvement.

$$t = \frac{39.11 - 0}{\frac{66.25}{\sqrt{245}}}$$

$t \approx 9.24$

d) H_0: The mean individual gain of Accelerated Math students is the same as the mean individual gain of traditional students. $(\mu_{dA} = \mu_{dT}$ or $\mu_{dA} - \mu_{dT} = 0)$

H_A: The mean individual gain of Accelerated Math students is greater than the mean individual gain of traditional students. $(\mu_{dA} > \mu_{dT}$ or $\mu_{dA} - \mu_{dT} > 0)$

Independent groups assumption: Individual gains of students from different classes should be independent.
Randomization condition: Although not specifically stated, classes in this experiment were probably randomly assigned to learn either Accelerated Math or traditional curricula.
10% condition: 231 and 245 are less than 10% of all students.
Nearly Normal condition: We don't have the actual data, so we can't check the distribution of the sample. However, the samples are large. The Central Limit Theorem allows us to proceed.

Since the conditions are satisfied, it is appropriate to model the sampling distribution of the difference in means with a Student's t-model, with 452.10 degrees of freedom (from the approximation formula).

We will perform a two-sample t-test. The sampling distribution model has mean 0, with standard error: $SE(\bar{d}_A - \bar{d}_T) = \sqrt{\dfrac{78.01^2}{231} + \dfrac{66.25^2}{245}} \approx 6.6527$.

The observed difference between the mean scores is 77.53 – 39.11 = 38.42

$t = \dfrac{(\bar{d}_A - \bar{d}_T) - 0}{SE(\bar{d}_A - \bar{d}_T)}$

Since the P-value is less than 0.0001, we reject the null hypothesis. There is strong evidence that the Accelerated Math students have an individual gain that is significantly higher than the individual gain of the students taught using traditional methods.

$t = \dfrac{38.42 - 0}{6.6527}$

$t \approx 5.78$

24. Pesticides.

H_0 : The percentage of males born to workers at the plant is 51.2%. ($p = 0.512$)
H_A : The percentage of males born to workers at the plant is less than 51.2%. ($p < 0.512$)

Independence assumption: It is reasonable to think that the births are independent.
Success/Failure Condition: $np = (227)(0.512) = 116$ and $nq = (227)(0.488) = 111$ are both greater than 10, so the sample is large enough.

The conditions have been satisfied, so a Normal model can be used to model the sampling distribution of the proportion, with $\mu_{\hat{p}} = p = 0.512$ and $\sigma(\hat{p}) = \sqrt{\dfrac{pq}{n}} = \sqrt{\dfrac{(0.512)(0.488)}{227}} \approx 0.0332$.

We can perform a one-proportion z-test. The observed proportion of males is $\hat{p} = 0.40$.

The value of $z \approx -3.35$, meaning that the observed proportion of males is over 3 standard deviations below the expected proportion. The P-value associated with this z score is approximately 0.0004.

$z = \dfrac{\hat{p} - p_0}{\sqrt{\dfrac{pq}{n}}}$

$z = \dfrac{0.4 - 0.512}{\sqrt{\dfrac{(0.512)(0.488)}{227}}}$

$z \approx -3.35$

$P = 0.0004$

0.4   0.512
$z = -3.35$

With a P-value this low, we reject the null hypothesis. There is strong evidence that the percentage of males born to workers is less than 51.2%. This provides evidence that human exposure to dioxin may result in the birth of more girls.

25. Dairy sales.

a) Since the CEO is interested in the association between cottage cheese sales and ice cream sales, the regression analysis is appropriate.

b) There is a moderate, linear, positive association between cottage cheese and ice cream sales. For each additional million pounds of cottage cheese sold, an average of 1.19 million pounds of ice cream are sold.

c) The regression will not help here. A paired *t*-test will tell us whether there is an average difference in sales.

d) There is evidence that the company sells more cottage cheese than ice cream, on average.

e) In part a, we are assuming that the relationship is linear, that errors are independent with constant variation, and that the distribution of errors is Normal.

In part c, we are assuming that the observations are independent and that the distribution of the differences is Normal. This may not be a valid assumption, since the histogram of differences looks bimodal.

f) The equation of the regression line is $Ice\hat{C}ream = -26.5306 + 1.19334(CottageCheese)$. In a month in which 82 million pounds of ice cream are sold we expect to sell:

$Ice\hat{C}ream = -26.5306 + 1.19334(82) = 71.32$ million pounds of ice cream.

g) Assuming the conditions for inference are satisfied, the sampling distribution of the regression slope can be modeled by a Student's *t*-model with (12 – 2) = 10 degrees of freedom. We will use a regression slope *t*-interval, with 95% confidence.

$b_1 \pm t^*_{n-2} \times SE(b_1) = 1.19334 \pm (2.228) \times 0.4936 \approx (0.09, 2.29)$

h) We are 95% confident that the mean number of pounds of ice cream sold increases by between 0.09 and 2.29 pounds for each additional pound of cottage cheese sold.

26. Infliximab.

H_0: The remission rates are the same for the three groups.

H_A: The remission rates are different for the three groups.

Counted data condition: The data are counts.
Randomization condition: Assume that these patients are representative of all patients.
Expected cell frequency condition: The expected counts are all greater than 5.

	Placebo (Obs / Exp)	5 mg (Obs / Exp)	10 mg (Obs / Exp)
Remission	23 / 38.418	44 / 39.466	50 / 39.116
No Remission	87 / 71.582	69 / 73.534	62 / 72.884

Under these conditions, the sampling distribution of the test statistic is χ^2 on 2 degrees of freedom. We will use a chi-square test for homogeneity.

$$\chi^2 = \sum_{all\,cells} \frac{(Obs - Exp)^2}{Exp} \approx 14.96,$$
and the *P*-value ≈ 0.0006.

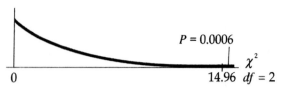

Since the *P*-value ≈ 0.0006 is low, we reject the null hypothesis. There is strong evidence that the remission rates are different in the three groups. Patients receiving 10 mg of Infliximab have higher remission rates than the other groups. These data indicate that continued treatment with Infliximab is of value to Crohn's disease patients who exhibit a positive initial response to the drug.

27. Weight loss.

Randomization Condition: The respondents were randomly selected from among the clients of the weight loss clinic.
10% Condition: 20 people are less than 10% of all clients.
Nearly Normal Condition: The histogram of the number of pounds lost for each respondent is unimodal and symmetric, with no outliers.

The clients in the sample had a mean weight loss of 9.15 pounds, with a standard deviation of 1.94733 pounds. Since the conditions have been satisfied, construct a one-sample *t*-interval, with $20 - 1 = 19$ degrees of freedom, at 95% confidence.

$$\bar{y} \pm t^*_{n-1}\left(\frac{s}{\sqrt{n}}\right) = 9.15 \pm t^*_{19}\left(\frac{1.94733}{\sqrt{20}}\right) \approx (8.24, 10.06)$$

We are 95% confident that the mean weight loss experienced by clients of this clinic is between 8.24 and 10.06 pounds. Since 10 pounds is contained within the interval, the claim that the program will allow clients to lose 10 pounds in a month is plausible. Answers may vary, depending on the chosen level of confidence.

28. Education vs. income.

a) **Straight enough condition:** The scatterplot is straight enough to try linear regression.
Independence assumption: The residuals plot shows no pattern.
Does the plot thicken? condition: The spread of the residuals is consistent.
Nearly Normal condition: The Normal probability plot is reasonably straight.

Since the conditions for inference are satisfied, the sampling distribution of the regression slope can be modeled by a Student's *t*-model with $(57 - 2) = 55$ degrees of freedom. We will use a regression slope *t*-test. The equation of the line of best fit for these data points is: *Income* $= 5970.05 + 2444.79(Education)$.

b) The value of $t \approx 5.19$. The *P*-value of less than 0.0001 means that the association we see in the data is unlikely to occur by chance. We reject the null hypothesis, and conclude that there is strong evidence of a linear relationship between education level and income. Cities in which median education level is relatively high have relatively high median incomes.

c) If the data were plotted for individuals, the association would appear to be weaker. Individuals vary more than averages.

d) $b_1 \pm t^*_{n-2} \times SE(b_1) = 2444.79 \pm (2.004) \times 471.2 \approx (1500, 3389)$

We are 95% confident that each additional year of median education level in a city is associated with an increase of between $1500 and $3389 in median income.

e) The regression equation predicts that a city with a median education level of 11 years of school will have a median income of $5970.05 + 2444.79(11) = \$32862.74$ $(t^*_{55} \approx 1.6730)$

$$\hat{y}_v \pm t^*_{n-2}\sqrt{SE^2(b_1) \cdot (x_v - \bar{x})^2 + \frac{s_e^2}{n}}$$

$$= 32862.74 \pm (1.6730)\sqrt{471.2^2 \cdot (11 - 10.9509)^2 + \frac{2991^2}{57}}$$

$$\approx (32199, 33527)$$

We are 90% confident that cities with 11 years for median education level will have an average income of between $32,199 and $33,527.

29. Diet.

H_0: Cracker type and bloating are independent.

H_A: There is an association between cracker type and bloating.

Counted data condition: The data are counts.
Randomization condition: Assume that these women are representative of all women.
Expected cell frequency condition: The expected counts are all (almost!) greater than 5.

	Bloat	
	Little/None (Obs/Exp)	Moderate/Severe (Obs/Exp)
Bran	11 / 7.6471	2 / 5.3529
Gum Fiber	4 / 7.6471	9 / 5.3529
Combination	7 / 7.6471	6 / 5.3529
Control	8 / 7.0588	4 / 4.9412

Under these conditions, the sampling distribution of the test statistic is χ^2 on 3 degrees of freedom. We will use a chi-square test for independence.

$\chi^2 = \sum\limits_{all\,cells} \frac{(Obs - Exp)^2}{Exp} \approx 8.23$, and the *P*-value ≈ 0.0414.

Since the *P*-value is low, we reject the null hypothesis. There is evidence of an association between cracker type and bloating. The gum fiber crackers had a higher rate of moderate/severe bloating than expected. The company should head back to research and development and address the problem before attempting to market the crackers.

30. Cramming.

a) H_0: The mean score of week-long study group students is the same as the mean score of overnight cramming students. $(\mu_1 = \mu_2 \text{ or } \mu_1 - \mu_2 = 0)$

H_A: The mean score of week-long study group students is the same as the mean score of overnight cramming students. $(\mu_1 > \mu_2 \text{ or } \mu_1 - \mu_2 > 0)$

Independent Groups Assumption: Scores of students from different classes should be independent.
Randomization Condition: Assume that the students are assigned to each class in a representative fashion.
10% Condition: 45 and 25 are less than 10% of all students.
Nearly Normal Condition: The histogram of the crammers is unimodal and symmetric. We don't have the actual data for the study group, but the sample size is large enough that it should be safe to proceed.

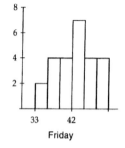

$$\bar{y}_1 = 43.2 \qquad \bar{y}_2 = 42.28$$
$$s_1 = 3.4 \qquad s_2 = 4.43020$$
$$n_1 = 45 \qquad n_2 = 25$$

Since the conditions are satisfied, it is appropriate to model the sampling distribution of the difference in means with a Student's *t*-model, with 39.94 degrees of freedom (from the approximation formula). We will perform a two-sample *t*-test. The sampling distribution model has mean 0, with standard error: $SE(\bar{y}_1 - \bar{y}_2) = \sqrt{\dfrac{3.4^2}{45} + \dfrac{4.43020^2}{25}} \approx 1.02076$.

The observed difference between the mean scores is 43.2 – 42.28 = 0.92.

Since the *P*-value = 0.1864 is high, we fail to reject the null hypothesis. There is no evidence that students with a week to study have a higher mean score than students who cram the night before.

$$t = \frac{(\bar{y}_1 - \bar{y}_2) - (0)}{SE(\bar{y}_1 - \bar{y}_2)}$$
$$t \approx \frac{0.92}{1.02076}$$
$$t \approx 0.90$$

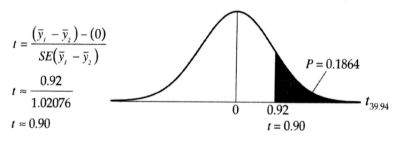

b) H_0: The proportion of study group students who will pass is the same as the proportion of crammers who will pass. $(p_1 = p_2 \text{ or } p_1 - p_2 = 0)$

H_A: The proportion of study group students who will pass is different from the proportion of crammers who will pass. $(p_1 \neq p_2 \text{ or } p_1 - p_2 \neq 0)$

Random condition: Assume students are assigned to classes in a representative fashion.
10% condition: 45 and 25 are both less than 10% of all students.
Independent samples condition: The groups are not associated.
Success/Failure condition: $n_1\hat{p}_1 = 15$, $n_1\hat{q}_1 = 30$, $n_2\hat{p}_2 = 18$, and $n_2\hat{q}_2 = 7$ are not all greater than 10, since only 7 crammers didn't pass. However, if we check the pooled value, $n_2\hat{p}_{pooled} = (25)(0.471) = 11.775$. All of the samples are large enough.

Since the conditions have been satisfied, we will model the sampling distribution of the difference in proportion with a Normal model with mean 0 and standard deviation

estimated by $SE_{pooled}\left(\hat{p}_1 - \hat{p}_2\right) = \sqrt{\dfrac{\hat{p}_{pooled}\hat{q}_{pooled}}{n_1} + \dfrac{\hat{p}_{pooled}\hat{q}_{pooled}}{n_2}} = \sqrt{\dfrac{\left(\frac{33}{70}\right)\left(\frac{37}{70}\right)}{45} + \dfrac{\left(\frac{33}{70}\right)\left(\frac{37}{70}\right)}{25}} \approx 0.1245.$

The observed difference between the proportions is:
$0.3333 - 0.72 = -0.3867.$

Since the *P*-value = 0.0019 is low, we reject the null hypothesis. There is strong evidence to suggest a difference in the proportion of passing grades for study group participants and overnight crammers. The crammers generally did better.

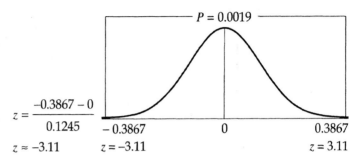

$z = \dfrac{-0.3867 - 0}{0.1245}$

$z \approx -3.11$

c) H_0: There is no mean difference in the scores of students who cram, after 3 days. $\left(\mu_d = 0\right)$

H_A: The scores of students who cram decreases, on average, after 3 days. $\left(\mu_d > 0\right)$

Paired data assumption: The data are paired by student.
Randomization condition: Assume that students are assigned to classes in a representative fashion.
10% condition: 25 students are less than 10% of all students.
Nearly Normal condition: The histogram of differences is roughly unimodal and symmetric.

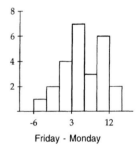

Since the conditions are satisfied, the sampling distribution of the difference can be modeled with a Student's *t*-model with 25 – 1 = 24

degrees of freedom, $t_{24}\left(0, \dfrac{4.8775}{\sqrt{25}}\right).$

We will use a paired *t*-test, with $\bar{d} = 5.04.$

Since the *P*-value is less than 0.0001, we reject the null hypothesis. There is strong evidence that the mean difference is greater than zero. Students who cram seem to forget a significant amount after 3 days.

$t = \dfrac{\bar{d} - 0}{\dfrac{s_d}{\sqrt{n}}}$

$t = \dfrac{5.04 - 0}{\dfrac{4.8775}{\sqrt{25}}}$

$t \approx 5.17$

d) $\bar{d} \pm t^*_{n-1}\left(\dfrac{s_d}{\sqrt{n}}\right) = 5.04 \pm t^*_{24}\left(\dfrac{4.8775}{\sqrt{25}}\right) \approx (3.03, 7.05)$

We are 95% confident that students who cram will forget an average of 3.03 to 7.05 words in 3 days.

e) H_0: There is no linear relationship between Friday score and Monday score. $\left(\beta_1 = 0\right)$

H_A: There is a linear relationship between Friday score and Monday score. $\left(\beta_1 \neq 0\right)$

Scatterplot	Residuals Plot	Normal Probability Plot	Residuals Histogram
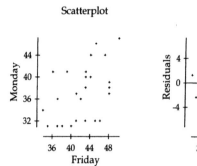	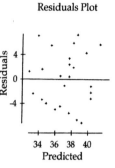	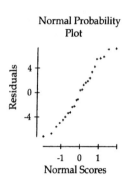	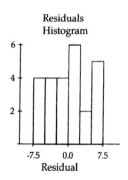

Straight enough condition: The scatterplot is straight enough to try linear regression.

Independence assumption: The residuals plot shows no pattern.

Does the plot thicken? condition: The spread of the residuals is consistent.

Nearly Normal condition: The Normal probability plot of residuals is reasonably straight, and the histogram of the residuals is roughly unimodal and symmetric.

Since the conditions for inference are satisfied, the sampling distribution of the regression slope can be modeled by a Student's t-model with $(25 - 2) = 23$ degrees of freedom. We will use a regression slope t-test.

Dependent variable is: **Monday**
No Selector
R squared = 22.4%   R squared (adjusted) = 19.0%
s = 4.518 with 25 - 2 = 23 degrees of freedom

The equation of the line of best fit for these data points is: $M\hat{o}nday = 14.5921 + 0.535666(Friday)$.

Source	Sum of Squares	df	Mean Square	F-ratio
Regression	135.159	1	135.159	6.62
Residual	469.401	23	20.4087	

Variable	Coefficient	s.e. of Coeff	t-ratio	prob
Constant	14.5921	8.847	1.65	0.1127
Friday	0.535666	0.2082	2.57	0.0170

The value of $t \approx 2.57$. The P-value of 0.0170 means that the association we see in the data is unlikely to occur by chance. We reject the null hypothesis, and conclude that there is strong evidence of a linear relationship between Friday score and Monday score. Students who do better in the first place tend to do better after 3 days.

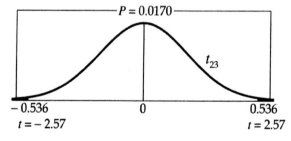